WindEurope Conference 2018

Journal of Physics: Conference Series
Volume 1102

Hamburg, Germany
25 - 28 September 2018

Editors:

Michael Muskulus
Carlo L. Bottasso

ISBN: 978-1-5108-7344-5
ISSN: 1742-6588

Printed from e-media with permission by:

Curran Associates, Inc.
57 Morehouse Lane
Red Hook, NY 12571

CURRAN ASSOCIATES INC.
proceedings.com

Some format issues inherent in the e-media version may also appear in this print version.

Printed by Curran Associates, Inc. (2018)

Institute of Physics
Dirac House, Temple Back
Bristol BS1 6BE UK

Phone: 44 1 17 929 7481
Fax: 44 1 17 920 0979

techtracking@iop.org

Additional copies of this publication are available from:

Curran Associates, Inc.
57 Morehouse Lane
Red Hook, NY 12571 USA
Phone: 845-758-0400
Fax: 845-758-2633
Email: curran@proceedings.com
Web: www.proceedings.com

Table of contents

Volume 1102

WindEurope conference 2018 within the Global Wind Summit

25–28 September 2018, Hamburg, Germany

Accepted papers received: 21 September 2018
Published online: 10 October 2018

Preface

Papers

Assessing the wind resource and turbine performance

Developments in turbine technology

Digitalisation

End of life issues

Environmental impacts, social acceptance & spatial planning

Health and safety

Integrating wind into the energy system

Operations and Maintenance

Preface

On behalf of the European Academy of Wind Energy (EAWE[1]) and WindEurope (formerly the European Wind Energy Association EWEA), we hereby have the pleasure of presenting the Scientific Proceedings of the WindEurope Conference 2018.

Global Wind Summit 2018 (WindEurope Conference & WindEnergy Hamburg) took place in Hamburg on 25-28 September 2018. It is one of the largest and most important meeting places of the wind energy community. This year's conference has featured 50 thematic sessions with a total of more than 500 speakers. As in previous years, the sessions mixed industrial content with scientific works, in order to increase the visibility of relevant scientific research and to facilitate an interchange of ideas between industry and academia.

In continuation of the successful agreement between WindEurope and IOP Publishing, the scientific contributions of the WindEurope Conference 2018 are published in the Journal of Physics: Conference Series, an Open Access journal managed by the Institute of Physics (IOP). All papers in this volume originate from scientific contributions presented at WindEurope Conference 2018, and have been peer-reviewed in order to ensure strict, international standards, coordinated by the Scientific Proceedings Editors and aided by the EAWE Wind Europe Scientific Committee.

New this year was the possibility for *every* accepted scientific abstract to submit a paper for the Scientific Proceedings. In previous years only the scientific papers presented in oral sessions were given this opportunity, but this year authors could submit a manuscript also for accepted scientific poster presentations. We believe this is an important development that benefits our community, and that greatly increases the attractiveness of WindEurope Conference. In particular, this guarantees long-term storage and easy electronic reference of conference papers, and we thank WindEurope for making this possible. Of course this positive development has also meant a much higher workload than previously. This year, a total of 46 original papers were submitted, of which 40 were eventually accepted for publication.

The papers were roughly organized in eight topics, which were handled by the following Section Editors:

- Ola Carlson, Chalmers University of Technology, Sweden:
  Integrating wind into the energy system
- Emilio Gomez-Lazaro, University of Castilla-La-Mancha, Spain:
  Digitalization
- Johan Meyers, KU Leuven, Belgium:
  Assessing the wind resource and turbine performance
- Michael Muskulus, Norwegian University of Science and Technology NTNU, Norway:
  End of life issues
  Environmental impacts, social acceptance & spatial planning
  Health & safety

[1] http://www.eawe.eu/

- Vasilis Riziotis, National Technical University of Athens NTUA, Greece: Developments in turbine technology
- Simon Watson, Delft University of Technology, The Netherlands: Operations & maintenance

The topics of the papers in these proceedings cover the whole spectrum of scientific interests in wind energy. The following selection demonstrates the variety of contributions and discusses some of the highlights.

The session on *Assessing the wind resource and turbine performance* contains works that discuss how extreme winds will be estimated in the New European Wind Atlas (Bastine et al.) and a novel idea for assessing atmospheric stability from measurements using a simple diagnostic plot (Svennigsen et al.). A number of other papers consider Lidar technology. For example, the paper by Würth et al. discusses how Lidar can be used to forecast the short-term power output of a wind turbine. The paper by Stock-Williams et al. demonstrates how a machine learning approach can be used to reconstruct wind fields more accurately from Lidar measurements, including an evaluation of the reconstruction uncertainty. Other prominent topics considered in the contributions are, for example, complex terrain and meso-scale models.

The session on *Turbine technology* has two interesting papers on wind turbine control, one dealing with optimal yaw under wake conditions (Urbán et al.) and the other developing observability analysis for analysing data-driven wind turbine models (Ritter et al.). The flight phases of ground-generating airborne wind energy systems are optimized in the paper by Warnock et al. The dynamics of a semi-submersible floating wind turbine are investigated with simulations (Galván et al.) and the fatigue in umbilical power cables is studied by Young et al.

In the *Operations & maintenance* session the paper by Rochat et al. demonstrates temperature sensing in a HVDC offshore cable. A general approach to anomaly detection, based on deep neural networks that predict probability distributions, is described by Vogt et al. An important tool in analyzing the reliability of wind turbine components is Failure Modes, Effects and Criticality Analysis (FMECA). The paper by Cevasco et al. demonstrates how this can help identify the most cost-critical components for maintenance actions.

The session on *Integrating wind into the energy system* features a study on a hybrid auxiliary power system that complements diesel generators with wind turbines and an energy storage solution, in order to provide emergency power to a large wind farm during islanding operation (Huang & Chen). The paper by Loukidis et al. is a careful study on the effect changes of financial support scheme in Greece can have on the uncertainty in wind farm revenues. Rebello et al. demonstrate how dynamic curtailment with a simple algorithm can, in principle, allow for participation in the secondary regulation market.

In the *Environmental impacts, social acceptance & spatial planning* session, only a few contributions were received. The paper by Ioannou et al. studies the effect uncertainty in electric market prices can have on the profitability of a wind farm, comparing different mathematical models for future electricity prices. Gonzalez et al. study the different raw materials needed for future renewable energy projects in a scenario for 2050. Although wind energy was the least affected, they identify certain risks for a successful transition to a low-carbon economy.

Only one paper was submitted to the *End of life issues* session. Pettas & Cheng discuss down-regulation and individual blade control for a 10 MW wind turbine as strategies for fatigue load reduction. It is shown that their combination is advantageous.

IOP Conf. Series: Journal of Physics: Conf. Series **1102** (2018) 011001 doi:10.1088/1742-6596/1102/1/011001

The session on *Digitalization* features one paper. Do & Berthaut-Gerentes provide a short study on how wind speed and power output depend on the time scale, reasoning that the current sampling rate for SCADA systems (aggregating 10 minutes of information by averaging) is close to optimal.

The *Health and safety* session also only features a single paper. Leimeister et al. discuss ideas on how to reduce health and safety hazards in lifting operations by employing automated, human-free hoisting systems and suitable fastening solutions for transport at sea.

As with previous events, the European Academy of Wind Energy is responsible for organizing the review process for scientific contributions, has contributed to the development of the sessions, and has provided chairs for sessions with scientific content. All papers were peer-reviewed by a Scientific Committee, consisting of scientists from EAWE member institutes and their associates.

Handling the many papers was challenging this year, as the paper review took place mostly within the academic summer holidays. We thank all authors for their understanding and willingness to take part in this procedure, including the short deadlines imposed by the tight publication schedule. We thank all reviewers for their engagement and entirely voluntary work, without which we could not guarantee the scientific quality and current interest of the accepted contributions. WindEurope is responsible for the organization and logistics of the conference, and we thank their highly professional staff and their associates for the excellent collaboration.

Michael Muskulus
Editor Scientific Proceedings WindEurope Conference 2018
Scientific Programme Coordinator WindEurope Conference 2018
Norwegian University of Science and Technology NTNU

Carlo L. Bottasso
Editor Scientific Proceedings WindEurope Conference 2018
President European Academy of Wind Energy EAWE
Technical University of Munich TUM

IOP Conf. Series: Journal of Physics: Conf. Series **1102** (2018) 011002 doi:10.1088/1742-6596/1102/1/011002

Peer review statement

All papers published in this volume of *Journal of Physics: Conference Series* have been peer reviewed through processes administered by the proceedings Editors. Reviews were conducted by expert referees to the professional and scientific standards expected of a proceedings journal published by IOP Publishing.

Evaluation of ERA-Interim and NCEP-CFSR Reanalysis Datasets against in-situ Measured Wind Speed Data for Keti Bandar Port, Pakistan

Zia ul Rehman Tahir[a,*], Muhammad Salman Sarfraz[a], Muhammad Asim[a], Muhammad Sajid Kamran[a], Shahid Imran[a], Nasir Hayat[a]

[a]Department of Mechanical Engineering, University of Engineering and Technology Lahore, Pakistan

ziartahir@uet.edu.pk

Abstract. The evaluation and comparison of reanalysis datasets (NCEP-CFSR and ERA-Interim) against mast measured wind data for Keti-bander port of Pakistan have been performed. In the last decade, wind farms have been developed on the land, but there is still need for the development of offshore wind farms on the 1100 km coast of Pakistan. The wind data measured at the Keti Bandar Port of Pakistan at different heights for the duration of 20 months has been used in this study. The evaluations of satellite datasets against measured data were done based on statistical analysis, monthly mean time series and Weibull probability distribution function. The wind data at 30 m, 60 m and 85 m was extrapolated using the Logarithmic law from 10 m height. At 30 m height both satellite datasets show very good agreement with measured data, the predictions of CFSR data are much better than ERA-Interim. The percentage difference for mean wind speed, wind power density and Weibull parameters at 30 m are less than 5%. Overall NCEP-CFSR gave better results compared to ERA-Interim. These results may be used for the offshore wind resource assessment for Pakistan due to unavailability of offshore wind speed data.

1. Introduction

Pakistan has been facing a shortage in the supply and demand of electricity over the past decade, so looking for alternative resources of energy like the wind is a good sustainable option. Due to the skewness of energy production systems towards fossil fuels, the economy of the country is badly affected because of high prices of fossil fuels and high dependence on oil price changes, leading to cost vulnerability [1-3]. The solution to reducing the dependency on fossil fuel i.e. residual furnace oil, diesel oil, etc. is to invest in renewable energy technologies. A share of 1.48% of the total energy supply is being produced from renewable energy sources in Pakistan[4]. Several countries are successfully dealing with growing energy demand by installing Renewable based energy resources[5, 6]. Fortunately, Pakistan is blessed with the plentiful potential of renewable energy sources. Pakistan's possibilities of using renewable energy resources that include Solar, Wind, Hydel and Biomass Energy, etc. [2, 7-12].

Pakistan Meteorological Department (PMD) has assessed the coastline of Pakistan during 2002, and it shows wind energy potential along the coastline of Sindh and Balochistan [13]. Pakistan has a total potential of wind energy about 346GW, out of which 120GW is feasible [14]. Pakistan has a total installed capacity of 789 MW out of which 199MW was added in 2017 [4]. Country's wind energy has a share of 1.48% to total energy generation till 2017. Based on available grid infrastructure in the country, the government has set a target for adding additional 1,200 MW of wind power by 2020, out of which 400MW will be added in 2018.[4]. There was not a single grid-connected wind farm exists in Pakistan until 2009, three grid-connected wind farms with total capacity of 150 MW had been added and 2.6 GW wind form will be added by 2030 [15]. The exponential increase in renewable energy projects especially wind projects is attracting attention day by day and for installation of such projects. An accurate wind potential estimate is of key importance. The installation of a wind mast to measure surface data is a costly process, satellite datasets can be used as an alternative to surface measured data.

Satellite reanalysis datasets are evaluated on the basis of the comparison performed with surface measured data. This method is most commonly used in recent studies for the different location [16-19]. The measured and satellite data was compared in three ways; using correlation by finding different statistical parameters, by monthly time series [19-22] and Weibull probability distribution function[23-25]. The statistical tools used in this analysis are; mean bias error (MBE), mean absolute error (MAE), root mean square error (RMSE) and correlation coefficient (R). For European region, Carvalho et al. [16, 25, 26] evaluated a number of satellite simulated datasets with surface measured datasets, for Iberian Peninsula analysis for offshore wind satellite data using ERA-Interim and NCEP-CFSR was reported. The evaluation of wind satellite data including ERA-Interim and NCEP CFSR for Portugal region was reported in Ref [26]. Staffell et al. [27] discussed the simulation of present and future wind speed for 23 European countries including ERA-Interim and NCEP- CFSR. Based on all these comparative and evaluative studies, this paper discussed the results and found quite reasonable comparisons.

Pakistan has a coastal line of around 1100 km, but neither an offshore wind farm developed nor offshore wind resource assessment was performed for Pakistan. The wind speed measured data port of Keti Bandar is compared with reanalysis datasets which could be used for the offshore wind resource assessment of Pakistan. The reanalysis datasets of Era-Interim and CFSR is evaluated for the coastal line of Pakistan, to determine which one of these satellite datasets shows better results with measured wind data and the results presented here can be of great value for the offshore wind resource assessment for Pakistan due to insufficient or unavailability of offshore wind speed data. The evaluation of NCEP CFSR and Era-Interim datasets has been performed using statistical parameters: MAE, MBE, STDE, RMSE and R. The comparison of these results with previously published results have been performed thus validating results and identifying satellite estimates variations for the South Asian region. This study is an initiative towards wind resource assessment for potential sites in Pakistan.

2. Wind Data

Alternative Energy Development Board of Pakistan (AEDB) along with United Nations Development Programme (UNDP) under the Global Environmental Facility (GEF) Wind Energy Program had installed three wind masts in the Gharo-Keti Bandar Wind Corridor [28], in order to facilitate the growth of wind energy. The AEDB-UNBDP Keti Bandar mast was installed in Keti Bandar town in the south-east of Karachi [28] at the start of 2009, the geographical location of the site is 24° 11′ 49.18″ N and 67° 37′ 38.67″ E. The location of the mast is about 30 km away from the shore. The mast consists of five anemometers at measurement levels of 85 m (two anemometers), 60 m, 30m, and 10m to measure wind speed and two wind vanes at 83.5 m and 28.5 m to measured wind direction. The data for 20 months duration from 1st March 2009 to 31st October 2010 was used to evaluate satellite reanalysis data for this study.

The monthly mean wind speed of the measured data is presented in Figure 1. There are two anemometers at 85 m height so the average wind speed from two anemometers is calculated. The mean wind speed at 85 m, 60 m, 30 m and 10 m are 6.93 m/s, 6.53 m/s, 5.55 m/s and 4.31 m/s respectively. The monthly mean wind speed at 85 m ranges from 5.49 m/s to 8.57 m/s, the maximum is in May and minimum in October. There is a general trend of increase in wind speed from January to May, then decrease from May to October, then a slight increase from October to December and finally decrease from December to January.

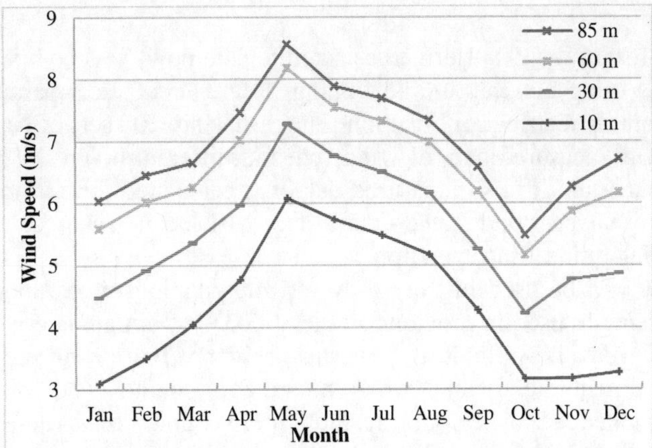

Figure 1. Monthly mean wind speeds

To extrapolate wind speed different methods have been developed over time, assuming wind shear profile follows either a logarithmic-linear law, a logarithmic law or a power law [29]. The vertical wind shear profile using simple mean wind speed is shown in Figure 2. The measured data is used to get the best fit for Power law [30] and Logarithmic law [31, 32] of vertical interpolation of wind speed. The power law exponent (alpha) used in Power law was calculated as 0.209 and surface roughness used in logarithmic law was calculated as 0.228 m. The logarithmic law has been commonly used in various recent studies for the validation of reanalysis data with surface measured data [33-38], as it provides the interpolation or extrapolation of wind speeds over a range of wind speeds, locations and altitudes [29, 39].

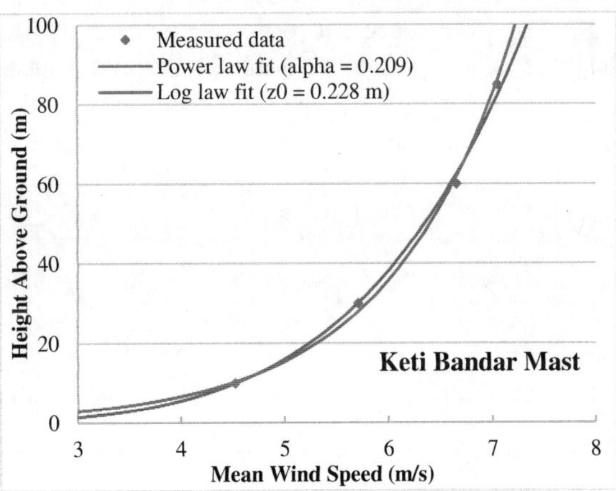

Figure 2. Vertical wind shear profile

The ERA-Interim reanalysis [40] is the third generation reanalysis developed by the European Centre for Medium-Range Weather Forecasts (ECMWF) released in 2006 is based on the calculations using an integrated forecast system model. It is a global reanalysis dataset having a spatial resolution of 0.75°×0.75° (~80 km) and a temporal resolution of six hours, the wind data is available from 1979 to onwards at 10m to 270 m height above ground. The ERA-Interim wind data was downloaded with a spatial resolution of 0.125°×0.125° (~13 km) at 10 m height above ground.

The Climate Forecast System Reanalysis (CFSR) [41] is the third generation reanalysis product developed by the National Centers for Environmental Prediction (NCEP) released in 2010. This is a global reanalysis dataset having a spatial resolution of 0.5°×0.5° (~55 km) and a temporal resolution of six hours. The wind data is available from 1979 to onwards at 10 m and 22 m height above ground.

3. Methodology

The measured data and reanalysis data (termed as satellite data now) was compared in three common ways; statistical analysis using correlation and finding different statistical parameters, monthly time series and Weibull probability distribution function. The reanalysis data sets of NCEP-CFSR and ERA-Interim are compared with measured data of wind speed at different heights. The reanalysis datasets were available at six hours durations so measured data was calculated for six hours duration from 10 minutes duration. The reanalysis wind speed data at the mast location was calculated from the four nearby grid locations using bilinear interpolation.

The Logarithmic law will be used in this study for interpolation or extrapolation of wind speed because it has been commonly used in previous studies [33-38] to extrapolate wind speed by satellite datasets. The ERA-Interim data is available at the vertical height of 10 m, wind speed data was vertically extrapolated using the Logarithmic law at 30m, 60 m and 85m height. The surface roughness used in Logarithmic law was downloaded for temporal resolution of six hours from the ECMWF website. The NCEP-CFSR data is available at the vertical height of 10 m and 22 m so satellite data was extrapolated using Logarithmic law from these two heights at 30 m, 60 m and 85 m. The measured data at 22 m height was interpolated using the logarithmic law, using surface roughness length for each time step.

4. Results

The evaluation of satellite wind speed was performed in three ways; monthly mean time series, statistical analysis and Weibull probability distribution function. This section presents the results obtained from three methods of comparison.

4.1. Time Series

The monthly mean wind speed at 10 m, 30 m, 60 m and 85 m from measured and ERA-Interim data is presented in Figure 3. There is a consistent overestimation of satellite wind speed at 10 m. At 30 m height, there is an overestimation of wind speed except for winter of 2009-10 so the over MBE is low. At 60 m and 80 m height, the trend is quite similar, but the satellite wind speeds are underestimated.

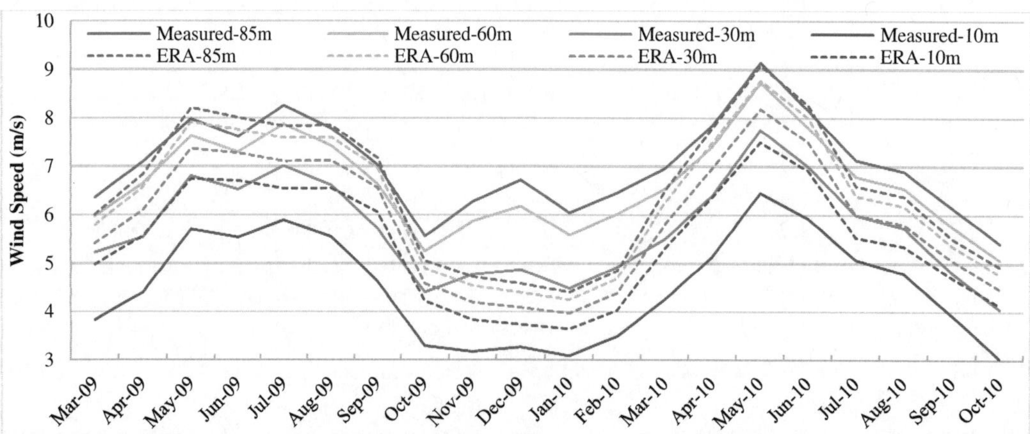

Figure 3. Monthly mean wind speed from measured and ERA-Interim data

The monthly mean wind speed at 10 m, 30 m, 60 m and 85 m from measured and NCEP-CFSR data is presented in Figure 4. There is a constant bias for 10 m height, satellite wind speed was overestimated. At 30 m, trends are closely related with minimum bias. At 60 m and 80 m the trend is of CFSR wind speed is like ERA-Interim. Both satellite data underestimate wind speed in winter, low wind speed in June 2009 was estimated well by CFSR, whereas not be ERA-Interim. For this location and duration discussed CFSR data gave a better estimation of wind speed compared to ERA-Interim.

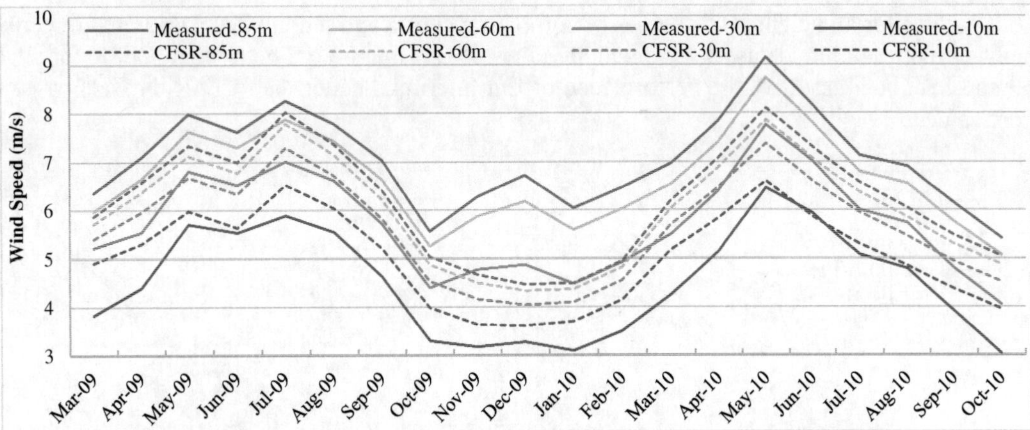

Figure 4. Monthly mean wind speed from measured and CFSR data

4.2. Statistical Analysis

The statistical parameters to compare the satellite data with measured data are; mean bias (MBE), mean absolute error (MAE), the root mean square error (RMSE), and correlation coefficient (R). A positive bias means that the satellite data is overestimated and vice versa. Bias, MAE, RMSE and R of both satellite datasets for six hours and 24 hours duration is presented in Table 1. The best results, characterised by higher values of correlation coefficient and lower values of errors, are highlighted in bold. The daily data for ERA-Interim shows better comparison compared to six hourly data whereas the best results are at 30 m height. Similarly, the daily data for NCEP-CFSR shows better comparison compared to six hourly data whereas the best results are at 30 m height. The NCEP-CSFR data shows better agreement with measured data in terms of errors compared to ERA-Interim but NCEP-CFSR has a lower value of R compared to ERA-Interim.

Table 1. Statistical analysis of the comparison between measured and satellite mean wind speed data

Satellite Data	Height	6 Hourly					Daily				
		MBE	MAE	RMSE	STDE	R	MBE	MAE	RMSE	STDE	R
ERA-Interim	10 m	0.897	1.295	1.600	**1.325**	**0.761**	0.897	1.057	1.237	**0.852**	**0.873**
	30 m	**0.192**	**1.157**	**1.506**	1.494	0.740	**0.192**	**0.811**	**1.050**	1.032	0.843
	60 m	-0.341	1.246	1.678	1.643	0.720	-0.341	0.899	1.275	1.229	0.805
	85 m	-0.513	1.340	1.846	1.773	0.700	-0.513	0.982	1.425	1.329	0.788
NCEP-CFSR	10 m	0.508	1.107	1.425	**1.331**	**0.739**	0.508	0.794	0.981	**0.839**	**0.835**
	22 m	**0.017**	**1.086**	**1.418**	1.418	0.719	**0.017**	**0.731**	**0.941**	0.941	0.812
	30 m	-0.075	1.151	1.516	1.514	0.699	-0.075	0.751	0.970	0.967	0.811
	60 m	-0.639	1.409	1.799	1.682	0.655	-0.639	1.001	1.313	1.147	0.764
	85 m	-0.842	1.580	2.032	1.849	0.612	-0.842	1.143	1.512	1.256	0.741

The scatter plots of measured wind speed and satellite estimates of wind speed for four heights (10 m, 30 m, 60 m and 85 m) are presented in Figure 5. At 10 m height, a best-fit line of ERA is parallel to the 1-1 line with an offset which shows a constant overestimation whereas CFSR line is intersecting 1-1 line which shows that lower winds are overestimated, and higher winds are underestimated. At 30 m height best-fit line form ERA is parallel and very close to 1-1 line whereas the behaviour of the CFSR line is like 10 m height. At 60 m height ERA line is parallel to 1:1 but higher winds are slightly underestimated, the trend is like ERA line at 85 m. The CFSR data is underestimating high wind speeds and overestimating lower wind speeds at all heights, although wind speed dispersion is not consistent

IOP Conf. Series: Journal of Physics: Conf. Series **1102** (2018) 012001 doi:10.1088/1742-6596/1102/1/012001

over the 1-1 line for all heights. The values of correlation coefficient tend to decrease with an increase in height for both satellite datasets. Correlation values of Era-Interim are better than CFSR for all 10, 30, 60 and 85m heights. Also, the performance of Era-Interim is better than CFSR datasets based on gradient values.

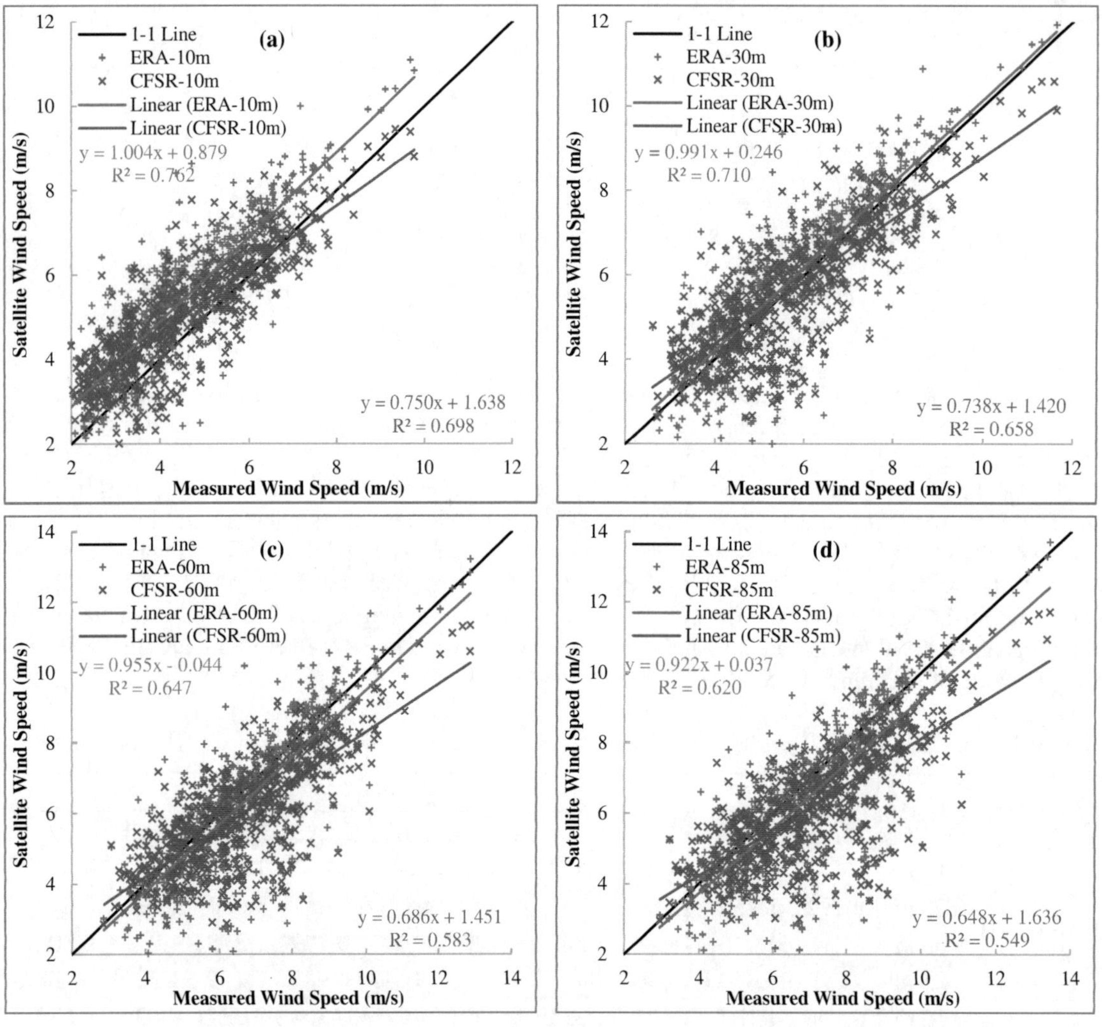

Figure 5. Correlation between surface measured and satellite wind speed

4.3. Weibull Probability Density Function

The Weibull probability density function is widely used to represents wind speed distribution due to its flexibility and simplicity. The scale factor A is related to mean wind speed while shape factor k is equivalent to the standard deviation [42]. The value of A and k for both measured and satellite data for six hours duration is given in Table 2, for satellite data percentage error is presented in parentheses next to the value. The measured mean wind speed at different heights and standard deviation in parenthesis are presented in Table 2. The mean wind speed from satellite data with a percentage error from measured speed in parenthesis is presented in Table 2. The mean wind speed for the complete duration from satellite data show good agreement with measured data at higher heights within 5% error, the same agreement is for Weibull factors A and k.

IOP Conf. Series: Journal of Physics: Conf. Series **1102** (2018) 012001 doi:10.1088/1742-6596/1102/1/012001

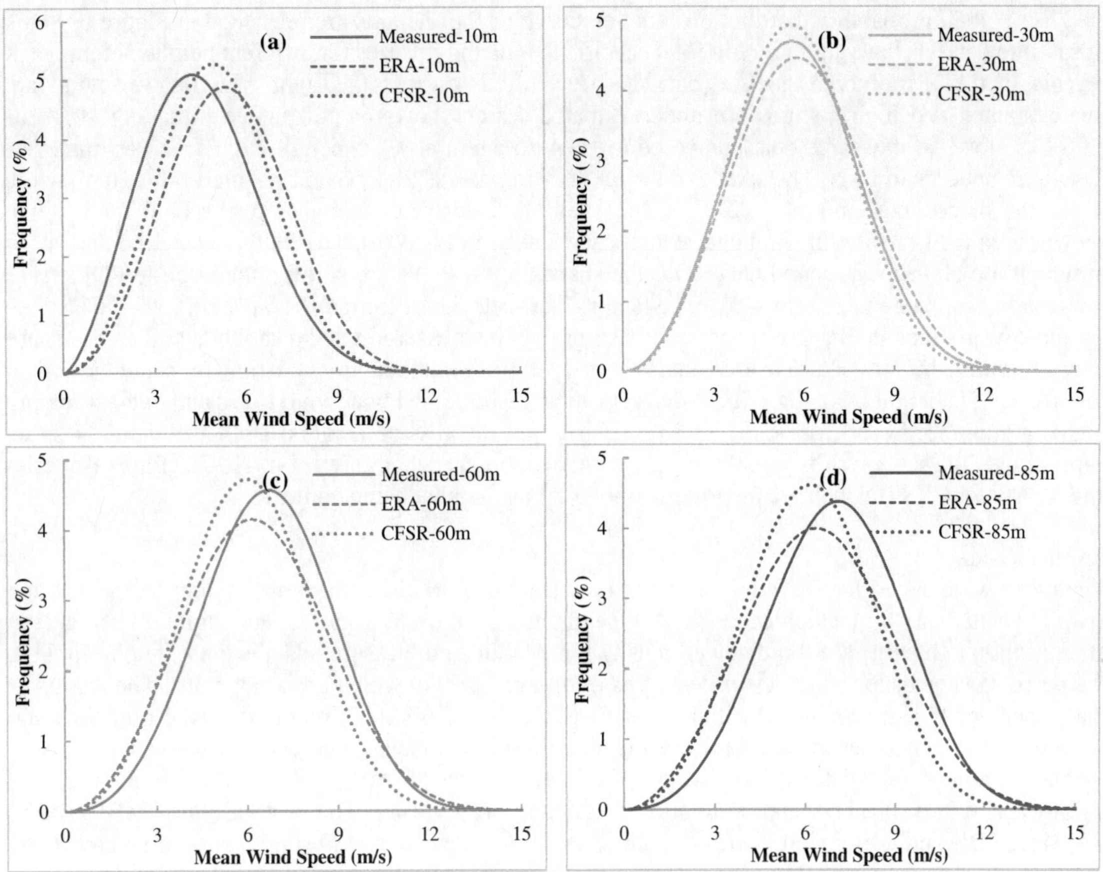

Figure 6. Weibull PDF between surface measured and satellite wind speed

Table 2. Statistics of the comparison between measured and satellite six-hourly mean wind data

Wind Data	Height	Mean wind speed (m/s)	Wind power density (W/m²)	Weibull scale factor *A* (m/s)	Weibull shape factor *k*
Measured	10 m	4.524 (1.850)	88	5.106	2.595
	22 m	5.444 (1.858)	136	6.089	3.071
	30 m	5.549 (1.960)	157	6.382	3.067
	60 m	6.529 (2.023)	232	7.384	3.502
	85 m	7.043 (2.133)	274	7.813	3.560
ERA-Interim	10 m	5.423 (19.87%)	127 (54.55%)	6.071 (18.90%)	3.032 (16.84%)
	30 m	5.901 **(6.34%)**	164 (12.10%)	6.606 **(3.51%)**	3.008 **(-1.92%)**
	60 m	6.313 (-3.31%)	202 **(-6.90%)**	7.068 (-4.28%)	2.993 (-14.53%)
	85 m	6.533 (-7.24%)	224 (-12.41%)	7.315 (-6.37%)	2.980 (-16.29%)
NCEP-CFSR	10 m	5.031 (11.21%)	108 (22.73%)	5.628 (10.22%)	3.039 (17.11%)
	22 m	5.460 **(0.29%)**	136 **(0.00%)**	6.101 **(0.20%)**	3.158 **(2.83%)**
	30 m	5.631 (1.48%)	148 (-5.73%)	6.287 (-1.49%)	3.191 (4.04%)
	60 m	6.012 (-7.92%)	178 (-23.28%)	6.704 (-9.21%)	3.264 (-6.80%)
	85 m	6.201 (-11.96%)	195 (-28.83%)	6.912 (-11.53%)	3.291 (-7.56%)

The Weibull probability distribution function curve of four heights are presented in Figure 6. PDFs of satellite datasets have similar trend for a specific height and it varies for different heights. Figure 5(a) reveals that PDFs of both satellite datasets are shifted towards the right which lower wind are overestimated and higher winds are underestimated which shows overall underestimations. PDF of CFSR is closer to measured data compared to ERA so again at 10 m height wind speed estimates of CFSR are better than ERA. The spread of the CFSR wind data is lesser than Era-Interim for 10 m, while it has the percentage error of 10.22% in the Weibull scale factor which is responsible for positive skewness in the Era-Interim wind data which can be seen in Fig 5 (a). The PDFs of satellite data at 30 m height are close to measured data PDF, the shapes of all PDFs are similar, the satellite PDFs have lower values of percentage errors of shape factor. The scale factor for the ERA is higher than CFSR.

The lower winds in Era-Interim are underestimating the measured wind data while higher winds are overestimating the surfaced measured data with an overall overestimation for 10 and 30m and vice versa for 60 and 85m height. Although CFSR dataset is underestimating lower wind speed and overestimating higher wind speeds for 10m only and vice versa for the rest of heights. Both the datasets show approximate the same results for 30m height with Weibull shape factor error of -1.92% for Era-interim and 4.04% for CFSR, which are the lowest among all the heights being compared.

5. Discussion

The mean wind speed for the whole duration is presented in Table 2, for both satellite the percentage error is positive at 10 m which decrease with height and changes to negative, the minimum percentage error is at 30 m height. The wind power density for satellite and measured data is presented in Table 2, the percentage error for wind power density is minimum at 60 m which are around 5%. The results of the present study were compared with the results of the previous studies reported for the European and Antarctic region by other researches to see the reliability of reanalysis datasets for Pakistan.

For six hourly wind data of ERA-Interim, the lowest values of MBE, MAE and RMSE are for 30m, and lowest values of STDE and correlation coefficient are for 10m. The best values of MBE, MAE, RMSE, STDE and correlation coefficient are 0.192 m/s, 1.157 m/s, 1.506 m/s, 1.325 m/s and 0.761 respectively. The values of MBE decrease up to 30m height with an overestimation of wind dataset and then decrease negatively for higher heights with an underestimation of the dataset. Garmashov et al. [17] compared Era-Interim over Black Sea speed at 10 m height MBE, RMSE and R values are 0.35 m/s, 1.90 m/s and 075. Carvalho et al. [26] compared Era-Interim for Portugal at 10m height, the values of statistical parameters MBA, RMSE, STDE and R were 0.34 m/s, 2.10 m/s, 2.02 m/s and 0.79 respectively. A comparison of the present analysis and previous analysis of wind speed reveals that Era-Interim data has better R values in Europe compared to Pakistan [16, 25] [17].

For six hourly data of NCEP-CFSR, the lowest values of MBE, MAE and RMSE are for 22m, and lowest values of STDE and correlation coefficient are for 10m. The best values of MBE, MAE, RMSE, STDE and correlation coefficient are 0.017 m/s, 1.086 m/s, 1.418 m/s, 1.331 m/s and 0.739 respectively. Carvalho et al. [43] evaluated satellite datasets (NCEP NCAR, ERA-interim and CFSR) for Iberian Peninsula and came at the conclusion that the results of CFSR are better than those shown by NCEP NCAR but worse than ERA-interim where the comparison has been performed for 5 potential sites and the values of MBE, RMSE and R for CFSR data range from 0.38 m/s to 0.82 m/s, 1.77 m/s to 2.33m/s and from 0.83 to 0.90 respectively for wind speed. Overall at 30m both CFSR and Era-Interim show very good results in term of statistical analysis, probability distribution function and time series comparison.

6. Conclusions

The evaluation of the NCEP-CFSR and ERA-Interim against the measured wind speed data has been presented for the first time for the coastline of Pakistan. The wind industry in Pakistan is under development stages with the possible potential of the offshore wind farm, this study will be an initiative toward the development of the offshore wind farms in Pakistan. The results presented here can be of excellent value for the offshore wind resource assessment for Pakistan due to insufficient or unavailability of offshore wind speed data.

The evaluation of satellite wind speed data against measured data was done based on monthly mean time series, statistical analysis, regression plots and probability distribution function. The monthly mean time series comparison reveals that CFSR data gave a better estimation of wind speed compared to ERA-Interim. The NCEP-CSFR data shows better agreement with measured data in terms of errors compared to ERA-Interim but NCEP-CFSR has a lower value of R compared to ERA-Interim. The performance of Era-Interim is better than CFSR datasets based on regression plots and a best-fit line. The comparison of PDFs of both satellites shows the similar trend at a different height. Overall NCEP-CFSR gave better results compared to ERA-Interim.

7. References

[1] Mahmood A, Javaid N, Zafar A, Ali Riaz R, Ahmed S and Razzaq S 2014 Pakistan's overall energy potential assessment, comparison of LNG, TAPI and IPI gas projects *Renewable and Sustainable Energy Reviews* **31** 182-93

[2] Hong S-Y and Pan H-L 1996 Nonlocal Boundary Layer Vertical Diffusion in a Medium-Range Forecast Model *Monthly Weather Review* **124** 2322-39

[3] Sher H A, Murtaza A F, Addoweesh K E and Chiaberge M 2015 Pakistan's progress in solar PV based energy generation *Renewable and Sustainable Energy Reviews* **47** 213-7

[4] Rehman S, Halawani T and Husain T 1994 Weibull parameters for wind speed distribution in Saudi Arabia *Solar Energy* **53** 473-9

[5] Chou M-D and Lee K-T 1996 Parameterizations for the absorption of solar radiation by water vapor and ozone *Journal of the atmospheric sciences* **53** 1203-8

[6] Parrish D F and Derber J C 1992 The National Meteorological Center's Spectral Statistical-Interpolation Analysis System *Monthly Weather Review* **120** 1747-63

[7] Mirza I A, Khalil M S, Amer M and Daim T U 2015 *Policies and Programs for Sustainable Energy Innovations: Renewable Energy and Energy Efficiency,* ed T U Daim, *et al.* (Cham: Springer International Publishing) pp 55-81

[8] Mirza U K, Ahmad N, Harijan K and Majeed T 2009 Identifying and addressing barriers to renewable energy development in Pakistan *Renewable and Sustainable Energy Reviews* **13** 927-31

[9] Tahir Z R and Asim M 2018 Surface measured solar radiation data and solar energy resource assessment of Pakistan: A review *Renewable and Sustainable Energy Reviews* **81** 2839-61

[10] Zameer H and Wang Y 2018 Energy production system optimization: Evidence from Pakistan *Renewable and Sustainable Energy Reviews* **82** 886-93

[11] Amjad M, Zafar Q, Khan F and Sheikh M M 2015 Evaluation of weather research and forecasting model for the assessment of wind resource over Gharo, Pakistan *International Journal of Climatology* **35** 1821-32

[12] Mabel M C and Fernandez E 2008 Analysis of wind power generation and prediction using ANN: A case study *Renewable energy* **33** 986-92

[13] Lun I Y and Lam J C 2000 A study of Weibull parameters using long-term wind observations *Renewable energy* **20** 145-53

[14] Bagiorgas H S, Mihalakakou G, Rehman S and Al-Hadhrami L M 2012 Wind power potential assessment for seven buoys data collection stations in Aegean Sea using Weibull distribution function *Journal of Renewable and Sustainable Energy* **4** 013119

[15] Siddique S and Wazir R 2016 A review of the wind power developments in Pakistan *Renewable and Sustainable Energy Reviews* **57** 351-61

[16] Carvalho D, Rocha A, Gómez-Gesteira M and Santos C S 2014 Comparison of reanalyzed, analyzed, satellite-retrieved and NWP modelled winds with buoy data along the Iberian Peninsula coast *Remote sensing of environment* **152** 480-92

[17] Garmashov A V, Kubryakov A A, Shokurov M V, Stanichny S V, Toloknov Y N and Korovushkin A I 2016 Comparing satellite and meteorological data on wind velocity over the Black Sea *Izvestiya, Atmospheric and Oceanic Physics* **52** 309-16

[18] Winterfeldt J 2008 Comparison of measured and simulated wind speed data in the North Atlantic. Universität Hamburg Hamburg)

[19] Song L, Liu Z and Wang F 2015 Comparison of wind data from ERA-Interim and buoys in the Yellow and East China Seas *Chinese journal of oceanology and limnology* **33** 282-8

[20] Rashmi R, Polnikov V, Pogarskii F, Gomorev I, Samiksha V and Vethamony P 2016 Long-Term Variability of the Wind Field over the Indian Ocean Based on ERA-Interim Reanalysis *Atmosphere-Ocean* **54** 505-18

[21] Kiss P, Varga L and Jánosi I M 2009 Comparison of wind power estimates from the ECMWF reanalyses with direct turbine measurements *Journal of Renewable and Sustainable Energy* **1** 033105

[22] Luo Y, Wang D, Gamage T P, Zhou F, Widanage C M and Liu T 2018 Wind and wave dataset for Matara, Sri Lanka *Earth System Science Data* **10** 131

[23] Chang T P 2011 Performance comparison of six numerical methods in estimating Weibull parameters for wind energy application *Applied Energy* **88** 272-82

[24] Werapun W, Tirawanichakul Y and Waewsak J 2015 Comparative study of five methods to estimate Weibull parameters for wind speed on Phangan Island, Thailand *Energy Procedia* **79** 976-81

[25] Carvalho D, Rocha A, Gómez-Gesteira M and Santos C S 2014 Offshore wind energy resource simulation forced by different reanalyses: comparison with observed data in the Iberian Peninsula *Applied Energy* **134** 57-64

[26] Carvalho D, Rocha A, Gómez-Gesteira M and Santos C S 2014 WRF wind simulation and wind energy production estimates forced by different reanalyses: comparison with observed data for Portugal *Applied Energy* **117** 116-26

[27] Staffell I and Pfenninger S 2016 Using bias-corrected reanalysis to simulate current and future wind power output *Energy* **114** 1224-39

[28] Memon N, Jafari A H, Yousuf I and Khan S A 2010 Analysis of data of AEDB-UNDP (WEP) wind masts installed in Gharo-Keti Bandar wind corridor. AEDB, Islamabad, Pakistan)

[29] Schallenberg-Rodriguez J 2013 A methodological review to estimate techno-economical wind energy production *Renewable and Sustainable Energy Reviews* **21** 272-87

[30] Hellmann G 1914 *Über die Bewegung der Luft in den untersten Schichten der Atmosphäre*: Königlich Preussischen Akademie der Wissenschaften)

IOP Conf. Series: Journal of Physics: Conf. Series **1102** (2018) 012001 doi:10.1088/1742-6596/1102/1/012001

[31] Manwell J F, McGowan J G and Rogers A L 2009 *Wind Energy Explained: Theory, Design and Application*: Wiley)

[32] Gualtieri G and Secci S 2011 Comparing methods to calculate atmospheric stability-dependent wind speed profiles: A case study on coastal location *Renewable Energy* **36** 2189-204

[33] Staffell I and Green R 2014 How does wind farm performance decline with age? *Renewable Energy* **66** 775-86

[34] Carvalho D, Rocha A, Gómez-Gesteira M and Silva Santos C 2014 Comparison of reanalyzed, analyzed, satellite-retrieved and NWP modelled winds with buoy data along the Iberian Peninsula coast *Remote Sensing of Environment* **152** 480-92

[35] Hawkins S, Eager D and Harrison G P 2011 Characterising the reliability of production from future British offshore wind fleets. In: *IET Conference on Renewable Power Generation (RPG 2011),* pp 1-6

[36] Ruti P M, Marullo S, D'Ortenzio F and Tremant M 2008 Comparison of analyzed and measured wind speeds in the perspective of oceanic simulations over the Mediterranean basin: Analyses, QuikSCAT and buoy data *Journal of Marine Systems* **70** 33-48

[37] Dhanju A, Whitaker P and Kempton W 2008 Assessing offshore wind resources: An accessible methodology *Renewable Energy* **33** 55-64

[38] Petersen E L, Mortensen N G, Landberg L, Højstrup J and Frank H P 1998 Wind power meteorology. Part I: climate and turbulence *Wind Energy* **1** 2-22

[39] Fox N I 2011 A tall tower study of Missouri winds *Renewable Energy* **36** 330-7

[40] Dee D P, Uppala S M, Simmons A J, Berrisford P, Poli P, Kobayashi S, Andrae U, Balmaseda M A, Balsamo G, Bauer P, Bechtold P, Beljaars A C M, van de Berg L, Bidlot J, Bormann N, Delsol C, Dragani R, Fuentes M, Geer A J, Haimberger L, Healy S B, Hersbach H, Hólm E V, Isaksen L, Kållberg P, Köhler M, Matricardi M, McNally A P, Monge-Sanz B M, Morcrette J J, Park B K, Peubey C, de Rosnay P, Tavolato C, Thépaut J N and Vitart F 2011 The ERA-Interim reanalysis: configuration and performance of the data assimilation system *Quarterly Journal of the Royal Meteorological Society* **137** 553-97

[41] Saha S, Moorthi S, Pan H-L, Wu X, Wang J, Nadiga S, Tripp P, Kistler R, Woollen J, Behringer D, Liu H, Stokes D, Grumbine R, Gayno G, Wang J, Hou Y-T, Chuang H-y, Juang H-M H, Sela J, Iredell M, Treadon R, Kleist D, Delst P V, Keyser D, Derber J, Ek M, Meng J, Wei H, Yang R, Lord S, Dool H v d, Kumar A, Wang W, Long C, Chelliah M, Xue Y, Huang B, Schemm J-K, Ebisuzaki W, Lin R, Xie P, Chen M, Zhou S, Higgins W, Zou C-Z, Liu Q, Chen Y, Han Y, Cucurull L, Reynolds R W, Rutledge G and Goldberg M 2010 The NCEP Climate Forecast System Reanalysis *Bulletin of the American Meteorological Society* **91** 1015-58

[42] Carvalho D, Rocha A, Gómez-Gesteira M and Silva Santos C 2014 WRF wind simulation and wind energy production estimates forced by different reanalyses: Comparison with observed data for Portugal *Applied Energy* **117** 116-26

[43] Carvalho D, Rocha A and Gómez-Gesteira M 2012 *Ocean surface wind simulation forced by different reanalyses: Comparison with observed data along the Iberian Peninsula coast* vol 56

Comparison of simulations and wind tunnel measurements for the improvement of design tools for Vertical Axis Wind Turbines

Sarah Barber and Henrik Nordborg

University of Applied Sciences Rapperswil, Oberseestrasse 10, 8640 Rapperswil, Switzerland

E-mail: sarah.barber@hsr.ch

Abstract. As wind turbines are getting larger and cost reductions through up-scaling are reaching a limit, there is increasing pressure to reduce their installation and operating costs, making the business case for Vertical Axis Wind Turbines (VAWTs) more interesting again. However, in order to properly assess the economic potential of large-scale VAWTs, improved design tools are required. As the research budget for VAWT projects is generally too small to carry out expensive large-scale wind tunnel tests and field measurements, alternative methods are required. In this work, the application of small-scale, lower budget methods for improving VAWT design tools is assessed. It is shown that (a) currently available tools for VAWT design have not been sufficiently validated, (b) lower budget, small-scale wind tunnel tests can be effective for examining VAWT performance in terms of average power coefficient vs. tip speed ratio as well as the dynamics of the torque, and (c) Detached Eddy Simulations on small-scale VAWTs can be effective for examining their performance in terms of average power coefficient vs. tip speed ratio as well as the forces on the blades. Based on these results, a process for the transfer of the results of lower budget, small-scale measurements and simulations into recommendations for improved design tools is being developed.

1. Introduction

Vertical Axis Wind Turbine (VAWT) technology has not yet managed to establish itself as a reliable technology for wind energy projects. This is mainly because VAWTs have several disadvantages over Horizontal Axis Wind Turbines (HAWTs), including a lower efficiency, higher bearing loads and lower average height above the ground. However, advantages include the positioning of the drivetrain at the base, leading to reduced installation and service costs, and the independency of operation to wind direction, resulting in less fatigue. As wind turbines are getting larger and cost reductions through up-scaling are reaching a limit, there is increasing pressure to reduce their installation and operating costs. This is making the business case for Vertical Axis Wind Turbines (VAWTs) more interesting again.

This increased interest is reflected in the recent research activities aimed at better understanding the dynamics and aerodynamics of VAWTs - for example from the DeepWind project ([1], [2] and [3]) and from Sandia National Laboratories ([4], [5] and [6]) as well as the development of new offshore floating concepts such as [7], [8] and [9].

However, in order to properly assess the economic potential of large-scale VAWTs, improved design tools are required. Usually, design tools are improved by comparing them to costly wind tunnel and full-scale field tests in which the shaft torque and blade loads are measured at a range of flow conditions, in order to better understand the physics. As the research budget for VAWT development projects is generally small and because very few large-scale VAWTs are actually in operation, it is currently not feasible to carry out such measurement campaigns.

In this work, the application of small-scale, lower budget methods for improving VAWT design tools is assessed. This is done by firstly reviewing the currently available design tools, then carrying out small-scale wind tunnel measurements and Detached Eddy Simulations (DES) and comparing the results to simulations using an open-source design tool, QBlade. Finally, the feasibility of these methods for improving design tools is assessed.

2. Review of currently available tools

Tools such as DNV GL's Bladed [10], DTU's HAWC2 [11] and NREL's Aerodyn [12] are typically used in the wind energy industry for HAWT design. These tools consist of various different modules in which, for example, the aerodynamics, structural dynamics and control systems can be modelled. The aerodynamics are typically modelled with a simple Blade Element Momentum code [13] including corrections for blade tip and root vortex losses [13], for high tip speed ratios [14] and for dynamic stall [15] as well as additions for dynamic wake modelling [16]. The models have been shown to be accurate enough for design purposes and are well-established in the industry.

However, far less research and development work has been done on VAWT design. As the aerodynamic behaviour of VAWTs is significantly different to HAWTs, these codes cannot be directly used for VAWTs. Some of the first in-depth studies on the aerodynamics of VAWTs were focused on understanding the dynamic stall effects and vortex shedding in the wakes ([17], [18], [19], [20]). More recent studies have focused on the aerodynamics and structural dynamics of large-scale floating VAWTs by developing and applying design tools ([3], [6], [21] and [22]), CFD ([23], [24], [25], [26] and [27]) and wind tunnel tests ([28] and [1]).

The validation status in terms of power coefficient (C_P) or average shaft torque of some aerodynamic codes is summarised in Table 1 on the next page. It can be seen that most of the tools have not been properly validated in terms of their shaft torque production (or C_P). It is concluded that more VAWT validation measurements are required in the community order to improve the design tools.

3. Assessment of lower budget methods for design tool improvement

3.1. Small-scale wind tunnel tests

Small-scale wind tunnel tests were carried out in an open-section wind tunnel consisting of a fan connected to a 3 m long pipe of diameter 1.4 m with a honeycomb grid to smooth the flow. A three-bladed H-rotor VAWT model with a diameter of 0.7 m and a height of 0.5 m was built out of an aluminium frame and 3D-printed NACA 0024 blades with a chord length of 0.07 m as shown in Figure 1. An in-house Labview programme was developed in order to hold the rotational speed of the wind turbine constant at values ranging from 50 to 600 rpm using a Maxon RE 50 motor with a 4.3:1 gearbox and ESCON 50/5 servo controller.

In order to check the quality of the wind inflow conditions, the wind tunnel was first calibrated at the centre of the inflow plane with a PeakTech 5060 anemometer mounted on a positioning frame. Stable conditions were found for wind speeds up to 7 m/s, giving a maximum Reynolds number of 130,000 at the available rotational speeds and the measured air conditions of 14.4 °C and 96 kPa. The wind field of the inflow plane was then measured for a range of wind speeds with the anemometer positioned at 35 locations in a grid of 7 × 5 points as shown in Figure 2 for 2.85 m/s at the centre (the rotor area is shown by the black rectangle). This results in an average wind speed across the rotor plane of 25% more than the wind speed in the centre of the channel.

IOP Conf. Series: Journal of Physics: Conf. Series **1102** (2018) 012002 doi:10.1088/1742-6596/1102/1/012002

Table 1. Validation status of main VAWT aerodynamic tools.

Tool name	Institute	Model	Validation status
-	TU Delft	Double multiple streamtube model [18]	Shown to be of poor quality
HAWC2	DTU	Modified linear actuator cylinder [3]	No published comparison to C_P measurements
CACTUS	Sandia	3D vortex code with lifting line approximation [6]	Over-prediction of C_P above max. C_P
QBlade	TU Berlin	Lifting-line free-vortex wake code [29]	Mediocre performance of C_P compared to RANS CFD
U2DIVA	TU Delft	2D unsteady multibody free-wake panel code [28]	No published comparison to C_P measurements
ARDEMA2D	University of Uppsala / Areva	Free-vortex particle wake [30]	Poor performance of C_P compared to wind tunnel tests
PHARWEN	Nenuphar	3D vortex panel [8]	Average blade forces compare well to field measurements
-	University of Glasgow	Vorticity transport model [21]	Blade forces and wake profiles compare fairly well to wind tunnel experiments

Figure 1. Wind tunnel set-up.

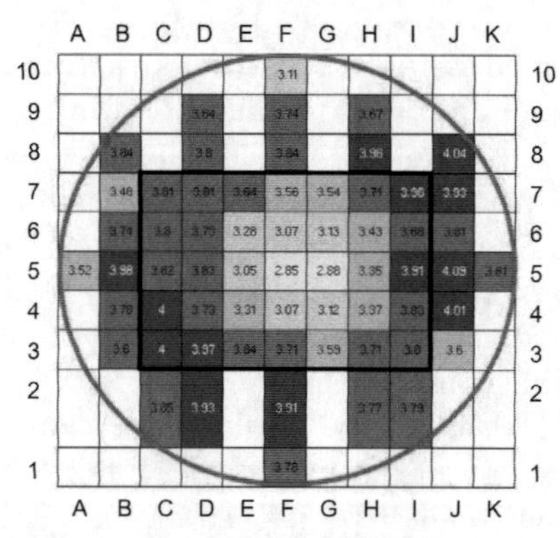

Figure 2. Measured inflow wind conditions at 2.85 m/s in the centre.

The key design parameter of a wind turbine is the shaft torque, allowing the rotor power coefficient (C_P) to be calculated as shown in Equation 1, where τ = rotor torque (Nm), ω = rotational speed (rad/s), ρ = air density (kg/m^3), V_0 = inflow wind speed (m/s) and A = rotor cross-sectional area (m^2).

$$C_P = \frac{\tau\omega}{\frac{1}{2}\rho V_0^3 A} \tag{1}$$

A Lorenz Messtechnik in-line torque sensor type DR-2112R was mounted between the motor and the rotor and measurements were carried out at a frequency of 2 kHz for at least 30 seconds per measurement point. Blade loads are also key to assessing design tools; however, they were not measured due to the expense and complexity of installing strain gauges in the rotating system.

The results in terms of power coefficient vs. tip speed ratio are shown in Figure 3 for wind speeds of 6.0 m/s, 6.5 m/s and 7.0 m/s, where the tip speed ratio (λ) is given by Equation 2 and R = rotor radius (m). They are compared to measurements on a similar-sized VAWT model from the University of New Hampshire [6], showing that the approximate magnitude is reasonable (the dimensions, aspect ratio, solidity and airfoil types are different). This shows that the chosen set-up could be used successfully to measure the full range of the power coefficient vs. tip speed ratio curve of the VAWT model.

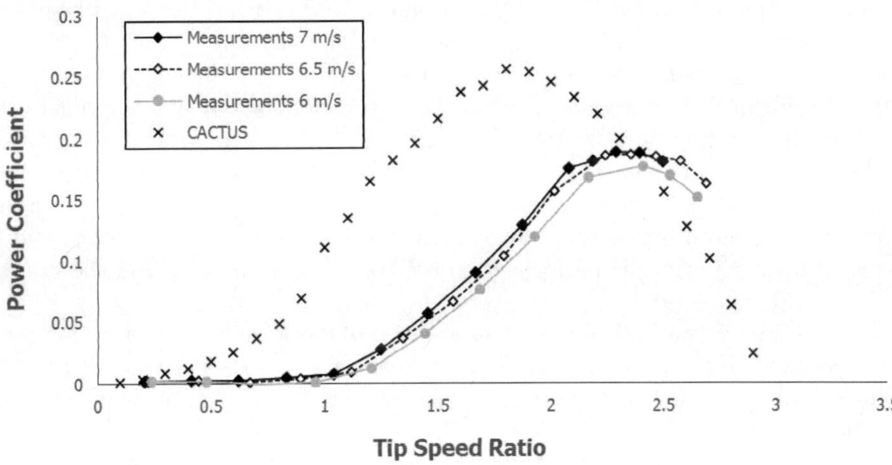

Figure 3. Wind tunnel measurement results.

$$\lambda = \frac{R\omega}{V_0} \tag{2}$$

Examination of the 2 kHz torque signal allows the periodical behaviour of the rotor to be analysed. Figure 4 shows the frequency content of three typical measurements carried out for 60 seconds each at tip speed ratios of 0.82, 1.65 and 2.47 (200, 400 and 600 rpm), obtained by carrying out Fast Fourier Transforms on the measurement data. As expected, clear peaks can be seen at the rotational speed (*1P*), at the blade passing frequency (three times the rotational speed = *3P*), at twice the blade passing frequency (*6P*) and at three times the blade passing frequency (*9P*). Large peaks at *2P* can also be seen for every rotational speed, indicating a rotor imbalance from an error in manufacture or pitch setting of one of the blades. Additionally, large peaks at *4.3P* can be seen. These are thought to be caused by the dynamics of the motor, which turns 4.3 times faster than the rotor. A small peak can be seen on each plot at a frequency of approximately 23 Hz, circled on the figure, indicating a dynamic that is independent of the rotor speed. This could be the first torsional natural frequency of the structure.

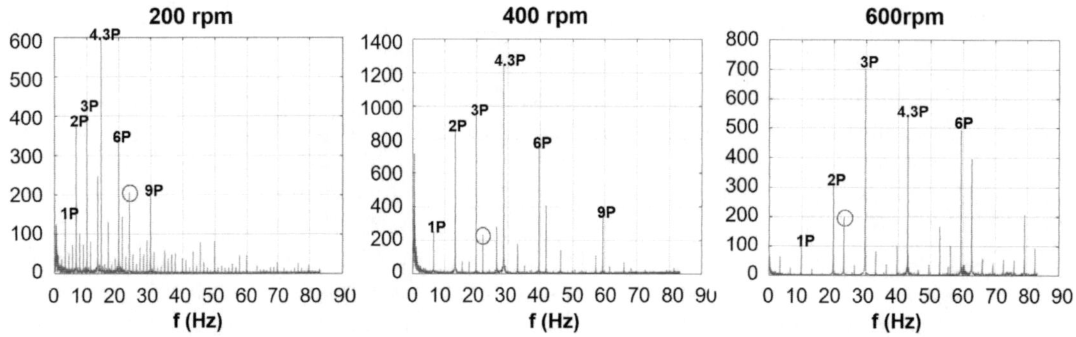

Figure 4. Frequency content of a three typical torque signals at tip speed ratios of 0.82, 1.65 and 2.47 (200, 400 and 600 rpm).

As the frequency content at *4.3P* is due to the motor and not to the wind turbine rotor, it needs to be removed in order to examine the dynamic behaviour of the rotor. On the left-hand graph of Figure 5, a typical torque signal at a tip speed ratio = 2 is shown with a low-pass filter of 100 Hz applied. The periodic nature of the torque can be seen; however, the *4.3P* content is dominating. Applying a 30 Hz low pass filter removes the *4.3P* frequency, but also the interesting *6P* and *9P* frequencies (middle graph). Here only the blade passing frequency can be seen, as well as the asymmetry due to the misalignment of one blade as discussed above. The *4.3P* peak was therefore removed manually from the Fourier transformed-data and then an Inverse Fast Fourier Transform was performed on the resulting dataset, giving the graph shown on the right-hand side of Figure 5.

This shows a distinctive periodical behaviour; the torque does not smoothly transition from the maximum to the minimum value, but it increases again slightly in the middle. This is thought to be caused by the interaction of the blades with the wake, and is examined further in the next section.

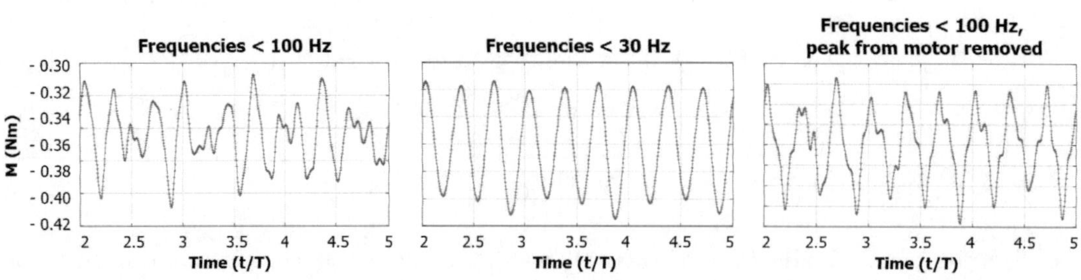

Figure 5. Typical torque signal at tip speed ratio = 2 (rotational speed = 435 rpm = 7.25 Hz).

In summary, it has been found that lower budget small-scale wind tunnel tests can be used to compare the performance in terms of power coefficient vs. tip speed ratio as well as to investigate

IOP Conf. Series: Journal of Physics: Conf. Series **1102** (2018) 012002 doi:10.1088/1742-6596/1102/1/012002

the natural frequencies and the dynamic behaviour of the total torque on the shaft within a rotor revolution. This has the potential for helping to understand the physics of VAWTs and thus improving design tools.

3.2. Detached Eddy Simulations on small-scale VAWTs

Computational Fluid Dynamics (CFD) can be used in order to gain detailed information of turbulent flows around objects. In this case, its application can potentially help to better understand the behaviour of VAWTs and therefore to improve design tools, either in combination with or instead of small-scale wind tunnel tests.

The basis of conventional CFD simulations are the Navier-Stokes equations, which define single-phase fluid flows and can only be solved for turbulent flows using turbulence modelling, as the resolution required to resolve all scales involved in turbulence is beyond what is computationally possible. The most common - and least computationally expensive - approach to turbulence modelling is to derive the Reynolds-Averaged Navier Stokes (RANS) equations from the instantaneous Navier-Stokes equations via a Reynolds decomposition. A more accurate - and more computationally expensive - approach is Large Eddy Simulations (LES), involving removal of the smallest scales of the through a filtering operation and modelling of their effect using subgrid scale models. This allows the largest and most important scales of the turbulence to be resolved, while greatly reducing the computational cost incurred by the smallest scales. A compromise between the two methods is Detached Eddy Simulations (DES), which uses RANS to resolve boundary layers and LES to resolve the larger eddies in the bulk flow. Therefore, the grid resolution for DES is not as demanding as pure LES, considerably reducing the cost of the computation.

As this work focuses on the application of small-scale, lower budget methods for improving VAWT design tools, it was decided to carry out Detached Eddy Simulations (DES) on the VAWT model in the small-scale wind tunnel. This was found to be the best compromise between computational expense and model accuracy, as the small scale of the simulations reduces the computational costs but retains the better accuracy of DES compared to RANS.

First of all, Detached Eddy Simulations were carried out using CFX ANSYS version 19.0 on one blade of the wind turbine for a Reynolds number of 88,000 as shown in Figure 6(a). The results in terms of lift coefficient (c_L) vs. angle of attack (α) are compared to the results of other turbulence models (SST k-omega, SST k-omega coarse, k-epsilon and Scale-Adaptive Simulation [31]) as well as to XFOIL simulations [32] with different values of *Ncrit* in Figure 6(b). *Ncrit* is the log of the amplification factor of the most-amplified frequency which triggers transition, and should be set depending on the ambient disturbance level in which the airfoil operates (= 4-8 for a dirty wind tunnel, 10-12 for a clean wind tunnel). First of all, it can be seen that the XFOIL results are highly dependent on the value of *Ncrit*. Also, it can be seen that DES is capable of reasonably predicting the flow over the blade.

Next, simulations were carried out on one half of the wind turbine and outlet pipe of the wind tunnel as shown in Figure 7. Two different meshes were used for the simulations, having 15 million and 44 million cells. The mesh around the structure and the blades was refined using inflation layers to reach a y+ of less than 30. Incompressible transient simulations were performed using the Detached Eddy Model for turbulence. The wall function is used to model the boundary layer very close to the blades. The time steps ranged from 0.5 to 1.0 ms, depending on the rotational speed of the turbine. The simulation results were checked for mesh and time-step independence. The simulations were run at the computing cluster at the University of Applied Sciences Rapperswil, using 80-120 cores per simulation. The input wind speed was set to a constant value of 6 m/s.

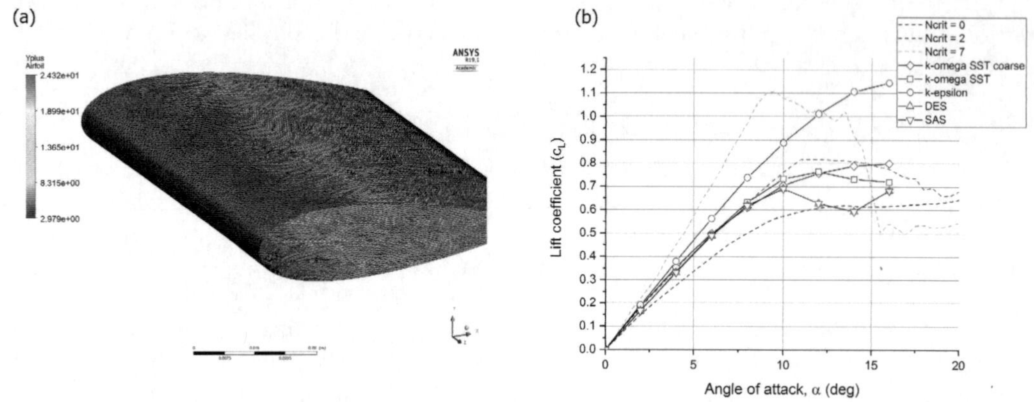

Figure 6. (a) Set-up of the Detached Eddy Simulations for one blade; (b) Results in terms of lift coefficient vs. angle of attack.

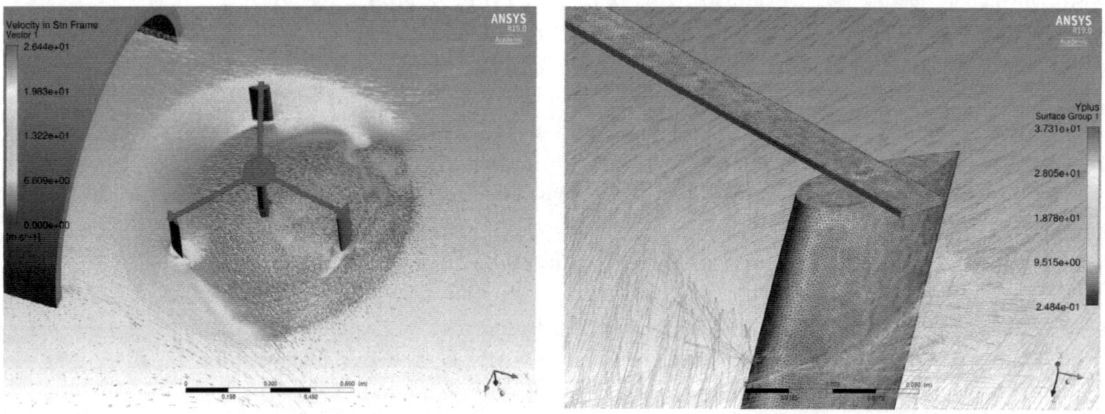

Figure 7. Set-up of the Detached Eddy Simulations.

The results in terms of power coefficient vs. tip speed ratio are shown in in Figure 8 compared to the wind tunnel measurements. At and above the maximum C_P, the results match fairly well. At lower tip speed ratios, the simulations match less well, due to the increasing average angle of attack (leading to more flow separation, which is difficult to simulate) and the decreasing Reynolds Number (reducing the likelihood of a turbulent flow, which is easier to simulate). Further differences are thought to be caused by the non-uniform wind input field in the wind tunnel, which is not accounted for in the simulations.

In Figure 9, the variation of torque with time for a tip speed ratio = 2 is examined. In part (a), the torque exerted on the shaft by one blade is shown. One large peak is visible per revolution, when the blade passes the front of its swept area and is operating at its optimal angle of attack and producing lift. A second, smaller peak is seen as the blade passes the back of its swept area, producing a small amount of lift. The torque even becomes negative at two azimuth locations, because here the drag dominates. The torque produced by the other two blades shows a similar pattern, offset by one-third of a revolution. The total torque, shown in part (b), is then given by the sum of these three torques at each azimuth position. Like the measurements shown in the previous section, this signal does not smoothly transition from the maximum to the minimum value, but it increases again slightly in the middle. This occurs due to the presence of the second small peak as each blade passes behind the tower. Additionally, it can be seen in part (b) that the magnitude of the torque fluctuations matches well with the measurements (both approximately 0.15 Nm), confirming the ability of the simulations to predict the physics correctly. Part (b) also shows a comparison between the positive torque produced by the blades and the resistance caused by the structure of the wind turbine, comprising approximately 10% of the total torque.

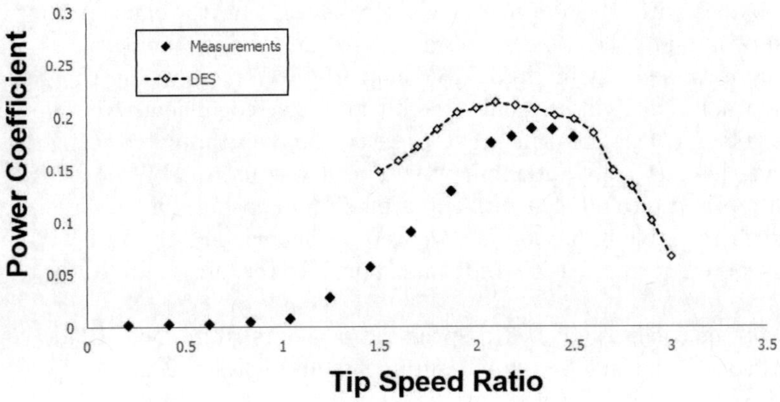

Figure 8. Comparison of measurements to DES.

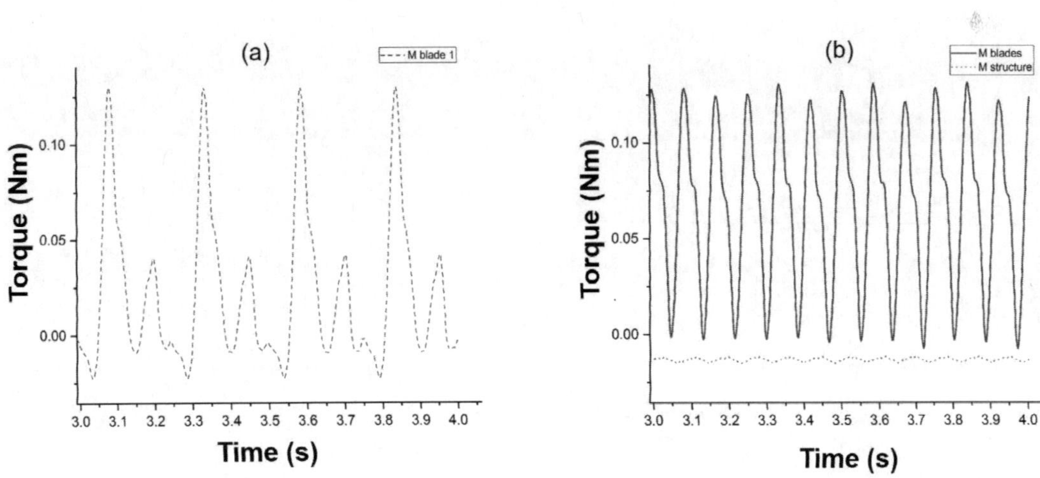

Figure 9. Variation of torque with time from DES: (a) For one blade; (b) For all blades (solid line) and for the structure (dashed line).

In summary, it has been found that lower budget Detached Eddy Simulations on small-scale VAWTs can be used to compare the performance in terms of power coefficient vs. tip speed ratio as well as the dynamic behaviour of the wind turbine. This has the potential for helping to understand the physics of VAWTs and thus improving design tools. Further analysis of these results is on-going.

3.3. Open-source design tool

The design tool that is applied in this work is the open-source software QBlade [29] from TU Berlin, which is capable of Double Multiple Streamtube Model (DMSM) and Lifting Line Free Vortex Wake (LLFVW) simulations. The DMSM [33] applies the actuator disc and blade element theory to VAWTs by splitting up the stream tube flowing through the VAWT into a set of smaller stream tubes that are passed through by each blade twice per revolution. The LLFVW method [34] models the blades with a single line of vortices located on the quarter chord of the blade and updates the position of the wake end nodes based on the local velocity.

Simulations were carried out in the DMSM and LLFVW mode using the geometry and air conditions of the small-scale wind tunnel. The lift and drag coefficients were obtained from the results of the one-bladed DES from the previous section and extrapolated using the Montgomery method in QBlade [29]. A constant input wind speed was used. For the DMSM simulations, the blade was discretised into 40 elements and a maximum epsilon for convergence of 0.001 was used. Tip losses and variable induction factors were neglected. For the LLFVW simulations, 10 rotor revolutions were simulated with 720 time steps. Tower effects and drag of the structure were neglected.

The average power coefficient vs. tip speed ratio results from the QBlade simulations are shown in comparison to the small-scale measurements and DES in Figure 10(a). It can be seen that the DMSM and the LLFVW simulations overestimate the power coefficient compared to the measurements and the DES at tip speed ratios above the peak power coefficient. This is expected from previous work showing that DMSM generally overestimates at high tip speed ratios due to the fact that backflow effects are not taken into account [33] and that LLFVW generally overestimates the torque [29]. In Figure 10(b), the familiar variation of torque with time can be seen as discussed in the previous sections.

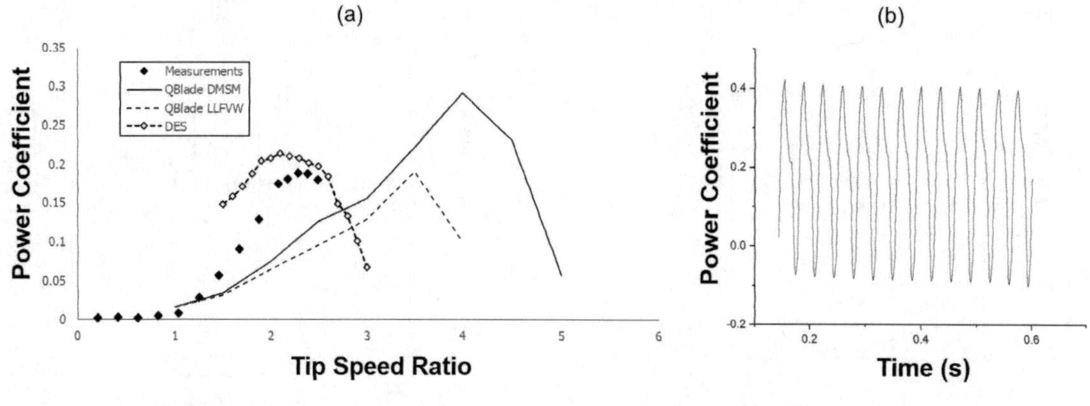

Figure 10. (a) Comparison of results from measurements, DES and QBlade in terms of power coefficient vs. tip speed ratio; (b) Power coefficient vs. time for the total torque from the QBlade LLFVW simulations for a tip speed ratio = 2.

3.4. Summary of lower budget methods

VAWT design codes were found to not match well with simulations and wind tunnel tests, both in the literature and in the present work. However, lower budget, small-scale wind tunnel tests and Detached Eddy Simulations were successfully used to examine the average torque as well as its variation with time in order to examine the physics of VAWTs. This has the potential for improving VAWT design tools and ultimately improving VAWT design and reducing wind farm operating costs. Based on these results, a process for the transfer of the results of lower budget, small-scale measurements and simulations into recommendations for improved design tools is currently being developed.

4. Conclusions

This work has shown the following:

- Currently available aerodynamic tools for VAWT design have not been sufficiently validated in terms of the shaft torque and blade loads, the most important design parameters.
- Lower budget, small-scale wind tunnel tests can be effectively used for examining VAWT performance in terms of average power coefficient vs. tip speed ratio as well as the dynamics of the torque.
- Lower budget Detached Eddy Simulations on small-scale VAWTs can be effectively used for examining their performance in terms of average power coefficient vs. tip speed ratio as well as the forces on the blades.
- A process for the transfer of the results of lower budget, small-scale measurements and simulations into recommendations for improved design tools is currently being developed.

Acknowledgments

Thanks to Lukas Schnider, Simon Boller and Roland Peterer for help with the wind tunnel measurements.

References

[1] Najafi N and Paulsen U S 2017 *Energy* 405–416
[2] Galinos C, Larsen T J, Madsen H A and Paulsen U S 2016 *Energy Procedia* **94** 319–328
[3] Madsen H A, Larsen T J, Paulsen U S and Vita L 2013 *Proceedings of 51st AIAA Aerospace Sciences Meeting including the New Horizons Forum and Aerospace Exposition (Chapter AIAA 2013-0913)* (AIAA)
[4] Owens B C and Griffiths D T 2014 *Journal of Physics: Conferrence Series* vol 012092 p 524
[5] Owens B C and Griffith T D 2014 *32nd ASME Wind Energy Symposium, AIAA SciTech Forum* vol 0518 (AIAA)
[6] Owens B C, Hurtado J E, Paquette J A, Griffith D T and Barone M 2013 *54th AIAA/ASME/ASCE/AHS/ASC Structures, Structural Dynamics, and Materials Conference* vol 1552 (AIAA)
[7] Borg M, Shires A and Collu M 2013 *Renewable and Sustainable Energy Reviews* **39** 1214–1225
[8] Dupont M, Pitance D, Kluczewska-Bordier J, Immas A and Silvert F 2017 *WindEurope Conference and Exhibition (Poster P.O. 262)*
[9] Pedersen T F, Paulsen U S, Madesn H A, Nielsen P H, Enevoldsen K, Tesauro A, Kragh K A, Vita L, Ritchie E, Leban K, Wedell-Heinen J and Larsen K H 2013 *European Wind Energy Conference & Exhibition 2013* (European Wind Energy Association (EWEA))
[10] 2018 Dnv gl bladed URL `https://www.dnvgl.com/energy/generation/software/bladed/index.html`
[11] 2018 Hawc2 URL `www.hawc.de`
[12] 2018 Nwtc information portal (aerodyn) URL `https://nwtc.nrel.gov/AeroDyn`
[13] Glauert H 1935 *Airplane propellers* vol 4 (Springer)
[14] Glauert H 1926 *A General Theory of the Autogyro* (ARCR R&M No. 1111)
[15] Leishman J G and Beddoes T S 1989 *Journal of the American Helicopter Society* **34**(3) 3–17
[16] Peters D A and He C J 1992 *Journal of American Helicopter Society*
[17] Ferreira C J S, van Zuijlen A, Bijl H, van Bussel G and van Kuik G 2010 *Wind Energy* **13** 1–17
[18] Ferreira C S, van Kuik G, van Bussel G and Scarano F 2009 *Experiments in Fluids* **46** 97–108
[19] Frounie P, Beguier C, Paraschivoiu I and Brochier G 1986 *Journal of Propulsion and Power* **5** 445–449
[20] Almohammadi K M, Ingham D B, Ma L and Pourkashanian M 2015 *Journal of Fluids and Structures* **57** 144–158

[21] Scheurich F, Fletcher T M and Brown R E 2011 *Wind Energy* **14** 159–177
[22] Mendoza V, Bachant P, Wosnik M and Goude A 2016 *Journal of Physics: Conference Series* (IOP Publishing) p 753
[23] Mendoza V and Goude A 2017 *IOP Conference Series: Journal of Physics* vol 012031 (IOP Publishing)
[24] Shamsoddin S and Port-Agel F 2014 *Energies* **7** 890–912
[25] Abkar M and Dabiri J O 2017 *Journal of Turbulence*
[26] Hezaveh S H, Bou-Zeid E, Lohry M W and Martinelli L 2017 *Wind Energy* **20** 713–730
[27] Bangga G, Lutz T, Dessoky A and Krmer E 2017 *Journal of Renewable and Sustainable Energy* **9**(5)
[28] Tescione G, Ragni D, Ferreira C J S and van Bussel G J W 2014 *Renewable Energy* **70** 47–61
[29] Marten D 2016 Qblade v0.95 - guidelines for lifting line free vortex wake simulations
[30] Eriksson S, Kjellin J and Bernhoff H 2013 *Energy Science Engineering* **1** 135–143
[31] Menter F and Egorov Y 2005 *AIAA Paper 2005-1095*
[32] Drela M 2000 Xfoil URL `http://web.mit.edu/drela/Public/web/xfoil/`
[33] Marten D and Wendler J 2013 Qblade guidelines v0.6
[34] Marten D, Pechlivanoglou G, Nayeri C and Paschereit C O 2016 *International Symposium on Transport Phenomena and Dynamics of Rotating Machinery*

IOP Conf. Series: Journal of Physics: Conf. Series **1102** (2018) 012003 doi:10.1088/1742-6596/1102/1/012003

Wind field reconstruction from lidar measurements at high-frequency using machine learning

Clym Stock-Williams[1], Paul Mazoyer[2] and Sébastien Combrexelle[3]

[1] TNO, Westerduinweg 3, 1755 LE Petten, The Netherlands
[2] Leosphere, 14-16 rue Jean Rostand, 91400 Orsay, France

Clym.Stock-Williams@tno.nl

Abstract. The increase in wind turbine size during recent years has led to a situation where the wind speed measured at hub height is no longer a sufficient representation of the power available to be extracted by a turbine. Light Detection and Ranging ("Lidar") wind sensing devices can measure wind speeds in a volume in front of the sensor, offering affordable and improved estimation of wind turbine power performance and loading. However, a Lidar can only measure the component of the wind velocity in the direction of the beam. This makes reconstruction of the wind velocity field from raw Lidar measurements a challenge, and current methods contain assumptions which cause inaccuracies in complex terrain and in wind turbine wakes. In addition, data availability is reduced in specific weather conditions, such as very cold climates and dense fog. A novel Machine Learning method is developed here, based on Gaussian Process regression, to remove these assumptions when producing full 3D wind fields from Lidar measurements. This approach is naturally robust to overfitting and predicts uncertainty derived from data density and machine error, without needing physical information about the local terrain. Initial validation by comparison of a Windcube Lidar with a meteorological mast, on a site with simple conditions, shows excellent performance which meets the threshold for commercial use. The method also infers data during measurement gaps, offering the potential for 100% data availability. Instantaneous 2D wake measurements in a plane can be made, which when validated and further developed will offer a breakthrough in understanding of wind farm interactions and turbine loading.

1. Introduction

The increase in size of wind turbines over the recent years has led to a situation where the wind speed measured at hub height is no longer a sufficient representation of the power available to be extracted by a turbine. The new version of IEC standard 61400-12-1 therefore allows for a new definition: Rotor Equivalent Wind Speed, which considers the variation of wind speed and angle across the rotor area.

Such a definition, however, requires data to calculate it. Light Detection and Ranging ("Lidar") wind sensing devices have emerged in recent years, which use lasers to measure wind velocity over a volume (a 3D "wind field") in front of the device. Ground-based Lidar measurements are approved in the new standard for power performance testing, although still when accompanied by a meteorological mast.

This advantage over traditional point-measurement anemometers, along with their portability, have enabled not only more accurate power performance measurements, but also several innovations which could reduce Cost of Energy ("CoE") significantly:

IOP Conf. Series: Journal of Physics: Conf. Series **1102** (2018) 012003 doi:10.1088/1742-6596/1102/1/012003

- *Yaw misalignment correction*. This has already demonstrated up to 5% improvement in power output.
- *Site suitability pre-construction studies*, to select appropriate turbines for the site, considering the wind speeds over the full area of the turbine. This reduces fatigue and lower maintenance costs, thereby increasing availability, as well as the use of Lidar reducing costs relative to meteorological masts.

There is a problem, however, with current methods used to reconstruct the wind field from the raw Lidar measurements. Simplifications are made to ensure reasonably accurate estimation of the 10-minute average wind speed and direction [1], resulting in the following limitations which are not inherent in the Lidar sensors themselves:

- lack of reliable information about the variability of the wind conditions at higher frequencies than 10 minutes;
- poor performance when the simplifying assumptions are not valid, particularly: in complex terrain; or when the sensors are used inside wind farms, since wind conditions then vary significantly across the Lidar measurement volume due to wakes.

Machine learning offers the potential to improve significantly the accuracy and volume of wind field information reconstructed from raw Lidar data, particularly in complex conditions.

Section 2 explains the methodology. First, the physical basis for the Lidar operation and measurements, so that the inputs and desired outputs of the wind field reconstruction algorithm are made clear. Second, the machine learning method applied: Gaussian Process ("GP") Regression. Third, the application of GPs to the wind field reconstruction algorithm.

Section 3 demonstrates the application of the developed algorithm to real wind measurements, to validate the method. Additional benefits over current methods are listed and examples given from real wind data.

Finally, section 4 draws conclusions and suggests future development priorities to realise the benefits fully and enable it to be widely used.

2. Approach

2.1. Lidar Physics

Lidar units point a laser beam (either pulsed or continuous wave) into the sky and analyse the backscatter signal to determine the radial wind speed through the Doppler effect.

Here, the Windcube Lidar unit from Leosphere is considered, which is a ground-based pulsed Lidar unit with 5 static beam directions. Four beams are oriented towards the four cardinal directions, angled by 28° from the vertical; while the fifth points vertically. Wind speeds can be measured in a 40 - 200m range, at up to 12 programmable heights (10 are generally used). All 5 beams are measured consecutively over a 4-second period.

Pulsed Lidar units have soft focus across all ranges simultaneously, since time of flight windowing can be used to detect the backscatter at a certain distance. However, since the pulse has a finite width, and because of spectral broadening, there is a measurement resolution of 25-30m in the location of the backscatter signal, which is almost constant with distance.

IOP Conf. Series: Journal of Physics: Conf. Series **1102** (2018) 012003 doi:10.1088/1742-6596/1102/1/012003

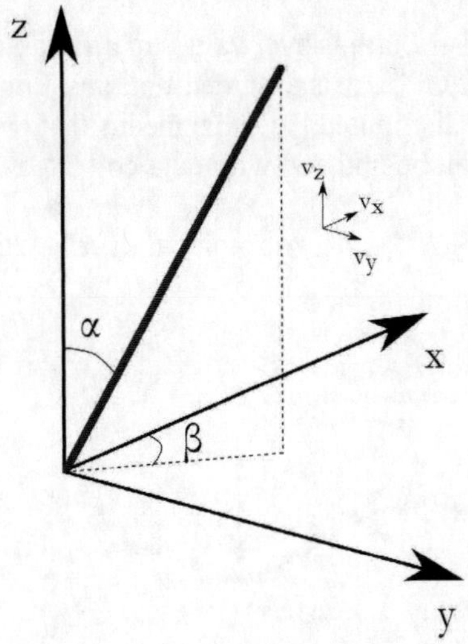

Figure 1. Spherical co-ordinate system for Lidar beam measurements.

Figure 1 shows the geometry. The component v_b of the wind velocity vector v measured by the beam (at any distance r) is the dot product of that velocity with the unit vector of the beam ($\hat{\boldsymbol{b}}$):

$$
\begin{aligned}
v_b &= \boldsymbol{v}^{\boldsymbol{T}} \cdot \hat{\boldsymbol{b}} \\
\therefore v_b &= -\left(v_x \sin\alpha \cos\beta + v_y \sin\alpha \sin\beta + v_z \cos\alpha\right)
\end{aligned}
$$

Equation 1. Radial flow speed measured by Lidar beam, due to true wind velocity.

Assumptions are therefore required to reconstruct the wind field, since only one resolved component of the wind field is measured at each point in space. This is known as the Cyclops Effect.

Currently, the standard wind field reconstruction algorithm supplied with many Lidar units assumes that all measurements within a certain time at the same height come from a stationary homogeneous process and therefore can be averaged together. For the WindCube that means averaging the North and South beams together to get v_x, and the East and West beams together to get v_y. More sophisticated methods include:

- Flow Complexity Recognition (FCR), Leosphere's patented method [2] for reproducing the wind field without homogeneity assumptions. This relies on fitting simplified Navier-Stokes equations and considering the topography of the ground around the Lidar.
- 10-minute averaged data using Taylor's "Frozen Turbulence" hypothesis [3];
- Real-time reconstruction by coupling a low-order dynamic wind model with a Kálmán filter [4];
- Reconstruction of a plane of wind velocity by using fast Navier-Stokes solver LINCOM [5];
- Use of a Wiener filter with various physics-based wind field assumptions [6];
- Use of ARMA-X modelling assuming Linear Time Invariance [7]; and
- A recent article which seems to retreat to a stationary homogeneous model [8].

The effort in [4] comes closest to using pure machine learning methods without imposing limiting assumptions on the wind field behaviour. There is however a gap for a new approach to reconstructing wind fields, minimising the number of site-specific or restrictive assumptions.

2.2. Gaussian Processes

Stochastic processes are an extension of the concept of probability distributions over values of a variable, to probability distributions over variables. In a Gaussian Process ("GP"), it is assumed that any finite number of variables described by that process are jointly Normally distributed [9]. This means that one can express a prior probability distribution over functions, which can be updated with data collected to provide predictions at any other point in the input space.

A GP is fully specified by its mean and covariance functions: $y \sim GP(\boldsymbol{\mu}, \boldsymbol{\Sigma})$. A graphical overview is given in Figure 2.

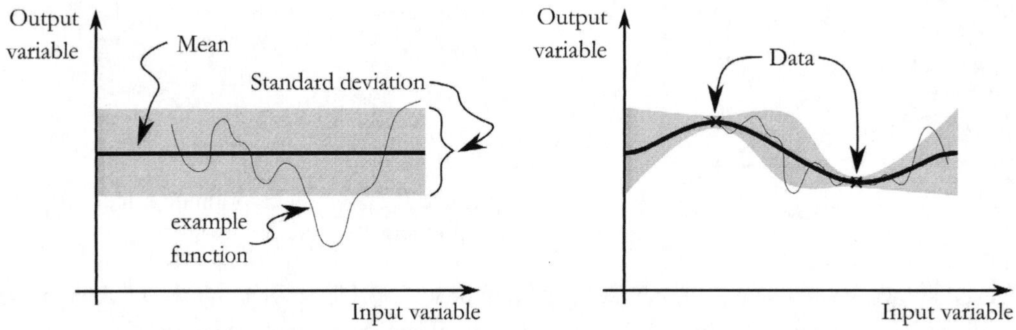

Figure 2. Basic overview of Gaussian Processes, after [9]: left, prior; right, posterior.

The covariance function k captures the relationship between the variables in the process:

$$k(x,x') = E[(f(x) - E[f(x)])(f(x') - E[f(x')])]$$

where x and x' denote values of the input variables and E is the expectation operator. Covariance "kernels" are often combined to create an appropriate model. The Mátern class of kernels provides functions which often realistically match natural processes. Each kernel is specified by a parameter ν, which is often set to half-integer values. For $\nu = {}^3/_2$, the kernel is:

$$k_{Mát3/2}(x,x') = \sigma_f^2 \left(1 + \frac{\sqrt{3}(x - x')}{\ell}\right) e^{-\frac{\sqrt{3}(x-x')}{\ell}}$$

Once the model is specified, the hyperparameter values (e.g. ℓ) should be optimised to fit the data. The data are still required to make predictions, since this is a non-parametric model.

GP regression works particularly well when inferring information from sparse, noisy data. This class of algorithms has been applied successfully to a number of problems, including the inference of meteorological data across time and space in a country [10] and the reconstruction of topographical surfaces from Lidar scans for robotics [11] and mapping [12]. However, it has not been applied to wind field reconstruction.

2.3. Wind Field Reconstruction using Gaussian Processes

In a similar manner to that used in [11], a GP is fitted to the aspect we do not understand – the speed measured by the Lidar beam in space and time – and geometric equations are then used, along with some physical assumptions, to convert these into wind velocities. The steps in the method are:

1. Create a stationary GP ("GP1") that can model beam velocities measured during only a small period of time (e.g. 1 minute).

2. Fit this model to a selection of periods in the data, then apply another GP ("GP2") over the hyperparameters of those models in order to smooth them.
3. To create predictions of beam speed for a given set of time stamps, we use a set of half-overlapping GP1s, with hyperparameters set from GP2 according to their time.
4. Predictions from each half-overlapping GP1 are combined through a form of Mixture of Experts to create the final beam velocity predictions: a "Virtual Lidar".
5. Geometrical, physical and statistical considerations are now used to infer the wind velocity vectors from the Virtual Lidar's beam speeds.

2.3.1. "GP1" for Short-Term Beam Speed Prediction. This GP takes measurements in space and time, in the co-ordinate system of the Lidar, and re-predicts a continuous beam speed field:

$$GP1: [\alpha_1, \alpha_2, r, t] \rightarrow v_b$$

Equation 2. Definition for GP1, mapping a static Lidar beam measurement location to beam speed.

Here, $\alpha_1, \alpha_2 \in [-\pi/2, \pi/2]$ describe the beam orientation in the well-behaved co-ordinate system $\alpha_1 = \alpha \cos\beta$, $\alpha_2 = \alpha \sin\beta$; $r \in [0, \infty]$ is the measurement distance from the Lidar unit; and $t \in \mathbb{R}$ is the time since a fixed reference. GP1 is defined by a zero mean function, Gaussian likelihood, Mátern 3/2 covariance kernel with Automatic Relevance Determination (ARD) [13] and box hyperpriors over the lengthscale hyperparameters for the variables α_1, α_2 and r with range [4,7].

Figure 3 shows an example output of instantaneous predictive mean and standard deviation of beam speed. As expected, a smooth variation of beam speed in space is seen, along with lower uncertainty where the measurements are taken along the beams.

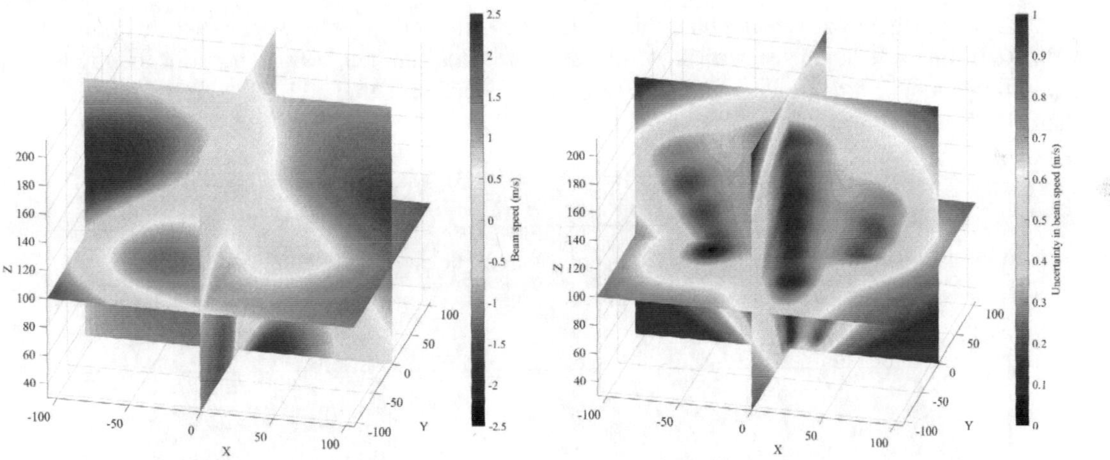

Figure 3. Plane slices across a 3D instantaneous wind field, showing mean beam speed (left) and standard deviation (right) as reconstructed by GP1.

2.3.2. "GP1" for Short-Term Beam Speed Prediction. The difficulty with GP regression in time (and in particular when fitting in real-time rather than to historical data) is that the predictions of the model are affected near the edges of the "window" by the lack of information about behaviour outside it. Wind conditions are expected to have turbulent length-scales that change in time, therefore fitting one GP for a whole day, or even 10-minute period, is not desirable.

The trade-off between amount of data available to gain a better fit and solution time is managed here by fitting one "locally-expert" GP1 to each "window" of 60 seconds of data, with a stride of 30 seconds.

As a result, each GP1 overlaps with its neighbours for 30 seconds. The individual predictions are then combined together through a weighting function, as follows:

$$v_{b*}(t) = \frac{\Sigma_i w_i(t) v_{b*_i}(t)}{\Sigma_i w_i(t)}$$

The weighting function chosen is a standard Normal distribution, centred on the middle of each GP Expert, with the time arbitrarily scaled such that the start and end are at ± 4. This is appropriate since, as previously mentioned, the GP has more information and therefore produces more accurate results near the centre of its window. Since the predictions of each GP are effectively fully correlated, the predictive uncertainties can be calculated in the following manner:

$$\delta v_{b*}(t) = \left(\sum_i w_i(t)\right)^{-1} \sum_i w_i(t) \delta v_{b*_i}(t)$$

Setting the hyperparameters of each GP1 by independent optimisation to the local data results in unphysical behaviour. Instead, the (five) hyperparameters from local optimisation of one GP1 model every 10 minutes are obtained. A multi-output GP ("GP2") is then fit to these values in order to smooth them, the predictions of which can then be used to set up a GP1 for any period in time covered by the data available (and predict into the future):

$$GP2: \tau \rightarrow \boldsymbol{h}$$

Equation 3. Definition for GP2, mapping the start time for GP1 to the hyperparameter values for GP1.

GP2 is defined by a zero mean function, Gaussian likelihood, squared exponential covariance kernel with ARD, multiple outputs specified by indicator variables, and a narrow hyperprior forcing the time length-scale to take a value of around e^{10} seconds (6 hours), creating a slowly varying function rather than one fit to the noise.

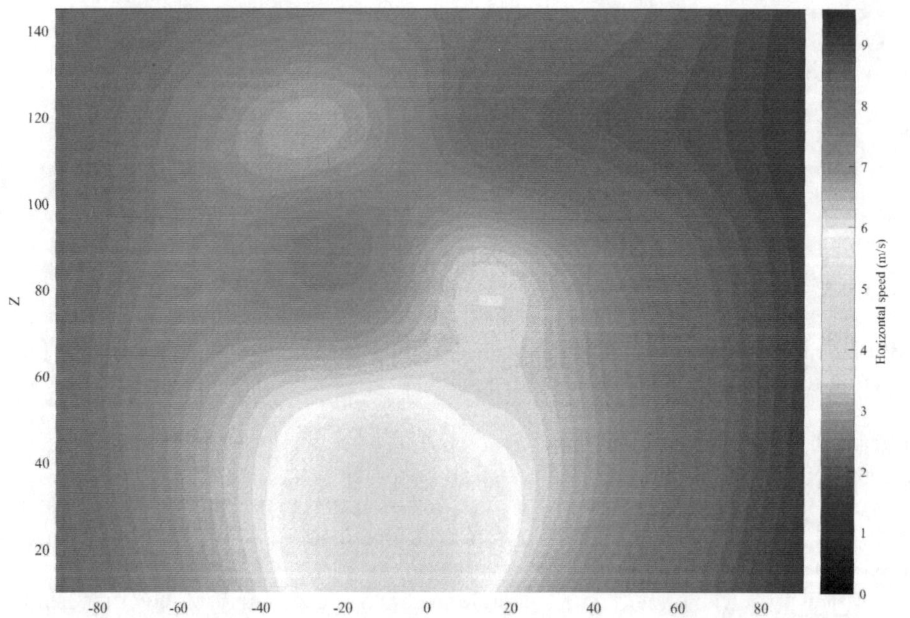

Figure 4. Vertical plane (metres East-West versus metres above ground) showing instantaneous horizontal wind speed which includes a wind turbine wake.

2.3.3. Conversion of Virtual Lidar to Wind Field. At this point, it is possible to predict (including uncertainty) the beam speed at any point in time and space, which we shall call a "Virtual Lidar". The GPs take care of spatial and temporal variability, leaving only the issue of removing the dimensionality reduction associated with the Lidar measurement:

$$V: v_b \rightarrow \boldsymbol{v}$$

The simplest approach is to return to the assumptions which were earlier discarded, and obtain five points at each desired height (although now with error reduction and uncertainty estimation) and calculate one wind velocity, assuming horizontal homogeneity. This is used in the next section.

A more computationally-expensive, but valuable, approach is to obtain a number of points in any area of interest (or in an entire 3D volume) and use maximum likelihood with Equation 1 to estimate the local velocity. This provides true instantaneous 3D wind fields, meaning turbine wakes can be reconstructed, such as shown in Figure 4.

3. Results

Data from a site known to have simple wind flow conditions and where a WindCube Lidar was co-located with a met mast, as shown in Figure 5.

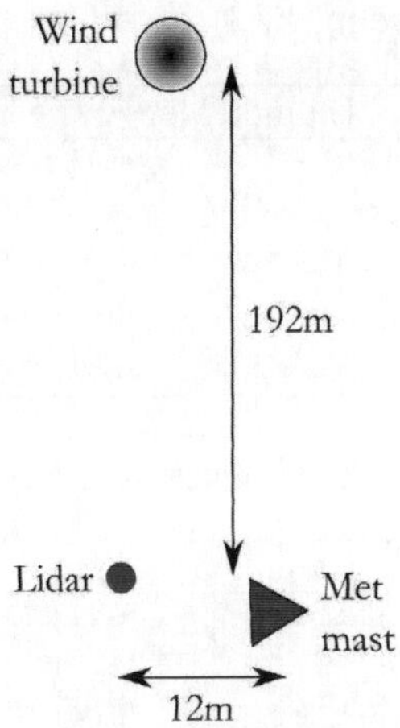

Figure 5. Site layout for validation.

The mast measurements are at 40, 60, 80, 100 and 116m, using cup anemometers and wind vanes. The mast data were only provided as 10-minute averages. The WindCube is configured to measure at exactly these heights, as well as additional heights up to 200m above ground. The time period when both Lidar and met mast data were available covers 44.6 days.

In order to resolve the Cyclops effect, beam speeds are here output from GP1 at $(x, y) = \left(\pm {}^{z}/_{2}, 0\right)$ to obtain v_x through averaging (and similarly for v_y) - very similar to the homogeneous assumptions currently made. The effect of the GP is still significant, however, since in this data set the WindCube v2 beams were not aligned with the North-South-East-West axes, but offset by 45°. The GP has therefore performed large spatial inference.

Example results are given in Figure 6 (wind speeds below 4m/s have been removed). The summary statistics for the two methods are given in Table 1 and Table 2. Bias is the mean and Scatter is the standard deviation of the residual errors.

Finally, evidence is shown in Figure 7 that GP inference during times of low data availability can result in reliable wind speed estimates.

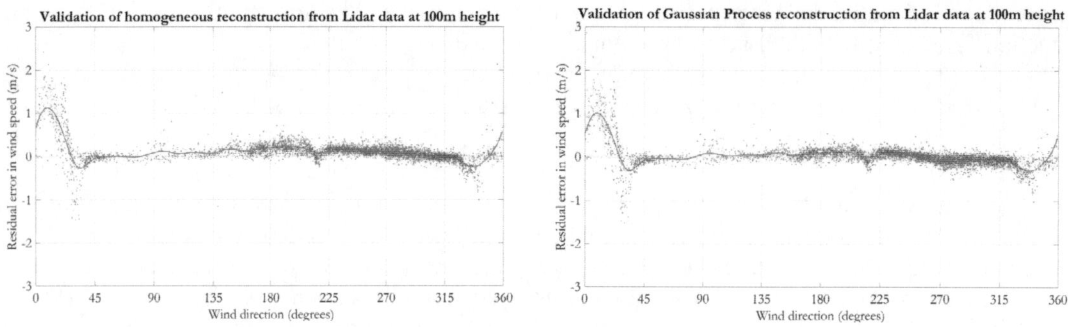

Figure 6. Diagnostic plot of residual errors versus direction at the 100m height. Left: Homogeneous (Leosphere); Right: GP. The undisturbed flow used for validation statistics are plotted in red. A smoothed average is also plotted as a line.

Table 1. Validation statistics for GP reconstruction (90% confidence interval)

	40m	60m	80m	100m	116m
Wind Speed Bias (m/s)	[−0.03, −0.02]	[−0.01, −0.00]	[−0.02, −0.01]	[−0.01, −0.01]	[−0.02, −0.01]
Wind Speed Scatter (m/s)	[0.12,0.12]	[0.11,0.12]	[0.11,0.11]	[0.11,0.12]	[0.10,0.10]
Wind Turbulence Bias (m/s)	[−0.16, −0.16]	[−0.13, −0.12]	[−0.07, −0.06]	[−0.03, −0.02]	[−0.01, −0.01]
Wind Turbulence Scatter (m/s)	[0.12,0.12]	[0.11,0.11]	[0.09,0.10]	[0.09,0.09]	[0.09,0.09]
Wind Direction Bias (°)	[−5.9, −5.9]	[−5.7, −5.7]	[−4.8, −4.7]	[−5.5, −5.4]	[−5.4, −5.3]

Table 2. Validation statistics for Homogeneous (Leosphere) reconstruction (90% confidence interval)

	40m	60m	80m	100m	116m
Wind Speed Bias (m/s)	[0.10,0.10]	[0.15,0.15]	[0.11,0.12]	[0.11,0.12]	[0.11,0.11]
Wind Speed Scatter (m/s)	[0.10,0.11]	[0.09,0.10]	[0.08,0.08]	[0.08,0.09]	[0.09,0.09]
Wind Turbulence Bias (m/s)	[−0.09, −0.09]	[−0.04, −0.04]	[−0.00, −0.00]	[0.03,0.04]	[0.04,0.04]
Wind Turbulence Scatter (m/s)	[0.09,0.10]	[0.09,0.09]	[0.08,0.08]	[0.08,0.08]	[0.09,0.09]
Wind Direction Bias (°)	[−5.9, −5.8]	[−5.7, −5.7]	[−4.7, −4.7]	[−5.4, −5.4]	[−5.3, −5.3]

4. Conclusions

Lidar can be used to measure wind speeds in a volume, offering the potential for more accurate estimation of wind turbine power performance and loading. However, a Lidar can only measure the component of the wind velocity in the direction of the beam, and suffers data loss caused by clouds and rain. This makes reconstruction of the wind velocity field from raw Lidar measurements a challenge, and current methods contain assumptions which cause inaccuracies in complex terrain and in wind turbine wakes.

A novel Machine Learning method has been developed, based on Gaussian Process regression, to remove these assumptions. This approach is naturally robust to overfitting, predicts uncertainty derived from data density and machine error, and – subject to further validation – is applicable to any site without tuning. Further, the method can be applied to most of the monostatic Lidar designs, with some adjustments for circular scanning patterns and in the future for floating Lidar.

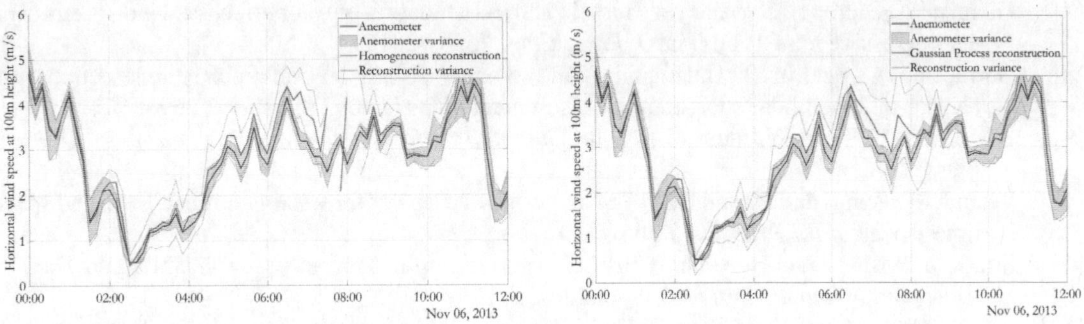

Figure 7. Comparison of 10-minute averaged wind speed from the met mast (blue) with reconstruction from Lidar (red) during a period of low data availability. Left: Homogeneous (Leosphere); Right: GP.

Validation at an onshore site with simple wind conditions over 44.6 days, against a met mast with cup anemometers and wind vanes at 5 heights covering 40-116m, gives the following results:

- The new method achieves an average wind speed bias of $-0.01m/s$ and a scatter of $0.11m/s$, meeting the threshold for commercial use.
- The new method infers data during Lidar measurement gaps, offering the potential for 100% data availability.
- Errors at one measurement height are automatically removed.
- Instantaneous 2D wake measurements in a plane can be made, which when validated and further developed will offer a breakthrough in understanding of wind farm interactions.

Despite these promising results, some priorities have been identified for future development:

- Conduct validation against 1-second sonic anemometer data.
- Further improve methods used for creating the Virtual Lidar.
- Adapt the model for floating Lidar by correcting for the rotation, and potentially adding additional terms for the translation, in GP1.
- Further improve methods used for reconstructing wind velocity from the Virtual Lidar.

References

[1] Bossanyi E, Kumar A and Hugues-Salas O 2014 Wind turbine control applications of turbine-mounted lidar *J. Phys. Conf.* **555** 012011

[2] Boquet M, Nibart M and Albergel A 2014 Method and device for determining the movements of a fluid from remote measurements of radial velocities *Europe Patent* EP2661636

[3] Raach S, Schlipf D, Haizmann F and Cheng P W 2014 Three dimensional dynamic model based wind field reconstruction from lidar data *J. Phys. Conf.* **524** 012005

[4] Towers P and Jones B 2014 Real-time wind field reconstruction from lidar measurements using a dynamic wind model and state estimation *Wind Energy* **19** 133–50

[5] Mikkelsen T, Herges T, Astrup P, Sjöholm M and Naughton B 2017 Wind field re-construction of 3d wake measurements from a turbine-installed scanning lidar *abstract from Int. Conf. on Future Technologies for Wind Energy* (Boulder: DTU)

[6] Guillemin F, Domenico D, Nguyen N, Sabiron G, Boquet M, Girard N and Coupiac O 2016

Nacelle lidar online wind field reconstruction applied to feedforward pitch control *J. Phys. Conf.* **753** 052019

[7] Givanani A, Bierbooms W and van Bussel G 2016 Estimation of rotor effective wind speeds using autoregressive models on lidar *J. Phys. Conf.* **753** 072018

[8] Borraccino A, Schlipf B, Haizmann F and Wagner R 2017 Wind field reconstruction from nacelle-mounted lidar short-range measurements *Wind Energy Science* **2** 269–83

[9] Rasmussen C E and Williams C K 2006 *Gaussian Processes for Machine Learning* (The MIT Press)

[10] Tastu J, Madsen H and Pinson P 2013 Short-term wind power forecasting: probabilistic and space-time aspects *DTU PHD-2013* **306**

[11] Smith M, Posner I and Newman P 2010 Generating implicit surfaces from lidar data *in Towards Autonomous Robotic Systems* (Plymouth, UK)

[12] Liu Y, Monteiro S T and Saber E 2015 An approach for combining airborne lidar and high-resolution aerial color imagery using gaussian processes *in Proc. Image and Signal Processing for Remote Sensing XXI* (Toulouse: SPIE) vol. 9643 ed. L Bruzzone

[13] Neal R M 1995 Bayesian Learning for Neural Networks *Lecture Notes in Statistics* vol. 118

IOP Conf. Series: Journal of Physics: Conf. Series **1102** (2018) 012004 doi:10.1088/1742-6596/1102/1/012004

Comparison of Reanalysis, Analysis and Forecast datasets with measured wind data for a Wind Power Project in Jhimpir, Pakistan

Zia ul Rehman Tahir[a,*], Muhammad Asim[a], Saqib Jamil[a], Rizwan Shad[b], Nasir Hayat[a], Ali Moaz[a], Muhammad Talha Akram[a], Muhammad Safyan[b]

[a]Department of Mechanical Engineering, University of Engineering and Technology Lahore, Pakistan
[b]Department of Mechanical Engineering, University of Central Punjab, Pakistan

ziartahir@uet.edu.pk

Abstract. This study aims to evaluate reanalysis, analysis and forecast data of wind speed and direction with the surface measured data at a site in Jhimpir region of Pakistan. Five *datasets* (reanalysis, analysis and forecast); NCEP-NCAR, 20th Century Reanalysis v2c (20C), CFSR, GFS and NCEP-FNL have been used for this study. The comparison has been made on six hourly and daily bases. The statistical parameters used are Mean Bias Error (MBE), Mean Absolute Error (MAE), Root Mean Square Error (RMSE) and correlation coefficient (R). The Weibull probability distribution has been drawn, the shape and the scale factors have been calculated. The CFSR data shows significantly good results for lower temporal resolution (daily) with a high correlation coefficient for a wind speed of 0.898 while at a higher temporal resolution of 20C shows significantly lower values of MBE of -0.87 m/s and CFSR shows lower values of RMSE of 2.09 m/s. CSFR wind direction data show good agreement with measured data in terms of statistical parameters The wind industry in Pakistan is under development stages with the possible potential of the offshore wind farm and expansion of land wind farms in other areas of the country, this study will be an initiative toward the initial wind resource assessment for potential sites in the wind corridor.

1. Introduction

Pakistan has been facing a shortage of about 4000 MW in the supply and demand of electricity over the past decade, so looking for alternative resources of energy, the wind is a good sustainable option. There was not a single grid-connected wind farm exists in Pakistan until 2007, but progress is being made in this regard in the past decade. The total capacity of operational wind projects at various locations is 308.2 MW according to the Alternate Energy Development Board (AEDB), 477MW projects are under construction and 663MW projects are in the pipeline. Pakistan Metrological Department (PMD) with financial assistance from the Ministry of Science and Technology brought about the Wind Power Potential Survey of Coastal Areas of Pakistan in June 2005, the estimated wind potential to be of 43000 MW in the Gharo-Jhimpir wind corridor [1], Jhimpir region has been identified as the most lucrative site for wind power plants in Pakistan by PMD.

The NCEP-NCAR *reanalysis* dataset [2, 3] jointly produced by the National Centers for Environmental Prediction (NCEP) and National Center for Atmospheric Research (NCAR) in the mid-nineties and this was the first generation of reanalysis datasets. The spatial resolution of wind speed data is 1.905°×1.875° (latitude × longitude) and temporal resolutions six hours (00, 06, 12, 18 UTC). The Twentieth Century *Reanalysis* Project (20C) was developed by the Physical Sciences Division of the Earth System Research Laboratory from NOAA and the CIRES/Climate Diagnostics Center of the University of Colorado [4]. The wind data of the 20th Century Reanalysis version 2c was used which has a temporal resolution of three hours with a spatial resolution of 1.905°×1.875°.

The Climate Forecast System Reanalysis (CFSR) is a third-generation *reanalysis* product by NCEP in released in 2010, which is available for from January 1979 [5]. The wind data of CFSR has a spatial resolution of 0.5°×0.5° and a temporal resolution of six hours. The Global forecast system (GFS) is a *forecast* data product prepared by NCEP [6, 7] available from December 2002. This wind data has a spatial resolution of 0.5° × 0.5° and a temporal resolution of six hours. The NCEP FNL (Final) is global operational *analysis* data produced by the Global Data Assimilation System (GDAS) [8]. This wind data has a spatial resolution of 1° ×1° and a temporal resolution of six hours and available from 1999.

The wind data measured by a private company at Jhimpir region is being compared data from *reanalysis*, *analysis* and *forecast* datasets (all termed in this paper as *datasets*) to analyze the resource potential. The site is located 114 km from Karachi and has been identified as a potential region for the wind power generation according to a resource potential estimation survey of Pakistan Metrological Department. The geographical location of the site where the mast was installed is 25.134° N and 67.996° E. The mast consists of five anemometers at measurement levels of 85 m, 60 m, 30m, and 10m to measure wind speed and two wind vanes at 83.5 m and 28.5 m to measured wind direction. The data for the duration of 36 months, from 1st January 2009 to 31st December 2011, was used to evaluate wind data from the datasets for this study.

The study is aimed to evaluate the reanalysis, analysis and forecast *datasets* for the site located Jhimpir region of Sind province. The evaluation or comparison of the datasets for wind resource assessment for South Asia has been performed to a very limited extent for few regions in Iran [9]. Amjad et al. [10] evaluated NCEP-NCAR data Gharo region using WRF simulated model at 10m and 30m heights using statistical parameters and time series. In this study, for the first time an extensive evaluation of five datasets: NCEP-NCAR, 20C, CFSR, GFS and NCEP-FNL against measured wind data at 10 m height has been performed using statistical parameters: MBE, MAE, RMSE and R. A comparison of these results with previously published results have been reported thus validating results and identifying the datasets variations for the South Asian region. The recent exponential increase in renewable energy projects especially wind projects are attracting attention and for the installation of such projects accurate potential estimate is of key importance. Since installing a wind mast and to measure surface data is a costly and time-consuming process, the datasets can be used as an alternative to surface measured data. This study is an initiative towards wind resource assessment of potential sites based on wind data from the datasets in Pakistan so that it can be sued to estimate the potential of the site without measured data.

2. Methodology

The quantitative comparison between the mast measured data and data datasets, and is of key importance to assess the potential of a site for the installation of a project and this comparison goes beyond having some graphs. Several methods had been used by researchers in literature for evaluation of the datasets against mast measured data. A comparison between the mast measured data and the datasets in literature has been made on the basis of daily time series [11], monthly mean wind speed [9, 11], sub-daily time series [12], annual average series [13], statistical analysis [14, 15], and probability distribution function[16, 17]. For evaluation of wind speed and direction from datasets, four methods have been used in this study. The evaluation of wind data from dataset against measured data was performed based on statistical analysis of wind speed and wind direction, monthly mean time series of wind speed, wind direction frequency in terms of wind rose and Weibull probability distribution function.

Statistical analysis includes calculating several parameters which provide an idea of how much a wind data from datasets is comparable. The statistical parameters include MBE (mean bias error), MAE (mean absolute error), RMSE (root mean square error), and correlation coefficient R, the details of calculations of these errors are presented in the Ref [17]. The wind rose represents the percentage of wind direction in sectors and for each sector and relative frequency of wind speed for each sector. The Weibull Probability Density Function (PDF) is an important tool to determine the resource probability density, cumulative frequency functions and subsequently to plot their respective graphs [18].

The datasets are available at a temporal resolution of three hours (20C) and six hours (NCEP-NCAR, CFSR, GFS and NCEP-FNL), 20th Century data had been converted to six hours. The reanalysis wind speed data at the mast location was calculated from the four nearby grid locations using bilinear interpolation. The data from datasets have been converted to daily mean as well for comparison. The surface measured data had a temporal resolution of ten minutes, so six hourly and daily mean data was calculated.

3. Results

In this section, results are presented in the form of statistical analysis, monthly mean time series of wind speed, the wind rose for wind direction and Weibull probability density function.

3.1. Statistical Analysis

MAE, MBE, RMSE and R computed from data from datasets and measured data of wind speed and wind direction for a temporal resolution of six hours and one day are presented in Table 1. A positive bias or error in the wind speed represents the data from datasets is overestimated and vice versa, whereas for wind direction a positive bias represents a clockwise deviation and vice versa. The lower values of errors and considerably higher value of correlation coefficient for wind speed and direction are of greater importance.

The statistical parameters for daily data are better compared to six hourly data. For six hourly wind speed data the lowest bias is for 20C but it shows worst value of R, CFSR has the lowest MAE, RMSE and the highest value of R. The six hourly wind direction data of CFSR has lowest MBE, MAE and RMSE and the highest value of R. The best values, higher correlation coefficient and lower errors for both six hourly and daily for all the datasets, are highlighted in bold. The overall wind speed biases are negative which indicate the underestimate of wind speed by satellite. The negative biases of wind direction indicate anticlockwise wind rotation from the datasets compared to measured data. For the site under study, CFSR data shows overall good comparison of both six hourly and daily data.

Table 1. Statistics of the comparison between measured and the wind data from datasets at 10 m

Temporal Resolution	Satellite Data	MBE		MAE		RMSE		R	
		Speed (m/s)	Dir'n (°)	Speed (m/s)	Dir'n (°)	Speed (m/s)	Dir'n (°)	Speed (m/s)	Dir'n (°)
Six Hourly	NCEP-NCAR	-1.52	-15.06	1.86	43.11	2.34	58.83	0.700	0.529
	20C	**-0.87**	-26.66	1.72	53.22	2.18	68.28	0.644	0.387
	CFSR	-1.44	**-11.13**	**1.69**	**33.65**	**2.09**	**51.85**	**0.785**	**0.689**
	GFS	-1.60	-19.25	1.92	37.45	2.41	54.56	0.683	0.593
	NCEP-FNL	-1.73	-19.57	1.99	37.63	2.49	54.52	0.679	0.579
Daily	NCEP-NCAR	-1.52	-21.18	1.61	39.45	1.96	53.14	0.824	0.565
	20C	**-0.87**	-32.78	**1.44**	50.99	1.83	66.18	0.718	0.464
	CFSR	-1.44	**-11.86**	1.48	**27.51**	**1.77**	**43.15**	**0.898**	**0.728**
	GFS	-1.60	-19.97	1.65	33.02	2.04	50.13	0.821	0.721
	NCEP-FNL	-1.73	-20.25	1.76	33.02	2.15	49.89	0.824	0.703

IOP Conf. Series: Journal of Physics: Conf. Series **1102** (2018) 012004 doi:10.1088/1742-6596/1102/1/012004

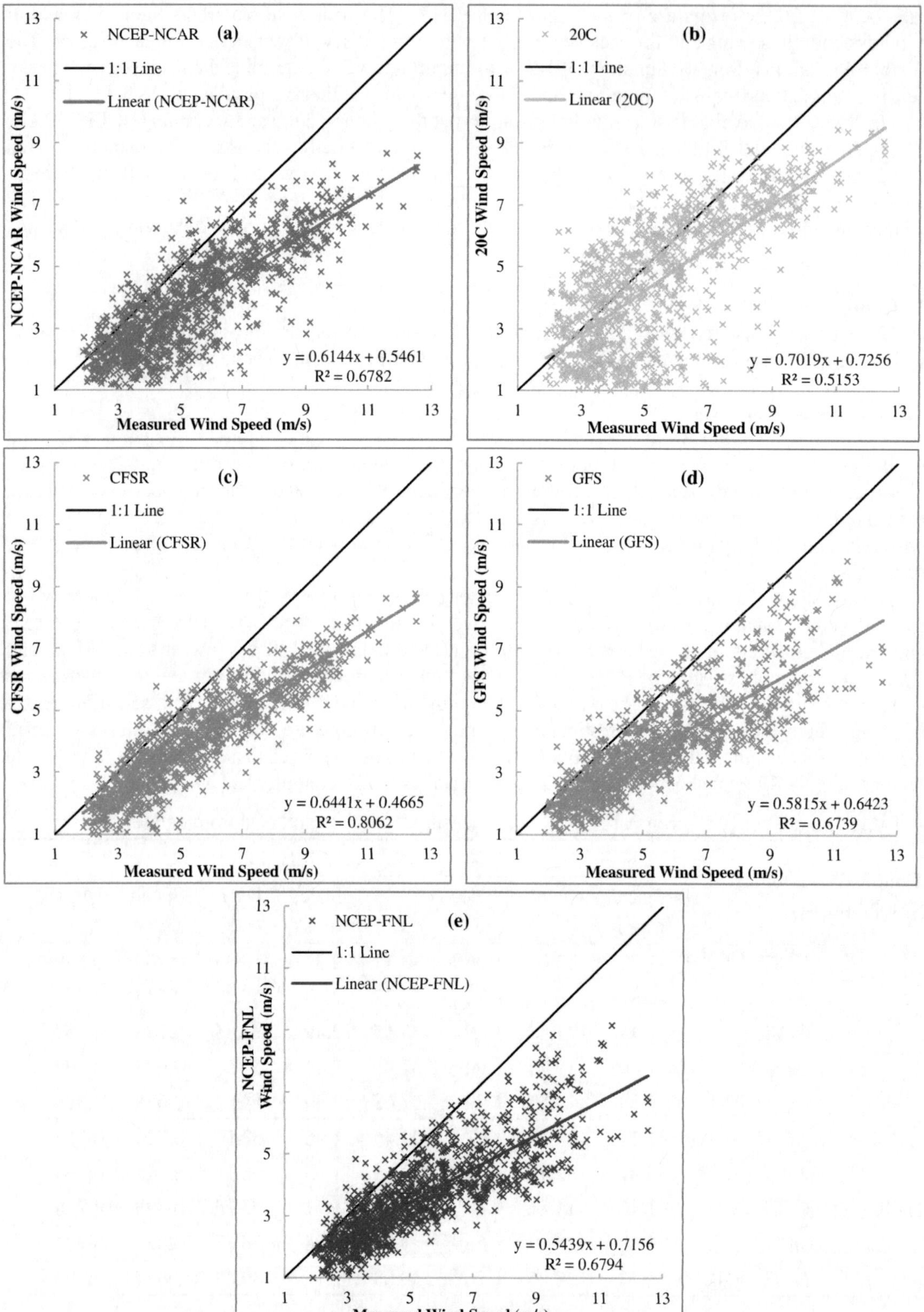

Figure 1. Correlation between surface measured and the datasets wind speed

The correlation between the measured data and data from the dataset mean daily wind speeds at 10 m height is shown in Figure1. The horizontal axis in the scatter plot is measured wind speed and the vertical axis is wind speed estimated by satellite. The best-fit line of correlation between both datasets along with 1:1 line is presented in Figure 1; the equation of the best-fit line is also presented. The equation of best-fit line has gradient close to one and intercept close to zero for better comparison. The highest gradient is for 20C and second best is for CFSR, the highest value of intercept is for 20C and lowest for CFSR. The CFSR data is in good agreement with measured data based on the equation of best fit and correlation coefficient.

3.2. Monthly Mean Time Series and Wind Rose

The monthly mean wind speed from measured and data from datasets for the duration of three years is presented in Figure 2. The general trends of the wind speeds for NCEP-CNAR and CFSR, GFS and NCEP-FNL are similar and closely related to each other. The trend of 20C is different from other satellites and measured data sets. The overall trend of all satellites except 20C is like measured with a consistent bias. All datasets underestimate wind speed in summer while only 20C overestimates from mid-August to mid-October. From April to October, 20C shows the least biased results while CFSR data represents the same trend as of surface measured data with a considerable non-varying bias, so its correlation coefficient has the highest value. Both NCEP-FNL and GFS show almost identical trend and that is the most biased from March to October. Although 20C shows overall good trend yet from November to February, it shows the most biased trends.

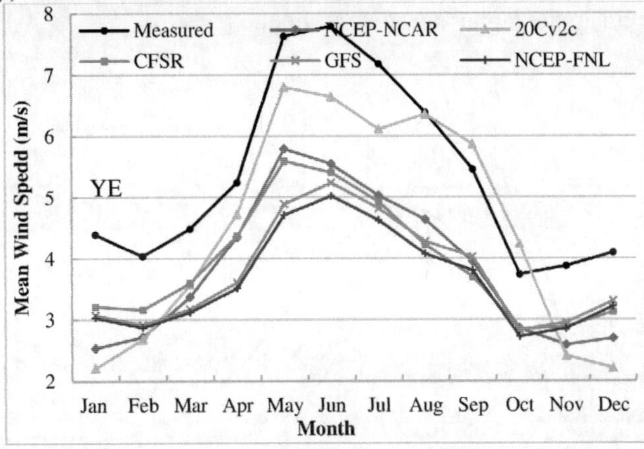

Figure 2. Measured and the datasets monthly mean wind speed

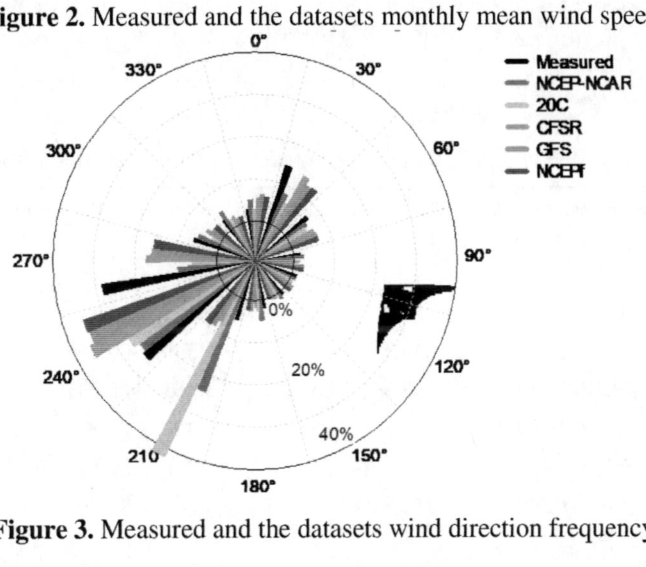

Figure 3. Measured and the datasets wind direction frequency

The wind rose in terms of wind direction frequency for measured data and data from the datasets for 12 sectors is presented in Figure 3. The dominant wind direction is southwest for measured data with 60% of wind. The dominant measured wind direction from March to October is towards southwest whereas form November to February (winter season) is towards the northeast. The wind rose shows that wind direction of CFSR, GFS and NCEP-FNL data has less rotation towards anticlockwise direction whereas NCEP-NCAR and 20C have a more anticlockwise rotation. The wind direction frequency for CFSR is closely related to measured data among all datasets.

3.3. Probability Distribution Function

The Weibull probability distribution function is widely used to represent wind speed distribution due to its flexibility and simplicity. The scale factor A is related to mean wind speed while shape factor k is equivalent to the standard deviation [19]. The Weibull probability distribution function (PDF) of measured data and data from datasets for the six hourly duration is presented in Figure 4. The datasets PDF's are shifted towards the left side of the wind speed axis relative to the measured PDF, this means low wind speed frequencies are overestimated whereas strong wind speed frequencies are underestimated by the datasets, which originates the overall wind speed underestimation. It is evident from the Figure 4 that data from 20C shows the closest trend to the surface measured data (underestimation of 16.34% in A) while on the other hand, NCEP-FNL shows the most deviated trend with the highest deviation in mean wind speed (underestimation of 32.20% in A). All datasets show a similar trend for Weibull scale factor except 20C. All datasets overestimate the lower speed wind frequencies and underestimate the higher speed wind frequencies.

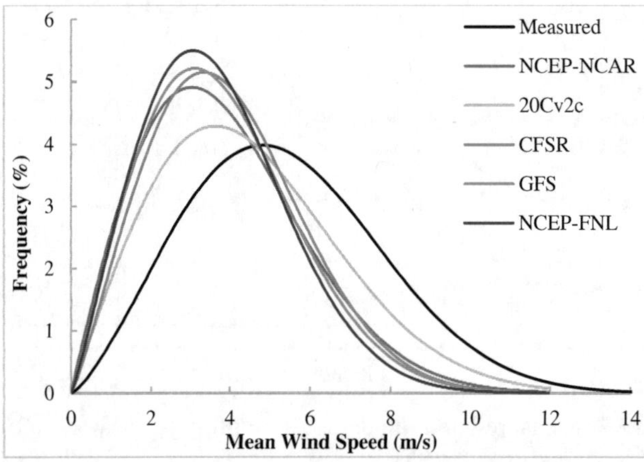

Figure 4. The Weibull probability distribution function of measured and the datasets wind speed

Table 2. Statistics of the comparison between measured and datasets six hourly wind data

Wind Data	Mean wind speed (m/s)	Mean wind direction (°)	Weibull Scale Factor A (m/s)	Weibull Shape Factor k
Measured	5.366	271.2	6.075	2.366
NCEP-NCAR	3.843 (-28.38)	233.2 (-14.01)	4.340 (-28.56)	1.979 (-16.36)
20C	**4.492 (-16.29)**	223.5 (-17.59)	**5.063 (-16.34)**	2.036 (-22.44)
CFSR	3.923 (-26.89)	**257.5 (-5.05)**	4.428 (-26.83)	**2.186 (-16.72)**
GFS	3.762 (-29.89)	253.4 (-6.56)	4.246 (-29.84)	2.095 (-20.19)
NCEP-FNL	3.634 (-32.28)	252.8 (-6.78)	4.103 (-32.20)	2.154 (-17.94)

The measured mean wind speed, direction, Weibull scale factor *A* and Weibull shape factor *k* are presented in Table 2. The mean wind speed and direction from datasets for three years duration with percentage error from measured speed is presented in parenthesis in Table 2. Mean wind speed from 20C is best with least percentage error and CFSR is the second best, mean wind direction from CFSR is best with least percentage error and GFS is second best. The Weibull scale factor for 20C is best with least percentage error and CFSR is the second best whereas the Weibull shape factor is best for CFSR and NCEP-FNL is the second best.

4. Discussion

The results of the statistical analysis between measured data and data from datasets are presented in Table 1. The MBE, RMSE and R for six hourly wind speeds from NCEP-NCAR are -1.52 m/s, 1.86 m/s and 0.700 respectively. A comparison of NCEP-NCAR wind speed data for five sites along the Iberian Peninsula coast was reported by Carvalho et al. [12], the MBE, RMSE and R for wind speed ranges from 0.03 m/s to 0.70 m/s, 2.28 m/s to 2.62 m/s and 0.73 to 0.82. Abderrahim et al. [20] compared NCEP-NCAR for five site in France, the MBE, RMSE and R for wind speed ranges from 1.49 m/s to 2.54 m/s, -0.62 m/s to -0.08 m/s and 0.71 to 0.84. A comparative study of present results with previous results revealed that for Pakistan the values of MBE, RMSE and correlation coefficient are in slight disagreement with the values obtained from the global analysis. However, the correlation coefficient is showing a relatively good estimate for Pakistan region. The MBE, RMSE and R for six hourly wind speeds from 20C are -0.87 m/s, 1.72 m/s and 0.644 respectively. A general observation about 20C is that there is a small MBE compared to other datasets but still the value of the correlation coefficient is not good.

The MBE, RMSE and R for six hourly wind speeds from CFSR are -1.44 m/s, 1.69 m/s and 0.785 respectively. A comparison of CFSR wind speed data was performed by Carvalho et al. [17] for five locations, the MBE and RMSE ranges from 0.42 m/s to 0.41 m/s and 2.06 m/s to 2.51m/s respectively. When location of Portugal was taken for analysis and evaluation purpose, it was seen that CFSR data shows ad results compared to NCEP FNL dataset with the values of MBE and RMSE for CFSR data range from 0.42 m/s to 0.41 m/s and 2.06 m/s to 2.51m/s respectively [17]. In another similar study for the Iberian Peninsula [12] the values of MBE, RMSE and R for range from 0.38 m/s to 0.82 m/s, 1.77 m/s to 2.33 m/s and from 0.83 to 0.90 respectively for wind speed. A comparison shows CFSR gives better estimation in Europe compared to Pakistan, the wind speed is mostly overestimated with higher values of correlation coefficients in Europe whereas in Pakistan it is underestimated with a comparatively lower value of correlation coefficient.

The MBE, RMSE and R for six hourly wind speeds from GFS are -1.60 m/s, 2.41 m/s and 0.683 respectively. Carvalho et al. [21, 22] compared GFS wind speed for five sites along the Iberian Peninsula coast at 10 m height, the MBE, RMSE and R ranging from 0.12 m/s to 0.4 m/s, 1.65 m/s to 2.26 m/s, and 0.81 to 0.90. The results obtained from the site in this study are comparable with data reported in the literature.

The MBE, RMSE and R for six hourly wind speeds from NCEP-FNL are -1.73 m/s, 2.49 m/s and 0.679 respectively. A comparison of NCEP-FNL with measured data reported in [23] had values of MBE, RMSE and R range from 0.24 m/s to 0.82 m/s, 1.70 m/s to 2.24 m/s and from 0.83 to 0.90 respectively. A similar study [21] reported had values of MBE, RMSE and R ranges from -2.02 m/s to 0.90 m/s, 1.82 m/s to 3.07 m/s and from 0.84 to 0.89 respectively. It is clear from above that statistical results of NCEP-FNL that for selected site in Pakistan are comparable with the previous studies, the values of correlation coefficients are higher for previous studies compared to the site for this study.

5. Conclusions

The evaluation of the NCEP-NCAR, 20C, CFSR, GFS and NCEP FNL datasets against the measured wind data has been presented for the first time for Pakistan. The wind industry in Pakistan is under development stages with the possible potential of the offshore wind farm and expansion of land wind farms in other areas of Pakistan, this study will be an initiative toward the initial wind resource assessment for potential sites in Pakistan.

The evaluation of datasets wind speed and direction data was performed in four ways; statistical analysis, monthly mean wind speed, wind direction frequency as the wind rose and Weibull probability distribution function. Based on MBE, MAE and RMSE CFSR data shows overall good agreement. CFSR data is in good agreement with the measured data based on scatter plot and value of correlation coefficient.

A comparison of monthly mean time series shows that 20C reanalysis data is good among all datasets. The wind direction frequency for CFSR is closely related to measured data among all datasets using wind rose comparison. A comparison of probability distribution function curves shows that 20C reanalysis data is good in terms of Weibull scale factor and CFSR is good in term of Weibull shape factor. Overall based on all parameters, CSFR reanalysis data gave good approximation and can be used as an alternative to surface measured data for initial site assessment for a wind farm.

6. References

[1] PMD 2005 WIND ENERGY PROJECT.

[2] Kalnay E, Kanamitsu M, Kistler R, Collins W, Deaven D, Gandin L, Iredell M, Saha S, White G, Woollen J, Zhu Y, Chelliah M, Ebisuzaki W, Higgins W, Janowiak J, Mo K C, Ropelewski C, Wang J, Leetmaa A, Reynolds R, Jenne R and Joseph D 1996 The NCEP/NCAR 40-Year Reanalysis Project Bulletin of the American Meteorological Society 77 437-72

[3] Kistler R, Kalnay E, Collins W, Saha S, White G, Woollen J, Chelliah M, Ebisuzaki W, Kanamitsu M, Kousky V, Dool H v d, Jenne R and Fiorino M 2001 The NCEP–NCAR 50-Year Reanalysis: Monthly Means CD-ROM and Documentation Bulletin of the American Meteorological Society 82 247-68

[4] Compo G P, Whitaker J S, Sardeshmukh P D, Matsui N, Allan R J, Yin X, Gleason B E, Vose R S, Rutledge G, Bessemoulin P, Brönnimann S, Brunet M, Crouthamel R I, Grant A N, Groisman P Y, Jones P D, Kruk M C, Kruger A C, Marshall G J, Maugeri M, Mok H Y, Nordli Ø, Ross T F, Trigo R M, Wang X L, Woodruff S D and Worley S J 2011 The Twentieth Century Reanalysis Project Quarterly Journal of the Royal Meteorological Society 137 1-28

[5] Saha S, Moorthi S, Pan H-L, Wu X, Wang J, Nadiga S, Tripp P, Kistler R, Woollen J, Behringer D, Liu H, Stokes D, Grumbine R, Gayno G, Wang J, Hou Y-T, Chuang H-y, Juang H-M H, Sela J, Iredell M, Treadon R, Kleist D, Delst P V, Keyser D, Derber J, Ek M, Meng J, Wei H, Yang R, Lord S, Dool H v d, Kumar A, Wang W, Long C, Chelliah M, Xue Y, Huang B, Schemm J-K, Ebisuzaki W, Lin R, Xie P, Chen M, Zhou S, Higgins W, Zou C-Z, Liu Q, Chen Y, Han Y, Cucurull L, Reynolds R W, Rutledge G and Goldberg M 2010 The NCEP Climate Forecast System Reanalysis Bulletin of the American Meteorological Society 91 1015-58

[6] National Centers for Environmental Prediction / National Weather Service / NOAA / U. S. Department of Commerce 2007 NCEP Global Forecast System (GFS) Analyses and Forecasts. (Boulder, CO: Research Data Archive at the National Center for Atmospheric Research, Computational and Information Systems Laboratory)

[7] 2007 NCEP Global Forecast System (GFS) Analyses and Forecasts. (Boulder, CO: Research Data Archive at the National Center for Atmospheric Research, Computational and Information Systems Laboratory)

[8] National Centers for Environmental Prediction / National Weather Service / NOAA / U. S. Department of Commerce 2000 NCEP FNL Operational Model Global Tropospheric Analyses, continuing from July 1999. (Boulder, CO: Research Data Archive at the National Center for Atmospheric Research, Computational and Information Systems Laboratory)

[9] Mostafaeipour A, Sedaghat A, Dehghan-Niri A and Kalantar V 2011 Wind energy feasibility study for city of Shahrbabak in Iran Renewable and Sustainable Energy Reviews 15 2545-56

[10] Amjad M, Zafar Q, Khan F and Sheikh M M 2015 Evaluation of weather research and forecasting model for the assessment of wind resource over Gharo, Pakistan International Journal of Climatology 35 1821-32

[11] Ozgener L 2010 Investigation of wind energy potential of Muradiye in Manisa, Turkey Renewable and Sustainable Energy Reviews 14 3232-6

[12] Carvalho D, Rocha A and Gómez-Gesteira M 2012 Ocean surface wind simulation forced by different reanalyses: Comparison with observed data along the Iberian Peninsula coast vol 56

[13] Kent E, Fangohr S and Berry D 2013 A comparative assessment of monthly mean wind speed products over the global ocean vol 33

[14] Satheesan K, Sarkar A, Parekh A, Kumar M R and Kuroda Y 2007 Comparison of wind data from QuikSCAT and buoys in the Indian Ocean International Journal of Remote Sensing 28 2375-82

[15] Ruti P, Marullo S, D'Ortenzio F and Tremant M 2008 Comparison of analyzed and measured wind speeds in the perspective of oceanic simulations over the Mediterranean basin: Analyses, QuikSCAT and buoy data vol 70

[16] Morales J M, Minguez R and Conejo A J 2010 A methodology to generate statistically dependent wind speed scenarios Applied Energy 87 843-55

[17] Carvalho D, Rocha A, Gómez-Gesteira M and Santos C S 2014 WRF wind simulation and wind energy production estimates forced by different reanalyses: comparison with observed data for Portugal Applied Energy 117 116-26

[18] Yissa Dawde O 2013 Wind Resource Data Analysis: The case of MYDERHU project site, Tigray regional state, Ethiopia.

[19] Carvalho D, Rocha A, Gómez-Gesteira M and Silva Santos C 2014 WRF wind simulation and wind energy production estimates forced by different reanalyses: Comparison with observed data for Portugal Applied Energy 117 116-26

[20] Abderrahim B, Croize-Fillon D, Queffeulou P, Liu C and Hervé R 2009 Evaluation of high-resolution surface wind products at global and regional scales vol 2

[21] Carvalho D, Rocha A, Gómez-Gesteira M and Santos C S 2014 Comparison of reanalyzed, analyzed, satellite-retrieved and NWP modelled winds with buoy data along the Iberian Peninsula coast Remote sensing of environment 152 480-92

[22] Carvalho D, Rocha A, Gómez-Gesteira M and Santos C S 2014 Offshore wind energy resource simulation forced by different reanalyses: comparison with observed data in the Iberian Peninsula Applied Energy 134 57-64

[23] Carvalho D, Rocha A, Gómez-Gesteira M and Silva Santos C 2014 Offshore wind energy resource simulation forced by different reanalyses: Comparison with observed data in the Iberian Peninsula vol 134

Wake detection in the turbine inflow using nacelle lidars

Dominique P Held[1,2], Nikolaos Kouris[2], Antoine Larvol[2] and Jakob Mann[1]

[1] Department of Wind Energy, Technical University of Denmark (DTU), Frederiksborgvej 399, 4000 Roskilde, Denmark
[2] Windar Photonics A/S, Helgeshøj Alle 16-18, 2630 Taastrup, Denmark

E-mail: domhel@dtu.dk

Abstract. In this study we present the ability to detect wakes in the inflow of turbines using nacelle-mounted continuous-wave lidar systems. Wake flows generate small-scale turbulence, which has significantly smaller length-scales than ambient turbulence. Due to the lidars large probe volume this turbulence is attenuated and will not be visible in the lidar's measurements. One approach to retrieve information about small-scale turbulence is by measuring the lidar Doppler spectrum width. Here we present an wake detection algorithm based on these measurements at two distinct locations in front of the turbine. By comparing the line-of-sight turbulence intensity and considering the instantaneous misalignment it is possible to detect half- and full-wakes. This has been tested during a 4.5 month long experiment and results show that situations where the wake affects the lidar measurements can be removed.

1. Introduction

Coherent Doppler wind lidars are a novel anemometry devices, which offer the possibility of remotely sensing wind speeds. These devices are especially attractive for wind energy applications where wind measurements within the first few 100 m of the boundary layer are required. Their spatial and temporal resolution allows measurements and flow characterization around utility-scale wind turbines [1, 2, 3]. Investigations of lidars mounted on top of the nacelle [4] or inside the spinner [5] to retrieve information about the turbine inflow have also been conducted. This led to a variety of proposed applications spanning from lidar-assisted yaw [6] and pitch control [7] over power curve validation [8] to wind farm control [9].

One of the problems associated with nacelle-mounted lidar systems is the limitation of only measuring the line-of-sight (LOS) component of the wind velocity. To obtain inflow information useful to control applications (like rotor-effective wind speed, yaw misalignment or vertical shear) reconstruction methods with appropriate wind models are necessary [10]. Usually the assumption of horizontal homogeneity is made. While being reasonable well satisfied in homogeneous, flat terrain, this assumption is heavily violated in complex terrain or when wakes impinge the field of view.

Thus it is necessary to detect flow situations where the assumptions of horizontal homogeneity is violated. An example of the severe influence of wakes on the misalignment measurement by a lidar system can be seen in fig. 1 and 2.

In fig. 1 three situations are illustrated. On the left horizontally homogeneous flow, which is aligned with the turbine, is shown. In this case the LOS components at each focus point are equal and the measured misalignment $\varphi = 0$. Once the flow is misaligned (the case of $\varphi > 0$ is shown in the center panel), the LOS component at the left focus point is reduced, while the right LOS component is increased. This leads to a measurement of $\varphi > 0$.

If the assumption of horizontal homogeneity is violated the wind vectors have different magnitudes even though the flow is aligned. A simplified situation is shown in the right panel of fig. 1, where a reduced speed is shown on the left side of the inflow. This can be caused by a wake, for example. Similar to the center panel this will cause a reduction of the LOS component on the left focus point and thus to a measured misalignment of $\varphi > 0$ despite an aligned flow.

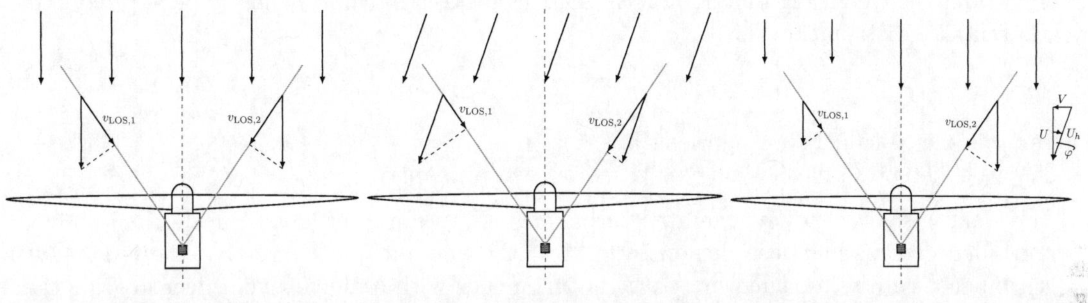

Figure 1. *Left*: Homogeneous inflow aligned with the turbine. *Center*: Homogeneous inflow misaligned with the turbine. *Right*: Impinging wake on the left part of the wind field.

The influence of wakes on experimental misalignment measurements can be seen in fig. 2. Here the 10 minute mean values are presented. The wake position of another upstream turbine is shown as a black vertical line. Outside the wake sector a average misalignment of around +5° can be observed. Inside the wake however, it can be seen that the misalignment varies substantially, reaching misalignments of up to ±35°. This implies that it is crucial to detect waked situations, because the biased lidar measurements can lead to incorrect turbine alignment and thus energy losses.

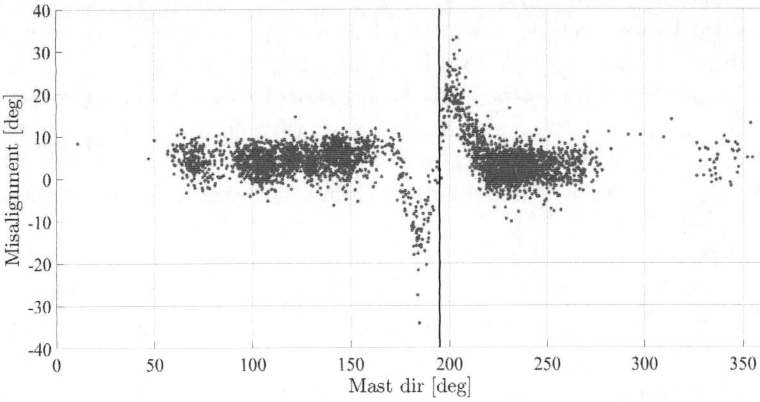

Figure 2. Turbine misalignment versus met mast wind direction; the vertical black line indicates a wake situation.

2. Methodology

In this study we want to show the possibility of detecting wake in the inflow of turbines using nacelle-mounted continuous-wave (cw) lidar systems.

The LOS wind speed can be expressed as a convolution of the LOS component of the wind vector and a weighting function φ:

$$v_r(\mathbf{r}) = \int_{-\infty}^{\infty} \varphi(s)\mathbf{n} \cdot \mathbf{u}(s\mathbf{n} + \mathbf{r})ds. \tag{1}$$

Here $\mathbf{n}$ is the beam unit vector, $\mathbf{r}$ is a vector pointing to the focus position and $\mathbf{u}$ is the 3-D wind vector field and s is the distance from the focus point along the beam. For a focused, cw, coherent lidar the weighting function (at large focus distances) can sufficiently be approximated by a Lorentzian distribution [11]:

$$\varphi(s) = \frac{1}{\pi}\frac{z_R}{z_R^2 + s^2}, \tag{2}$$

where z_R is the so-called Rayleigh length.

The wake detection methodology is based on two principles:

(i) The length-scale of wake-generated turbulence is of significant smaller scale than ambient turbulence. This has been demonstrated by [12], where it was shown that the wake-added turbulence can be modeled by synthetic turbulence with a much smaller length scale than ambient turbulence. The models fit experimental observations well.

(ii) Small-scale turbulence is responsible for increasing the width of a Doppler spectrum. Due to the large probe volume of a cw lidar small-scale fluctuations will be attenuated. The filtered turbulence will widen the Doppler spectrum. This has been experimentally verified by [13].

Combining these two principles, it is possible to define a wake detection parameter. Here we use the second centralized statistical moment v_s^2 of the Doppler peak, which is the spectrum varaince, to estimate its width

$$v_s^2(t) = \frac{\int (v - v_r(t))^2 W(t,v)dv}{\int W(t,v)dv}, \tag{3}$$

where $v_r(t)$ is the wind speed measured by the lidar (see eq. 1), $W(t,v)$ is the Doppler spectrum as a function of time and LOS speed and v is an integration dummy.

To facilitate the detection the spectra variance is normalized by the LOS speed v_r to arrive at the LOS TI. By comparing these values from the two different beams, wakes in the inflow can be detected. The sign of the difference in LOS TI shows which half of the rotor is impinged by the wake. When high instantaneous LOS TI values compared to the averaged LOS TI are encountered a full wake is flagged.

An example of two different spectra, where one laser beam is measuring inside a wake, is shown in 3. Both spectra were obtained within 5 s. It can be seen that the blue spec has a much larger width indicating the presence of a significant amount of small-scale turbulence.

3. Experimental setup

A cw nacelle-mounted lidar by Windar Photonics is mounted on a Vestas V80 turbine. The device is a short-range lidar with a range of 80 m and can focus sequentially at two different focus locations with a angle of 60° in between. The measurement period spanned from beginning of November 2014 to mid of March 2015. A total of approximately 10,000 10-minute periods have been acquired. The site can be seen in fig. 4, where the turbine with the lidar systems is indicated by the marker. The surrounding turbine can also be seen.

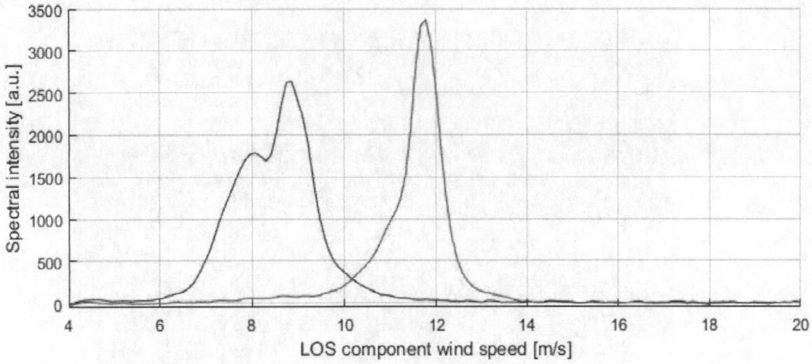

Figure 3. Example of two spectra where the blue spectrum show an enhanced peak width due to a measurement inside a wake.

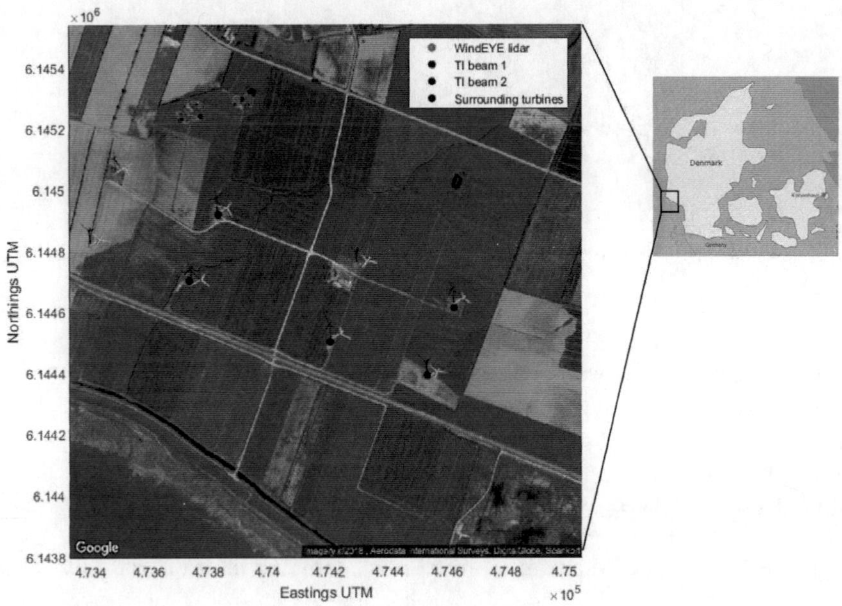

Figure 4. Google earth screen shot of the test site

4. Results

The results for the experimental campaign mentioned in the previous section can be found in fig. 5 and 6. The measurements are taken at a focus distance of 80 m. Firstly, the LOS TI as a function of yaw angle are shown. Here we have considered the wakes of the 5 closest turbines to the lidar-equipped turbine.

It can be observed from fig. 5 that the LOS TI is substantially higher at yaw angles when the turbine is pointing towards one of the surrounding turbines. Further when rotating clockwise, it can be seen that beam 2 will be affected first by the wake as this beam points to the right as seen from the turbine (compare fig. 1 again). So by comparing the difference in LOS TI can give information which side of the rotor is affected by the wake. Also, the LOS TI values for beam 2 have in general a lower value as for beam 1. The reason lies in the fact that the turbine is on average misaligned. The LOS wind speed on beam 2 is higher and thus reduces the LOS TI.

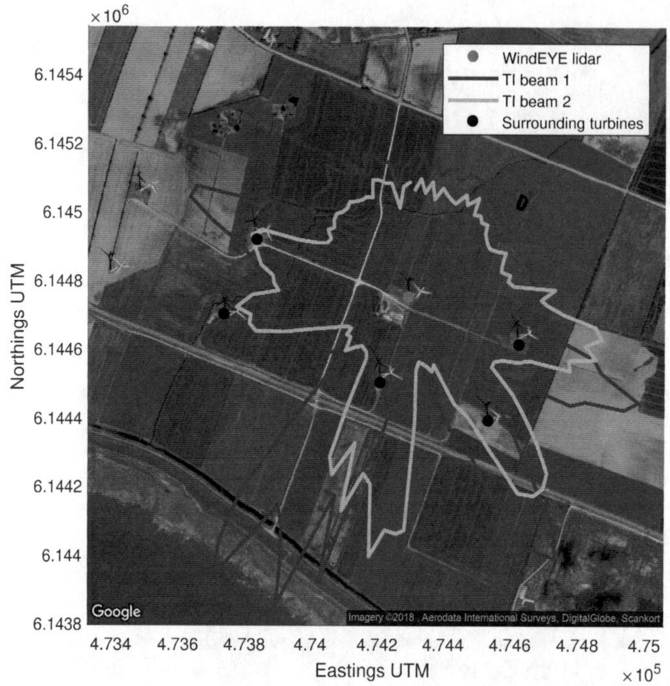

Figure 5. Map illustration of the LOS TI for beam 1 and 2. Solid points indicate the positions of the turbines.

Next, the performance of the wake detection algorithm is presented in fig. 6. The top panel shows the 10-minute mean misalignment of the turbine against its yaw position. The sinusoidal curves indicate the influence of wakes affecting the lidar measurements (compare to fig. 2). The surrounding turbines are shown as colored vertical lines and the sinusoidal influences of the wake are centered around them. Closer turbine tend to give more severe wake effects as the wind speed deficit has been recovered less compared to turbines further away.

The red, magenta and black scatter points show detected half- and full-wake situations, respectively. Here *Wake 1* refers to wake on the left half of the rotor (seen from the lidar) and *Wake 2* on the right half. It can be seen that the respective wake situations can be captured and a distinction between half- and full-wake is possible. There are some outliers, especially for the half-wake cases. It can probably attributed to non-wake sources of additional turbulence, like buildings or trees.

The bottom panel show the misalignment signal after all wake situations have been removed. Here it can be seen that it was able to remove the sinusoidal wake influences. Such a signal can be used for wind turbine yaw alignment.

5. Conclusion

Here we presented a methodology to detect wake situations in the inflow of turbines using 2-beam nacelle lidars. This is of importance since the assumptions made to derive wind field information from lidar measurements imply that a wake in one of the two beams is interpreted as a large misalignment. The Doppler peak width can be used to measure the small-scale turbulence found in wake flows. We showed results of a measurement campaign, where a wake detection algorithm has been tested. The results are promising and imply that it is possible to detect wake-affected turbine inflow. Thus, nacelle lidars can be used as turbine misalignment sensors. Future work

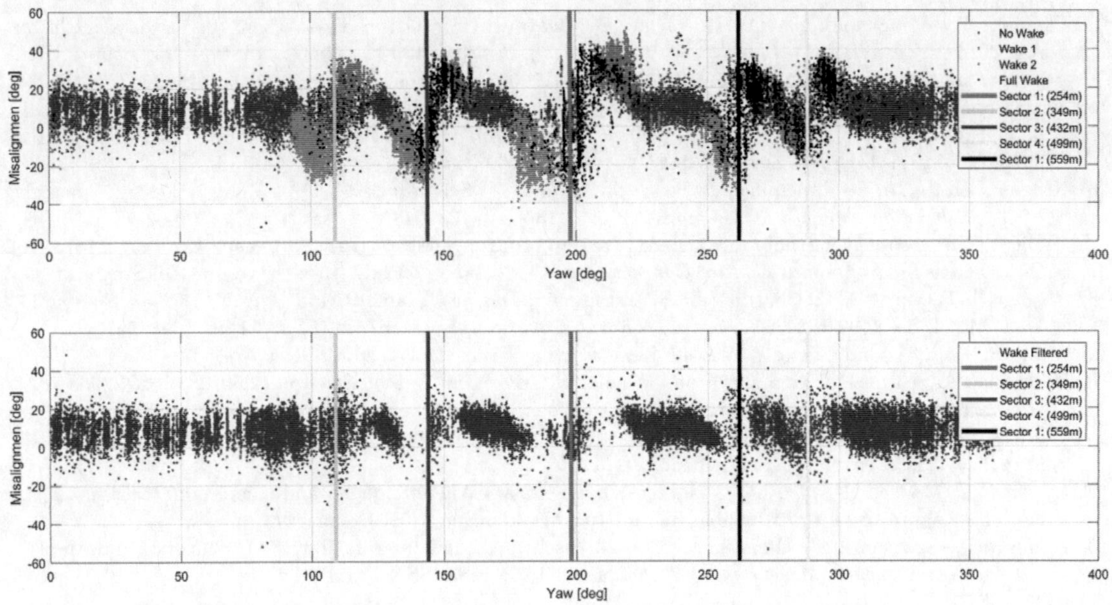

Figure 6. 10-minute mean turbine misalignment versus yaw position. Vertical lines indicate the surrounding turbines.

includes the analysis of different ambient turbulence levels. Interest lies in testing the algorithm in situations of high ambient turbulence, where the difference in peak width might not be as significant. Also we will investigate the possibility of not only detecting wake but to estimate the wake deficit simultaneously.

The learning objectives can be summarized as follows:

- Nacelle-mounted lidars can measure wind turbine misalignment, but wake situations include a bias in the measurements
- Wake turbulence is increasing the Doppler peak width of lidar measurements.
- By comparing the lidar Doppler peaks second centralized statistical moment wake flows can be detected in the inflow of turbines.
- A detection algorithm can be used to flag inflows affected by wakes and thus allowing a nacelle lidar to be used as a turbine misalignment sensor.

Acknowledgments

This study was supported by Innovationsfonden Danmark in form of an industrial PhD stipend (project number: 5016-00182). The authors want to thank Vattenfall A/S for providing the wind turbine SCADA data.

References

[1] Trujillo J J, Bingöl F, Larsen G C, Mann J and Kühn M 2011 *Wind Energy* **14** 61–75 ISSN 1099-1824 URL `http://onlinelibrary.wiley.com/doi/10.1002/we.1608/full`

[2] Machefaux E, Larsen G C, Troldborg N, Hansen K S, Angelou N, Mikkelsen T and Mann J 2016 *Wind Energy* **19** 1535–1551 ISSN 1099-1824 URL `http://onlinelibrary.wiley.com/doi/10.1002/we.1608/full`

[3] Barthelmie R, Doubrawa P, Wang H and Pryor S C 2016 *Journal of Physics: Conference Series* **753** 032034 ISSN 1742-6588 URL `http://stacks.iop.org/1742-6596/753/i=3/a=032034?key=crossref.22fd57bb37636861e3aff6c3189b12d7`

[4] Fleming P A, Scholbrock A K, Jehu A, Davoust S, Osler E, Wright A D and Clifton A 2014 *Journal of Physics: Conference Series* **524** 012002 ISSN 1742-6596 URL `http://stacks.iop.org/1742-6596/524/i=1/a=012002?key=crossref.efcc12681c6b46c26d377e1e2f7672c7`

[5] Mikkelsen T, Angelou N, Hansen K, Sjöholm M, Harris M, Slinger C, Hadley P, Scullion R, Ellis G and Vives G 2014 *Wind Energy* **16** 625–643 ISSN 1099-1824 URL `http://onlinelibrary.wiley.com/doi/10.1002/we.1608/full`

[6] Kragh K a, Fleming P A and Scholbrock A K 2013 *Journal of Solar Energy Engineering* **135** 031018 ISSN 0199-6231 URL `http://solarenergyengineering.asmedigitalcollection.asme.org/article.aspx?doi=10.1115/1.4023971`

[7] Schlipf D 2015 *Lidar-Assisted Control Concepts for Wind Turbines* Ph.D. thesis University of Stuttgart

[8] Wagner R, Pedersen T F, Courtney M S, Antoniou I, Davoust S and Rivera R L 2014 *Wind Energy* **17** 1441–1453 ISSN 1099-1824 URL `http://onlinelibrary.wiley.com/doi/10.1002/we.1608/full`

[9] Raach S, Schlipf D and Cheng P W 2017 *Wind Energy Science* **2** 257–267 ISSN 1742-6588

[10] Borraccino A, Schlipf D, Haizmann F and Wagner R 2017 *Wind Energy Science* **2** 269–283 ISSN 2366-7621 URL `http://www.wind-energ-sci-discuss.net/wes-2017-10/`

[11] Sonnenschein C M and Horrigan F A 1971 *Applied Optics* **10** 1600–4 ISSN 0003-6935 URL `http://www.ncbi.nlm.nih.gov/pubmed/20111170`

[12] Larsen G C, Madsen H A, Torben J and Troldborg N 2008 Wake modeling and simulation Tech. rep. Risø National Laboratory for Sustainable Energy Technical University of Denmark

[13] Branlard E, Pedersen A T, Mann J, Angelou N, Fischer A, Mikkelsen T, Harris M, Slinger C and Montes B F 2013 *Atmospheric Measurement Techniques* **6** 1673–1683 ISSN 18671381

Extreme Winds in the New European Wind Atlas

David Bastine[1,2], Xiaoli Larsén[3], Björn Witha[4], Martin Dörenkämper[2], Julia Gottschall[2]

[1] Jade University of Applied Sciences, Ofener Straße 16/19, 26121 Oldenburg, Germany
[2] Fraunhofer Institute for Wind Energy Systems IWES, Am Seedeich 45, 27572 Bremerhaven, Germany
[3] DTU Wind Energy, Frederiksborgvej 399, 4000 Roskilde, Denmark
[4] ForWind, Institute of Physics, University of Oldenburg, Küpkersweg 70, 26129 Oldenburg, Germany

E-mail: julia.gottschall@iwes.fraunhofer.de, david.bastine@jade-hs.de

Abstract. As a part of the New European Wind Atlas project, we investigate the estimation of extreme winds from mesoscale simulations. In order to take the smoothing effect of the simulations into account, a spectral correction method is applied to the data. We show that the corrected extreme wind estimates are close to the values obtained from offshore met masts. Hence, after further investigations we plan to use the examined approach as a basis for the calculation of extreme winds on the complete New European Wind Atlas, which will be publicly available at the end of the project.

1. Introduction

The European scale project New European Wind Atlas (NEWA) [1] aims at developing a New European Wind Atlas as a standard for site assessment in wind energy. The atlas is being developed on the basis of mesoscale simulations using the Weather Research and Forecasting model (WRF). The simulations will be supported and validated by a large number of field measurements stemming from various met masts and measurement campaigns across Europe, e.g. [2].

One specific feature of the atlas will be the estimation of extreme winds, which play a decisive role for the design of wind turbines and wind farms. In order to make sure that the wind does not exceed a wind turbine's design specification, reliable estimates of extreme winds are crucial. Following the IEC-61400-1:2005 standard [3], we are interested in the 10-min-average wind speed at hub height, which has a recurrence period of 50 years. This wind is often referred to as the 50-year wind.

Estimating the 50-year wind for a site using mesoscale simulations is a very challenging task. Firstly, the mesoscale models describe the orography and surface roughness around the site on relatively coarse grids. Thus, for example, local, small-scale speed-up effects through hills or surface roughness changes may not be captured. Secondly, smoothing is embedded in the mesoscale simulations in order to ensure numerical stability leading to an effective resolution several times coarser than the spatial resolution setup. This smoothing effect leads to a reduced variability of the wind and consequently to an underestimation of extreme winds.

This article focuses on correcting the smoothing effect in the mesoscale modeled winds using the spectral correction method (SCM) [4, 5, 6]. The SCM has shown some promising results

IOP Conf. Series: Journal of Physics: Conf. Series **1102** (2018) 012006 doi:10.1088/1742-6596/1102/1/012006

in connection with the use of reanalysis data [5]. However, its performance is expected to vary with different model outputs and the simulations from NEWA are of a much finer resolution (3 km) than the modeled data that have been previously used in connection with the SCM (about 50 km).

Hence, the major goal of this work is to examine the performance of the SCM in combination with the simulations used for the New European Wind Atlas. In this article, we present a first study based on simulations covering a period of 10 years in central Europe. At the end of the project 30 years of simulated data will be available and used for the extreme wind estimation. We compare our results to measurements from four different locations. Three of the considered sites are offshore while the last one is on relatively simple and flat terrain. We expect the local effects of orography and surface roughness to be negligible for the offshore sites and partially also for the simple onshore case. This is expected to allow an isolated investigation of the smoothing effect.

The methodology is introduced in Section 2 including the modeling (Section 2.1), measurements (Section 2.2), the estimation of extreme wind (Section 2.3, 2.4). The SCM will be applied to the simulations and the resulting extreme winds will be compared to the estimates from measurements. The results are presented and discussed in Section 3. In the end, final conclusions are drawn in Section 4.

2. Methodology

2.1. Mesoscale Modeling with WRF

The mesoscale wind atlas, as part of the NEWA project, is simulated with the widely used open-source mesoscale model WRF [7] version 3.8.1. For our study, we used simulations of all available ten years (2008-2017) from one sub-domain of the wind atlas covering Central Europe (see Figure 1). The horizontal resolution of the innermost model domain is 3 km. The wind atlas consists of week-long simulations, driven by the new ERA5 climate reanalysis dataset and using the MYNN PBL scheme [8, 9]. The output contains a number of atmospheric variables at several heights from 10 m to 500 m with an output time step of 30 minutes corresponding to an output frequency of 48 day^{-1}. A snapshot of the simulations showing the instantaneous wind speed at 100 m height can be found in Figure 1 for one selected point in time.

2.2. Measurements

In order to validate our estimations of extreme winds, time series of wind speed from four different measurement sites are used, which are briefly described in Table 1. Their locations are shown in Figure 1. FINO1 and FINO3 are located in the North Sea, FINO2 in the Baltic Sea and the met mast Cabauw onshore Western of the Netherlands.

All measurements are performed by cup anemometers and the considered data consists of ten-min-average wind speeds given every ten minutes. Details can be found in [10, 11]. Measurement gaps smaller than six hours have been linearly interpolated. Longer gaps were omitted for the calculation of the spectra. We only use time series which overlap with the completed simulations even though longer measurement time series are partially available.

	FINO1	FINO2	FINO3	Cabauw
height [m]	100	102	90	80
period [years]	10	9	8	10
availability [%]	95	94	83	99
terrain	offshore	offshore	offshore	flat

Table 1. Description of the measurement sites.

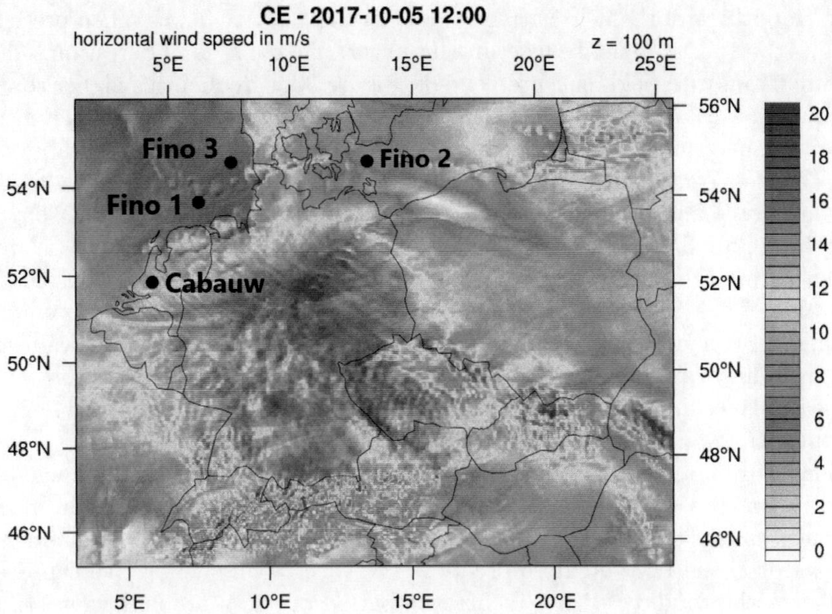

Figure 1. Simulation snapshot of the WRF model domain for central Europe. The color denotes the wind speed in $\mathrm{m\,s^{-1}}$ at a height of 100 m. The black dots represent the approximate positions of the measurement sites.

2.3. Extreme Wind Estimation

We define $u_{max}^{(i)}$ as the annual maximum of the i-th year, while $\langle u_{max} \rangle = \langle u_{max}^{(i)} \rangle_i$ denotes the average annual maximum, which we estimate by the sample mean of the annual maxima. Following e.g. [12, 13] , we assume that the annual maxima are Gumbel-distributed. We estimate the corresponding parameters α and β with the annual maximum method (AMM) [14], which is based on a probability weighted moment procedure. The estimated parameters can be used to estimate the T-year wind u_T, which is defined as the wind occurring with a probability of $\frac{1}{T}$ in one year. It is estimated by

$$u_T = \beta - \alpha \ln\left[\ln\left(\frac{T}{T-1}\right)\right] \approx \beta + \alpha \ln T \ (\text{for}\, T >> 1) \tag{1}$$

The statistical uncertainties, which can be found e.g. in [14, 6], increases with T and is quite high for $T = 50$ years for only 10 years of measurements and simulations. Hence, before the 30-year simulation is ready at the end of NEWA project, we investigate not only the desired 50-year wind but also the average annual maximum, in order to compare quantities with less uncertainty and obtain a more significant validation. Additionally, in the context of the SCM we will also consider the wind $\tilde{u}_{T=1}$ with a recurrence period of $T = 1$ years, representing the wind which is exceeded once a year on average. Note that $\tilde{u}_T$ approximately equals u_T only for $T >> 1$.

2.4. Spectral Correction Method

As already mentioned in the introduction, mesoscale simulations underestimate the extreme winds due to a smoothing of the field variables such as the velocity field. When presented in the spectral domain, the spectral energy level in the mesoscale range is underestimated as can be seen by the comparison with field measurements in Figure 3 for frequencies higher than 1 day^{-1}. It has been shown, for example by *Larsén et al.* [15, 16], that the power spectral density (PSD), also simply called spectrum in the following, usually follows $S(f) \propto f^{-5/3}$ above a certain frequency f_c. The frequency f_c is related to the integral time scale, which is often found to be in the order of 1 day in the mid-latitude here. Setting $S(f) \propto f^{-5/3}$ for $f > f_c = 1.5\,\text{day}^{-1}$ yields a "corrected", hybrid PSD S_{cor}, which is shown as the red dashed line in Figure 3. Note also that the corrected spectrum is extrapolated up to the output frequency of the 10 min-measurements. Choosing the value of f_c is a crucial step in the SCM and can lead to some uncertainty, as will be discussed further in Section 3. In order to reduce the sensitivity to f_c, the exact value of $S_{cor}(f_c)$ is determined by a linear regression (in the log-log framework) for $1.3\,\text{day}^{-1} < f < 1.7\,\text{day}^{-1}$. For spectral peaks exceeding the used extrapolation procedure the original value of the WRF spectrum is kept.

The main idea of the SCM is to estimate a correction factor for the extreme winds based on the original and corrected spectrum following *Larsén et al.* [4, 5]. Under certain mathematical assumptions, such as stationarity and Gaussian-distributed u and $\dot{u} = \frac{du}{dt}$, extreme winds can be expressed based on the PSD or more precisely the spectral moments. We can even find an explicit formula for the wind $\tilde{u}_T$ with a recurrence period of one year. Following [4], it is given by

$$\tilde{u}_T = \langle u \rangle + \sqrt{m_0}\sqrt{2\ln\left(\frac{1}{2\pi}\sqrt{\frac{m_2}{m_0}}T\right)}\ , \tag{2}$$

where $\langle u \rangle$ is the mean velocity and

$$m_j := 2\int_0^\infty d\omega\ \omega^j S(\omega) \tag{3}$$

are the spectral moments. For discrete signals, we have to estimate m_2 by integrating up to only half the sampling frequency of the corresponding signal.

Due to these assumptions made, we expect high uncertainties when using Equation (2) to estimate $\tilde{u}_T$. However, it has been shown that the ratio of the corrected and uncorrected values of $\tilde{u}_T$ can lead to a good description of the smoothing effect and hence a possible correction factor [4, 5]. More precisely, we can estimate the spectral moments for the corrected PSD ($m_j^{(corr)}$) and uncorrected PSD (m_j) and define a correction factor

$$C := \frac{\tilde{u}_T(m_0^{(corr)}, m_2^{(corr)})}{\tilde{u}_T(m_0, m_2)} \tag{4}$$

We choose $T = 1$ year and apply the resulting factor to the extreme wind estimates directly obtained from the original simulations. Hence, the SCM-corrected values $\langle u_{max} \rangle$ and u_T are $C \cdot \langle u_{max} \rangle$ and $C \cdot u_T$, respectively, where u_T has been estimated applying the AMM method to the original WRF simulations (see Section 2.3). Note that in some articles C is also called smoothing effect denoted as SE.

It should be noted again that the SCM method is based on many assumptions and not all of them are fulfilled. Thus, its performance has to be assessed by the comparison with measurements. One reason for the choice of $T = 1$ year to calculate the correction factor, is that some of the assumptions made become problematic for higher T. For example, the non-Gaussian tail of the velocity distribution plays a more important role. Hence, we calculate the correction factor for $T = 1$ year and assume that the smoothing effect on the 50-year wind can be taken into account by the same factor. Numerically, we can also estimate $\tilde{u}_T$ or even u_T for different assumptions such as a non-Gaussian probability density function (pdf) of u, but this is beyond the scope of this paper and will be investigated in a future article.

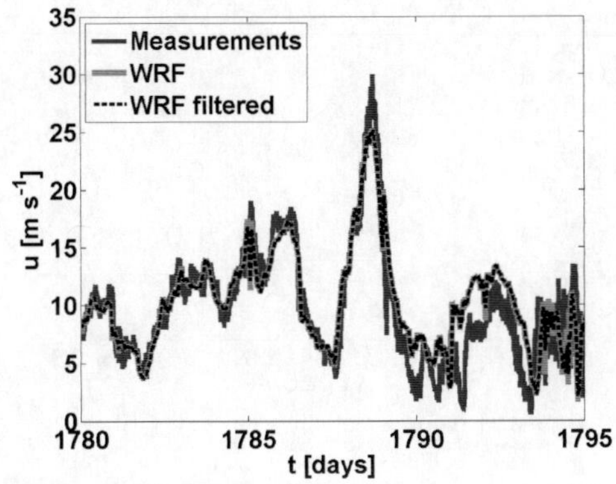

Figure 2. Wind speed time series at the location of FINO1.

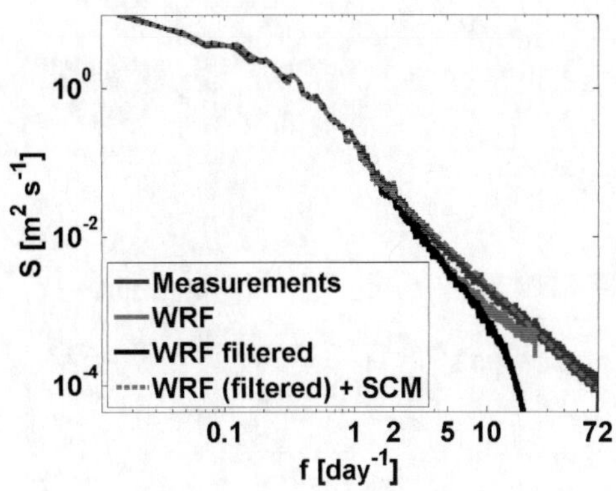

Figure 3. PSD of measurements (blue) and simulations (gray) and filtered simulations (black). The red dashed line shows the corrected PSD based on the filtered spectrum.

3. Results and Discussion

3.1. Properties of the WRF Time Series

Comparing the simulated time series (grey line in Figure 2) with the measured one (blue line in 2), we can clearly see the smoothing effect of the simulations. The measured time series contains a lot more high-frequency dynamics. Consequently, the highest wind speed in the presented time period seems to be around $5\,\mathrm{m\,s^{-1}}$ lower for the simulated time series. The missing fluctuations also manifest themselves in the lower spectral energy level of the WRF simulations for frequencies higher than approximately $1\,\mathrm{day}^{-1}$, as shown in Figure 3.

At the high-frequency end of the original WRF-PSD an upward bend can be observed indicating the presence of high-frequency dynamics on scales even finer than the output frequency of $48\,\mathrm{day}^{-1}$. Since we do not see this behavior in the measurements, such a bend indicates non-physical dynamics at the corresponding short time scales, as pointed out by [17] (see e.g. Figure 10). The observed high-frequency fluctuations can be artificial and numerical noises as a consequence of the high spatial resolution [18]. Thus, the study of [18] shows that a higher resolution does not necessarily lead to a more accurate simulation on small scales. In our case, however, the upward bend is weak enough that no energy is aliased to the low frequencies $f < 3\,\mathrm{day}^{-1}$. Hence, we still expect a good agreement with measurements on larger time scales.

3.2. Sensitivity of the SCM to the Output Time Step of the Simulations

For the extreme winds and the SCM, on the other hand, the higher energy on short time scales might be problematic since the SCM strongly depends on the tail of the spectrum due to the important second-order spectral moment m_2 weighting the spectrum with f^2 (see Equation (3)). Due to the high energy in the tail, the value of m_2 and consequently the SCM method and the resulting extreme wind estimates can be very sensitive to the output time step of the simulation, as illustrated in Figure 4. Choosing the output time step between 0.5 and 2 hours leads to an uncertainty caused by the time step choice in the order of 0.8 ms$^-1$ for the annual average maximum and $2\,\mathrm{m\,s^{-1}}$ for u_{50}.

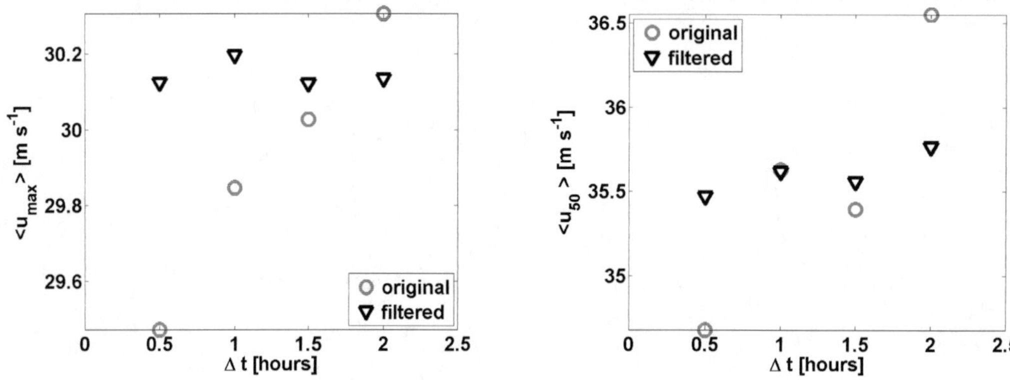

Figure 4. Estimated extreme winds using SCM depending on the output time step of the simulations for the location of FINO1. The gray circles show the estimates based on the original WRF time series, while for the black triangles a running 4-point average has been applied to the WRF data.

Usually, when undesired high-frequency dynamics are present in a signal, this energy is filtered out by a low pass filter, if possible before discretization. However, in our case we could not obtain a low-pass filtered output of the simulations, such as 10- or 30-min-averages but are bound to the saved data of the model which are instantaneous values every 30 minutes. Hence, we decided to apply a simple running two-point average (averaging two consecutive values) to this output corresponding to a filtering time scale of 60 minutes. The applied filter leads to a strong drop in the spectrum, as shown in Figure 3. For such a filtered spectrum, the integrals in Equation (3) are now well-defined since the spectrum approaches zero very fast. Consequently and most importantly, the SCM is now relatively insensitive to the choice of output time steps below 60 minutes. This insensitivity is hard to show for the two-point average since we only have data with an output time time step of 30 minutes. Thus, Figure 4 illustrates this phenomenon for a 4-point running average, reducing energy on time scales shorter than 2 hours. Consequently, the extreme winds estimates are almost constant when applying the SCM to the filtered data showing the filtering the data can lead to a more consistent SCM procedure. In order to keep as much realistic low frequency information as possible but to remove enough non-physical high frequency energy, we apply the SCM to the WRF data filtered with the running two-point average. The filtered data will be simply referred to as the WRF data in the following.

3.3. Application of the SCM

We now apply the SCM method to the filtered WRF time series at the different measurement sites. As a first step, the PSDs of measurements and simulations are estimated at all sites (Figures 3 and 5). After correction, the spectra seem to match the measured ones relatively well for the offshore cases and the uniform terrain in Cabauw. It is remarkable that the diurnal peaks at $f = 1, 2, 3\,\mathrm{day}^{-1}$ are matched really well for Cabauw.

In addition to the corrected spectrum using $f_c = 1.5\,\mathrm{day}^{-1}$, the thin red lines in Figure 5 represent corrected spectra for $f_c = 1.2\,\mathrm{day}^{-1}$ and $f_c = 1.8\,\mathrm{day}^{-1}$. Except for FINO3 these alternative corrected spectra do not differ strongly from the $f_c = 1.5\,\mathrm{day}^{-1}$ case resulting in a small sensitivity of the SCM to f_c. For FINO3, a relatively high sensitivity to f_c is found since at $f_c = 1.2\,\mathrm{day}^{-1}$ is the slope of S still differs strongly from $-5/3$. This leads to an overestimation of the small-scale energy for $f_c = 1.2\,\mathrm{day}^{-1}$ and consequently to a sensitive dependence of the correction factors on f_c for FINO3, as discussed further in the next paragraph.

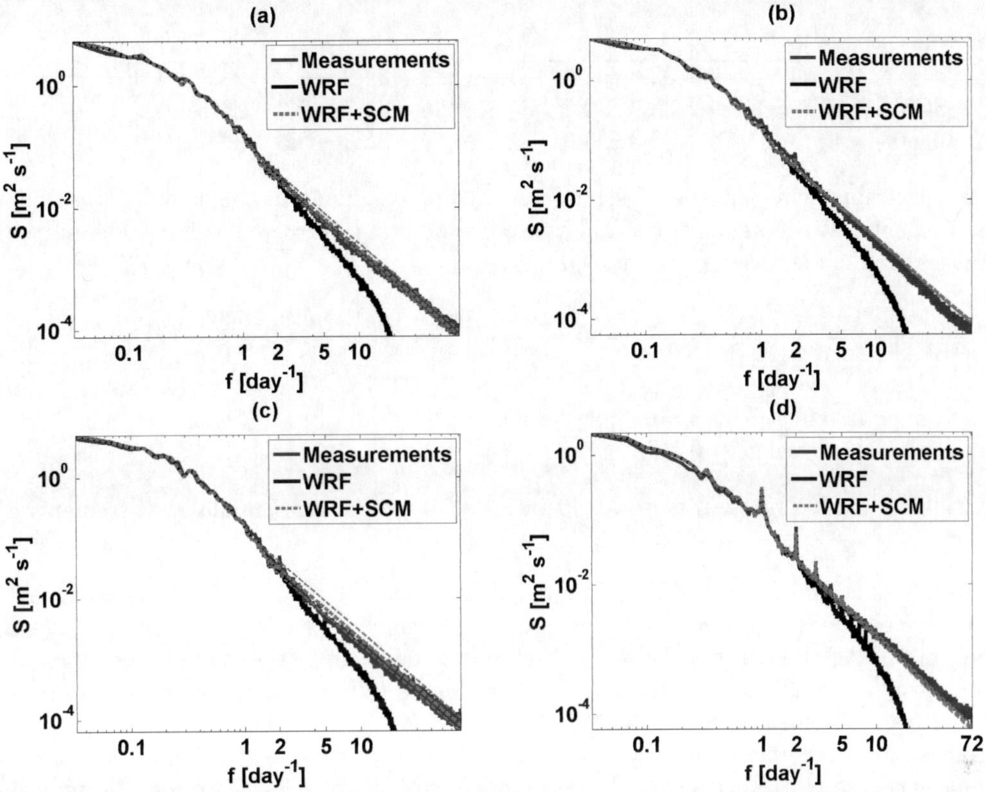

Figure 5. Power Spectral Density of measurements and simulations for all measurement sites: (a) FINO1 (b) FINO2 (c) FINO3 (d) Cabauw. The thick red dashed line shows the corrected PSD for $f_c = 1.5\,\mathrm{day}^{-1}$. In order to investigate the sensitivity on the value of f_c, all figures also include thin red dashed lines showing the corrected PSD using $f_c = 1.2\,\mathrm{day}^{-1}$ and $f_c = 1.8\,\mathrm{day}^{-1}$. Since these lines are often really close to each other, they are not always easy to see indicating a low sensitivity to f_c.

The estimated extreme winds can be found in Table 2. Additionally, we compare the estimated correction factors with the actual ratios of the extreme winds for WRF and measurements in Table 3.

For all three offshore sites the estimated annual average maxima are remarkably close to the measured ones indicating that the SCM seems to be able to take the smoothing effect successfully into account. For FINO1, for example, we find $\langle u_{max} \rangle = 27.8\,\mathrm{m\,s^{-1}}$ for the WRF simulations and SCM-corrected values of $\langle u_{max} \rangle = 30.0\,\mathrm{m\,s^{-1}} \pm 0.7\,\mathrm{m\,s^{-1}}$ close to the measured value $\langle u_{max} \rangle = 30.7\,\mathrm{m\,s^{-1}} \pm 1.1\,\mathrm{m\,s^{-1}}$. The given uncertainty is one standard deviation of the sample mean. The correction factor is approximately $C = 1.08$ for all three offshore cases very close to the ratios of measurements and simulations (1.10, 1.08, 1.12). We estimate the uncertainty of the correction factor caused by the choice of f_c, which has been discussed above, to be around 0.004 for FINO1 and FINO2 and 0.008 for FINO3. These uncertainty needs to be

IOP Conf. Series: Journal of Physics: Conf. Series **1102** (2018) 012006 doi:10.1088/1742-6596/1102/1/012006

Annual average max. $[m\,s^{-1}]$	FINO1	FINO2	FINO3	Cabauw
WRF	27.8 ± 0.6	26.2 ± 0.6	27.4 ± 0.6	20.8 ± 0.4
WRF + SCM	30.0 ± 0.7	28.3 ± 0.7	29.6 ± 0.6	22.3 ± 0.4
Measurements	30.7 ± 1.1	28.4 ± 0.8	30.8 ± 1.0	24.4 ± 1.1
50-year wind $[m\,s^{-1}]$				
WRF	32.8 ± 2.3	31.6 ± 2.3	32.2 ± 2.2	24.3 ± 1.5
WRF + SCM	35.9 ± 2.5	34.8 ± 2.5	34.7 ± 2.3	26.1 ± 1.6
Measurements	39.3 ± 3.6	35.3 ± 3.0	38.5 ± 3.5	34.0 ± 3.9

Table 2. Extreme wind speeds: Estimates of average annual maxima and 50-year winds for measurements, WRF simulations and the corresponding corrected value. The statistical uncertainties are denoted as one standard deviation.

	FINO1	FINO2	FINO3	Cabauw
WRF + SCM (Correction factor)	1.08	1.08	1.08	1.07
Annual average maxima (Measurements/WRF)	1.10	1.08	1.12	1.17
50-year wind (Measurements/WRF)	1.20	1.12	1.20	1.38

Table 3. Estimated correction factors and ratios of measured and simulated extreme winds.

investigated further and has to be kept in mind when using the SCM without measurements.

For the 50-year winds, the SCM also leads to values much closer to the measured ones than for the original WRF simulations (Table 2). However, in contrast to the annual average maxima, the 50-year winds are underestimated for all offshore sites but due to the high uncertainties, it is hard to assess if this is a significant result.

One reason for a possible underestimation of the 50-year winds might be that the used WRF setup could have difficulties with the accurate modeling of storms. Furthermore, the smoothing embedded in the simulations might have a stronger effect on the extreme wind estimates for higher T. This will be examined further in the future.

For the onshore site Cabauw, the SCM still leads to an improvement but it is not as good as in the offshore cases. Particularly, the 50-year wind seems to be strongly underestimated. The reason for this underestimation in the estimation and correction of the spectrum, since the corrected spectrum matches the measured one remarkably well. Former results in [6] show that combined with a microscale modeling approach, the SCM can yield a relatively a good estimation at this site. Thus, the small-scale roughness and orography might play a more important role than we expected despite the simplicity of the site. This will be investigated further, when taking small-scale effects are taken into account in a future step of of the project, which will be based on the WAsP methodology.

4. Conclusions and Outlook

The simulations set up for the New European Wind Atlas are based on the open source model WRF and the ERA5 reanalysis dataset. In this article, we examined the combination of these simulations with the SCM, in order to estimate extreme winds.

A spectral comparison of the WRF simulations with the measurements indicates non-physical behavior of the simulations in the high-frequency region, a known phenomenon for mesocale simulations with a relatively high spatial resolution. We illustrated that these high-frequency dynamics can be a significant source of uncertainty for the extreme wind estimation but that in the case of the NEWA data the application of a very simple low-pass filter leads to a robust estimation procedure.

Using the filtered WRF data, the SCM clearly improves the extreme wind estimates of the WRF simulation. Particularly offshore, estimates of the annual average maxima very close to the measured values have been found. The SCM also leads to valuable estimates of the 50-year wind but our results indicate a moderate systematic underestimation, which we will investigate further when more simulated data is available.

For the onshore site Cabauw, the extreme wind estimates also lead to a clear improvement but still show a strong underestimation. It is likely that the small-scale orography and roughness plays a more significant role than expected.

Thus, in a following step, we will combine the SCM with a micro-scale modeling approach based on the WAsP Engineering technique (see e.g. [14, 19, 20]). This way the small-scale orography and roughness can be taken into account. The resulting extreme wind estimates will be compared to measurements of many different met masts placed in sites of different complexity The uncertainties of the combined approach will be investigated in great detail.

We plan to offer two different ways to present the extreme winds in the new European wind atlas. First, the "coarse extreme wind", as estimated in this article, will be given. Despite its uncertainties onshore, it still offers a good first impression of the extreme wind climate in Europe, particularly offshore. Additionally, we plan to add a flag denoting the level of uncertainty corresponding to the specific site. Second, we are going to offer the SCM-corrected generalized extreme wind climate (GEWC) [6], which the user can use as an input to micro-scale modeling such as WAsP Engineering. This way site-specific extreme wind estimates all over Europe can be obtained without having to perform a single measurement. The new European wind atlas will therefore be of great value to the wind energy industry, particularly in the field of resource assessment.

Acknowledgements

The NEWA project is funded by the Federal Ministry of Economic Affairs and Energy under Grant number 0325732. The simulations were performed at the HPC cluster EDDY, located at the University of Oldenburg and funded by the Federal Ministry of Economic Affairs and Energy under Grant number 0324005.

References

[1] New european wind atlas (newa) URL `http://www.neweuropeanwindatlas.eu/`
[2] Mann J, Angelou N, Arnqvist J, Callies D, Cantero E, Arroyo R C, Courtney M, Cuxart J, Dellwik E, Gottschall J *et al.* 2017 *Phil. Trans. R. Soc. A* **375** 20160101
[3] International Electrotechnical Commision 2005 International standard iec 61400-1:2005 Tech. rep. ISBN 2-8318-8161-7
[4] Larsén X G, Ott S, Badger J, Hahmann A N and Mann J 2012 *Journal of Applied Meteorology and Climatology* **51** 521–533 ISSN 1558-8424
[5] Larsén X G and Kruger A 2014 *Journal of Wind Engineering and Industrial Aerodynamics* **133** 110–122 ISSN 0167-6105
[6] Hansen B, Larsn X, Kelly M, Rathmann O, Berg J, Bechmann A, Sempreviva A and Ejsing Jrgensen H 2016 *Extreme Wind Calculation Applying Spectral Correction Method Test and Validation* (Denmark: DTU Wind Energy)
[7] Skamarock W, Klemp J B, Dudhia J, Gill D O, Barker D M, Duda M G, Huang X Y and Wang W 2008 A description of the advanced research wrf version 3 Tech. rep. National Center for Atmospheric Research Boulder, USA
[8] Nakanishi M and Niino H 2006 *Boundary-Layer Meteorology* **119** 397–407
[9] Nakanishi M and Niino H 2009 *Journal of the Meteorological Society of Japan. Ser. II* **87** 895–912
[10] Fino research platforms URL `https://www.fino-offshore.de/en/`
[11] Cabauw experimental site for atmospheric research (cesar) URL `http://www.cesar-observatory.nl/`

[12] Mann J, Kristensen L and Jensen N O 1998 *Bridge Aerodynamics* A/S Storebaelt; Oresundskonsortiet; COWI Consulting Engineers & PlannersEOLEOLAS; COWI Fdn; DTU Tech Univ Denmark; Danish Grp of IABSE; Permanent IntEOLEOLAssoc Nav Congresses (PIANC); DANHIT High Performance Comp & Networking;EOLEOLInt Bridge, Tunnel & Turnpike Assoc (IBTTA); World Rd Assoc (PIARC)
[13] Coles S, Bawa J, Trenner L and Dorazio P 2001 *An introduction to statistical modeling of extreme values* vol 208 (Springer)
[14] Abild J 1994 Application of the wind atlas method to extremes of wind climatology Tech. Rep. Risoe-R-722 (EN) Risoe National Laboratory Roskilde, Denmark
[15] Larsén X G, Vincent C and Larsen S 2013 *Quarterly Journal of the Royal Meteorological Society* **139** 685–700
[16] Larsén X G, Larsen S E and Petersen E L 2016 *Boundary-Layer Meteorology* **159** 349–371 ISSN 1573-1472 URL `https://doi.org/10.1007/s10546-016-0129-x`
[17] Skamarock W C 2004 *Monthly weather review* **132** 3019–3032
[18] Floors R, Hahmann A and Peña A *Journal of Geophysical Research: Atmospheres*
[19] Larsén X G and Mann J 2009 *Wind Energy* **12** 556–573
[20] Wasp engineering URL `http://www.wasp.dk/weng`

Accuracy quantification of a CFD model in complex site by cross-checking wind measurements

G Crasto[1], L Terzi[2] and A Lombardi[2]
[1] via Gaspara Stampa 7, 09131, Cagliari, IT
[2] Renvico srl, via San Gregorio 34, 20124, Milano, IT
E-mail: giorgio.crasto@gmail.com, {ludovico.terzi, andrea.lombardi}@renvico.it

Abstract. In the present study it is discussed a sensitivity analysis on input parameters of a CFD resource assessment methodology built on top of OpenFOAM 4.1, a well-known open source library for numerical fluid dynamics. RANS are solved for dry uncompressible air, the Atmospheric Boundary Layer is calculated without thermal effects (neutral conditions). It is highlighted how the control of a metric based upon a weighted root mean square error (wRMSE) of the local accelerations (speed ups) can guide the wind analyst in the design of the computational mesh and further numerical settings. Moreover, it will be quantified the wRMSE as a function of the distance to the measurement points. The parameters analysed are the horizontal extension of the model, the horizontal resolution within the refined area, which is covering the layout of the wind farm, three different versions of k-ε turbulence closure and the number of directional sectors considered.

1. Objectives

The objective of the present work is to define a metric that can be used to evaluate the uncertainty of the wind resource assessmenmt model, hence to guide the wind analyst to further improve his/her estimations. The chosen metric is a weighted root mean square error ($wRMSE$), of the sector wise errors $su_{CFD,i} - \overline{su_i}$ on speed-ups as in the **eq. 1**:

$$wRMSE = \sqrt{\sum_{i=1}^{ns} [f_i(su_{CFD,i} - \overline{su_i})^2]} \quad (1)$$

$$su_{CFD,i} = \frac{uCFD_{target,i}}{uCFD_{ref,i}} \quad (2)$$

$$\overline{su_i} = \frac{1}{nt_i} \sum_{t=1}^{nt_i} \frac{u_{target,t,i}}{u_{ref,t,i}} \quad (3)$$

$$\hat{\sigma}_{su_i} = \sqrt{\frac{1}{nt_i - 1} \sum_{t=1}^{nt_i} \left(\frac{u_{target,t,i}}{u_{ref,t,i}} - \overline{su_i}\right)^2} \quad (4)$$

Where the wRMSE is calculated, in turn, for each pair of measurement points; considering one as the reference and a second one as the target all the possible combinations are accounted for in the cross-checking procedure. In the present study, having a total of 4 measurement levels, deployed on three masts, led to 12 combinations of reference-target points.

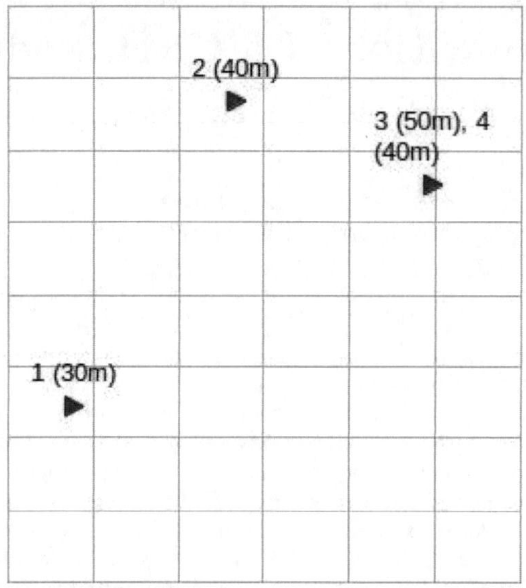

Figure 1. Layout of masts (top view) with numbered measurement levels, height in parentheses.

The wind rose at the reference point is discretized with ns directional sectors, the frequency f_i is computed at the reference for the i-th direction, calms are disregarded by filtering-out wind speeds smaller than 2 m/s. The speed-up for the i-th direction defined in **eq. 2** is obtained from the ratio of the horizontal wind speed from the CFD database at the target point divided by the same variable at the reference point. A number ns of CFD models are run, equal to the number of sectors ns, for a central value of the directional sector, imposed at the inlet(s), therefore the horizontal wind speeds appearing in **eq. 2** are obtained by weighting two sets of numerical results, i.e. the wind speeds components are linearly interpolated in order to reconstruct the chosen wind direction at the reference point. The speed-up ratio from the measurements $\overline{su_i}$ is instead obtained, as noted in **eq. 3** by averaging the concurrent ratios, filtered at reference by wind direction and wind speeds for calms, with the chosen threshold of 2 m/s.

In **Figure 2** is given an example of sector wise speed-ups from concurrent measurements and predicted by the CFD model. Once the $wRMSE$ is computed for each combination of measurement level (considering a complete measurement level formed by at least an anemometer coupled with a wind vane, i.e. capable of measuring both wind speed and direction), a reference and a target, they can be plotted against the distance between sensors. The $wRMSE$ on the speed-ups defines an estimate of the goodness of the CFD prediction only by considering the magnitude of the velocity vector; while indeed a possible misprediction of the velocity vector at the target point it is also in the direction. In analogy with the $wRMSE$ on the speed-up it could be defined a metric on the directional shift between reference and target points. In **Figure 3** an example of directional shifts, sectorwise, measured and predicted by the CFD.

2. Approach

The Digital Terrain Model of a location in Europe has been loaded from the EU-DEM v1.1 [1], the map of use-of-land from the Corine Land Cover (CLC) 2012; the former with 25m resolution and the latter having a 100m resolution. The EU-DEM is a hybrid product based on SRTM

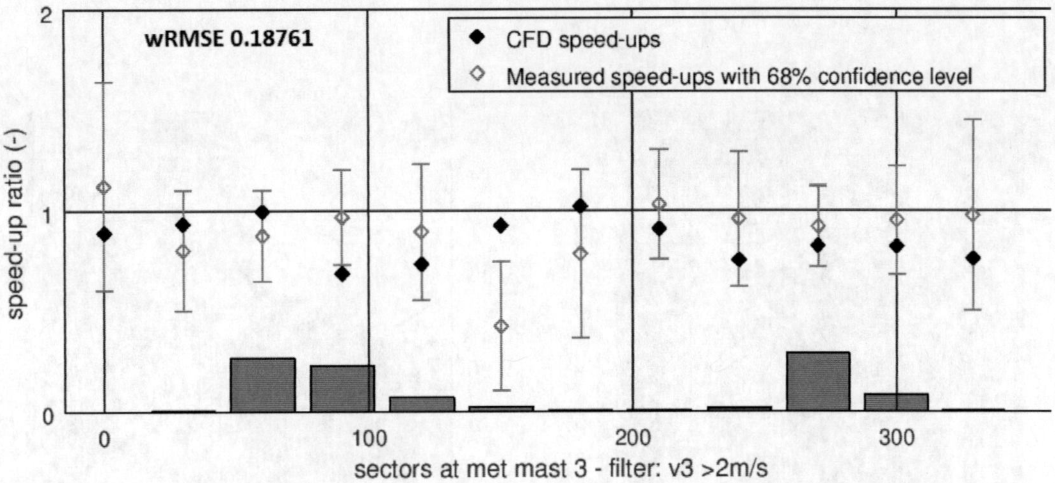

Figure 2. Example of sectorwise errors on speed-ups (sector width 30°). From reference 3 to target 2. CFD speed-ups (*black diamonds*) as for **eq. 2**, speed-ups from measurements (*empty diamonds with barred confidence level*) the 68% confidence level is estimated by $\pm\ \hat{\sigma}_{su_i}$ as for **eq. 4**; frequencies f_i at reference point (*green columns*).

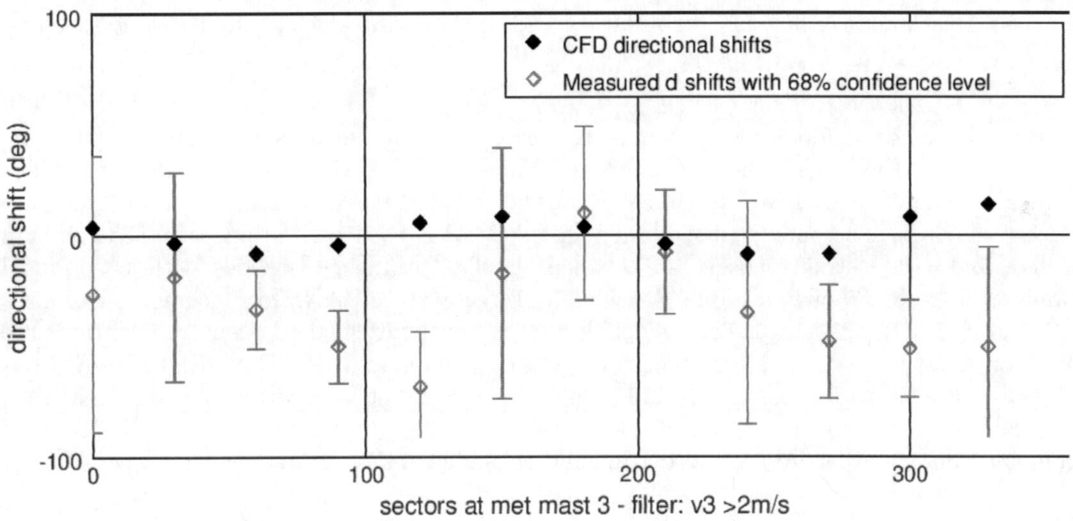

Figure 3. Example of sectorwise errors on directional-shifts. From reference 3 to target 2.

and ASTER GDEM data fused by a weighted averaging approach. The statistical validation of EU-DEM v1.0 documents a relatively unbiased (-0.56 meters) overall vertical accuracy of 2.9 meters RMSE, which is fully within the contractual specification of 7m RMSE (European Commission 2009). The upgrade to version 1.1 contains the correction of geo-positioning issues, a reduced number of artefacts and an improved vertical accuracy.

From the use-of-land codes of the CLC it is derived a roughness map, according to a conversion table which has been mainly derived from [3].

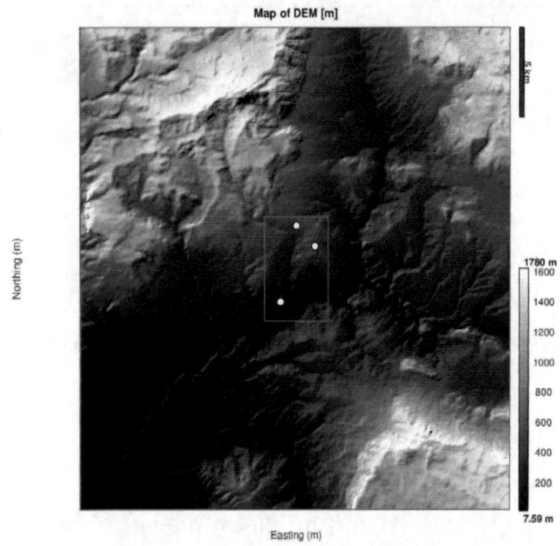

Figure 4. Digital Elevation Map (DEM) from EU-DEM v1.1 [1]. White circles indicate the met-masts, black circles the wind turbines.

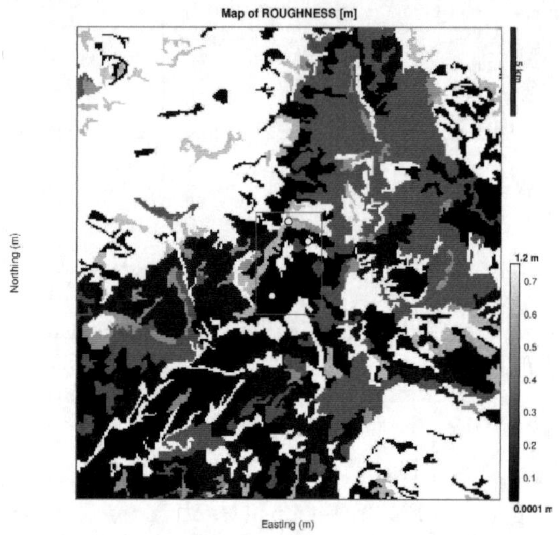

Figure 5. Roughness map from the Corine Land Cover 2012 [1]. White circles indicate the met-masts, black circles the wind turbines.

Several CFD simulations of the wind are carried out on the area of study, by changing the extension of the model, the resolution and number of directional sectors simulated. The Reynolds-Averaged Navier-Stokes (RANS) equations are therefore integrated over a domain covering the area. The open source library OpenFOAM is used for the purpose. The domain is discretized with a structured grid of hexahedral cells, turbulence is closed with a k-ε model (standard [6], RNG [7] or Realizable [8]), while the steady state RANS for uncompressible flows are integrated with the `simpleFoam` algorithm; only continuity and momentum equations are solved for this study so there are no thermal effects considered in the CFD models. All equations are firstly solved with a 1st order upwind scheme (`bounded Gauss upwind`), up to a prescribed level of convergence, and secondly the solution is refined by switching the numerical scheme of the momentum equations to a 2nd order upwind (`bounded Gauss linearUpwind grad(U)`).

2.1. Boundary conditions

The computational domain is bounded on the lowest surface by the terrain's orography, the upper boundary is a horizontal plane set at an automatically calculated height, function of the terrain complexity, lateral boundaries are vertical planes oriented as the Cartesian directions North-South, West-East. The lateral faces orientation implies that when the simulated wind is blowing from one of the Cartesian directions there is one inlet, one outlet and two no-friction walls; in all the other cases there are two inlets and two outlets.

Analytical expressions for the solved variables are assigned to the inlets, the outlets are set with a hydrostatic distribution of pressure, the upper boundary has a no-friction wall condition. A wall-function for fully-rough regimes is prescribed at terrain adjacent cells. Further information on the model is given in [4, 5].

IOP Conf. Series: Journal of Physics: Conf. Series **1102** (2018) 012007 doi:10.1088/1742-6596/1102/1/012007

3. The methodology

The present study is a sensitivity analysis of CFD models to some of the most important settings: the extension of the model, the turbulence closure of the RANS, the number of wind sectors modeled and the resolution of the computational mesh in the area of interest. Over the area studied there were deployed three met-masts, having respectively 30, 40 and 50m height and sensors mounted at several levels. The three masts were installed at the site with consistent overlapping time, 293 days of concurrent data for the first two masts erected in the area (1 and 2), few months of overlapping between the third mast (3 and 4) and the previous two. The first mast is located in a valley and the wind, by observing the wind-rose, looks quite channeled by it, the second mast is located on the highest part of the valley, almost on a saddle point between a ridge, at south-east, where it's located the third mast, and a group of mountains on the north-west, with a summit at 1780 m above sea level, see **Figure 4** for a top-view of the orography. Two main directions are identifed: wind from north-east and from south-west. In turn a measurement point is selected as reference and all the remaining points as predicted ones. The measurements at the prediction points, considered as ground truth, are used to evaluate the CFD error. In this fashion it's possible to plot the error in predicting the local acceleration, in terms of differences between measured speed ups and predicted by the CFD model. The CFD results are interpolated in order to reproduce the wind direction at reference before computing the speed-ups. In order to have one single metric for couple of measurement points a Root Mean Square Error ($wRMSE$) is computed against the directional sectors, weighting them by their frequency. It is finally studied how the $wRMSE$ error of the speed-ups varies with the model settings such as the extension of the CFD model, the horizontal resolution and the number of wind sectors modeled. The study is quantifying the accuracy of the CFD model through the chosen metric in order to provide guidelines for future CFD micrositing analysis.

In the **Figures 6**, **7** and **8** it is shown how the $wRMSE$ of speed-ups varies in the horizontal extrapolation with the distance between measurements and predicted points (targets). The general case in the legend ttt xx yy zz is characterized by a ttt turbulence model, xx directional sectors, yy meters of horizontal resolution, zz kilometers of distance between the refinement area and the frontier of the CFD domain. The refinement area, carachterized by a uniform horizontal grid resolution, includes all turbines and met-masts, framing them with 500 meters; its borders are highlighted in orange in **Figures 4** and **5**.

4. Results

The run test cases are summirized in **Table 1**, the varying parameters are: the turbulence model (standard k-ε, SKE; the Re-Normalization Group k-ε, RNG; the Realizable k-ε, Real.) the number of directional sectors (12, 24 or 36); the horizontal resolution in the refinement area; the width of the buffer around the area of refinement.

The $wRMSE$ metric is plotted against the distance between the sensors: in **Figure 6** it can be observed how the $wRMSE$ varies in a significant manner with model extension, while in **Figure 7** it can be observed how the turbulence closure did not play an important role for the selected project, it has to be emphasized how a finer mesh could generate different scenarios with the three turbulence models. With 100 m of horizontal resolution the RNG and Realizable k-ε gave almost identical results, while a little improvement was observed with the standard k-ε. In **Figure 8** it's instead shown the variation of the chosen metric by changing the number of sectors with which the wind roses is interpolated, and the number of CFD simulations, one per sector. It can be observed as an increasing number of sectors do not improve so much the accuracy of the prediction, with a unexpected tendency of increasing the $wRMSE$ with the number of sectors considered. This result should be investigated further, eventually also by decoupling the number of CFD models and the number of sectors used to discretized the wind roses, i.e. more CFD simulations per directional sector.

Table 1. List of cases run, classified by input parameters.

Turbulence model	SKE	SKE	SKE	RNG	Real	RNG	Real
Number of sectors	12	12	24	12	12	12	12
Resolution [m]	30	100	30	30	30	100	100
Buffer width [km]	1.4	3	1.4	3	3	3	3

Turbulence model	RNG	SKE	SKE	SKE	SKE	SKE	SKE	SKE
Number of sectors	12	12	12	24	36	12	12	12
Resolution [m]	100	100	100	100	100	30	180	100
Buffer width [km]	5	5	7	5	5	5	11	11

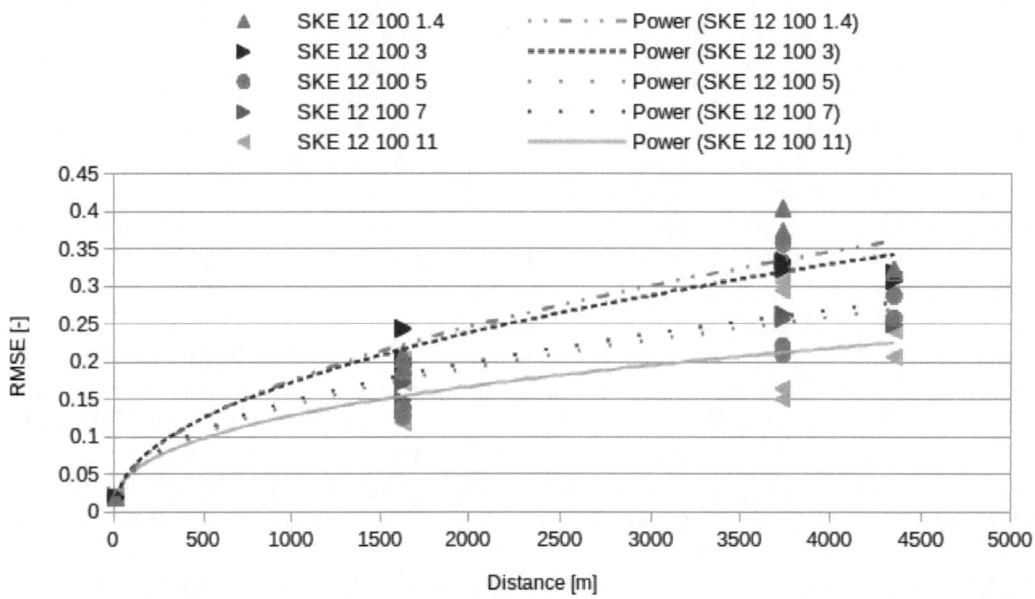

Figure 6. Dependence of the metric $wRMSE$ upon the distance to the reference met-mast. Regression with a parabolic function. Variation with buffer width, i.e. extension of the model.

For this complex site a buffer around the area of interest (the wind farm layout) of 11 km was providing a significant reduction of the target metric to be reduced, the $wRMSE$. The turbulence models were showing much less impact on the $wRMSE$ on the speed-ups. It is yet to be investigated a horizontal resolution smaller than 100 m coupled with 11 km of buffer width.

5. Conclusions

The weighted RMSE of CFD speed-ups are evaluated for a number of test cases, providing guidance for future projects in terms of model extension, resolution of the grid, number of sectors to model. The methodology is useful to quantify the accuracy of the CFD model, in predicting the wind speed, and in the design phase of the model to keep the CFD accuracy above

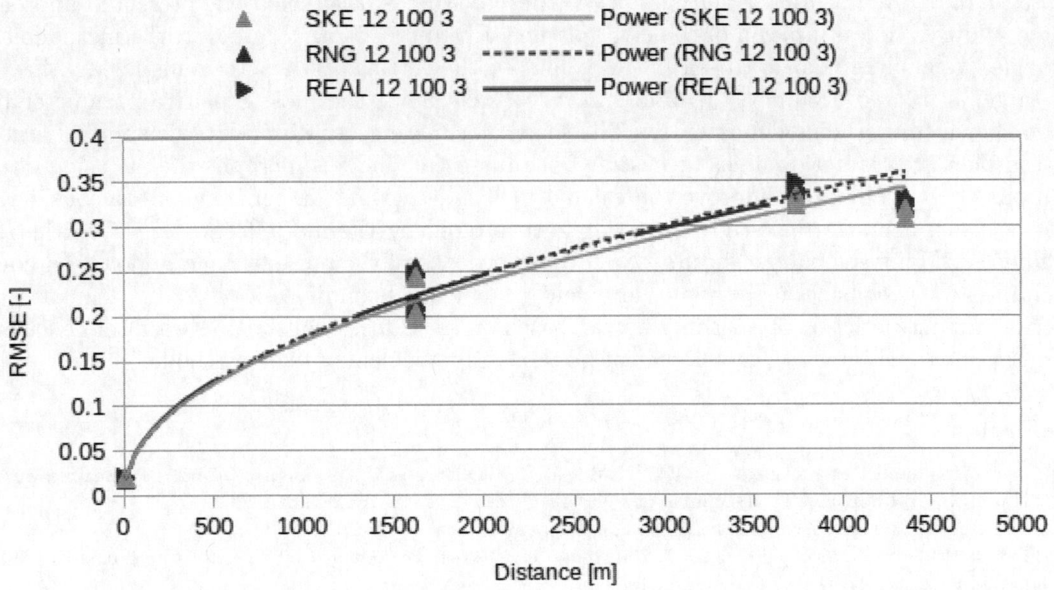

Figure 7. Dependence of the metric $wRMSE$ upon the distance to the reference met-mast. Regression with a parabolic function. Variation with tested turbulence models.

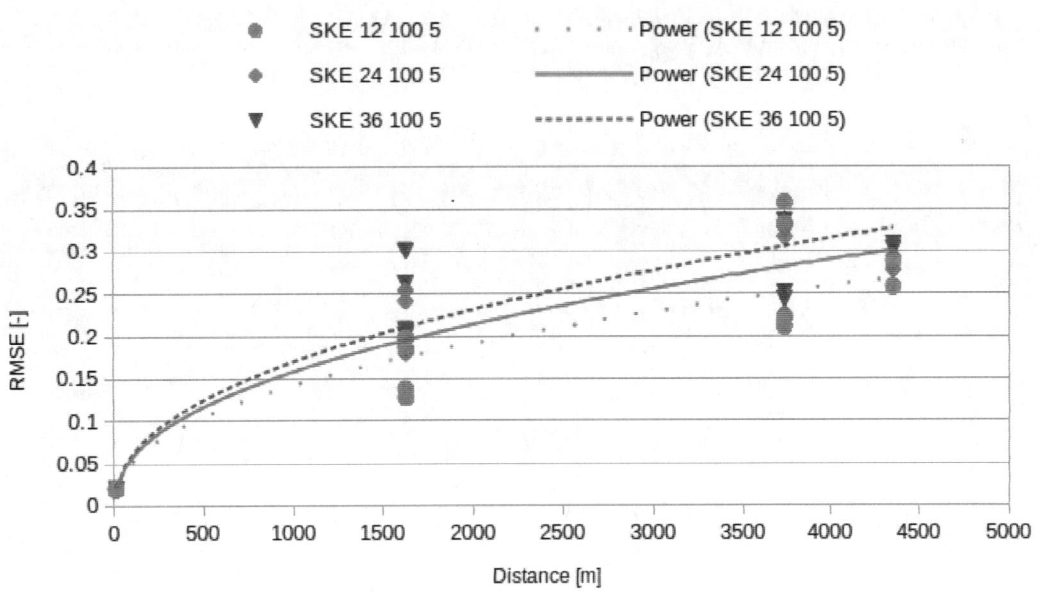

Figure 8. Dependence of the metric $wRMSE$ upon the distance to the reference met-mast. Regression with a parabolic function. Variation with number of sectors.

a minimum requirement. The authors of the present studies retain that the presented approach could allow the quantification of the error of the CFD micro-siting. Moreover, the methodology provides valuable guidance to select the values of the settings in a deterministic way, having as target a desired accuracy, or minimum $wRMSE$. Since the most sensitive parameter has been shown to be the model extension, the search for more accurate preditiction should aim at considering even larger domains or nesting with mesoscale models, peraphs the inclusion of the Coriolis effects could also improve the calculations as larger and larger domain are considered. The measurements in this study have not been filtered by thermal effects, e.g. with thermal gradient or Monin-Obukhov length, so, a future approach of the presented methodology it could be improved by considering only measurement near neutral conditions.

Also the introduction of a second metric, based on the directional shift between reference and target points will be very likely object of possible improvements of the methodology.

References

[1] Copernicus Land Pan-European, for EU-DEM and CORINE Land Cover. `https://land.copernicus.eu/`.

[2] Schuren E 2014 `https://land.copernicus.eu/user-corner/technical-library/eu-dem-2013-report-on-the-results-of-the-statistical-validation`.

[3] Silva J, Ribeiro C and Guedes R 2007 *Roughness length classification of Corine Land Cover classes* EWEC 2007 Proceedings.

[4] Crasto G, Segalini A, Astolfi D and Castellani F 2018 *Assessment of simulation and modelling errors for three CFD wind models* 6th Symposium on OpenFOAM in Wind Energy (SOWE), Visby June 2018

[5] Crasto G and Castellani F 2016 *Energy Yield Assessment and Site Suitability using OpenFOAM* 4th Symposium on OpenFOAM in Wind Energy (SOWE), TU Delft 2-4 May 2016

[6] Launder BE and Spalding DB 1974 *The numerical computation of turbulent flows* Computer Methods in Applied Mechanics and Engineering. 3 (2): 269-289 doi:10.1016/0045-7825(74)90029-2

[7] Yakhot V, Orszag SA, Thangam S, Gatski TB and Speziale CG 1992 *Development of turbulence models for shear flows by a double expansion technique* Physics of Fluids A, Vol. 4, No. 7, pp1510-1520.

[8] Shih TH, Liou WW, Shabbir A, Yang Z and Zhu J 1995 *A new k-e eddy viscosity model for high:reynolds number turbulent flows* Computers and Fluids, 24(3):227238.

Wind power prediction in complex terrain using analog ensembles

Yngve Birkelund[1], Stefano Alessandrini[2], Øyvind Byrkjedal[3] and Luca Delle Monache[2]

[1] Department of Engineering and Safety, The Arctic University of Norway, Tromsø, Norway
[2] Research Application Lab, National Center for Atmospheric Research, Boulder, CO, USA
[3] Kjeller Vindteknikk AS, Norway

E-mail: yngve.birkelund@uit.no

1. Introduction

The intermittent power generation from wind energy industry influence both power grid stability, energy system operators and the power marked economics. Accurate wind power predictions may reduce the cost impact of wind energy, and several forecasting techniques and methods have been proposed during the last decades [1, 2]. The performance of different forecasting methods is normally evaluated using root mean square error (RMSE) or mean absolute error (MAE), and the performance in general decreases for wind sites in complex terrain [3, 4].

In this paper we investigate the use of analog ensembles (AnEn) for wind power prediction [5] with focus on five wind power parks in northern Norway. This cold climate region can in general be characterized with good wind resources, but is challenging with the complex terrain consisting of large mountains, valleys and fjords [6, 7]. As input to the AnEn forecast method, we have used wind data from the open and operational weather forecast provided by the Norwegian Meteorological Institute [1]. The AnEn wind power prediction is compared to downscaling by the high-resolution WindPRO park modelling, using the same weather forecast as input. In addition, the naive persistence output model is included as a reference.

The paper is organized as follows: Section 2 describe the wind parks, the wind power forecasting methods and the verification statistics used. Results are shown and discussed in Section 3, followed by recommendations and conclusions in Section 4

2. Method

2.1. Wind parks

The location of the five wind parks in this study are shown in Fig. 1. The overview shows the terrain elevation in northern Norway in gray tones from black to white, corresponding to 0m and 2000m height above sea level, respectively, and the ocean is also shown in white. Nygårdfjellet wind park is found south west in this figure, quite close to the border to Sweeden. This wind park is located at an elevation of about 400m, in a valley going west-east and with high mountains close by to the north, and where high wind from east is common during the winter season [7]. Fakken wind park is located on a small island on the coast, on a small hill at 40-200m elevation, with open sea to north, close by mountains to the west and a mountain range to the south dividing with two large fjords. Havøygavlen is located on a flat low island, at about 200m

[1] http://thredds.met.no

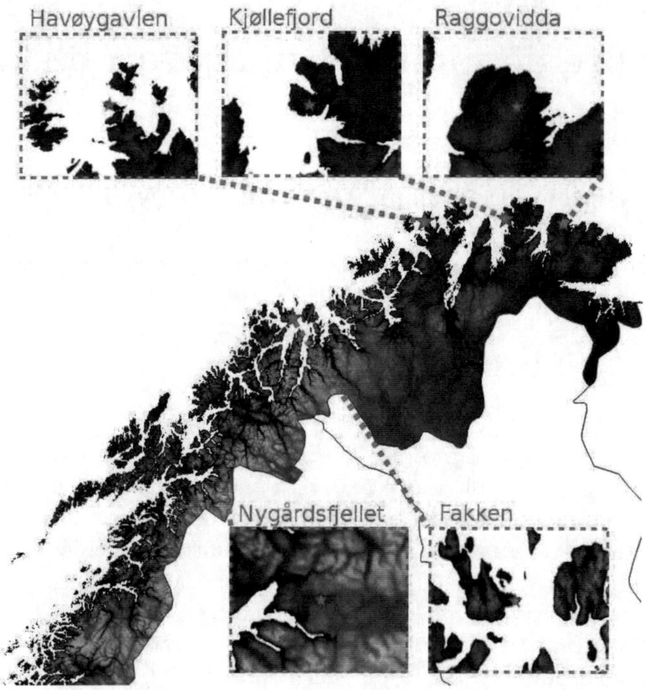

Figure 1. Northern Norway. Ocean shown in white. Gray tones from black to white shows terrain height from 0m to 2000m, respectively.

Table 1. Description of wind park sites.

Name	MW	Location °N/°E	RIX	Terrain features
Fakken	54.0	70.098 / 20.081	5-10	Small hill, high steep mountains and fjords
Havøygavlen	40.5	71.012 / 24.589	10-20	Flat low island, steep cliffs and fjords
Kjøllefjord	39.1	70.922 / 27.268	0-5	Low flat mountains and fjords
Nygårdfjellet	32.2	68.504 / 17.879	5-10	Inland valley, high steep mountains
Raggovidda	45.0	70.769 / 29.094	0-5	Flat inland mountain, close to coast

elevation. No nearby high mountains, but the island has steep cliffs down to the ocean and there are several fjords and narrow straits close by. Kjøllefjord is located at a low flat mountain area, at about 260m elevation, and with a large fjord to the west going north-south and some smaller fjords north and south going in the east-west direction. Raggovidda is the newest wind park in this region, is located at about 430m elevation on a large flat mountain area, with good wind resources and a capacity factor above 45% the last three years.

These wind parks are compared and summarized in terms of location, turbine types, site ruggedness (RIX)[8] and nearby terrain features in Table 1. The close by and steep cliffs down to the ocean make a large impact on the site ruggedness for Havøygavlen. The large mountains and fjords around Fakken and Nygårdsfjellet could also be challenging for wind prediction using numerical weather modelling.

2.2. Wind power forecasting

The weather forecast used in this paper is the Meteorological Cooperation on Operational numerical weather prediction (MetCoOp) run by Norwegian Meteorological Institute and Swedish Meteorological and Hydrological Institute. The forecast covers both countries with a horizontal resolution of 2.5km, using a horizontal grid of 739x949 points centered at 63.5°N and 15°E, and is issued at 00, 06, 12 and 18UTC with a 66 hour lead time. The MetCoOp is a branch of HARMONIE model developed within the High Resolution Limited Area Model consortium with several European meteorological institute, and was the main operational forecast in Norway from 2014 to 2016. This high resolution model has been shown to improve the wind, temperature and precipitation forecast in complex terrain [9]. The MetCoOp data set had some missing data in 2016, and in those cases, the succeeding MEPS or the similar Arctic AROME forecast data is used as a replacement.

Based on the current MetCoOp weather forecast F_t, at lead time t, the AnEn method search through historical weather forecasts to find similarities based on a set of physical variables. The ranking of similarities between current forecast and a historical forecast A_t is found following the distance metric [10]

$$||F_t, A_t|| = \sum_{i=1}^{N_v} \frac{w_i}{\hat{\sigma}_i} \sqrt{\sum_{j=-\tau}^{\tau} \left(F_{i,t+j} - A_{i,t+j}\right)^2} \tag{1}$$

where N_v is the number of physical variables to consider, w_i is the weighting chosen for the i-th parameter, $\hat{\sigma}_i$ is the estimated standard deviation of the variable, $F_{i,t}$ is the forecast for the i-th variable at lead time t, and τ describe the window of smoothing along the lead times of the forecasts. The N_A highest ranked historical forecast at that particular lead time t are the set of analog ensembles A_{t_n}, and the mean of the observed analog ensembles power output is the AnEn power forecast $PW_t^{\{anen\}}$ for lead time

$$PW_t^{\{anen\}} = \frac{1}{N_A} \sum_{n=1}^{N_A} PW^{\{anen\}} \left[A_{t_n}\right]. \tag{2}$$

The $PW^{\{anen\}} \left[A_{t_n}\right]$ are the power forecast ensemble which can be used to investigate the statistical properties of the forecast, e.g. standard deviation or interquartile range, and it is also possible to weight the analog outputs differently and/or use Kalman filtering to improve the forecast [10, 5].

The physical variables in the AnEn power forecast using Eq. (1) are 10m W_t and direction D_t interpolated to the wind park position shown in Table 1. The AnEn power forecast can use wind speed from 10m directly, even if the wind turbine nacelle typically are placed around 80m above the terrain, as the observed historical output from the nearest historical forecast is used as predicted forecast. The AnEn power estimate have 66 hourly forecast lead times, as this is the underlying wind forecast form the MetCoOp model.

Our AnEn wind power forecast is the mean value using $N_A = 20$ analogs, with equal weights on wind speed W_t and direction D_t, with window smoothing $\tau = 1$ and without bias correction [11].

We will compare the results from AnEn power forecast method with the naive persistence model (PRE), where the current power output as the forecast for all lead times, PW_t^{pre}. PRE is often used as a starting point both in operational use and for comparing different forecast techniques. We have this method as a base line, as the persistence methods performs well for short time periods [1].

The second type of forecasting we have included in this study uses power curves from the park wind turbines and numerical weather prediction for the wind estimate (PNWP). This method use the same 10 meter wind speed and direction data as the AnEn method, but now in combination with a wind park model from WindPRO software. As the WindPro park model assumes wind speed at turbine height, we have included a park depended wind scaling constant c giving park power output as

$$PW_t^{pnwp} = ppc\left[cW_t, D_t\right], \tag{3}$$

where the park power curve function $ppc[\cdot]$ is a linear interpolation from the park specification covering wind ranges $0,5$-$29,5$ m/s and 12 wind directions (N, NNE, NE, ...) of the WindPRO Park Power Verification Model (PPV) [12]. The input to the PPV model is based on N50 maps of Norway from Norwegian Mapping Authority, with 20 meter elevation levels and terrain roughness index, and the PPV model takes into account the power curve for each turbine, the placement of the turbine, the local topography and the wake effects between turbines [12]. The wind scaling constant c is chosen for each wind park site by optimizing the model output RMSE using the training data set, as described in Eq. (5).

Using the same input data for both PNWP and AnEn, allows for a fair comparison between the models as they share the same inherent variability from the numerical modeling of a changing weather pattern.

2.3. Forecast performance

The wind power data in this study are measured hourly total output, PW_t^o, during 2016 for each site listed in Table 1 provided by the Norwegian Water Resources and Energy Directorate (NVE)[2]. The power production time series are normalized with the corresponding wind park installed capacity, as listed in Table 1, as this allows a direct comparison between the different wind parks.

The wind forecast data is chosen so that all lead times are within 2016, creating a total of 1456 power forecast time series each of 66 lead times, $PW_{t,i}^m$, where m corresponds to the *anen*, *pnwp* or *pre* forecast method, t is the lead time and i is the forecast time series. The forecast data is split into a training set consisting of the first $N_T = 1200$ times series, and the last $N_P = 200$ time series are used to calculate the prediction performance for each method.

Based on the error $e_{t,i}^m$ between the predicted and observed power production

$$e_{t,i}^m = PW_{t,i}^m - PW_{t,i}^o \tag{4}$$

the root mean square error (RMSE) at each lead time t is defined as [13]

$$RMSE_t^m = \sqrt{\frac{1}{N_P}\sum_{i=1}^{N_P}\left(e_{t,i}^m\right)^2}. \tag{5}$$

The training set consisting of the first $N_T = 1200$ time series, corresponds to the period from January 1st to October 26th, is used to find the optimal RMSE scaling factor c for each wind park combining Eq. (3) and Eq. (5). The verification period runs from November 9th to December 29th, and the no overlap period between those ensure that there is no overfitting for the scaling factor.

The different power forecast methods is further analyzed using the bias b_t^m, or the mean error $\overline{e_t^m}$,

$$b_t^m = \overline{e_t^m} = \frac{1}{N_P}\sum_{i=1}^{N_P} e_{t,i}^m, \tag{6}$$

[2] https://www.nve.no/energiforsyning-og-konsesjon/vindkraft/

standard deviation of the error,

$$\sigma_t^m = \text{std}\left(e_t^m\right) = \sqrt{\frac{1}{N_P - 1}\sum_{i=1}^{N_P}\left(e_{t,i}^m\right)^2} \tag{7}$$

and the skill score ss_t^m based on the mean square error and the *pre* basic persistence forecast

$$ss_t^m = 1 - \frac{(RMSE_t^m)^2}{(RMSE_t^{pre})^2}. \tag{8}$$

Note that $ss_t^{pre} = 0$, and that the skill score can be interpreted as the effective improvement in other two method compered to the basic persistence forecast. A value above zero shows that the method is better than the PRE method, and the value can also be interpreted as the improvement in percent [14].

The last statistical quantity used in the comparison of the methods is the Pearson correlation coefficient defined as [15]

$$r_t^m = \frac{1}{N_P - 1}\sum_{i=1}^{N_P}\left(\frac{PW_{t,i}^m - \overline{PW_t^m}}{\text{std}\left(PW_t^m\right)}\right)\left(\frac{PW_{t,i}^o - \overline{PW_t^o}}{\text{std}\left(PW_t^o\right)}\right), \tag{9}$$

where $\overline{PW_t^m}$ and $\text{std}(PW_t^m)$ is the mean and standard deviation of the forecast methods, using the notations and definitions in Eq. (6) and Eq. (7), respectively. The correlation coefficient $r_t^m \in [-1, 1]$ describe the linear relationship between the forecast and measured value, where a value close to 1 (or -1) indicates a strong relationship and a value close to 0 means that these to quantities are uncorrelated [16].

3. Results

The RMSE for each power forecast method as a function of lead time are shown in Fig. 2, where the number in the parenthesis is the lead time averaged RMSE of the corresponding forecast method. The persistence method (PRE), shown as black solid lines, performs as expected with low RMSE at low lead times, but RMSE increases rapidly as t increases for all wind parks. Comparison between the wind parks shows that Nygårdsfjellet has the highest RMSE values, which may indicate a more rapid change of wind pattern at this park. The differences between the parks are in general quite small, and the average RMSE across wind parks, shown in lower right plot Fig. 2, shows that RMSE increases the first 20 hours and then stabilizes at RMSE $0.45 - 0.5$.

The numerical weather forecast results, PNWP, are shown as a red dashed-dotted lines in Fig. 2. For this forecast method, the RMSE starts around 0.2-0.25 at $t = 1$ hour, and increases almost linearly with lead time for all parks up to lower average RMSE value 0.35-0.4 than the other three, which indicates that the meteorological forecast is more precise here. The PNWP results are significantly lower than PRE for $t > 2$ for all wind parks, but at shortest lead time the persistence method still preforms better. Note also that the PNWP seems to increase slower in the 12-36 lead time region for Fakken and Nygårdsfjellet wind parks compared to the three other wind parks, indication that this method could provide good input for day-a-head trading.

The AnEn forecast results, shown as a blue dashed line in Fig. 2, provides a lower RMSE than the PNWP in all case. The overall results shows a RMSE with more than 10% improvement for all lead times. The improvement is highest and lowest for Havøygavlen and Nygårdsfjellet, respectively. The AnEn method shows in overall a consistent linear increase in RMSE from 0.2 to 0.3 for lead time 1 to 66 hours, and the performance variation is very small between the different parks.

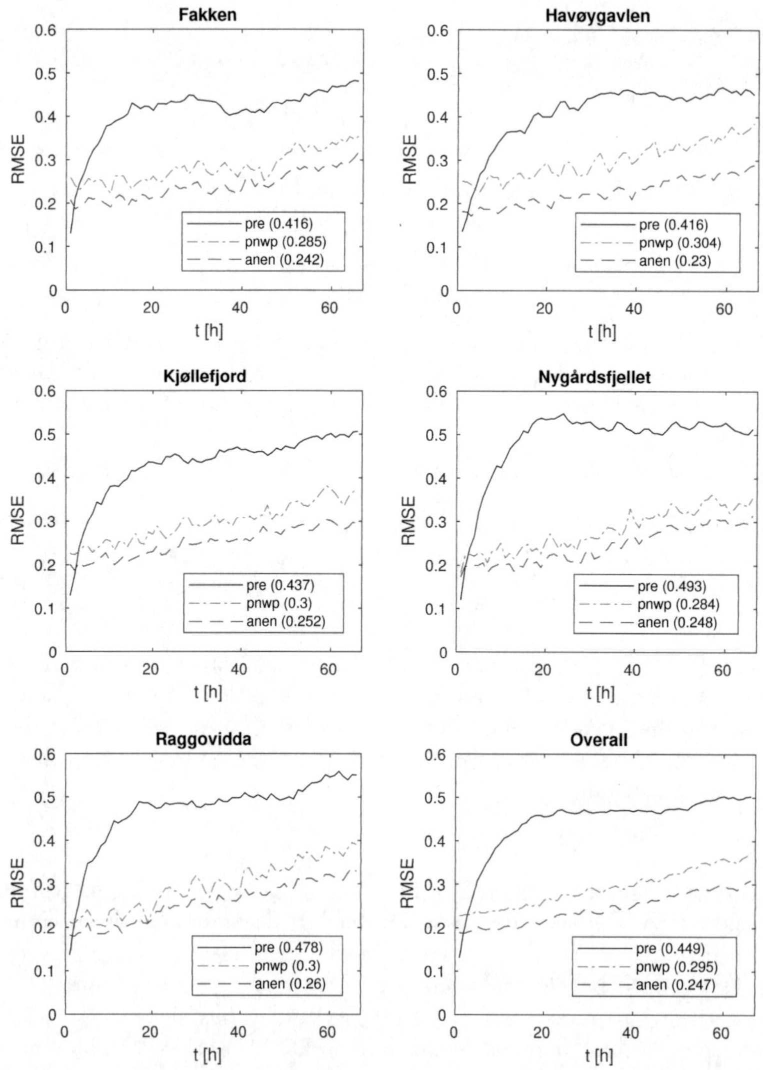

Figure 2. RMSE results for each wind park as function of lead time, with the lead time averaged RMSE in legend parenthesis. Lower right plot shows the overall RMSE averaged all wind parks.

The upper left plot in Fig. 3 shows the overall bias of the different forecast methods. PRE has a bias close to zero, while PNWP and AnEn are slightly over- and underestimating the power output, respectively. The bias increases for higher lead times, but the lead time averaged bias is under 2%. This bias in the AnEn case might come because of yearly variation of wind speeds, as the verification period at the end of the year might include more high wind events.

The overall standard deviation of the errors are shown in the upper right plot of Fig. 3. The standard deviation for all methods are almost identical to the RMSE results plotted in the lower right corner of Fig. 2. This shows that the prediction error is mainly caused by high variation in all cases, and not from bias.

The PRE method has a high correlation coefficient close to 0.92 for $t = 1$, but the correlation drops rapidly and falls below 0.50 at $t = 7$ hours. The PNWP and AnEn forecast methods, on

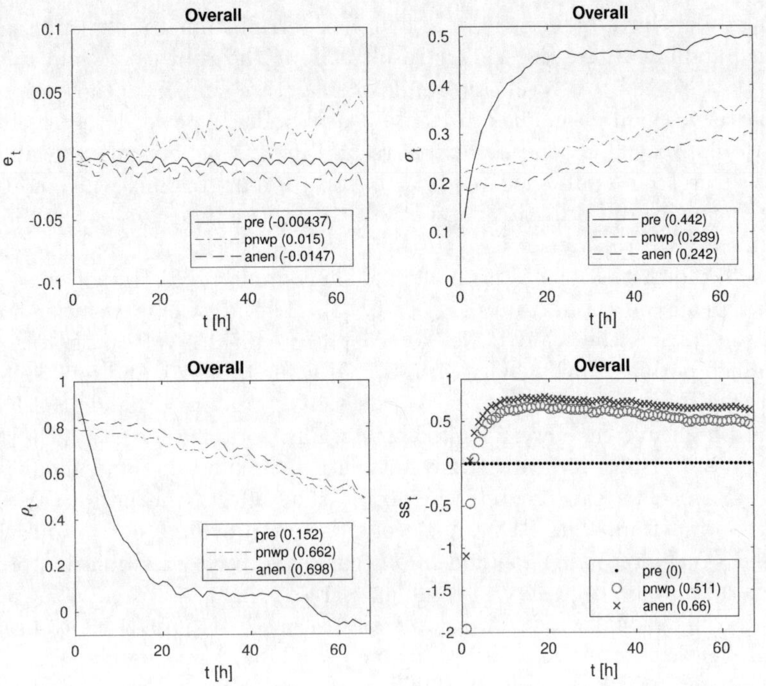

Figure 3. Overall performance of PRE, PNWP and AnEn methods. Upper left: Bias. Upper right: Standard deviation of error. Lower left: Correlation. Lower right: Skill score

the other hand, starts with a correlation close to 0.8 but the decay is significantly slower and the correlation does not drop below 0.5 even for the longest lead time $t = 66$. The AnEn method has a slightly higher correlation with measurement compared to the PNWP method.

The last plot is the skill score results, shown in the lower right corner of Fig. 3. To allow for easier readout for each lead time, the hourly results are shown as black dots, red circles and blue crosses, for PRE, PNWP and AnEn forecasting method, respectively. This figure shows that PRE performers best for $t = 1$, for $t = 2$ the PRE and AnEn methods performs equally good as $SS_1^{anen} = 0$, but for all other lead times the AnEn forecast performs best. The average improvement of AnEn and PNWP compared to PRE is 66% and 51%, respectively, and for lead times $t \in [12, 36]$ relevant for day ahead energy trading, these numbers increases to 75% and 65%.

4. Conclusions

In this paper we have investigated five different wind parks in northern Norway with respect to power prediction using three different prediction methods: PRE, PNWP and AnEn.

We have shown that the AnEn method provides a robust power forecast at all wind parks in this study. Overall, the RMSE for this method is 44% lower than the persistence method and 16% lower than optimized wind speed from numerical weather prediction and power curves. The performance of all methods is also investigated using standard deviation of error, skill score and Pearson correlation coefficient, and the AnEn method perform better than persistence for $t \geq 2$ hours lead time and better than PNWP for all lead times.

The large improvement in RMSE from PNWP to AnEn method in the case of Havøygavlen wind park might be explained by the high ruggedness index for this site. It is well known that numerical weather models have higher errors in complex terrain, but the AnEn method seems to be successful in handling those cases. Our results about the robustness and improvement using AnEn method compared to PNWP in complex terrain are similar to those represented in [4], although the latter uses all ensembles to create a probabilistic power forecast and we have used a deterministic approach. Furthermore, the results in this paper expands the use of the AnEn method into several wind parks located in a coastal complex terrain, which add some new insight into the performance of both AnEn and PNWP based methods.

The site ruggedness (RIX) has been used in the literature to classify the complexity of the terrain, and it is well known that the performance of short-time forecasting of wind speed and direction strongly depends on the terrain complexity [3]. The MetCoOp weather forecast has been used in this study, and the wind power forecast from the PNWP method have almost identical large variance both at Havøygavlen with a RIX value of 10-20, and for Kjøllefjord and Raggovidda where the RIX values are in the 0-5 class. On the other hand, the Fakken and Nygårdsfjellet wind park have the lowest RMSE errors, although these parks have the highest nearby mountains. Our results indicates that both nearby mountains and mountain range, valleys, fjords and sharp cliffs can create local events that affects the performance, in both negative and positive direction. The RIX value alone do not provide a clear indication to the trustfulness of the PNWP forecasting method in our case, and a closer evaluation of the NWP performance in general may be necessary in complex terrain.

When it comes to operational wind power forecast systems, it is important to keep in mind that the numerical weather prediction results are not available immediately. The MetCoOp forecast at initiation time (00, 06, 12 and 18 UTC), uses some time both to collect receive and process all observations and to do the numerical simulation, which implies that the forecast typically is available for public 2.5 hours after the initiation time. The AnEn method is not computer intensive and does not add extra time in this setting, but in practice both AnEn and PNWP method results should be shifted two leads hours forward because of the NWP processing time for a fair comparison with the PRE method. Thus, we could expect that the PRE method performers best up to $t = 3$ lead hours, and AnEn best after this. As day-a-head power trading typically has to be submitted before 12:00 the day before, this corresponds to lead time in the 12-36 hour region. For these lead times, the wind power forecast system is less effected by the NWP processing time and the AnEn method should be a natural choice among the methods investigated in this paper.

In this paper, the input wind parameters to the PNWP and the AnEn power prediction models are limited to a numerical weather prediction model with 2.5 km horizontal resolution taken 10 meter above the ground. If the input came from a higher resolution model, and/or the wind parameters at turbine hub heights were used, we would expect a decrease in RMSE for the PNWP. Changing the input parameters could also improve the AnEn performance, and it would have been interesting to quantify the ratio on improvements in these models.

Our results are limited to five wind parks and about one year of data, and some forecasts time series from other models had to be included. With longer training data, using a single operational forecast model and more advanced versions of analog ensembles, we can expect increased performance in AnEn.

5. Learning objectives

The paper introduces different classes of short-time power prediction, and explain an experimental setup for the investigation of model performance. The wind parks included represents different complex terrain types, including high steep mountains, valleys, fjords and sharp cliffs. This might create significant localized speed up or wind shadow area effects. Out results indicates that the analog ensembles method provide an robust and efficient method to short-time power production in complex terrain.

References

[1] Wu Y K and Hong J S 2007 *Power Tech, 2007 IEEE Lausanne* pp 504 –509
[2] Giebel G, Brownsword R, Kariniotakis G, Denhard M and Draxl C 2011 The state-of-the-art in short-term prediction of wind power - a litterature overview, delivery anemos.plus, Risø DTU, Denmark
[3] Costa A, Crespo A, Navarro J, Lizcano G, Madsen H and Feitosa E 2008 *Renewable and Sustainable Energy Reviews* **12** 1725 – 1744
[4] Kariniotakis G, Marti I, Casas D, Pinson P, Nielsen T, Giebel G, Usaola J, Sanchez I, Palomares A, Brownsword R, Tambke J, Focken U, Lange M, Louka P, Kallos G, Lac C, Sideratos G and Descombes G 2004 *Proc. European Wind Energy Conference*
[5] Alessandrini S, Monache L D, Sperati S and Nissen J 2015 *Renewable Energy* **76** 768 – 781
[6] Byrkjedal Ø and Åkervik E 2009 Wind map of norway (in norwegian) Report 9 The Norwegian Water Resources and Energy Directorate
[7] Bilal M, Birkelund Y and Homola M 2015 *Clean Energy Technologies* **3** 106–109
[8] Bowen A J and Mortensen N G 1996 *Proc. European Union wind energy conference* O15.2 pp 584–587
[9] Mller M, Homleid M, Ivarsson K I, Kltzow M A , Lindskog M, Midtb K H, Andrae U, Aspelien T, Berggren L, Bjrge D, Dahlgren P, Kristiansen J, Randriamampianina R, Ridal M and Vignes O 2017 *Weather and Forecasting* **32** 609–627
[10] Delle Monache L, Nipen T, Liu Y, Roux G and Stull R 2011 *Monthly Weather Review* **139** 3554–3570
[11] Delle Monache L, Eckel F A, Rife D A, Nagarajan B and Searight K 2013 *Monthly Weather Review* **141** 3498 – 3516
[12] Weir D E 2014 Wind power - production 2013 (in norwegian) Tech. Rep. 20 The Norwegian Water Resources and Energy Directorate
[13] Madsen H, Pinson P, Kariniotakis G, Nielsen H A and Nielsen T S 2005 *Wind Engineering* **29** 475–489
[14] Murphy A H and Epstein E S 1989 *Monthly Weather Review* **117** 572–582
[15] Dunn O J and Clark V A 1986 *Applied Statistics: Analysis of Variance and Regression (2nd ed.)* New York, NY, USA: John Wiley & Sons, Inc.
[16] Rodgers J L and Nicewander W A 1988 *The American Statistician* **42** 59–66

A novel method to quantify atmospheric stability

Lasse Svenningsen[1], René M. M. Slot[1,2], and Morten L. Thøgersen[1]

[1]EMD International A/S, Niels Jernes Vej 10, Aalborg, Denmark
[2]Aalborg University, Dep. of Civ. Eng., Thomas Manns Vej 23, Aalborg, Denmark

E-mail: ls@emd.dk

Summary. Atmospheric stability significantly influences the characteristics of a wind resource and strongly affects wind turbine power production and structural loads. Stability is governed by the thermal structure of the atmospheric boundary layer (ABL). Unstable ABLs are convective and increase turbulence, but reduce vertical wind shear, while stable ABLs reduce turbulence and increase wind shear. In neutral ABLs mechanical effects of terrain and roughness dominate.

The Monin-Obukhov length (MOL) and Richardson numbers are the recommended methods to quantify atmospheric stability. These methods require advanced and expensive measurements of temperature differences or 3D-sonic covariances, not installed on standard wind power masts.

This study presents a novel methodology to quantify stability based only on standard wind measurements. At high wind speeds turbulence and wind shear converge to the neutral ABL response in a given direction while the relative deviations of shear and turbulence at lower wind speeds strongly correlate with stability. When normalized by the neutral shear and turbulence, these deviations directly quantify stability. Here we present a novel method to estimate MOL directly from the ratio of normalized shear and turbulence.

The proposed method to quantify stability from standard wind measurements, provides an important input to advanced wind flow modelling of stability effects to improve the accuracy of predicted power production and structural loads.

1. Introduction - Importance and effects of atmospheric stability

Atmospheric stability is fundamental to understanding and characterizing the wind climate at a potential wind farm location as it significantly influences the wind turbine power production [1,2] and the structural loads [3–5]. Stability is determined by the buoyancy forces in the atmospheric boundary layer (ABL) and, hence, the acceleration or attenuation of vertical air movements. These buoyancy effects lead to convective mixing in the unstable ABL (UABL) or attenuation of vertical movements and stratification in the stable ABL (SABL). In the UABL turbulence is increased above the background level determined by the mechanical effects of terrain and roughness and the convective mixing reduces the vertical variation of wind speed (i.e. wind shear). The SABL has the opposite effect on wind shear and turbulence, and reduces turbulence to below the mechanical level resulting in stratified flow and increased wind shear. The neutral ABL (NABL) has neutral buoyancy and the wind flow is dominated by the mechanical effects of surface terrain and roughness.

The well-established standard methods to quantify atmospheric stability are based on measurements of vertical temperature differences or turbulent fluxes estimated from 3D-sonic covariance measurements. These methods are the Richardson numbers Ri (gradient or bulk) and the Monin-Obukhov length (MOL) L, typically given as $1/L$ [2,3,6]. Ri and $1/L$ estimate the ratio of buoyancy effects and mechanical effects and tend to zero in the NABL, where mechanical effects dominate. Both Ri and $1/L$ are negative in UABLs and positive in SABLs.

Alternative methods have been proposed to categorize stability directly from wind shear [7]. Such methods neglect the important effect that shear is not only controlled by stability, but to a large degree also by terrain and roughness. Neglecting these effects leads to wrong stability classification in directions with very high or low surface roughness or significant terrain-induced wind speed-up.

In this paper we present a novel method to quantify stability based on measurements from a standard mast setup typical for wind power projects. Such a mast only logs 10min mean and st. dev. of wind speed, and 10min wind direction mean at multiple measuring heights (z). This setup allows calculation of the 10min values of the wind shear exponent, $\alpha(t)$, and turbulence intensity, $TI(t) = \sigma(t)/u(t)$, which both depend strongly on stability. Section 2 describes the theoretical basis of the proposed method. Section 3 presents the four steps of the method and section 4 demonstrates the application of the method to observations from the Cabauw tower. Section 5 analyses the validity of the main assumption of the proposed method.

2. Monin-Obukhov similarity theory for shear and turbulence

The Monin-Obukhov Similarity Theory (MOST) provides a theoretical basis to quantify the expected deviations of wind shear as a function of stability, i.e. MOL. MOST is valid in the surface layer where fluxes are close to constant, typically covering the lower 10% of the ABL or approximately up to 50-100m above ground level [8,9]. The basic governing equation is the logarithmic wind profile with the MOST stability correction, Ψ_m.

$$u(z) = \frac{u_*}{\kappa}\left\{\ln\left(\frac{z}{z_0}\right) - \Psi_m\left(\frac{z}{L}\right)\right\} \tag{1}$$

Ψ_m is an integral of the non-dimensional wind shear function, φ_m, which is the 'universal scaling function' for momentum [8,9]. u_* is the friction velocity, κ is Von Karman's constant typically taken to be 0.4, z is the height above ground, and z_0 is the roughness length. In the limit of neutral stability $\varphi_m = 1$, and $\Psi_m = 0$. The vertical wind shear exponent, α, is typically estimated using multiple heights on the wind profile, but may also be calculated directly from the wind speed derivative at a single height using the non-dimensional wind shear function, φ_m, according to [4]:

$$\alpha \equiv \frac{du/dz}{u/z} = \frac{u_*}{u \cdot \kappa}\varphi_m\left(\frac{z}{L}\right) \tag{2}$$

The longitudinal component of turbulence ($\sigma_u = TI \cdot u$) does not fully follow the universal scaling laws of MOST as it does not entirely scale with height. However, σ_u does scale with u_* and depend on L, and non-dimensional scaling functions, φ_1, have been published in [8,10] covering the stable and unstable side, respectively. In the neutral limit φ_1 approaches a constant value. On the stable side φ_1 depends on z/L [11,12], whereas on the unstable side φ_1 depends only on $1/L$ and slightly on the boundary layer depth, z_i [8,10,13,14] as summarized in equations (3) and (4).

$$\sigma_u = u_*\varphi_1\left(\frac{z}{L}\right) \quad \text{(stable)} \tag{3}$$

$$\sigma_u = u_*\varphi_1\left(\frac{z_i}{L}\right) \quad \text{(unstable)} \tag{4}$$

The typical theoretical expressions published for Ψ_m, φ_m and φ_1 are summarized in Table 1 with the relevant references. The constants a, b and c are general and vary across the φ-functions and stability categories in the table. However, in the cited references the typical values for Ψ_m are b=4.7 (or 5.0) for stable conditions and b=16 for unstable conditions where c=1/4. Similarly, b=4.7 for φ_m under stable conditions and b=16, c=-1/4 for unstable conditions. For φ_1 a=1.7, b=0.31, c=0.63 under stable conditions and a=2.3, b=0.042, c=1/3 for unstable.

The phi-functions must be continuous across the stability spectrum, hence, their stable and unstable expressions must tend to the same neutral limit for large L. This is automatically fulfilled for φ_m as the published expressions are 'pre-normalized' to unity in the neutral limit. For φ_1 this continuity requires the constant a to be identical across the neutral, stable and unstable categories, but published values vary in the range of 1.7-2.5 [8,11]. For the purpose of this study we use the typical and intermediate value a=2.3 from [10]. As we shall see later the proposed method will only utilize ratios of φ_1, so the assumed value of a has limited significance except to ensure continuity across the stability categories.

Table 1. *Summary of published expressions for the universal functions, Ψ_m, φ_m and φ_1, under neutral, stable and unstable conditions. Relevant references are placed besides each expression. The constants for the neutral limit may be inferred from letting $|L| \to \infty$.*

	Neutral	Stable		Unstable	
Ψ_m	0	$-b\frac{z}{L}$	[8,10]	$\ln\left[\left(\frac{1+x^2}{2}\right)\left(\frac{1+x}{2}\right)^2\right] - 2\tan^{-1}(x) + \frac{\pi}{2}$ $x = \left(1 - b\frac{z}{L}\right)^c$	[8,10]
φ_m	1	$1 + b\frac{z}{L}$	[8–10]	$\left(1 - b\frac{z}{L}\right)^c$	[8–10]
φ_1	a	$a + b\left(\frac{z}{L}\right)^c$	[11,12]	$a\left(1 - b\frac{z_i}{L}\right)^c$	[10,13]

MOST makes several assumptions about the characteristics of the surface layer and their validity influences the accuracy of the presented method. In addition to constant fluxes, MOST assumes horizontal homogeneity e.g. of terrain and roughness as well as quasi-stationary conditions [9]. Additionally, according to [8] MOST is not valid at low wind speeds. For wind power applications the latter is not a problem as wind turbines do not operate below 3-5m/s. The other MOST assumptions are neither fulfilled for most real wind turbine locations nor for most 10min intervals during turbine operation. However, as we shall see in the next section describing the method and in section 5 validating the method, these assumptions are partially alleviated by using of ratios of the universal φ-functions. These ratios do not require MOST to be accurate in an absolute sense, only in a relative sense, to accurately predict the stability induced deviations of shear and turbulence from their neutral level.

3. Description of the proposed method

A typical wind power mast setup with wind speed and direction measurements at multiple levels, allows calculation of 10min values of the wind shear exponent, $\alpha(t)$, and turbulence intensity, $TI(t)$. The proposed method seeks to quantify stability in each 10min data interval via quantifying the deviations of shear and turbulence from their neutral levels in the wind direction of each 10min interval. The method has four processing steps to quantify stability in each 10min interval described in the following.

***Step 1** – estimate neutral levels of shear and TI versus direction*

First step of the method is to estimate the neutral levels of turbulence, $\widehat{TI}_N(\theta)$, and wind shear, $\hat{\alpha}_N(\theta)$, as a function of direction. These are estimated from the highest 1-3% of wind speeds using robust median statistics applied to a 10-20° wide rolling window of wind directions ensuring a smooth and robust estimation in 1° direction steps.

***Step 2** – estimate normalized deviations of shear and TI*

Second step is to calculate the normalized shear and turbulence and their deviations from unity ($\delta\alpha$ and δTI) for each 10min measurement. The normalization is relative to the neutral estimates from 'step 1', as shown in equation (5).

$$\delta TI(t) = \frac{TI(t)}{\widehat{TI}_N(\theta)} - 1 \quad \& \quad \delta\alpha(t) = \frac{\alpha(t)}{\hat{\alpha}_N(\theta)} - 1 \tag{5}$$

***Step 3** – apply smoothing kernel to enhance stability signal*
Third step, which can be omitted, is to apply a smoothing kernel to the time series of normalized shear and TI deviations to enhance the stability signal and reduce spurious variations. A centered median filter with a 1-3 hour window is a very effective compromise between retaining the signal and reducing the noise considerably.

***Step 4** – make diagnostic plot and estimate Monin-Obukhov lengths*
In the final step the normalized shear and TI deviations $\{\delta TI(t), \delta\alpha(t)\}$ are used to make a diagnostic plot to visually check the data. In the plot the magnitude of the $\{\delta TI(t), \delta\alpha(t)\}$ deviation represents the strength of the stability effects and the direction indicates the type of stability:

- Neutral: low magnitude, i.e. $\sqrt{\delta TI(t)^2 + \delta\alpha(t)^2} \rightarrow 0$
- Stable: significant magnitude in 2. quadrant, $\delta TI(t) < 0$ & $\delta\alpha(t) > 0$
- Unstable: significant magnitude in 4. quadrant, $\delta TI(t) > 0$ & $\delta\alpha(t) < 0$

The normalized shear and TI directly quantify stability and can be converted to an estimate of MOL based on theoretical considerations presented later in this section.

A large magnitude of deviations in the first and third quadrants of the diagnostic plot in 'step 4' can represent more ambiguous stability effects where the data deviate from the basic assumptions. Such data could result from systematic biases in the estimation of the neutral levels at high wind speeds due to complex meteorological effects. Such an effect could be the Charnock effect where roughness increases with wind speed for water surfaces, leading to increasing shear and turbulence in the high wind speed limit, introducing a bias. Similarly, data with significant magnitude in first quadrant ($\delta TI(t) > 0$ & $\delta\alpha(t) > 0$) could result from stable conditions coinciding with a ramp in wind speed or unstable conditions with large scale convective perturbations to the profile. Data in third quadrant ($\delta TI(t) < 0$ & $\delta\alpha(t) < 0$) could result from stable conditions with a shallow nocturnal ABL with measurement heights in the residual layer.

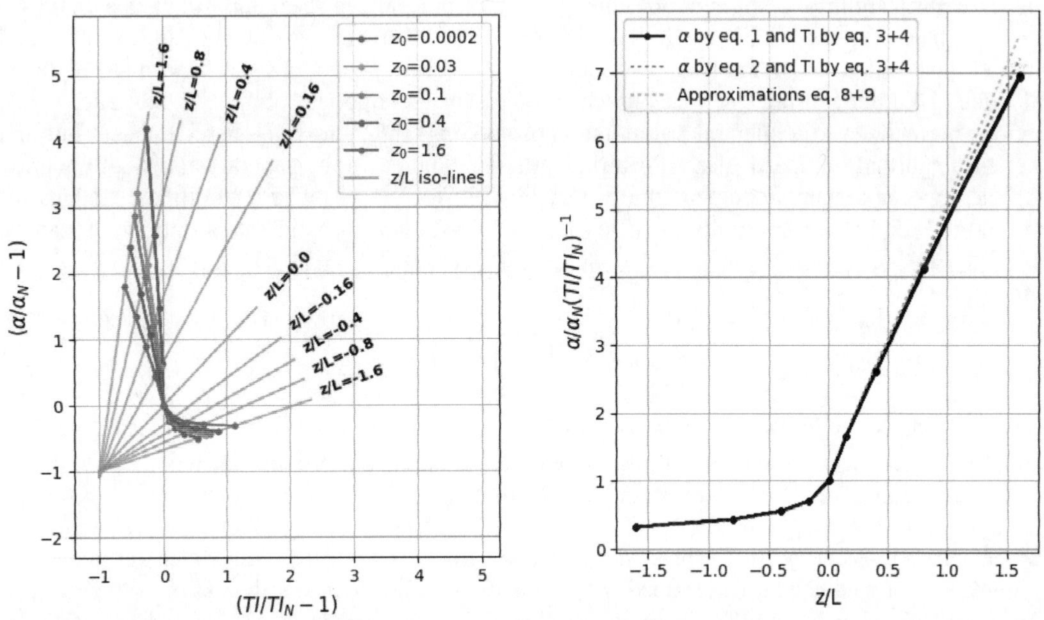

***Figure 1.** Left: theoretical normalized shear and turbulence deviations for five roughnesses (coloured lines) and nine stabilities (grey z/L-'iso-lines'). Right: ratios of α/α_N $(TI/TI_N)^{-1}$ versus z/L for the same data, where the five roughness curves collapse to a single (black) curve across the stabilities.*

To study the proposed method we generate theoretical wind speed profiles and turbulence data and apply steps 1 and 2 of the proposed method to calculate theoretical normalized deviations of shear and turbulence. The results are shown in Figure 1 (left) and the data are generated using the Ψ_m and φ_1 expressions in Table 1 with the constants outlined above the table, as well as equations (1) and (3)+(4) to calculate the mean wind speed profile and turbulence intensity. These calculations are performed for nine stabilities ranging from very unstable, over neutral to very stable represented by values of L from -50m over ±10,000m to 50m. Height above ground is set to $z = 80$m with 40m and 140m included for shear estimation to match the observations presented later in section 4. The ABL height is set to $z_i = 2000$m, but the results have a very low sensitivity to this assumption. This calculation setup is further repeated for five different roughness values ranging from $2 \cdot 10^{-4}$m to 1.6m. In Figure 1 (left) the solid lines show the normalized shear and turbulence deviations for these five roughness values across the nine stabilities. Points representing same z/L (stability) are connected to form z/L-'iso-lines'. Note that these z/L-'iso-lines' are all straight and converge to the same point {-1,-1}.

The peculiar convergence to {-1,-1} of the z/L-'iso-lines' in Figure 1 (left) indicate that the ratios of the normalized shear and turbulence are independent of roughness. If instead just the normalized shear and turbulence are plotted the z/L-'iso-lines' converge to {0,0}. As a consequence the five curves for different roughnesses collapse to a single curve in Figure 1 (right) when plotting the normalized shear divided by the normalized turbulence versus z/L. This single curve strongly resembles the shape of the universal function for momentum, φ_m, but does not match it exactly. To further analyse this result, equations (2), (3), and (4) are combined to investigate the theoretical expectation for the ratio of the normalized shear and turbulence. It turns out that in this ratio the wind speeds and friction velocities all cancel out as they appear both in equation (2) and equations (3)+(4), hence, yielding the result for a particular z/L:

$$\frac{\alpha}{\alpha_N}\left(\frac{TI}{TI_N}\right)^{-1} = \frac{\varphi_m}{\varphi_{m,N}}\left(\frac{\varphi_1}{\varphi_{1,N}}\right)^{-1} \tag{6}$$

This is a significant result. Not only does the ratio of the normalized shear and turbulence annul the effect of roughness as seen in Figure 1. Equation (6) shows that the ratio of normalized shear and turbulence equals the ratio of the two corresponding universal φ-functions, each normalized to its neutral value. All other constants or variables cancel out. This is important because in the ratios of the φ-functions normalized to its neutral value, the constants in Table 1 to some degree cancel out and become less important – in particularly for turbulence. This increases the general validity of equation (6) with the important example being complex terrain with steep slopes where modified constants for φ_1 are reported in [13]. Their results indicate that the adjustment of φ_1 in steep topography can be treated as factor $f(\hat{\theta}, z/L)$ as summarized in equation (7), where $\hat{\theta}$ is a characteristic terrain slope in a particular direction.

$$\varphi_{1,topo} \cong f\left(\hat{\theta}, \frac{z}{L}\right) \varphi_{1,flat} \tag{7}$$

If $f(\hat{\theta}, z/L)$ does not depend too strongly on z/L, it will mostly cancel out when φ_1 is normalized to its neutral value (cf. equation 6), making the φ_1 ratios more valid in non-flat topography. These arguments seem plausible also for normalized wind shear, φ_m, for moderate deviations from neutral stability in steep terrain, which also appears to be indicated by the results in [15] for unstable conditions. Similar considerations apply to the bias introduced by a Charnock effect for water surfaces. The increasing water roughness with wind speed will positively bias the neutral estimates of both shear and turbulence, slightly skewing a $\{\delta TI, \delta\alpha\}$ plot like Figure 1 (left), but the bias will, at least partially, cancel out in the estimated ratios $\alpha/\alpha_N\,(TI/TI_N)^{-1}$ damping the effect of the bias.

An important consequence of Figure 1 (right) is that it can be used to estimate L by inversely looking-up values of z/L from observed ratios of $\alpha/\alpha_N\,(TI/TI_N)^{-1}$. Taking a closer look at Figure 1 (right) the stable part is close to linear in z/L and the unstable part is close to linear in $-\ln(-z/L)$. This conclusion can also be reached by analysing equation (6) inserting the φ-expresssions in Table 1 using the values outlined above the table. The linear and log-linear approximations to the stable and unstable regimes are summarized in equations (8) and (9) and shown in Figure 1 together with the exact model.

$$\text{Stable:} \quad \frac{\alpha}{\alpha_N}\left(\frac{TI}{TI_N}\right)^{-1} > 1, \quad \frac{\alpha}{\alpha_N}\left(\frac{TI}{TI_N}\right)^{-1} \cong 4.1\frac{z}{L} + 1 \tag{8}$$

$$\text{Unstable:} \quad \frac{\alpha}{\alpha_N}\left(\frac{TI}{TI_N}\right)^{-1} < 1, \quad \frac{\alpha}{\alpha_N}\left(\frac{TI}{TI_N}\right)^{-1} \cong -0.15\ln\left(\frac{-z}{L}\right) + 0.4 \tag{9}$$

This result derived from equation (6) states that MOL can be estimated directly from measurements of wind speed and turbulence, although MOL itself depends on heat flux. However, the effect of heat flux enters implicitly via the estimation and normalization with the neutral shear and TI levels to quantify the deviations from the neutral state. Hence, this normalization accounts for the effect of non-zero heat flux and non-neutral temperature profile.

4. Demonstration of proposed method at Cabauw tower

The proposed method is demonstrated using data from the Cabauw meteorological research tower [16] situated in quite homogenous terrain in the Netherlands. The mast is equipped with high grade research instruments and the data are carefully filtered for faulty measurements. Wind speed measurements from $z \in \{40\text{m}, 80\text{m}, 140\text{m}\}$ are used for the shear estimates and the 80m data for turbulence and wind directions. Figure 2 shows the result of 'step 1' estimation of the neutral TI and shear levels for the direction 220°±10°. Estimated neutral levels are shown as green lines and the 2% highest wind speeds utilized for the estimation are shown in magenta. The trumpet shape of the data points towards lower wind speeds (grey samples) reflects the increasing importance of stability leading to larger deviations and increased variability around the neutral levels.

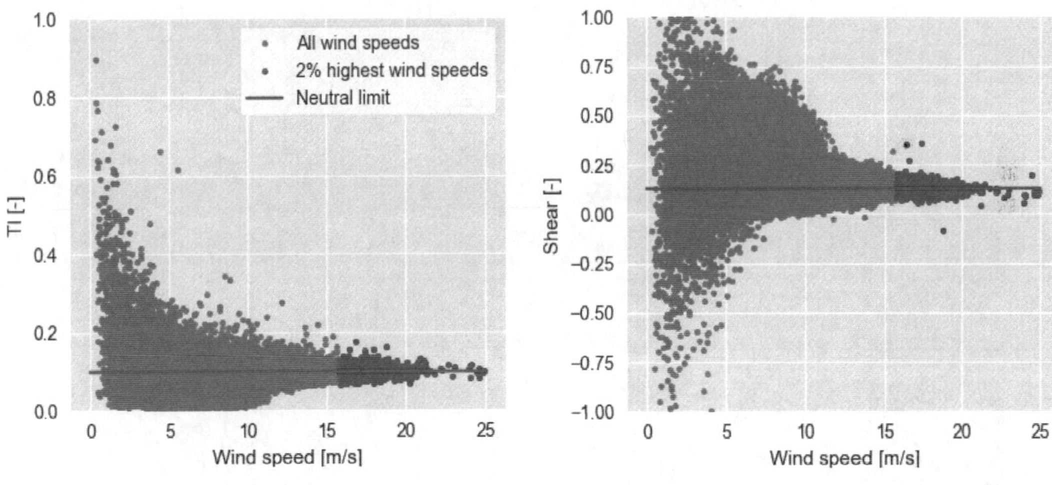

Figure 2. *All TI (left) and shear (right) 10min data for a 20° window around 220°. Values for the 2% highest wind speeds are shown in magenta. Green lines show the 'neutral estimates' of TI and shear.*

Figure 3 shows the result of applying steps 2+3 to estimate the normalized shear and turbulence deviations and how smoothing enhances the stability signal. The plot shows eight days covering the rise and fall of a high wind speed event where both shear and turbulence deviations are close to zero during the highest wind speeds. As the winds level off a clear diurnal stability variation emerges with positive shear deviations and negative TI deviations during night and vice versa during day (vertical grid lines are 00:00h). Note the increased noise of the normalized deviations for wind speeds below 5m/s and note also how the smoothing effectively reduces the spurious high frequency variations.

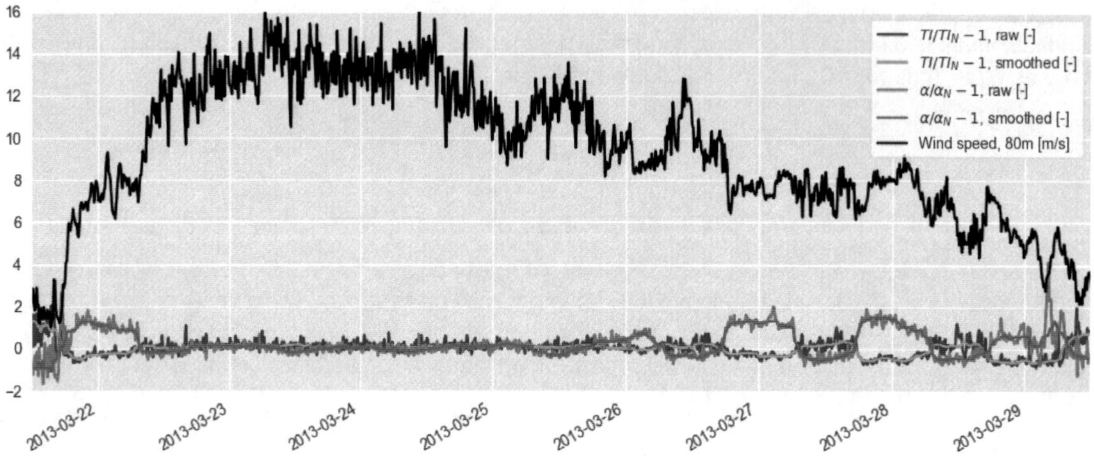

Figure 3. *Estimated normalized shear and turbulence deviations and wind speed at 80m for a period of eight days around a high wind speed event. Smoothed versions of the shear and turbulence deviations are overlaid for comparison of the effect of smoothing.*

Figure 4 shows the final result of steps 1-4 for the Cabauw data plotting the smoothed normalized deviations $\{\delta TI(t), \delta\alpha(t)\}$ for all the data. The colours indicate different wind speed thresholds and the superimposed z/L-'iso-lines' are based on equations (8) and (9). The z/L values are chosen to represent typical L thresholds of different stability regimes [e.g. 6]: neutral ($|z/L| \leq 80\text{m}/500\text{m} = 0.16$), over near neutral ($|z/L| \leq 80\text{m}/200\text{m} = 0.4$) to stable/unstable ($|z/L| \leq 80\text{m}/100\text{m} = 0.8$) and very stable/unstable ($|z/L| \leq 80\text{m}/50\text{m} = 1.6$).

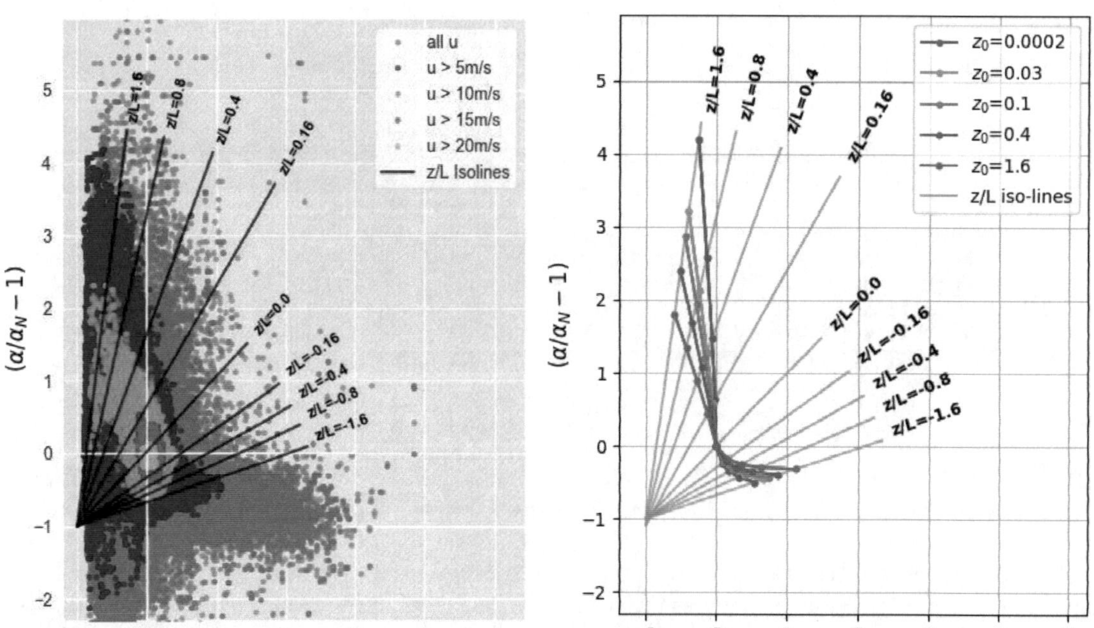

Figure 4. *Left: Estimated $\{\delta TI(t), \delta\alpha(t)\}$ data for Cabauw for different wind speed thresholds clearly showing data gradually moving towards the centre (neutral stability) for higher wind speeds. 'Iso-lines' for z/L from equations (8) and (9) are superimposed. Right: Figure 1 (left) for comparison.*

Notice the clear bulls-eye like centering of the data for the higher wind speed thresholds and how the observed $\{\delta TI(t), \delta\alpha(t)\}$ data resembles the theoretical data in Figure 1, albeit with some scatter. Figure 1 is inserted (right) for easy comparison. The grey data points represent wind speeds below 5m/s where the MOST starts to break down [8] and show a considerably larger scatter than the coloured points at higher wind speed. However, such low wind speeds are not relevant as wind turbines stop below 3-5m/s.

5. Validation of main assumption of the proposed method

The foundation of the proposed method is the existence of the 'universal' relationship illustrated in Figure 1 (right) between MOL and the ratio of normalized shear and normalized TI. This universal relationship is derived in equation (6) as the ratio of MOST universal φ-functions, which are only a function of MOL. Equations (8) and (9) approximate equation (6) for the φ-expressions summarized in Table 1. The existence of this universal relationship can be analysed and validated when concurrent measurements of both MOL and estimates of normalized shear and turbulence are available for a mast. We perform this validation analysis for data from two very diverse research masts both measuring MOL with 3D-sonic equipment. The two masts are high quality research grade masts operated by DTU at the Høvsøre test site [17] on the Danish west coast, and at Ryningsnäs [18] situated in a forest in south-western Sweden. Consistent sensor heights are available for the masts and 80m is used for MOL, wind direction and TI and 80m and 100m are used for the shear estimates. Both data sets have been carefully filtered for faulty data such as icing and for wake directions from operating turbines. Only relatively infrequent wind sectors are influenced by wakes, approximately 340° to 20° for Høvsøre and 30° to 70° as well as 160° to 200° for Ryningsnäs. For consistency with the proposed method the MOL measurements and the estimated concurrent normalized shear and turbulence data are both smoothed as described in 'step 3' of the proposed method (cf. section 3). Finally, to ensure similar appearance of the plots only data for one year is included for each mast, although multiple years are available for Høvsøre. No further filtering has been performed except excluding data for wind speeds below 3m/s. The results of this validation analysis are shown in Figure 5.

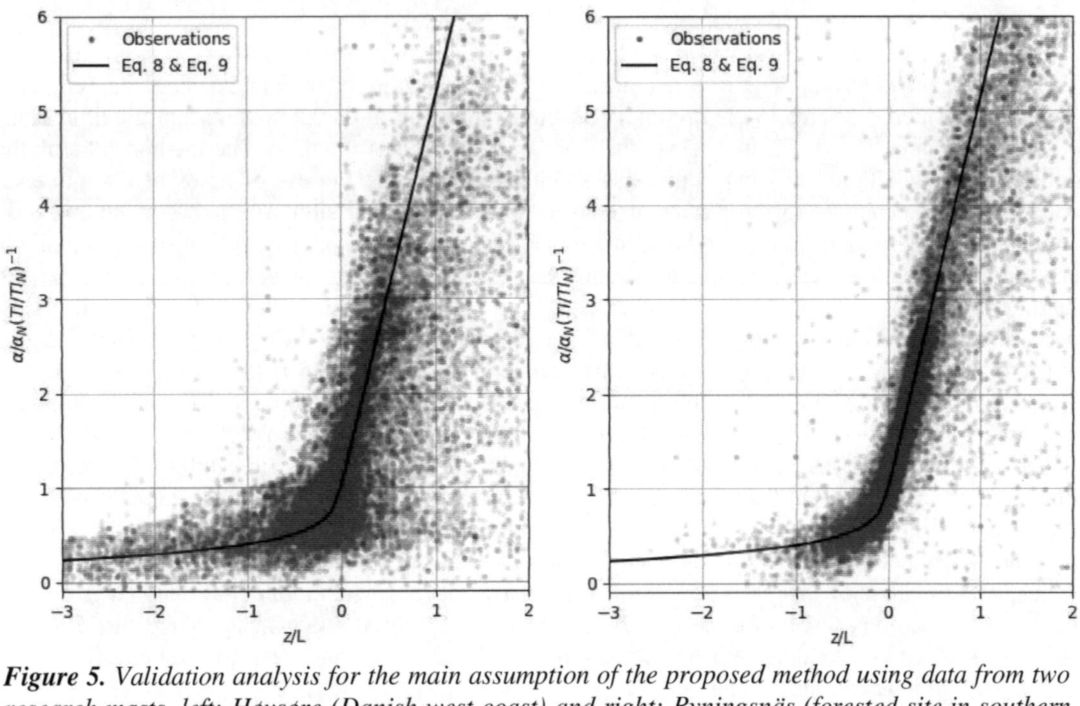

Figure 5. *Validation analysis for the main assumption of the proposed method using data from two research masts, left: Høvsøre (Danish west coast) and right: Ryningsnäs (forested site in southern Sweden). The theoretically derived expressions eq. (8) and (9) are superimposed as the black curve.*

Note the relatively high consistency between the approximate expression and the observed data points of z/L versus $\alpha/\alpha_N\,(TI/TI_N)^{-1}$. The data points in Figure 5 clearly follow the tendency of the superimposed black curve from the equations 8 and 9, which approximate the main assumption of the proposed method, equation (6). For Ryningsnäs there are very few data for $z/L < -1$, indicating that significantly unstable conditions are rare on this forested site. Høvsøre on the other hand has more data representing unstable conditions, and these scatter nicely around the theoretical expectation. In general, the observed scatter can to some degree be explained by the uncertain nature of the measurements. The MOL data are based on 3D-sonic covariances which estimate the heatflux and friction velocity and is a relatively noisy point-measurement itself. Similarly, shear estimates are calculated from wind speed measurements across multiple heights and is often prone to spurious variation and accumulation of error across the sensors. Or the upper shear height may be above ABL in very stable conditions. Furthermore, wind speed trends present within some of the 10min data samples (i.e. non-stationarity) will introduce a bias in the turbulence data. Considering these multiple and significant error sources the main assumption of the proposed method appears to hold surprisingly well for the two analysed masts, clearly reproducing the theoretically derived expressions in equations 8 and 9.

6. Conclusions, recommendations and discussion

This paper presents a novel method to quantify atmospheric stability using measurements from wind power masts with standard instrumentation. The method quantifies stability effects via normalizing observed shear and turbulence by their neutral levels, estimated as the high wind speed limit in each wind direction. A theoretical basis for the proposed method has been established using similarity theory, and describes how the ratio of normalized shear and turbulence only depends on z/L, i.e stability, and is independent of roughness and wind speed. Hence, a ratio of normalized shear and turbulence can be directly converted to a Monin-Obukhov length using the derived universal relationship (equation 6) which itself is a function of universal similarity functions for momentum and turbulence. This universal relationship is approximately linear for the stable part and log-linear for the unstable part. Convenient approximations have been derived (equations 8 and 9) which allow estimation of Monin-Obukhov lengths directly from observed ratios of normalized shear and turbulence.

The method has been demonstrated on data from the Cabauw tower in the Netherlands situated in almost homogenous terrain. The results quite closely resemble the theoretical expectations, but show some scatter which is related to the accumulated uncertainties in the measurements and the limitations of the Monin-Obukhov similarity theory, the basis of the proposed method. The method inherits the limitations of MOST, but this paper presents solid arguments why the use of ratios of the universal functions is less sensitive to the particular constants in the adopted similarity expressions and only requires the similarity expressions to be accurate in a relative sense. Similarly, the ratios of normalized shear and turbulence are less sensitive to potential biases from complex meteorological effects like the Charnock effect for water surfaces. However, the approximate expressions in equation (8) and (9) are based on equation (2) estimating wind shear at a single point of the main height. Using multiple heights for the shear estimate will introduce a small bias relative to this assumption as shown in Figure 1 (right). Thus, it is recommended not to base the shear estimate on heights too far from the main height, contrary to common practice in wind energy applications seeking to estimate the full rotor shear.

The validity of the main assumption of the method, the universal relationship approximated by equation (8) and (9) is tested using measured values of Monin-Obukhov lengths at two research masts in a coastal setting and in a forest. The results of the validation analysis show that the main assumption of the proposed method appears to hold well for the two analysed masts at 80m above ground.

Further validation of the proposed method against measured Monin-Obukhov lengths from 3D-sonics is required, in particular for masts in complex terrain and for multiple measuring levels. This will allow a better understanding of the limitations of the method and test the generality of the approximations in equations (8) and (9).

Acknowledgements
This work is part of the windPROSPER and InnoWind projects, funded by the national Innovation Fund Denmark. Cabauw data are provided via the Cesar data portal and data for DTU masts Høvsøre and Ryningnäs are available via the Rodeo data portal. The authors are very grateful for the support and data contributions.

References
[1] Wagner R, Courtney M, Larsen T J and Paulsen U S 2010 *Simulation of shear and turbulence impact on wind turbine performance* vol Risø-R-172
[2] Newman J and Klein P 2014 The Impacts of Atmospheric Stability on the Accuracy of Wind Speed Extrapolation Methods *Resources* **3** 81–105
[3] Holtslag M C, Bierbooms W A A M and Van Bussel G J W 2014 Estimating atmospheric stability from observations and correcting wind shear models accordingly *J. Phys. Conf. Ser.* **555** 1–10
[4] Kelly M, Larsen G, Dimitrov N K and Natarajan A 2014 Probabilistic meteorological characterization for turbine loads *J. Phys. Conf. Ser.* **524**
[5] Dimitrov N, Natarajan A and Kelly M 2014 Model of wind shear conditional on turbulence and its impact on wind turbine loads *Wind Energy* **17** 657–69
[6] Archer C L, Colle B A, Veron D L, Veron F and Sienkiewicz M J 2016 On the predominance of unstable atmospheric conditions in the marine boundary layer offshore of the U.S. northeastern coast *J. Geophys. Res.* **121** 8869–85
[7] Wharton S and Lundquist J K 2012 Assessing atmospheric stability and its impacts on rotor-disk wind characteristics at an onshore wind farm *Wind Energy* **15** 525–46
[8] Stull R B 1988 *An Introduction to Boundary Layer Meteorology.* (Dordrecht: Kluwer Academic Publishers)
[9] Kaimal J C and Finnigan J J 1994 *Atmospheric Boundary Layer Flows - Their Structure and Measurement* (New York: Oxford University Press)
[10] Panofsky H A and Dutton J A 1984 *Atmospheric Turbulence - Models and Methods for Engineering Applications* (New York: John Wiley & Sons)
[11] Brooks I M, Söderberg S and Tjernström M 2003 The turbulence structure of the stable atmospheric boundary layer around a coastal headland: Aircraft observations and modelling results *Boundary-Layer Meteorol.* **107** 531–59
[12] Pahlow M, Parlange M B and Portè-Agel F 2001 On Monin- Obukhov similarity in the stable atmospheric boundary layer *Boundary-Layer Meterology* **99** 225–48
[13] Park M S and Park S U 2006 Effects of topographical slope angle and atmospheric stratification on surface-layer turbulence *Boundary-Layer Meteorol.* **118** 613–33
[14] Højstrup J 1982 Velocity Spectra in the Unstable Planetary Boundary Layer *J. Atmos. Sci.* **39** 2239–48
[15] Nadeau D F, Pardyjak E R, Higgins C W and Parlange M B 2013 Similarity Scaling Over a Steep Alpine Slope *Boundary-Layer Meteorol.* **147** 401–19
[16] Monna W A A and Van Der Vliet J G 1987 *Facilities for research and weather observations on the 213 m tower at Cabauw and at remote locations. KNMI Scientific Rep. WR-87-5*
[17] Peña A, Floors R, Sathe A, Gryning S E, Wagner R, Courtney M S, Larsén X G, Hahmann A N and Hasager C B 2016 Ten Years of Boundary-Layer and Wind-Power Meteorology at Høvsøre, Denmark *Boundary-Layer Meteorol.*
[18] Arnqvist J, Segalini A, Dellwik E and Bergström H 2015 Wind Statistics from a Forested Landscape *Boundary-Layer Meteorol.* **156** 53–71

IOP Conf. Series: Journal of Physics: Conf. Series **1102** (2018) 012010 doi:10.1088/1742-6596/1102/1/012010

IEA Wind Task 32: Best Practices for the Certification of Lidar-Assisted Control Applications

David Schlipf[1], Nikolai Hille[2], Steffen Raach[3], Andrew Scholbrock[4], Eric Simley[4*]

[1] Stuttgart Wind Energy, University of Stuttgart, Germany
[2] DNV GL Renewables Certification, Hamburg, Germany
[3] sowento & University of Stuttgart, Germany
[4] National Renewable Energy Laboratory, Golden, CO, USA
[*] with Envision Energy during the IEA Wind Task 32 workshop in January 2018

Abstract. Lidar-assisted control is a promising technology for reducing structural loads on wind turbines. Guidelines for certifying wind turbines with this technology are important for its widespread adoption. As a first step, an IEA Wind Task 32 workshop was held, bringing together wind turbine manufacturers, lidar suppliers, certifiers, consultants, and researchers to identify barriers to certification with lidar-assisted control and ideas for mitigating the barriers. This paper builds on the outcome of the workshop by providing an analysis of the areas affected and the remaining challenges. Further, initial best practices are given to address the challenges, covering the lidar system, control and protection systems, design loads, and type testing.

1. Introduction

Lidar-assisted control (LAC) of wind turbines has been an active area of research for over a decade [1]. Both simulation-based studies and field-testing campaigns show promising reductions in structural loads as well as increases in energy production using feedforward control relying on preview measurements of the approaching wind field. However, LAC has not become widely adopted by the wind industry yet. One of the most significant barriers to the widespread adoption of the technology is that guidelines for certifying wind turbines with LAC have not been established. Without clear design standards for wind turbines using LAC, it is difficult for wind turbine manufacturers to fully assess the value creation of the control technology and move forward with including it in a commercial turbine design.

The general objective of the International Energy Agency (IEA) Task 32 extension is to identify and mitigate barriers related to the use of lidar for wind energy applications including loads and control by organizing topical workshops with tangible outcomes. While the first workshop focused on how to optimize lidar systems for LAC [2], the workshop "Certification of Lidar-Assisted Control Applications" of Task 32 in January 2018 at the DNV GL office in Hamburg brought together lidar manufacturers, wind turbine manufacturers, research organizations, and certification bodies to develop suggestions for guidelines for type certification of wind turbines with LAC. This paper builds on the main results of the workshop to provide a first step toward the development of standards for LAC.

First, an overview of type certification and an analysis of the areas affected by LAC is provided, along with a discussion of how these areas are related to existing standards. The

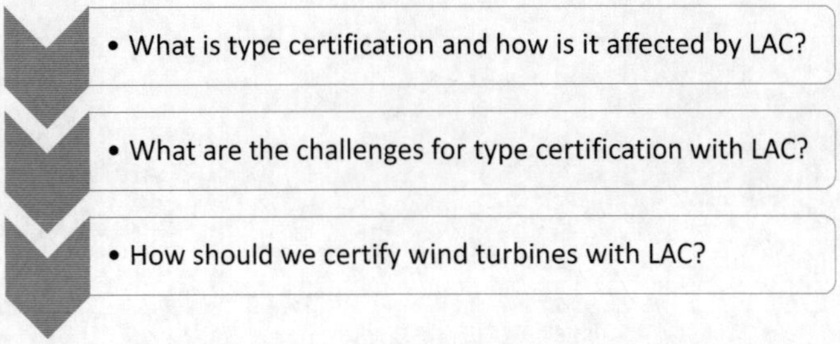

Figure 1: Key questions for the certification of lidar-assisted control, which served as the session themes for the IEA Wind Task 32 workshop and represent the main sections of this paper.

second subsection is devoted to the feedback from the four relevant stakeholder groups (lidar manufacturers, turbine manufacturers, consultants, and researchers), all answering the question: What are the challenges for type certification with LAC? This section is intended to alert the community to open issues. In the final section, initial best practices are given, based on information on how to certify wind turbines with LAC in the identified areas gathered during a discussion round with experts at the workshop. This section is intended to serve as the starting point in the development of a dedicated standard to certify turbines with LAC.

The work is structured into three main sections similar to the workshop organization (see Figure 1).

2. What is type certification and how is it affected by LAC?

Certification is the confirmation of design requirements by an accredited party (certification body) to a third party (authorities, financiers, insurers, etc.) based on accepted standards. This process supports the confidence in technical integrity, assists in satisfying health and safety requirements, and helps with securing investment.

Type certification of wind turbines is typically performed according to well-known standards, such as International Electrotechnical Commission (IEC) 61400-22 [3] or DNV GL-SE-0441 [4]. Beside further details both standards have in common that the following three modules form the basis for a wind turbine type certificate:

- Design evaluation
- Manufacturing evaluation
- Type testing.

During design evaluation, it is confirmed that the respective wind turbine design complies with requirements set up by applicable design standards. These design standards address all components of the wind turbine and represent the state of the art in wind turbine technology. Manufacturing evaluation intends to ensure that the production of components proceeds under consideration of the requirements assumed for the component design. Type testing closes the link between the real wind turbine and the theoretical design by confirming the model assumptions using measurements from the wind turbine in the field. LAC will mainly affect the design evaluation and type testing modules, with no significant relevance for manufacturing evaluation.

Today, wind turbines with LAC can be certified based on individualized interpretation of the standards and assessment of the specific lidar system and wind turbine type. However, existing standards do not contain explicit guidance and requirements for wind turbines with LAC. Certification of wind turbines with LAC thus currently relies on an interpretation of the standards and employs risk-based mechanisms, such as technology qualification DNVGL-RP-A203 [5].

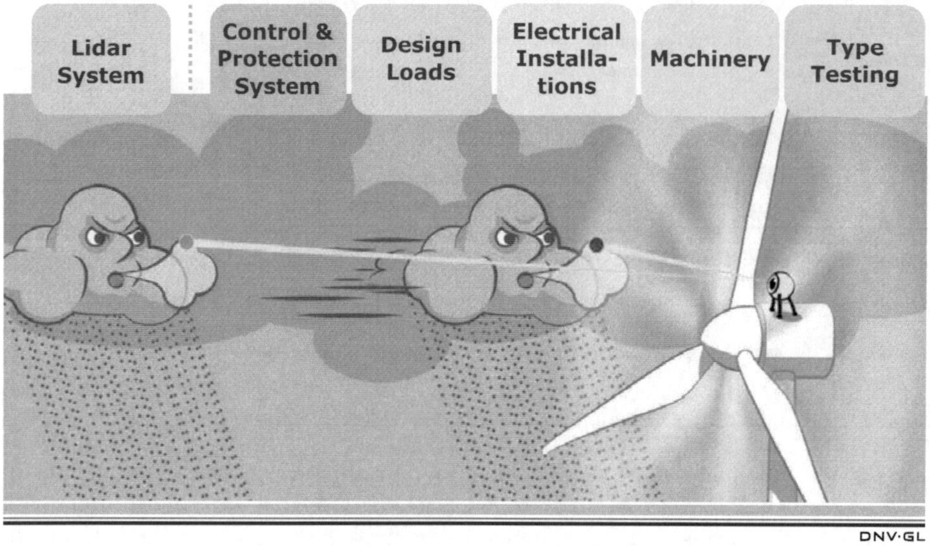

Figure 2: Lidar-assisted control in certification context.

Developing new standards considering LAC involves identifying requirements that ensure that a wind turbine with LAC meets the same safety level as the design standards for customary wind turbines. LAC uses a feedforward control loop for load reductions. Undetected faults in the lidar or in other parts of the LAC may cause the load reduction to fail or result in a load increase, thus possibly causing damage to the wind turbine. As a consequence, appropriate requirements need to be set up for a wind turbine with LAC to ensure safe operation. Proposals for best-practice requirements will be described in Section 4.

To comply with these requirements, the lidar manufacturer has to provide documentation on the lidar system to the wind turbine manufacturer. This documentation is the basis for integrating the lidar system with the wind turbine and its control and protection system. The wind turbine manufacturer analyzes the integrated design and documents the design and test results for the certification body. For LAC, the assessment by the certification body will typically focus on the areas given in Figure 2.

2.1. Lidar system

For the lidar system, the following features are considered relevant. They require documentation by the lidar manufacturer:

- Measurement uncertainty
- Data processing
- Data availability
- Failure mode and effects analysis (FMEA) or similar
- Lightning protection
- Electromagnetic compatibility (EMC)
- Personal safety.

Although it may be difficult to determine measurement uncertainty, it is important to do so when the measurement uncertainty could influence the loads. Depending on the integration of the lidar system and wind turbine controller, the measurement uncertainty needs to be determined by the lidar and wind turbine manufacturer together. For the lidar system, no specific design standard exists. For FMEA and personal safety, recourse can be taken to DNVGL-ST-0438 [6], whereas lightning protection and EMC is addressed in DNVGL-ST-0076 [7].

2.2. Control and protection system
For the control and protection system, the following features are considered relevant. They require documentation by the wind turbine manufacturer on the integrated system wind turbine with LAC:

- Control system
- FMEA or similar
- Manuals
- Analysis of failure probability
- Sanity checks
- Maintenance plan.

Within the documentation of the control and protection system, clarification is required regarding how the turbine reacts to unavailability and malfunctioning of the lidar. A special focus on sanity checks is needed as they ensure that a malfunction—which is undetected by the lidar—is detected by the wind turbine's control and protection system. For further guidance, refer to the design standard DNVGL-ST-0438 [6].

2.3. Design loads
For the calculation of the design loads a simulation model needs to be set up. It appropriately has to reflect the dynamic behaviour of the wind turbine with LAC regarding loads. The model requires input according to the following aspects:

- Wind evolution model
- Lidar model
- Controller model
- Load case definition.

The load case definition must consider cases of lidar system unavailability and malfunctioning within fatigue and extreme load calculations. Careful interpretation of deterministic design load cases is required to ensure that they match well with LAC and still fulfil the targeted safety level. For further guidance, refer to the design standards IEC 61400-1 [8] and DNVGL-ST-0437 [9].

2.4. Electrical installations and machinery
The lightning protection of the lidar system requires integration with the wind turbine's lightning protection system (see DNVGL-ST-0076 [7]). The mounting of the lidar system is assessed in context with the machinery components (see DNVGL-ST-0361 [10]).

2.5. Type testing
Type testing is of special importance because the simulation model of a new turbine type with LAC features several additional model uncertainties compared to a customary wind turbine. These model uncertainties are addressed by load measurements and the safety and function test. General validation is performed by direct one-to-one comparison of measurements with simulations (refer to DNVGL-ST-0438 [6] and DNVGL-ST-0437 [9]).

Besides fatigue load measurements, the extreme load behaviour must be evaluated, especially if the LAC design reduces extreme loads. During the safety and function test, it is relevant to investigate additional stops as well as error cases that are evoked by the lidar system. If the wind turbine with LAC changes to a baseline control mode when the lidar system is not available, then transitions between these control modes require testing, too. If extreme loads are reduced significantly, example events must be investigated where the load-reducing feature is active.

3. What are the challenges for type certification with LAC?

Building on the general requirements for certification with LAC proposed in Section 2, challenges identified by the four stakeholder groups at the workshop (lidar manufacturers, turbine manufacturers, consultants, and researchers) are discussed here. Note that the challenges presented by the stakeholders are occasionally augmented with the authors' observations.

3.1. Challenges for lidar manufacturers

From the lidar manufacturers' perspective, the two main challenges for certification with LAC are (1) properly assessing and simulating the atmospheric conditions and other variables that affect availability and measurement accuracy, and (2) including potentially proprietary lidar and wind field reconstruction methods in the certification process. As explained by Davoust *et al.* [11], atmospheric conditions such as very clean air, snowfall and extreme fog can cause availability to decrease due to either low aerosol density or high optical attenuation. Furthermore, variables such as the lidar sampling time and rotor speed affect availability due to blade blockage. Therefore, (1) the atmospheric conditions that will be experienced by a wind turbine should be characterized so that lidar availability can be estimated, and (2) the impacts of atmospheric conditions and blade blockage on availability and measurement noise should be included when simulating lidar measurements during the design evaluation stage. Two strategies for reconstructing wind field variables from raw lidar measurements (e.g., accounting for the induction zone [12]) can be used in LAC. Line-of-sight velocities can be used to reconstruct the wind field variables of interest in the controller, giving the wind turbine manufacturer more freedom in the design process, or the reconstructed wind field variables can be output directly by the lidar. If the latter strategy is adopted, methods for including the lidar manufacturer's wind field reconstruction algorithms in controller simulations while respecting their potentially proprietary nature need to be devised. One solution is compiled software with a defined interface (raw lidar data to derived signals) e.g., in the form of a dynamic link library (DLL) similar to the controller DLL usually used for aeroelastic simulations. This software should be identical to the software used on the real lidar system.

3.2. Challenges for turbine manufacturers

For wind turbine manufacturers, the primary challenge in certifying turbines using LAC lies in the application of LAC to extreme load reduction. In contrast to fatigue load assessment, many design load cases (DLCs) used to determine extreme loads during design evaluation include unrealistic deterministic wind conditions (e.g. gusts, direction changes) that present challenges when incorporating lidar measurements [13]. Depending on one's interpretation of the design standards, these extreme wind events could be measured almost perfectly by lidar—but this is clearly not justified. However, it is not reasonable to exclude the use of LAC for these DLCs completely. Therefore, clear guidelines for simulating lidar measurements in extreme wind fields should be established. An additional challenge for wind turbine manufacturers is proving that the extreme loads made possible by the use of LAC can be achieved even when the lidar is unavailable (for example, by detecting when lidar measurements are unavailable and switching to a derated mode of operation).

3.3. Challenges for consultants

For consultants, there are several open aspects that exist regarding the type certification of LAC. Consultants mainly provide their knowledge to lidar and turbine manufacturers to accompany the process of type certification, deriving their own recommended practices for realizing LAC. Thus, consultants can play a significant part in helping shape future design standards for type certification with LAC (for example, by helping determine if additional DLCs are needed). Another challenge identified by consultants is the need to develop a standard realistic lidar simulator that is general enough to be used with a variety of wind turbine simulation tools in the certification process. In doing so, consultants can help standardize elements of lidar simulation including the range-weighting functions used for spatial averaging [14], wind evolution modeling, the influence of nacelle motions on measurement locations and measured velocities, and the impact of the induction zone via the reduced velocities upstream of the turbine [15].

3.4. Challenges for researchers

The research community has identified LAC certification challenges based on field testing of LAC as well as theoretical work on wind field simulation. From field experiments performed by the research community using LAC, reliability was found to be an issue. In order to ensure the safety of the turbine, a robust control design should not look solely at the normal operation of the wind turbine, but also consider fall-back solutions for when lidar signals are lost or determined to be of poor validity. In general, LAC methods need to be developed, which account for the changing measurement quality. In order to account for this, a FMEA should be included in the certification process for LAC. On the simulation side, an important challenge for researchers is the creation of realistic wind field models through the integration of wind evolution into load simulations and the addition of turbulence to extreme wind event simulations. Although several wind evolution models exist, additional validation and re-tuning of the models is needed [16, 17]. Additionally, there are several open research questions with implications for the effectiveness of LAC for extreme load reduction. For example, how can gusts, direction changes, and other extreme deterministic wind events be embedded in turbulence to allow for the realistic simulation of lidar measurements while preserving the intent of the DLC? How does a change in wind direction evolve and from which direction does it propagate? In essence, can it be detected by a forward-measuring lidar? Further, wind field reconstruction methods for complex flow situations (complex terrain, measurements within wakes) and a simulation environment for testing these methods need to be developed.

4. How should we certify wind turbines with LAC?

During the workshop, the participants were split into four working groups based on the areas identified in Section 2: lidar system, control and protection system, design loads, and type testing. Electrical installations and machinery have been considered to be already sufficiently covered by existing standards. The participants were rotating through four different stations, each hosted by an expert moderator. This section summarizes and elaborates on the discussion results.

4.1. Lidar system

The lidar system is the key measurement device in LAC. Hence, it must be considered as part of the safety and control system and therefore treated according to the regulations of the safety and control system. Quantifications on mean time to failure, the probability of failure, and the probability of failure on demand need to be provided for the lidar sensor to analyze the safety system properly. Furthermore, the following guidelines on quantifying data availability and measurement uncertainty are needed. Practical aspects like personal safety in use, maintenance, and mounting procedures of the lidar as well as electrical and lightning protection standards must be ensured. In the following, the different aspects are discussed in more detail.

4.1.1. Failure mode and effects analysis for LAC In order to certify wind turbines with LAC, the theoretical consideration of the failures at the lidar sensor need to be considered. The FMEA method is an analytical methodology that takes the reliability of a system into account and displays the impact on other components and the overall system. The general intention in implementing security features that observe the lidar sensor is that a failure in the sensor should not lead to a worse situation than without it. Self-monitoring capabilities of the lidar system can help to minimize the impact of a failure to the overall safety chain. However, not every failure mode can be covered with a self-monitoring. Altogether, the FMEA serves as an analysis tool but can be used to quantify a critical limit that the system should not exceed.

4.1.2. Personal safety Personal safety is a necessary aspect that needs to be answered before realizing LAC. Mainly, three aspects are included: (1) eye safety, (2) electrical safety, and (3) handling safety.

Eye safety needs to be ensured. A guided use regarding installation, operation, and failure cases is necessary and should be provided in the documentation. Here, an adaption of a standardized certification procedure can be used.

Electrical safety is a requirement. Ensuring safety to the equipment from a lightning hazard is necessary as well.

Installation of lidar sensors on the nacelle offers the potential of higher risk for the technicians. Here, safety procedures for installation, service, and dismounting are needed to ensure that turbine manufacturers comply with requirements.

Altogether, adhering to the personal safety aspect is crucial. With clear procedures and the adoption of standardized processes, the health and safety of all personnel can be ensured.

4.1.3. Data availability Data availability is an important aspect when considering the lidar sensor in the safety and control system. There is no standardized definition of data availability of lidar systems for LAC. This challenges the comparability of lidar systems for LAC. A request would be to deliver a meaningful availability measure for each provided signal. Here, a clear link to measurement uncertainty is given, which will be discussed in Subsection 4.1.4.

4.1.4. Measurement uncertainty Several aspects influence the measurement uncertainty of a lidar system (e.g., external aspects like atmospheric and weather conditions, handling aspects like the exact orientation and alignment of the system and its components, or internal aspects like the uncertainty of the data processing algorithm or real-time properties). Altogether, a clear guideline is needed for a specified uncertainty analysis that standardizes the quantification. Furthermore, a guideline is needed on how to perform or verify the uncertainty analysis in the type certification.

4.1.5. Data processing For data processing, the turbine manufacturer asks for a standardized metric of signals that are provided by the lidar sensor. This implies that dedicated signals, such as the rotor-effective wind speed, need to be defined and specified in the guidelines. Then, they can be directly used in the controller. If unspecified signals are used or the data processing is included in the wind turbine controller, this aspect needs to be included in the type certification by the certifying body.

IOP Conf. Series: Journal of Physics: Conf. Series **1102** (2018) 012010 doi:10.1088/1742-6596/1102/1/012010

4.2. Control and protection system
During the workshop, ideas were generated for how LAC should be addressed in the certification of the control and protection systems, which keep the turbine within its operating limits and protect it when safe operating conditions are exceeded. Here, the contribution to both systems needs to be distinguished, because different requirements need to be fulfilled. In the near future, lidar will probably mainly improve control functions. Later, lidar assistance might be used to reduce extreme loads and will become part of the protection function.

Further, a turbine with LAC should have a traditional feedback control system to rely on when lidar measurements are not available. Due to the importance of reliable wind measurements for LAC to properly regulate and protect a wind turbine, a lidar quality signal should be required from the lidar, indicating the trustworthiness of the measurement so the turbine can switch to feedback control when necessary. Additionally, "sanity checks" should be performed by the controller to determine lidar measurement accuracy; for example, by comparing the lidar measurement with the wind speed estimated from turbine sensors. To accomplish this, a possible method is the calculation of the measurement coherence bandwidth (wavenumber at which the coherence between both wind speed estimates drops below 0.5).

It was concluded that for LAC no special requirements on a maintenance plan are necessary. A standard operation and maintenance manual that includes a maintenance plan is sufficient.

4.3. Design loads
Recommendations for including LAC in load simulations during the design evaluation stage were made in the following four categories: lidar modeling, wind field modeling, fatigue load calculations, and extreme load calculations.

4.3.1. Lidar modeling
It is generally agreed that lidar models should include the limitation to line-of-sight velocity measurements (i.e., the ability to measure only the wind velocity aligned with the beam direction), spatial averaging (weighting of the wind speeds along the beam according to an appropriate range-weighting function [14]), and realistic modeling of the scan pattern measurement sequence. Additional details that should be modeled include the impact of nacelle motions on the measurement locations and the relative velocities measured by the lidar, the consequences of blade blockage on lidar availability, and the effects of the induction zone. The induction zone model should include the reduced velocities upstream of the turbine where the lidar measures the wind as well as the additional preview time afforded by the reduced velocities.

Two methods for realistically simulating lidar measurements were discussed. The first approach is to include blade blockage and atmospheric conditions, such as visibility and aerosol density, in a detailed simulator meant to recreate the impacts of these parameters on availability and measurement noise [18]. This would require site-specific knowledge of the atmospheric conditions or at least a conservative estimate of them. The second approach uses knowledge of the scan pattern, range-weighting functions, wind evolution and spectral properties of the wind field to generate random realizations of the lidar measurements that adhere to the calculated theoretical measurement coherence (e.g., see [19, 20]). While this approach misses the direct impact of variables, such as visibility or aerosol density on measurement error, it can capture their impact in a statistical sense through the identified measurement coherence. Regardless of the method used, there should be a clearly defined interface between the lidar model and the controller/wind turbine simulator so that lidar manufacturers can supply potentially proprietary lidar models that can be used with a variety of simulation codes (e.g., as a DLL).

Because of the importance of atmospheric conditions, such as visibility and aerosol density, in determining lidar availability and measurement noise, a suggestion was made to create different atmospheric condition "classes" that are analogous to the mean wind speed and turbulence classes that currently exist in the standards [8]. This atmospheric condition class system would provide standardization in the determination of lidar availability, although work is needed to define a small representative set of atmospheric conditions that are likely to be encountered. Well-known methods for including wind evolution in lidar measurement simulations for stochastic turbulent wind fields have been established, relying on longitudinal spatial coherence functions to describe the evolution of turbulence as it advects downstream [19, 21]. While several longitudinal coherence models for wind evolution have been used in the LAC community [22–24], more work is needed to validate the models against field observations [16, 17].

4.3.2. Load simulations

Participants agreed that methods for determining fatigue loads using LAC are relatively straightforward. When calculating fatigue loads, imperfect lidar availability can be accounted for by performing separate load simulations with and without LAC, weighting the resulting damage based on the percentage of time the lidar is assumed to be available. However, loss of the lidar signal due to unfavorable atmospheric conditions is expected to occur frequently enough that DLCs capturing the transition from LAC to baseline control (or vice versa) should factor into the fatigue load calculations.

Calculating extreme loads using LAC is more complicated, especially if the controller relies on lidar measurements to mitigate the impact of extreme gusts or other events on loads. A chief concern is the possibility of lidar unavailability when extreme events occur. Therefore, extreme load DLCs should be simulated with and without LAC, including operation in a safe mode if necessary when the lidar is unavailable. Furthermore, DLCs should be developed to simulate the transition to a safe mode of operation when a lidar fault occurs. For rare events, such as extreme gusts and direction changes, it may be determined that the likelihood of lidar unavailability or of a lidar fault occurring simultaneously with an extreme event is low enough that it does not need to be considered or that the safety factor can be reduced. However, first it must be determined if such extreme events occur during weather conditions that increase the probability of unavailability (e.g., heavy rain).

While performing extreme load simulations with LAC in extreme turbulence (DLC 1.3 in the IEC 61400-1 Ed. 3 standard [8]) is relatively straightforward, deterministic extreme events, such as the extreme operating gust, extreme coherent gust with direction change, and extreme wind shear [8] are problematic because (1) they do not contain turbulence (and can therefore be measured accurately by a lidar, which is unrealistic) and (2) they are only defined at the turbine location (the speed and direction of propagation toward the turbine is unclear).

A solution to the first obstacle is to redefine the extreme events as stochastic turbulent wind fields where the rotor-averaged wind parameters of interest match the intended deterministic event. Wind evolution can then be introduced to the turbulent realizations of the extreme wind fields [21]. At least two methods have been developed for generating stochastic turbulent wind fields with constraints imposed on rotor-averaged quantities. First, Bierbooms and Cheng [25] discuss a statistical method for combining a random unconstrained stochastic field with the most likely residual field such that constraints imposed on a linear combination of points in the field are met. Alternatively, Schlipf and Raach [13] present an engineering approach in which the phase angles of the Fourier coefficients of a stochastic wind field are optimized so that the resulting rotor-averaged time series matches the intended extreme event. Dimitrov and Natarajan [26] developed a method, which fits turbulent wind fields to lidar measurements and which might be used for extreme event simulations as well.

Finally, uncertainty in the propagation speed and direction of extreme events can be addressed by including several different realizations of the parameters in the appropriate DLCs. For example, the extreme operating gust and extreme wind shear events can be simulated with a range of advection speeds toward the turbine while the extreme coherent gust with direction change can be simulated with the wind direction change propagating from a range of directions, from straight ahead to approaching from the side.

4.4. Type testing

Groups also met to discuss the topic of type testing, and their feedback is summarized here. In considering type testing for certifying LAC, a load performance test should be performed on a wind turbine with LAC. The following guidelines were produced from the four working groups.

4.4.1. Test matrix

A test matrix should include wind speed bins over the operational range of LAC. Typically, this is a subset of the operational range of a wind turbine as LAC has been used in above-rated wind conditions. The test matrix should also look to include various atmospheric conditions (e.g., various turbulence intensities, shears/veers, lidar backscatter conditions, etc.) and weather conditions (e.g., sunny skies, cloudy, fog, rain, snow, etc.) in order to show robust performance. In order to evaluate the LAC performance, data should be collected without the LAC portion of the controller as a baseline to compare against.

4.4.2. Instrumentation

An independent meteorological mast and/or lidar would be useful for the type testing to better quantify the wind speed and atmospheric conditions. Additionally, the wind turbine would need to be outfitted with strain gauges in the wind turbine tower at multiple stations/directions to assess the tower loading. Similarly, strain gauges would need to be installed on the wind turbine blades, again at multiple stations/directions in order to assess the blade loading. Instrumentation to measure the nacelle motion and blade tip deflection would also be useful.

4.4.3. Other considerations

Fault cases should also be included in the test matrix to show the control design performs well in the event of the loss of LAC. The type testing should only consider simple terrain and simple flow conditions where the wind can be considered homogeneous. Finally, in assessing the performance of LAC, 10-minute-averaged data points may be insufficient to effectively evaluate the performance as the control happens in real time and the wind disturbance can vary greatly over a 10-minute period. Instead, smaller averaged data points (~1 minute average) may be of more use. Data analysis in the frequency domain would also be useful to show the LAC performance at different operational frequencies of the wind turbine.

5. Conclusions

The lack of standards for the certification of wind turbines using lidar-assisted control (LAC) is a main barrier for the wide adoption of the technology by the wind industry. This paper presents the results of an IEA Wind Task 32 workshop that brought together experts from research and industry and thus is a first step toward a "best practices" document, providing solutions both to practical and research questions.

First, the areas of type certification impacted by LAC are identified to be the lidar system itself, the control and protection system, design loads, and type testing. Although turbines with LAC can be already certified by interpretation of existing standards, there are still several challenges to overcome for lidar and turbine manufacturers, consultants, and researchers before a dedicated standard can be issued. These challenges are described in this paper. Despite these challenges, the paper already presents discussions and first ideas on how to certify wind turbines with LAC.

The IEA Wind Task 32 will continue to bring experts together to foster lidar-assisted control and help make it an accessible and beneficial technology for the wind community.

6. Acknowledgements
This work was authored, in part, by the National Renewable Energy Laboratory, operated by Alliance for Sustainable Energy, LLC, for the U.S. Department of Energy (DOE) under Contract No. DE-AC36-08GO28308. Funding provided by the U.S. Department of Energy Office of Energy Efficiency and Renewable Energy Wind Energy Technologies Office. The views expressed in the article do not necessarily represent the views of DOE or the U.S. government. The U.S. government retains and the publisher, by accepting the article for publication, acknowledges that the U.S. government retains a nonexclusive, paid-up, irrevocable, worldwide license to publish or reproduce the published form of this work, or allow others to do so, for U.S. government purposes.

References
[1] Scholbrock A, Fleming P, Schlipf D, Wright A, Johnson K and Wang N 2016 Lidar-enhanced wind turbine control: Past, present, and future *Proceedings of the American Control Conference* pp 1399–1406

[2] Simley E, Fürst H, Haizmann F and Schlipf D 2018 *Remote Sensing* **10**

[3] International Electrotechnical Commission (IEC) 61400-22 2010 Wind turbines – Part 22: Conformity testing and certification

[4] DNVGL-SE-0441 2016 Type and component certification of wind turbines

[5] DNVGL-RP-A203 2017 Technology qualification

[6] DNVGL-ST-0438 2016 Control and protection systems for wind turbines

[7] DNVGL-ST-0076 2015 Electrical Installations

[8] IEC 61400-1 2005 Wind turbines - Part 1: Design requirements

[9] DNVGL-ST-0437 2016 Loads and site conditions for wind turbines

[10] DNVGL-ST-0361 2016 Machinery for wind turbines

[11] Davoust S, Jehu A, Bouillet M, Bardon M, Vercherin B, Scholbrock A, Fleming P and Wright A 2014 Assessment and optimization of lidar measurement availability for wind turbine control *Proceedings of the European Wind Energy Association annual event* (Barcelona, Spain)

[12] Borraccino A, Schlipf D, Haizmann F and Wagner R 2017 *Wind Energy Science* **2** 269–283

[13] Schlipf D and Raach S 2016 *Journal of Physics: Conference Series* **753** 052011

[14] Simley E, Pao L Y, Frehlich R, Jonkman B and Kelley N 2014 *Wind Energy* **17** 413–433

[15] Simley E, Pao L Y, Gebraad P and Churchfield M 2014 *Journal of Physics: Conference Series* **524** 012003

[16] Schlipf D, Haizmann F, Cosack N, Siebers T and Cheng P W 2015 *Meteorologische Zeitschrift* **24** 565–579

[17] Davoust S and von Terzi D 2016 Analysis of wind coherence in the longitudinal direction using turbine mounted lidar *Proceedings of The Science of Making Torque from Wind* (Munich, Germany)

[18] Mazoyer P and Marchta M 2014 Lidar simulator: a coupled wind and lidar simulator *European Wind Energy Association annual event* (Barcelona, Spain)

[19] Laks J, Simley E and Pao L Y 2013 A spectral model for evaluating the effect of wind evolution on wind turbine preview control *Proceedings of the American Control Conference* (Washington, USA)

[20] Raach S and Schlipf D 2018 Cross-tool realistic lidar simulations *Presentation at the IEA Wind Task 32 workshop on Certification of Lidar-Assisted Control Applications* URL `https://www.ieawindtask32.org/ws08-internal-documents/`

[21] Bossanyi E 2012 Un-freezing the turbulence: improved wind field modelling for investigating lidar-assisted wind turbine control *Proceedings of the European Wind Energy Association annual event* (Copenhagen, Denmark)

[22] Pielke R A and Panofsky H A 1970 *Boundary-Layer Meteorology* **1**(2) 115–130 ISSN 0006-8314

[23] Kristensen L 1979 *Boundary-Layer Meteorology* **16** 145–153

[24] Simley E and Pao L Y 2015 A longitudinal spatial coherence model for wind evolution based on large-eddy simulation *Proceedings of the American Control Conference* (Chicago, USA) pp 3708–3714

[25] Bierbooms W and Cheng P 2002 *Journal of Wind Engineering and Industrial Aerodynamics* **90** 1237–1251

[26] Dimitrov N and Natarajan A 2017 *Wind Energy* **20** 79–95

Evaluation of the Weather Research and Forecasting (WRF) model with respect to wind in complex terrain

Kine Solbakken and Yngve Birkelund
Department of Physics and Technology, UiT, The Arctic University of Norway
E-mail: kine.solbakken@uit.no

Abstract. In this study the performance of the Weather Research and Forecast (WRF) model in a complex and coastal terrain has been evaluated with focus on wind resource assessment. The study area is a small community on the northern part of the island Senja, Norway. The community, with fishery and seafood as its main industry, is being limited by poor grid connection. One of the solutions is to increase the production of local power from wind energy. There are no in-situ wind measurements in the area, and therefore numerical weather prediction models, namely the WRF model, is being evaluated as a method for wind resource assessment. The WRF model has been run for the whole of 2017 with high resolution covering an area large enough to include the three closest weather stations. The model is compared to the observed wind speed and direction. It is found that the model is able to reproduce the average wind speed and wind direction quite well for two of the locations, while for the third location the average wind speed is considerably overestimated compared to the observations. The Root Mean Square Error (RMSE) and Mean Absolute Error (MAE) found are larger than in other comparable studies.

1. Introduction

The aim of this study is to evaluate how well the Weather Research and Forecasting (WRF) model is able to reproduce the wind speed and the wind direction in a specific area in Northern Norway characterized by a complex and coastal terrain. A full year of WRF simulation results have been compared to in-situ wind observations retrieved from three weather stations provided by the Norwegian Meteorological Institute (MET).

The area of interest is a small community with a population of about 5000, located on the northern part of the island Senja, Norway. The main industry is fishery and seafood production. To be able to keep competing on a global market, the industry is currently facing large changes in terms of a more efficient and automated production line. The new technology, combined with a need for larger cooling and freezing facilities, will increase the energy demand. However, the grid capacity and the quality of the power supply have become limiting factors for economical growth. Already, the total power and energy load have, on several occasions, been higher than the grid capacity.

There are three suggested solutions for this problem. The first is to strengthen the regional power grid with a submarine power cable from the main land to the island. The second option is to supplement with power from a diesel generator, a solution that has been working for one of the neighboring islands.The third option is to introduce new local production of renewable energy and the use of a micro power grid to ensure efficient use of the energy in the community.

The local industry and the community want to keep a sustainable profile. Together with the grid owner, they are investigating the possibility of increasing the energy efficiency as well as the use of local renewable sources such as solar, wind and geothermal energy. In this study the focus will be on wind power resource assessment. Wind energy is considered a mature and economical technology and is currently one of the fastest growing renewable energy sources, both worldwide and in Norway [1, 2].

Some of the standard methods for assessing the wind energy potential are by local wind measurements, either wind measurement campaigns or from meteorological weather stations; extrapolation of free atmospheric wind from global data sets; or by use of wind atlases [3]. For the specific area of interest in this study, there are no in-situ measurements, and because of the complex terrain, use of global data sets or wind atlases might give insufficient results [4].

Another method, that has proven to give a good first approximation of the wind energy resource, is Numerical Weather Prediction (NWP) models. NWP models have certain advantages compared to real measurements, such as low cost, high resolution both horizontally and vertically over a large area, they do not have missing data and can provide long time series of data in a short amount of time [1]. However, NWP models are only an approximation to the real atmosphere due to simplification of the topography and the physical processes [5]. Therefore, for this study, performed in an area characterized by coastal and mountainous topography, it is important to emphasize that the observed wind can be affected by local conditions that the model is not able to capture. The wind retrieved from the model is generated over large grid cells with an average topography and roughness that can be very different from the real conditions [6]. Methods to achieve a better comparison of the simulations and the observations could for instance be by coupling of the NWP model with a micro scale model [5]; or by generalizing both the observed and simulated wind to a common elevation and roughness level as described in [6]. Further down scaling is consider to beyond the scope of this paper, and therefore the observations and the simulations will be compared directly.

The NWP model chosen for this study is the WRF model. Several studies have evaluated the performance of WRF with a special focus on the model's ability to reproduce the wind speed [7, 8, 9, 10]. A good agreement between the WRF model simulations and wind speed measurements for areas characterized by a complex terrain, are reported in [11, 12, 13]. Typically for these studies are low biases between the observed wind speeds and the simulated wind speed. When also the temporal simultaneity is considered, the deviation is larger. It is also found that the WRF model tend to overestimate the low wind speed and underestimate the higher wind speeds. In the study by [10], special attention is given to the WRF models ability to reproduce the wind direction.

The focus of this paper is the performance of the WRF model's ability to reproduce the wind speed and direction. In section 2 the method for the study is presented, in section 3 the results are presented and discussed and section 4 gives the conclusion.

2. Approach

2.1. Study area and the measured wind data

The study area is located at the northern part of the island Senja, shown in figure 1 at $69°18''$ N and $17°40''$ E. The area covers approximately $800km^2$ and consists of complex terrain with fjords and steep mountains. The elevation ranges from sea level to 1000 meters above mean sea level (a.m.s.l.). The main wind pattern along the coast of Northern Norway is dominated by both westerlies and polar fronts.

The focus of this study is the northern part of the island Senja. However, there are no available wind observations within this area. Therefore, the WRF model has been set up to

IOP Conf. Series: Journal of Physics: Conf. Series **1102** (2018) 012011 doi:10.1088/1742-6596/1102/1/012011

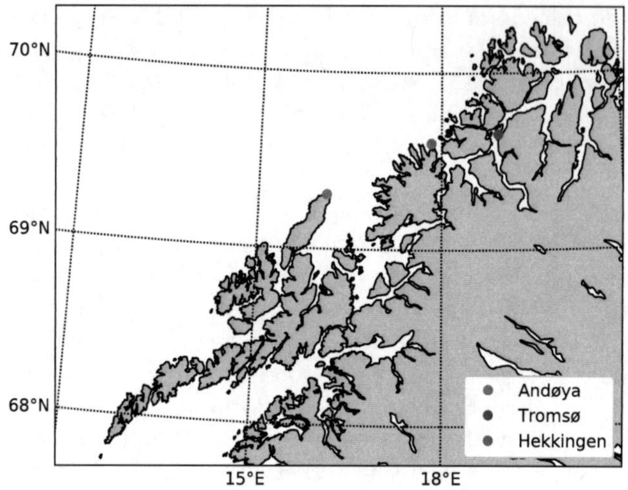

Figure 1: Map of the study area including the locations of the three meteorological stations.

Station name	Station number	Latitude (°N)	Longitude (°E)	Elevation (m a.s.l.)	Elevation in WRF (m a.s.l.)
Andøya	87110	69.3073	16.1312	10	7.6
Tromsø	90450	69.6536	18.9368	100	52
Hekkingen	88690	69.6013	17.8303	14	8

Table 1: Summary of the characteristics of the meteorological weather stations the wind observations are retrieved from. The elevation is given in meters above sea level (m a.s.l.). In addition, the model terrain heights of the closest grid point to each of the locations are also listed here.

cover an area of 18 000 km^2 to also include the three closest weather observation stations. The extended area has similar topography to that found on Senja. The three weather observation stations are operated by the Norwegian Meteorological Institute (MET) and the exact location and elevation above sea level can be found in table 1. All the wind measurements are retrieved from 10 meters above ground.

The quality of the data from 2017 retrieved from these station are good in terms of few missing data points. For Tromsø and Hekkingen less than 1% of the data for 2017 are missing. For Andøya, 2.3% of the data are missing, mainly in the short time periods 13th to 18th of July, 25th of July to 4th of August and 28th to 29th of August. For the time steps where observational data are missing, the corresponding data has been excluded from the model results.

2.2. Model set up and numerical simulations

The NWP model used in this study is the Advanced Research WRF (ARW) version 3.9.1. A detailed description can be found in [14]. The 20-category Moderate Resolution Imaging Spectroradiometer (MODIS) land use data and the Global Multi-resolution Terrain Elevation Data 2010 (GMTED2010) have been selected. Both the datasets were retrieved from the National Center for Atmospheric Research (NCAR) and have a resolution of 30 arc-seconds. The numerical simulations were initialized by the ERA-Interim reanalysis data provided by the

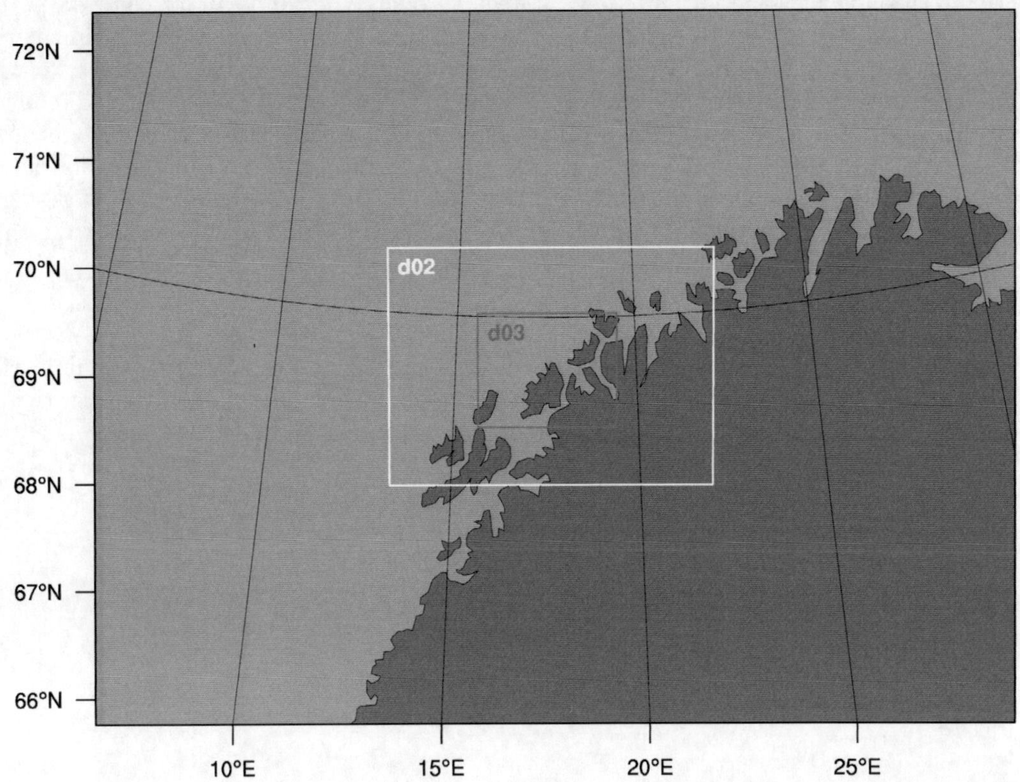

Figure 2: WRF domain d01, d02 and d03 with 25 km, 5 km and 1 km resolutions respectively.

European Center for Medium-Range Weather Forecast (ECMWF). The spatial resolution of the input data is approximately 80 km over 38 vertical pressure levels with a temporal resolution of 6 hours. The ERA-Interim data have been down scaled by the WRF model through three domains as illustrated in figure 2. The outermost domain is hereafter referred to as d01, the second as d02 and the innermost as d03. The grid size of the domains are reduced by a factor of 5 from 25 km in d01 to 1 km in d03. A one-way nesting strategy is applied, were the flow of information goes from the coarser domains to the finer domains, with no feedback. The center of d01 is at the latitude $69°154''42'$ N and longitude $17°66''00'$ E and corresponds to the north end of the island Senja. Polar stereographic projection is used as recommended for high latitudes. Domain d01 covers a large area of 775 km in north-south direction and 1025 km in east-west direction to ensure all large scale events are included. The inner domain, d03, covers an area that includes the locations of the three meteorological weather stations and stretches 121 km in north-south direction and 151 in east-west direction. The vertical structure consists of 50 terrain following vertical coordinates. To ensure numerical stability a time step of 5 seconds is used in d03.

For the physical configuration, the model offers a wide range of options for how to parametrize the physical sub-grid processes. This allows an optimal set up for the study case. In this study the CONUS configuration has been adopted, which consists of the following parametrization schemes: Thompson micro physics scheme, the Mellor–Yamada–Janjic scheme (MYJ) scheme to represent the planetary boundary layer physics, the Tiedtke scheme for the cumulus parametrization, the Monin-Obukhov Eta scheme for the surface layer processes, the Noah

IOP Conf. Series: Journal of Physics: Conf. Series **1102** (2018) 012011 doi:10.1088/1742-6596/1102/1/012011

Land Surface Model and the RRTMG schemes for short- and long-wave radiation [14].

The simulations have been initialized every week through 2017, and run for eight days. The first 24 hours of each week are considered to be spin up time for the simulations. A time period of one year is considered to be sufficient to determine diurnal and seasonal variations for wind resource assessment purposes [1]. Therefore, the same period has been considered to be sufficient also for this study. All the results discussed in this paper are 10 minute average values retrieved from the inner most domain at 10 meter height above ground from the same position as the meteorological stations. This is achieved by linear interpolation between the two closest grid points in the domain. The model terrain height for the closest grid point to each of the three stations have been listed in table 1

2.3. Evaluation methods

The following statistical measures are used to compare the deviation θ' between the numerical simulations and the measurements; root mean square error (RMSE), mean absolute error (MAE) and bias and are given by the following equations:

$$RMSE = \left[\frac{1}{N}\sum_{i=1}^{N}(\theta_i')^2\right]^{\frac{1}{2}} \tag{1}$$

$$Bias = \frac{1}{N}\sum_{i=1}^{N}\theta_i' \tag{2}$$

$$MAE = \frac{1}{N}\sum_{i=1}^{N}|\theta_i'| \tag{3}$$

Where for the wind speed, the deviation θ' is given by

$$\theta' = \theta^{sim} - \theta^{obs} \tag{4}$$

This is also true for the wind direction when the deviation is less than or equal to 180°. If $|\theta'| > 180°$ [12], then

$$\theta_i' = (\theta_i^{sim} - \theta_i^{obs}) \times \left(1 - \frac{360}{|\theta_i^{sim} - \theta_i^{obs}|}\right) \tag{5}$$

The MAE gives the mean of the deviation θ' between corresponding observation and simulation pairs. The RMSD gives the mean of the square of the deviation θ'. Due to the square operation, the RMSD emphasizes instances with large deviation. The bias gives the deviation between the observation sample mean and the simulated sample mean.

Another statistical measure often used to characterize the distribution of wind speed is the Weibull probability distribution function (PDF) [15]. The Weibull PDF is given by the following equation

$$f(U;\beta,\alpha) = \beta\alpha^{-\beta}U^{(\beta-1)}e^{-(\frac{U}{\alpha})^{\beta}} \tag{6}$$

The shape parameter β controls the shape of the function and the scale parameter α controls the width of the function.

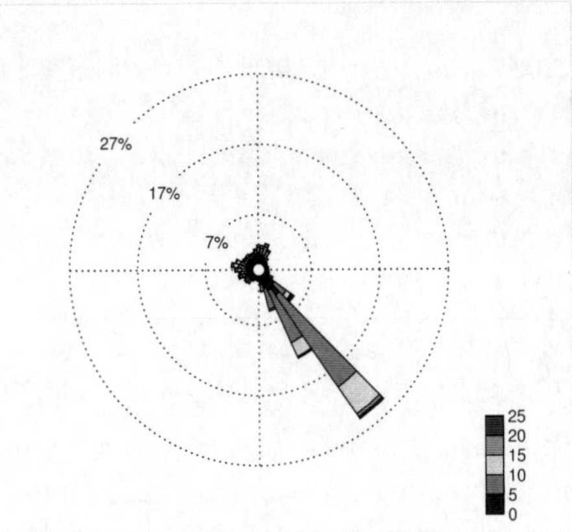

Figure 3: Observations, Hekkingen.

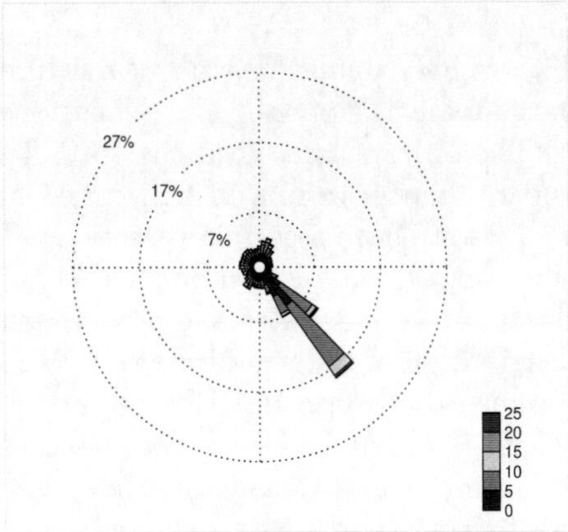

Figure 6: WRF, Hekkingen.

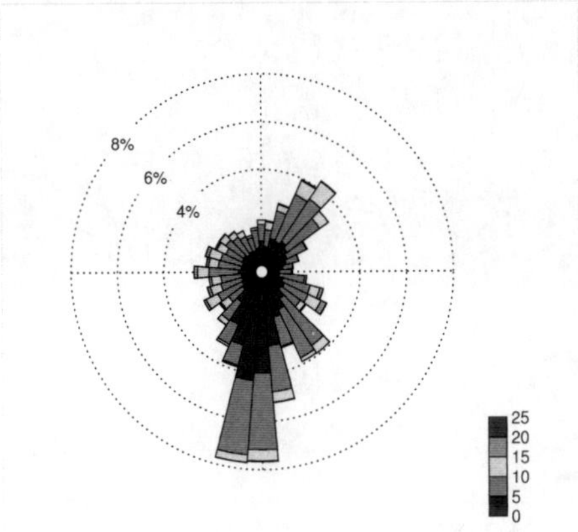

Figure 4: Observations, Andøya.

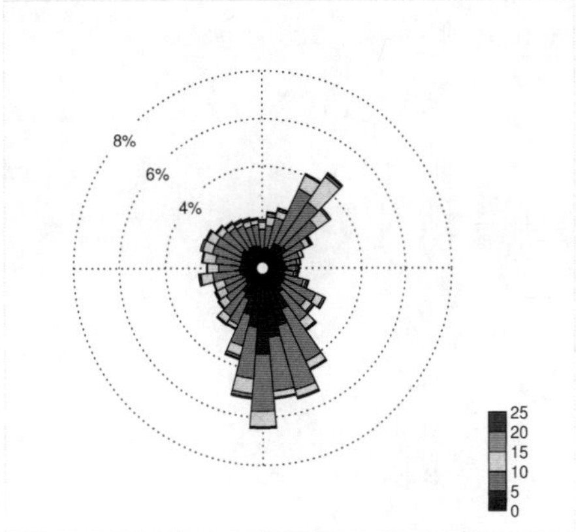

Figure 7: WRF, Andøya.

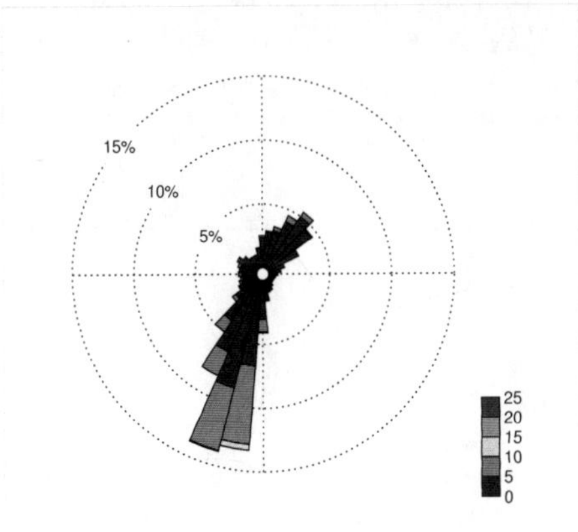

Figure 5: Observations, Tromsø.

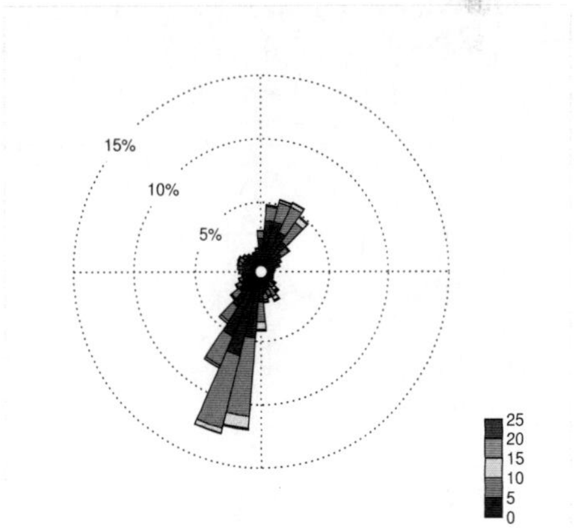

Figure 8: WRF,Tromsø.

3. Results

The WRF simulations are carried out over the year 2017 and compared to the in-situ measurements from the three locations Hekkingen, Andøya and Tromsø.

The observations from the three locations are presented as wind roses in figure 3, 4 and 5 and the model results in figure 6, 7 and 8. The circles represent in percentage how often the wind blows from a certain direction. The color bar gives the magnitude of the wind, where blue is the lowest wind speed and red the highest wind speeds. When comparing the wind roses for the observed data with the model results, it is evident that the model is able to reproduce the distribution of the wind directions well for all locations.

The observations for Hekkingen show a clear main wind direction from the south-east. The model is able to capture this but has some higher wind speeds than the observations. For Andøya the observations show a diverse distribution of wind direction with two main directions, one from the north-east and one from the south. The model show a similar distribution. Also for Tromsø the model is able to reproduce the main wind direction from south south-west. For all locations the model captures the main wind direction well, but with a slightly lower frequency compared to the observations.

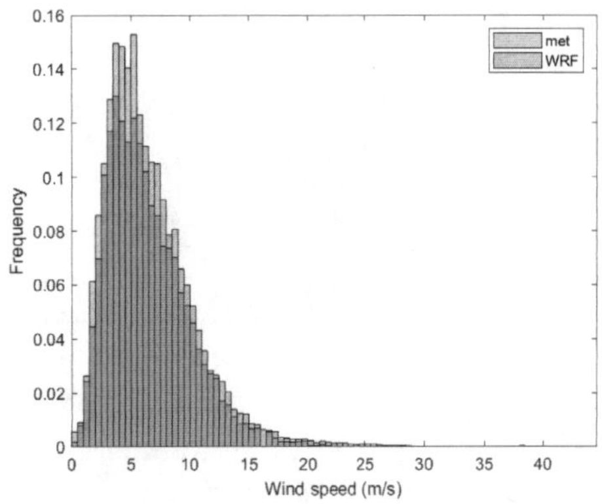

Figure 9: Wind speed PDF for Andøya.

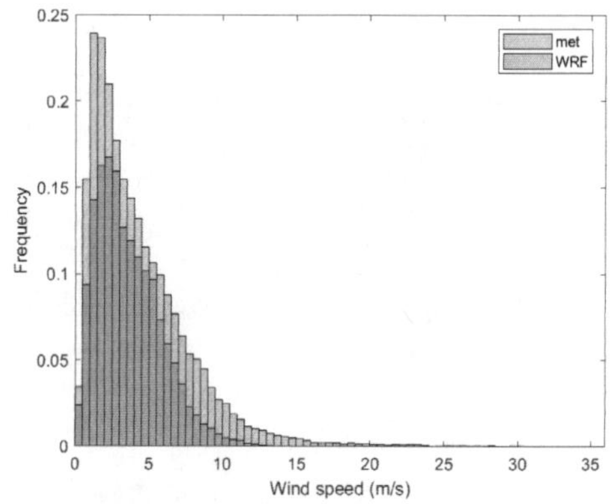

Figure 10: Wind speed PDF for Tromsø.

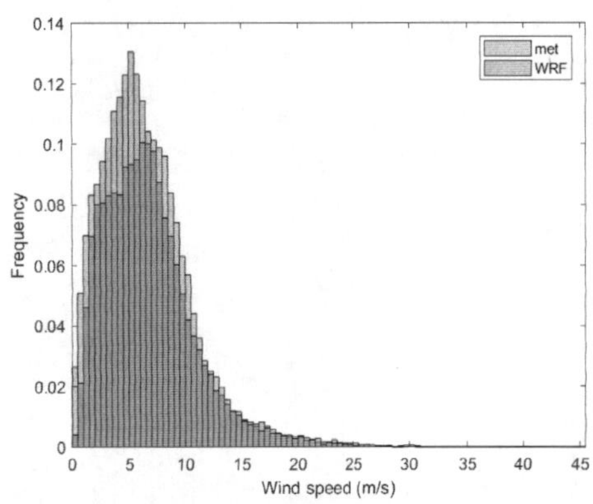

Figure 11: Wind speed PDF for Hekkingen.

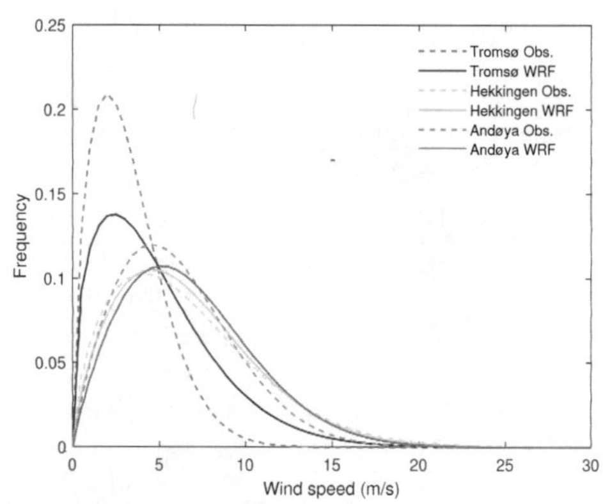

Figure 12: Weibull PDF plot for Hekkingen, Tromsø and Andøya.

The wind roses give a good indication on the model's ability to reproduce the observed wind field, especially with respect to the wind direction. To gain a better understanding of the model's ability to reproduce the magnitude of the wind, the distributions of the wind speeds are presented in figure 9, 10 and 11 for Andøya, Tromsø and Hekkingen respectively.

	Mean		Scale		Shape	
	Observations	WRF	Observations	WRF	Observations	WRF
Tromsø	3.26	4.84	1.61	1.44	3.65	5.36
Hekkingen	6.70	6.67	1.61	1.74	7.45	7.52
Andøya	6.11	6.81	1.88	1.88	6.88	7.70

Table 2: The mean of the observed and simulated wind speeds and the Weibull PDF scale and shape parameters.

For both Andøya and Tromsø the model tends to overestimate the frequency of the wind speeds above ~6 m/s and underestimate for wind speeds below this. For Hekkingen, on the other hand, the model underestimates the frequency of wind speeds above ~7 m/s and wind speed below ~2 m/s, while it underestimates for the moderate wind speeds. For all locations, the difference in frequency is largest for the moderate wind speeds. The model performs better for Hekkingen than it does for Andøya and especially when compared to Tromsø. The wind observations are done at 10 meters above ground. It is therefore not unlikely that the observations are influenced by obstacles in close vicinity. The weather station Tromsø is located within the city area, while Hekkingen is a small unpopulated island in the ocean.

The Weibull PDFs in figure 12 derived from the observations and the simulations show similar results. For Hekkingen the Weibull PDFs are in good agreement with a slight shift to the right. Andøya have a larger deviation between the two Weibull PDFs but still comparable. For Tromsø on the other hand, the underestimation of the frequency of low to moderate wind speeds and the overestimation for the high wind speed is evident.

The corresponding scale and shape parameters are presented in table 2. The best results are found for Hekkingen, with only small differences between the scale and the shape parameters derived from the observed and the model results. However, when comparing to the study by [12] the difference between these parameters are larger. In the study by [12] the wind measurements the WRF simulation was compared with was done at 60 m agl., while in this study it is done 10 m agl. In the study by [11] the performance of WRF was evaluated against observations done at several heights. It was found that the biases were considerable larger at the lowest measuring point at 40 m agl. than for the measurements done higher up.

In table 3 a statistical evaluation of the performance of the model for both wind speed and direction is summarized in terms of RMSE, bias and MAE.

The wind speed bias gives an indication of the model's tendency to over- or underestimate the results compared to the observations. The wind speed bias for Andøya indicates a slight overestimation of the mean wind speed. The bias for Tromsø shows a considerable overestimation. The bias for Hekkingen is close to zero and therefore shows no clear over or under estimation of the model simulations. The larger biases for Tromsø and Andøya might be related to the model's representation of the topography. Due to the resolution of the model, the complex topography is smoothed. While the observed wind speed is slowed down due to the topography, the smoothed representation will result in a lower friction and therefore higher wind speed [12]. For the wind direction, the biases show a good agreement between the simulated and the observed wind direction for all locations. Andøya has the lowest bias, where the simulated direction is slightly rotated clockwise compared to the observations. For Tromsø and Hekkingen the biases are larger and show a counter clockwise rotation.

The RMSE and the MAE show a quite high deviation between the model and the observations both for the wind speed and direction. Compared to similar studies [9, 12] the RMSE is larger. Especially the RMSE for wind direction are larger than what is found in [10]. It is evident that even though the wind roses and the biases show a good agreement between the observed and

	RMSE		Bias		MAE	
	Speed (ms^{-1})	Dir.(°)	Speed (ms^{-1})	Dir.(°)	Speed (ms^{-1})	Dir.(°)
Tromsø	3.57	112	1.58	-5.25	2.40	72
Hekkingen	4.10	101	-0.02	-3.21	3.02	60
Andøya	3.50	92	0.69	0.21	2.43	52

Table 3: The RMSE, bias and MAE for Tromsø, Hekkingen and Andøya. The speed is given in m/s and the direction (Dir) is given in degrees.

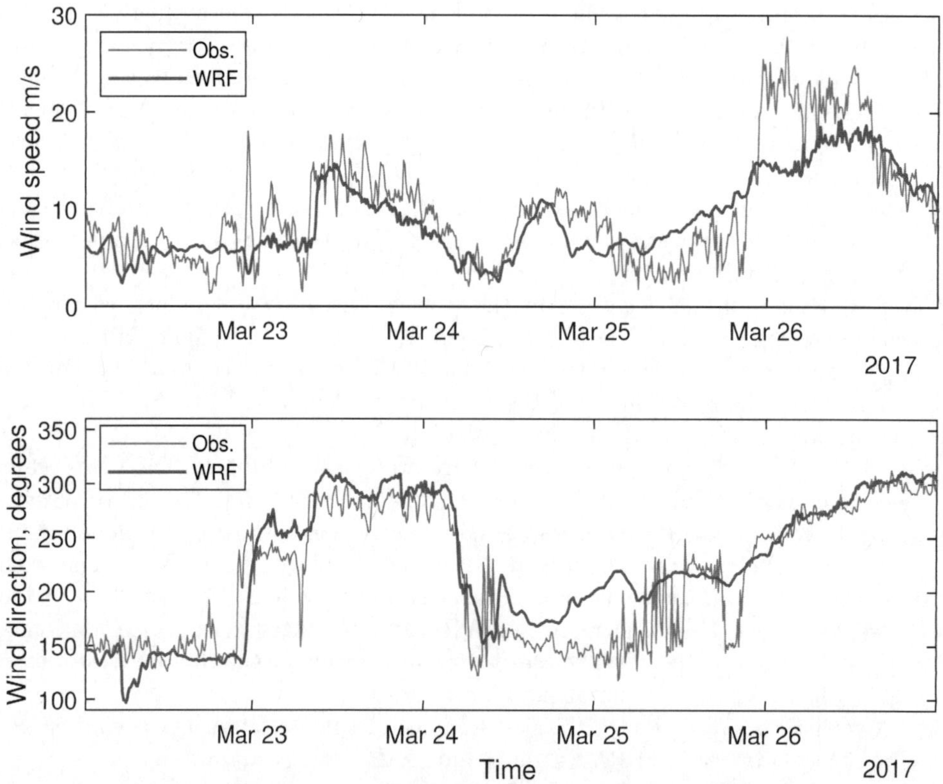

Figure 13: An example period of the observed and the simulated wind speed and direction for Hekkingen. The upper plot show the wind speed and the bottom plot show the wind directions for the observations (in grey) and the WRF simulations (red).

simulated wind direction, the RMSE and MAE indicates that the model has some challenges. The wind roses and biases show that the model is able to reproduce the annual average wind directions quite well, but when the temporal similarities is considered, the deviation between the simulations and the observations are more severe.

A short period of the wind speed time series are shown in the upper plot in figure 13 and the wind direction for the corresponding time in the plot under. The WRF wind speed and the direction follow the general trend seen in the measurements, but the model is not capable of reproducing the rapid variations seen in the measurements. An example of overestimation of low wind speed, and underestimation of high wind speed, can also be identified at March 25th and in the start of March 26th, respectively.

4. Conclusion

The aim of this study is to evaluate the performance of the WRF model in a complex and coastal terrain with respect to wind speed and direction. WRF simulations have been run for one year and compared to wind observations from three weather stations located close to the area of interest. A good agreement is found for two of the locations when comparing the annual averages of the WRF simulations and the observations. At the third location the model overestimates the wind speed but is able to capture the wind direction quite well.

The RMSE and the MAE indicate that there are still challenges when it comes to accurately reproducing the wind field in a complex terrain. Reasons for the high values for RMSE and MAE should be investigated further. Some of the deviations found can be related to the fact that the model results are compared to wind measurements done at 10 m agl. where obstacles in close vicinity might influence the measurements. A further study should therefore also include wind observations at wind turbine hub height to ensure a more accurate evaluation of the model for wind resource assessment purposes.

Meso-scale numerical weather prediction models have proven to give satisfactory first estimates of the wind fields and can be used to locate promising sites with respect to wind power. WRF is a widely used and tested model and gives results in good agreement with observations. However, near-surface wind modeling remains challenging, especially for areas with complex terrain. Further research must therefore be done to identify weaknesses and improve the models.

References

[1] Al-Yahyai S, Charabi Y and Gastli A 2010 *Renewable and Sustainable Energy Reviews* **14** 3192 – 3198 ISSN 1364-0321 URL `http://www.sciencedirect.com/science/article/pii/S1364032110001814`

[2] Weir D E and Aksnes N 2018 Vindkraft -produksjon i 2017/Wind Power -Production in 2017 Tech. Rep. 10 Norges vassdrag - og energidirektorat

[3] Landberg L, Myllerup L, Rathmann O, Petersen E L, Jørgensen B H, Badger J and Mortensen N G *Wind Energy* **6** 261–271 (*Preprint* `https://onlinelibrary.wiley.com/doi/pdf/10.1002/we.94`) URL `https://onlinelibrary.wiley.com/doi/abs/10.1002/we.94`

[4] Kunz M, Mohr S, Rauthe M, Lux R and Kottmeier C 2010 *Natural Hazards and Earth System Sciences* **10** 907–922 URL `https://www.nat-hazards-earth-syst-sci.net/10/907/2010/`

[5] Carvalho D, Rocha A, Santos C S and Pereira R 2013 *Applied Energy* **108** 493 – 504 ISSN 0306-2619 URL `http://www.sciencedirect.com/science/article/pii/S0306261913002705`

[6] Larsén X G, Badger J, Hahmann A N and Mortensen N G *Wind Energy* **16** 1167–1182 (*Preprint* `https://onlinelibrary.wiley.com/doi/pdf/10.1002/we.1544`) URL `https://onlinelibrary.wiley.com/doi/abs/10.1002/we.1544`

[7] Caroline D, N H A, Alfredo P and Gregor G *Wind Energy* **17** 39–55 (*Preprint* `https://onlinelibrary.wiley.com/doi/pdf/10.1002/we.1555`) URL `https://onlinelibrary.wiley.com/doi/abs/10.1002/we.1555`

[8] El-Samra R, Bou-Zeid E and El-Fadel M 2018 *Atmospheric Research* **203** 68 – 82 ISSN 0169-8095 URL `http://www.sciencedirect.com/science/article/pii/S0169809516305798`

[9] Giannaros T M, Melas D and Ziomas I 2017 *Renewable Energy* **102** 190–198 URL `http://dx.doi.org/10.1016/j.renene.2016.10.033`

[10] Jiménez P A and Dudhia J 2013 *Journal of Applied Meteorology and Climatology* **52** 1610 URL `https://journals.ametsoc.org/doi/10.1175/JAMC-D-12-0266.1`

[11] Fernández-González S, Martín M L, García-Ortega E, Merino A, Lorenzana J, Sánchez J L, Valero F and Rodrigo J S 2018 *Journal of Applied Meteorology and Climatology* **57** 733–753 (*Preprint* `https://doi.org/10.1175/JAMC-D-17-0121.1`) URL `https://doi.org/10.1175/JAMC-D-17-0121.1`

[12] Carvalho D, Rocha A, Gómez-Gesteira M and Santos C S 2014 *Applied Energy* **117** 116 – 126 ISSN 0306-2619 URL `http://www.sciencedirect.com/science/article/pii/S0306261913009847`

[13] Byrkjedal Ø and Berge E 2008 *Kjeller Vindteknikk AS*

[14] Skamarock W C, Klemp J B, Dudhia J, Gill D O, Barker D M, Duda M G, Huang X Y, Wang W and Powers J G 2008 1–113 URL `http://www2.mmm.ucar.edu/wrf/users/docs/arw`$_v$*3.pdf*

[15] Morgan E C, Lackner M, Vogel R M and Baise L G 2011 *Energy Conversion and Management* **52** 15 – 26 ISSN 0196-8904 URL `http://www.sciencedirect.com/science/article/pii/S019689041000227X`

IOP Conf. Series: Journal of Physics: Conf. Series **1102** (2018) 012012 doi:10.1088/1742-6596/1102/1/012012

Investigation of the validity of the Blade Element Momentum Theory for wind turbine simulations in turbulent inflow by means of Computational Fluid Dynamics

S. Ehrich[1], **C.M. Schwarz**[1], **H.Rahimi**[1,2], **B. Stoevesandt**[2] **and J. Peinke**[1,2]

[1] ForWind, Institute of Physics, University of Oldenburg, Oldenburg, 26129, Germany

[2] Fraunhofer IWES, Oldenburg, 26129, Germany

E-mail: sebastian.ehrich@uni-oldenburg.de

Abstract. In this work three different numerical methods are used to simulate a MW class wind turbine under turbulent inflow conditions. These methods are a blade resolved Computational Fluid Dynamics (CFD) simulation, an actuator line based CFD simulation and a Blade Element Momentum (BEM) approach. For all three methods sectional and integral forces are investigated in terms of mean, standard deviation, power spectral density and fatigue loads. It is shown that the power spectral densities of integral forces are in good agreement in the low frequency range for all methods. However, deviations are observed from 2 Hz onward. Although the fatigue loads of the CFD based methods are similar for the torque, the loads from BEM differ by 22%. For the thrust BEM only deviates by 4.5% from the blade resolved case. Nevertheless the actuator line based simulation differs by up to 7.3%. The standard deviations of sectional forces for all three methods relate very well to the standard deviation of integral forces. Nevertheless, a similar relation for the fatigue loads could not be observed.

1. Introduction

One of the most common tools in wind turbine design and load calculation is the Blade Element Momentum method (BEM). Although BEM based tools are very efficient, they are based on many simplifications like stationary environment, 2D airfoil characteristics and inviscid fluid behavior [1]. While preserving its efficiency, over the years a lot of correction models have been developed to improve the accuracy, e.g. by using tip loss models or dynamic stall models. However, especially for turbulent inflow scenarios the reliability of such predictions is still unknown. This is mainly due to the relatively coarse discretization of wind fields and rotor blades, the dynamic aerodynamic response to complex inflow, the unresolved flow field but also the assumption of inviscid and stationary flow [1]. Nevertheless, turbulent inflow scenarios are relevant to the design and certification process, e.g. to the design load cases 1.1-1.3 of the IEC standard [2]. One of the limited work in this field is done by Madsen et al. [3] where for turbulent inflow conditions differences for the spectra of sectional forces between BEM and Blade Resolved (BR) Computational Fluid Dynamics (CFD) simulations were shown. However, global quantities like thrust and torque, and the corresponding fatigue loads were not investigated.

This work aims to assess the validity of BEM based simulations for turbulent inflow conditions comparing them to two higher fidelity CFD methods, namely the BR simulation and the Actuator

Line (AL) model. New in this comparison is the investigation of spectra of integral forces, the analysis of fatigue loads and a bridge between BR and BEM calculations, namely the AL method. This work is organized as following. The basic setup of the turbine, the turbulent wind fields, the CFD and BEM simulations, and the fatigue load estimation are addressed in section 2. In section 3 key aerodynamic quantities of the turbine such as power, thrust and sectional forces and the corresponding equivalent fatigue loads as well as spectral properties are investigated. Finally a conclusion including future work is presented in section 4.

2. Methodology

In this section the basic methodology used in this work as well as the fatigue load estimation is presented.

2.1. The turbine

The numerical studies are performed on the NREL-5MW reference wind turbine with a rotor diameter of 126m [4]. This three bladed horizontal axis wind turbine with a blade length of 61.5m was designed by the National Renewable Energy Laboratory. It is close to the current average rotor size of newly installed wind turbines which accounts to 100-120m. More details on the reference turbine can be found in Table 1. Because there is a lack of experimental data for large wind turbines, and the NREL-5MW turbine has been investigated many times, it can be considered as one of the most important wind turbine design which are available within the wind energy research community.

Table 1. Overview of NREL-5MW turbine.

Parameter	Value
Number of rotor blades	3
Rotor diameter	126m
Rated aerodynamic power	5.3 MW
Rated wind speed	11.4 m/s
Rated rotational speed	12.1 rpm
Blade cone angle	2.5°
Shaft tilt angle	5.0°
Blade length	61.5m
Blade mass	17740 kg

2.2. The wind fields

As the findings are believed to be general and only weakly dependent on wind modelling details a simplified model for the wind fluctuations is used. For the sake of simplicity the wind profile is uniform, i.e. shear is not taken into account. Velocity fluctuations are represented by a Gaussian process realized at the inflow patch of the CFD domain. In particular, coupled stochastic Ornstein-Uhlenbeck processes of the form

$$\frac{\mathrm{d}u_i{}^{(k)}(t)}{\mathrm{d}t} = -\gamma\left(u_i{}^{(k)}(t) - \bar{u}^{(k)}\right) + \sqrt{D}\sum_j^N H_{ij}\Gamma_j{}^{(k)}(t) \tag{1}$$

with an exponentially decaying spatial and temporal correlation function for all wind components are used. In this stochastic differential equation each subscript $i, j = 1\ldots N$ denotes a specific cell on the inlet patch where the turbulence is generated and each superscript $k = x, y, z$ corresponds to the longitudinal, lateral or vertical velocity component. $\gamma = 0.97\ s^{-1}$ is a damping factor, $\bar{u}^{(y,z)} = 0\ ms^{-1}$

and $\bar{u}^{(x)}$ the mean velocities in x, y and z direction, $D = 0,62\ m^2 s^{-3}$ a diffusion constant, H the lower Cholesky-decomposed correlation matrix and Γ a Gaussian, delta-correlated random noise vector with variance 2.
Fields with a mean wind speed of $\bar{u}^{(x)} = 11.4\ ms^{-1}$ corresponding to the rated operating condition of the wind turbine and a Turbulence Intensity (TI) of 20% are fed into the CFD domain. This relatively high TI had to be chosen, because of the natural decaying process leading to much smaller turbulence intensities in the vicinity of the turbine in the absence of shear. To make the comparison between different codes in a similar manner, the pitch and rotor speed are fixed and the effect of structural deformations, tilt and cone as well as the tower and nacelle are neglected.

2.3. Numerical methods
In the following the BR, AL and BEM method are described in more detail:

- BR: The blade resolved simulations are performed with the open-source CFD software OpenFOAM V4.1 [5], which is an open-source CFD package with a collection of modifiable libraries written in C++. The simulation domain is discretized using the bladeblockMesher [6] and windTurbine mesher tools [7] with a combination of structured and unstructured grids consisting of a total of 36M cells. The mesh of the complete simulation domain and the rotating part are shown in Figure 1 with a sectional view on the blade mesh in Figure 2. In order to limit the cell aspect ratio in radial direction, adaptive wall functions are being applied. The solution for the flow field is achieved on solving the incompressible Delayed Detached Eddy Simulations (DDES) equations where the flow is assumed to be fully turbulent, i.e. the laminar-turbulent transition in the boundary layer is not considered. The closure problem is treated by use of the Sparlart-Allmaras (SA) turbulence model [8]. In order to solve the pressure-velocity coupling, the PIMPLE algorithm, is used, which is a combination of the loop structures of SIMPLE [9] and PISO [10]. Integration in time was performed using an implicit second order backward scheme together with a second order linear upwind scheme for discretizing the convection terms. For these simulations 360 cores were used for approximately 20 days resulting in 600 seconds simulation time.

- AL: The rotor is simplified by an AL model [11], where all rotor blades are reduced to rotating lines with a radial distribution of body forces at 60 positions along the lines. A Gaussian filtering kernel with smoothing parameter ε equal to 4 was used to obtain well fitting sectional forces to the BR case. Based on the angle of attack at the given positions on the line, body forces are gained from lookup tables containing polars of two dimensional airfoils. Except for a coarser resolution in the vicinity of the rotor the domain is exactly the same as for the BR case leading to a domain size of 22M cells. The flow field is simulated by Large Eddy Simulations (LES) with a Smagorinsky subgrid-scale model [12]. These simulations have been conducted on 360 cores for 25 hours real time reaching 600 seconds simulation time.

- BEM: This method uses wind fields recorded in the CFD domain on an equidistant 31×31 grid with a grid size of 4.5m. In this case the fields are obtained at the position of the rotor from the same mesh as for the AL method but in an empty CFD domain. This seems to be a reasonable choice, because the TI at the rotor is expected to be comparable for the CFD and BEM simulations. The BEM based tool FAST v8 [13] in combination with AeroDyn v15 [14] with the Beddoes Leishman type unsteady aerodynamics model [14, 15] is used in this work. The BEM model relies on the very same airfoil data as the AL.

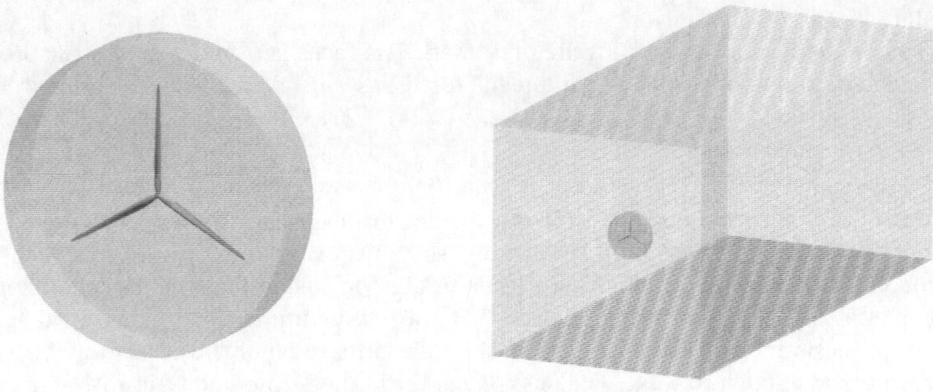

Figure 1. Rotating part and full domain of BR simulation.

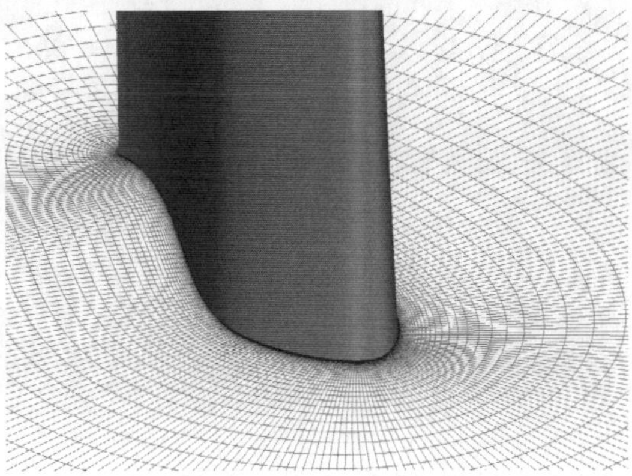

Figure 2. View on pressure side of the rotor blade.

2.4. Fatigue load estimation

One of the most important quantities relevant for wind turbines are the fatigue loads on different parts of the turbine measured by damage equivalent loads also known as Equivalent Fatigue Loads (EFL) which will play a role in the load analysis of the three methods analysed here. In general EFL are defined as

$$EFL = \left(\frac{\sum n_i r_i^m}{N}\right)^{\frac{1}{m}} \tag{2}$$

with the specific load ranges r_i, the corresponding amount n_i of load ranges, the reference number of ranges N and the Wöhler exponent m. More information on the estimation of the EFL and the underlying rainflow algorithm can be found in [16] and [17].

3. Results

In this section the results of the study are presented. The focus lies on the wake structure, average and standard deviation of sectional and integral forces, as well as fatigue loads and power spectral densities.

3.1. Wake dynamics

Snapshots of the velocity magnitude for the BR and AL cases are shown in Figure 3. Despite the differences in the methods both simulations show very similar fields in terms of the velocity magnitude upstream of the turbine. This leads to the conclusion that the induction zone of the turbine is also similar for both simulations. The wake structure is also in very good agreement. However, the BR simulation is resolving the tip and root vortex in a more physical way which results in smaller structures in the near wake. The missing nacelle in the AL method leads to a jet like behavior at the center of the rotor, but the influence on the far wake is neglegible.

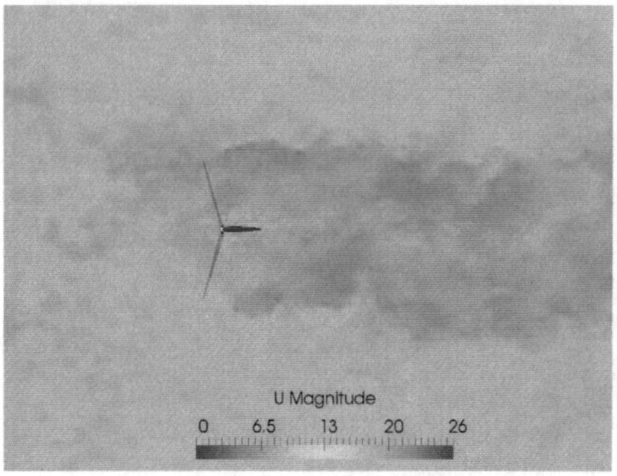

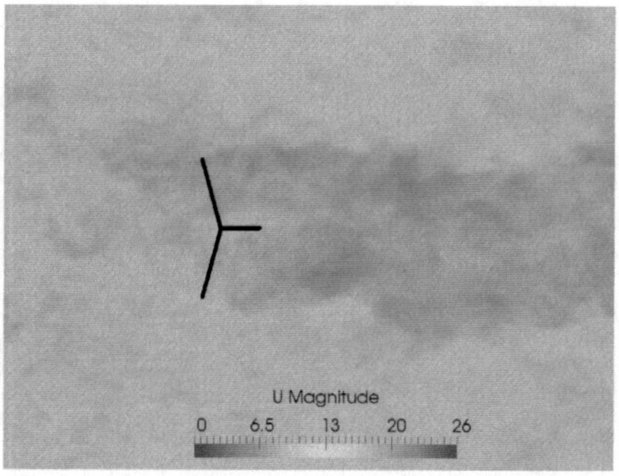

Figure 3. Snapshots of the velocity magnitude for BR (left) and AL (right) simulations.

3.2. Sectional forces

All three simulation methods (BR, AL and BEM) are compared based on averaged sectional quantities and their standard deviation. As most important aspects we select sectional axial and tangential forces on the blade, which do not reveal big differences for the average between all cases in the mid-span as shown in Figure 4). However at the root (up to 30% of span) and the very tip (from 80% of span onwards) some differences can be observed. BEM and AL rely on similar root and tip correction models and very same airfoil data which explains their good agreement. But in contrast to the BR case, where 3D effects are modelled physically, the forces at the root section are far off supporting the well known issues for BEM and AL.

The standard deviations for the tangential and axial force distributions, explicitly shown in Figure 5, are very similar for the BEM and AL methods but with a bit larger values for the BEM case in most sections. With the BR case as the reference the standard deviations of sectional forces tend to be lower at the root of the blade for the BEM and AL methods whereas they are larger in the mid-span. Close to the tip of the blade the standard deviations for the AL method are much smaller while BEM is in good agreement with the BR simulation at the outermost section.

IOP Conf. Series: Journal of Physics: Conf. Series **1102** (2018) 012012 doi:10.1088/1742-6596/1102/1/012012

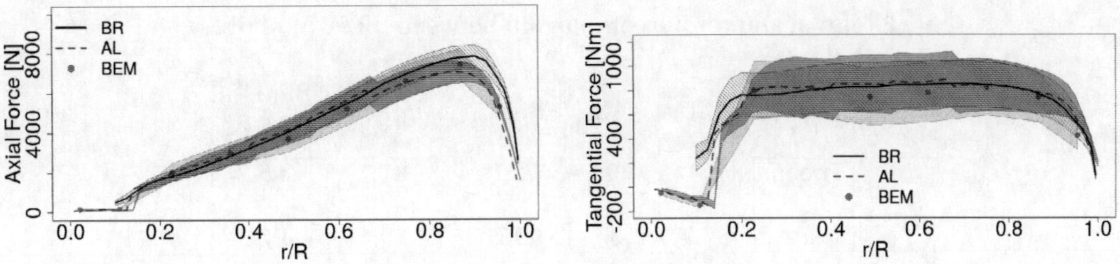

Figure 4. Radial distribution of axial (left) and tangential (right) forces for BR, AL and BEM cases. The shaded areas illustrate one standard deviation of the fluctuations.

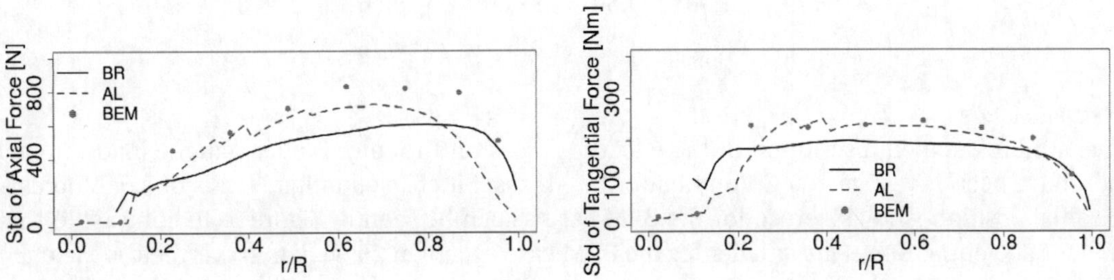

Figure 5. Radial distribution of standard deviation of axial (left) and tangential (right) forces for BR, AL and BEM cases.

3.3. Integral forces

As presented in Table 1, the integral aerodynamic thrust and torque, show a very close agreement between all the methods. The AL method shows up to 8.3% difference in thrust and 2.5% in torque with respect to BR method. The BEM simulation shows 4.2% and 1.8% difference in the thrust and torque compared to BR method, respectively,

Analog to the relation between mean sectional forces and mean integral forces, the standard deviation of sectional forces corresponds to a standard deviation in the thrust and torque. The standard deviation for the AL case is for both quantities approximately 10% smaller than for the BR simulation. In contrast to that, out of the BEM simulation an approximately 20 to 30% increase in the standard deviation is obtained.

Table 2. Thrust and torque comparison between BR, AL and BEM.

		BR	AL	BEM	AL/BR [%]	BEM/BR [%]
Thrust [kN]	mean	761	698	729	91.7	95.8
	std	37.9	33.9	49.7	89.4	131.1
	EFL	147.6	136.8	141.0	92.7	95.5
Torque [MNm]	mean	4.43	4.32	4.35	97.5	98.2
	std	0.495	0.452	0.599	91.3	121.0
	EFL	1.904	1.897	1.486	99.6	78.0

3.4. Fatigue loads

To give a more detailed discription of the effect of forces on the rotor blades, fatigue loads at each section have been investigated. Compared to the AL case, increased fatigue loads of axial forces at each radial position were observed for the BEM case as can be seen in Figure 6. In contrast to that, the EFL of tangential forces are smaller for the BEM case except at 20% span. Nevertheless the effect on integral forces like thrust and torque are different to what one would expect out of the radial force distributions as seen in Table 2. On the one side, the EFL of thrust are in good agreement for all three methods, while we would expect a higher value for the BEM case. On the other side, the loads of torque are almost the same for the BR and AL method, but the BEM approach gives a 22% lower estimation. It can be concluded that in contrast to the standard deviation, the EFL of thrust and torque can not be estimated directly out of the EFL of sectional forces. This highlights the fact that the dynamics of global forces are in general different to the dynamics of sectional forces and the output of the rainflow counting algorithm is not easily predictable for complex time series.

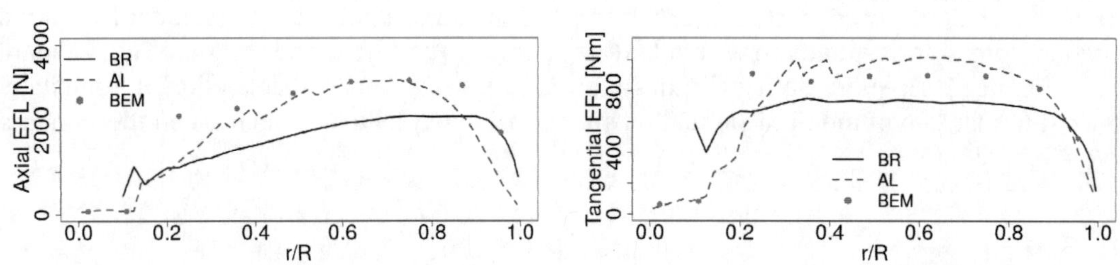

Figure 6. Radial distribution of axial (left) and tangential (right) equivalent fatigue loads for BR AL and BEM methods.

Further analysis of EFL by the underlying range histograms in Figure 7 shows that for the thrust force all three methods have in general a similar behaviour. At low ranges BEM is following the AL trend and in the mid range BEM has a bit higher counts as AL and BR. Because of their large weight the most important part for the EFL are the high thrust ranges which almost equally often occur for all three cases leading to similar EFL.

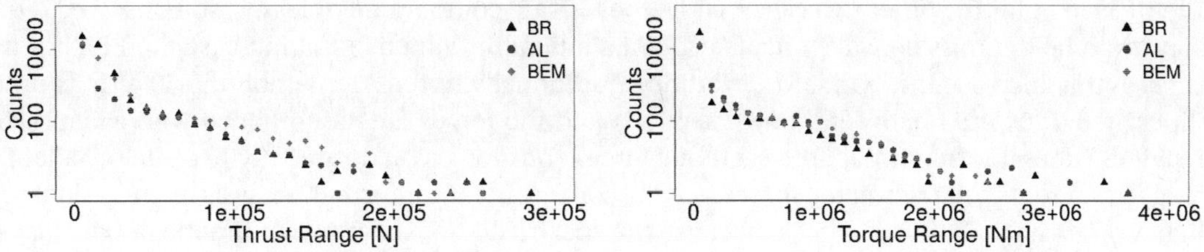

Figure 7. Range histograms for thrust (left) and torque (right).

The range histogram of the torque in Figure 7 is also very comparable for all three cases but the AL as well as the BR cases show ranges much higher than the highest one for BEM at approximately $2.4 \cdot 10^6$ Nm. This leads to the 22% difference in the EFL of torque shown in Table 2.

3.5. Power spectral densities

For a analysis of the dynamics of torque and thrust power spectral densities are investigated in Figure 8. All three methods show a very good agreement for low frequencies up to 2 Hz including the 3P frequency and the first two harmonics. For higher frequencies up to approximateley 5 Hz BEM and AL are very similar but the BR method drops up to this point. The reason for the power spectral density drop might be the RANS model working in the boundary layer of the blade while the AL uses LES everywhere and therefore does not smooth out the wind field in the vicinity of the blade. Additionally, the time step could be too large for the fine grid resolution in the blade region for the BR simulation. The difference between BEM and AL in the high frequency range could be according to the missing damping and filtering effects in the BEM case, while AL uses a Gaussian filter to smooth out the forces along the blade. The noisy behaviour of the power spectral density of the AL method at high frequencies is part of every AL simulation and is also a result of the convolution of the forces given at the airfoil sections with a Gaussian kernel.

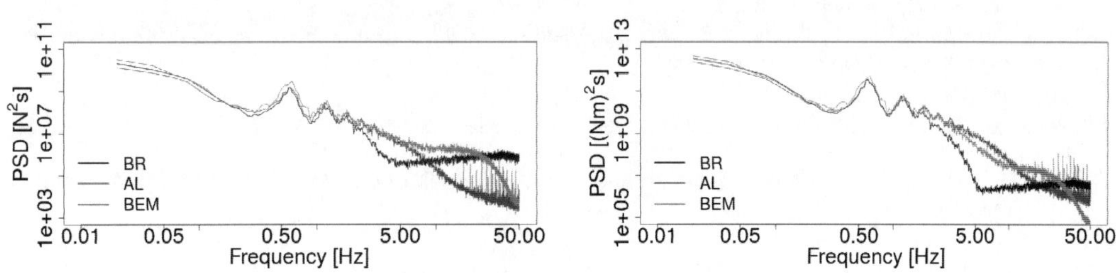

Figure 8. Power spectral densities of thrust (left) and torque (right) for BR, AL and BEM cases.

4. Conclusions

Out of this work it can be concluded that the AL, BR and the BEM method with extracted wind fields at the rotor plane from an empty domain are very similar with respect to averaged tangential and axial forces, except close to the root and the tip. However differences are observed when looking at the standard deviations of the forces. For the integral quantities thrust and torque BEM tends to overestimate the standard deviation compared to the other two methods. This leads to the conclusion that either the wind fields have to be extracted downwind of the rotor position from the empty domain simulation or a correction of wind fields for the BEM case has to be considered.

The question, why the standard deviation of the global quantitities investigated here are much higher for the BEM than for the other two cases, while the EFL are comparable or less, is still unsolved. One reason could be the missing wake effects of BEM simulations which could increase the EFL but do not necessarily increase the standard deviation. Additionally in the AL method the 60 airfoils used are linearly interpolated between known sections and the forces are smoothened with a Gaussian kernel. FAST does not smooth out the sectional forces and along a blade are only 13 airfoils available, leading to different dynamics in the forces. A good starting point for further analysis might be looking at higher order statistics of the forces, because the spectra do not seem to capture the fatigue loads. Further studies on this point have to be done in the future work to find the reason for this behavior. Nevertheless, BR and AL simulations result in very similar results for the avarage, standard deviation and EFL of thrust and torque, which is also the result of tuning the parameters of the AL to the BR results. The power spectral densities of thrust and torque for all three methods show good agreement on lower frequencies but it is still not very clear what leads to the large differences on high frequencies. Finally it is also not clear which of all the results are most realistic, because no experiments for comparison were performed.

Acknowledgments
The simulations were performed at the HPC Cluster EDDY, located at the University of Oldenburg (Germany) and funded by the Federal Ministry for Economic Affairs and Energy (Bundesministeriums für Wirtschaft und Energie) under grant number 0324005.

References

[1] Schepers J G 2012 *Engineering Models in Wind Energy Aerodynamics* Ph.D. Thesis Delft University of Technology URL `https://repository.tudelft.nl/islandora/object/uuid:92123c07-cc12-4945-973f-103bd744ec87/datastream/OBJ/download`
[2] International Electrotechnical Commission 2005 IEC 61400-1 Standard International Electrotechnical Commission
[3] Madsen H A, Sørensen N N, Bak C, Troldborg N and Pirrung G 2018
[4] Jonkmann J, Butterfield S, Musial W and Scott G 2009 Definition of a 5-MW Reference Wind Turbine for Offshore System Development Tech. Rep. TP-500-38060 National Renewable Energy Laboratory 15013 Denver W Pkwy, Golden, CO 80401, USA
[5] OpenCFD 2007 *OpenFOAM - The Open Source CFD Toolbox - User's Guide* OpenCFD Ltd. United Kingdom 1st ed
[6] Rahimi H, Daniele E, Stoevesandt B and Peinke J 2016 *Wind Engineering* **40** 148–172 ISSN 0309524X URL `https://doi.org/10.1177/0309524X16636318`
[7] Rahimi H 2018 *Improving Aerodynamic Engineering Models for Wind Turbines by means of Computational Fluid Dynamics* Ph.D. Thesis University of Oldenburg
[8] Spalart P and Allmaras S 1992 **439**
[9] Patankar S and Spalding D 1972 *International Journal of Heat and Mass Transfer* **15** 1787 – 1806 ISSN 0017-9310 URL `http://www.sciencedirect.com/science/article/pii/0017931072900543`
[10] Issa R 1986 *Journal of Computational Physics* **62** 40 – 65 ISSN 0021-9991 URL `http://www.sciencedirect.com/science/article/pii/0021999186900999`
[11] N Sørensen J and Shen W Z 2002 **124** 393
[12] Smagorinsky J 1963 *Monthly Weather Review* **91** 99–164
[13] Jonkman B and Jonkman J 2016 Fast v8.16.00a-bjj Tech. rep. NREL URL `https://wind.nrel.gov/nwtc/docs/README_FAST8.pdf`
[14] Jonkman J, Hayman G, Jonkman B and Damian R 2016 AeroDyn v15 User's Guide and Theory Manual (draft version) Tech. rep. NREL URL `https://wind.nrel.gov/nwtc/docs/AeroDyn_Manual.pdf`
[15] Leishman J G and Beddoes T S 1989 *Journal of the American Helicopter Society* **34** 3–17
[16] Madsen P H 1990 *Recommended practices for wind turbine testing and evaluation. 3. Fatigue loads*
[17] Rychlik I 1987 *International Journal of Fatigue* **9** 119 – 121 ISSN 0142-1123 URL `http://www.sciencedirect.com/science/article/pii/0142112387900545`

Forecasting wind ramps: can long-range lidar increase accuracy?

I Würth, S Ellinghaus, M Wigger, M J Niemeier, A Clifton, and P W Cheng

Stuttgart Wind Energy (SWE), University of Stuttgart, Allmandring 5B , 70569 Stuttgart

E-mail: wuerth@ifb.uni-stuttgart.de

Abstract. The national funded WindforS project VORKAST investigated the use of a long-range lidar for very short-term power forecasting of a wind turbine in complex terrain. This ability is essential for the grid integration of large amounts of wind energy. This paper describes the process of setting up the lidar, data handling, and wind field reconstruction. A process based on Taylor's frozen turbulence hypothesis is used to propagate wind speeds to the turbine and to forecast wind ramps. The lidar-based forecast is currently less accurate than persistence in these conditions. It is expected that the use of a more realistic propagation model will improve the forecasts in such complex terrain.

1. Introduction

Germanys goal for the Energiewende is to have a share of renewable energies in total electricity consumption of 35 % by 2020, by 2050 even 80 % [1]. To achieve these goals, the energy system is being reformed and the process will likely lead to a decentralized system with many interconnected power generating plants that need to be coordinated. Along with solar and hydro power, wind represents the largest renewable power source within this system. One of the main challenges for the electrical grid operator is to match an electric load profile to the power fluctuations of these weather dependent sources. Electric grid system operators must instantaneously and continuously balance supply and demand to maintain the reliability of the grid. Therefore, predictions of the power from different sources are necessary [2]. Current wind power predictions are based on numerical weather prediction (NWP) models, which use mathematical models of the atmosphere and ocean to predict the weather based on current weather conditions. The horizontal resolution of these models ranges from 1 to 10 kilometers (km) [3]. Short-term forecasts of up to one hour are usually implemented using the persistence method. For this method it is assumed, that there will be no changes in wind speed within the next designated time period, meaning the upcoming wind speed is equal to the present one. While the wind forecast for horizons of more than an hour using NWP models can be regarded as sufficiently precise, the fact that persistence is the state-of-the-art prediction method for horizons of less than an hour can be problematic for locations in which wind conditions frequently change and are therefore not accurately predicted using the conditions at the time of the forecast. This is particularly the case for a phenomenon called wind ramps, also referred to as wind ramping event or wind power ramp, which are defined as a sudden and strong change in wind power output from a wind turbine or wind park [4].

The University of Stuttgart proposes a novel approach for short-term forecasting of the wind power, which is based on wind speed measurements upwind of the wind turbines. Within the national funded WindForS research project VORKAST the use of a long-range wind lidar for wind power forecasting for horizons up to 60 minutes (min) is being investigated. Wind lidar devices measure the wind speed along the line-of-sight (LOS) direction of a laser beam. Modern scanning wind lidars, can measure the LOS wind speed up to a distance of 10 km in a hemisphere centered on the device. The wind speed is measured in multiple range gates along the laser beam and thus measurements with a very high temporal and spatial resolution are possible. In contrast to NWP models and persistence also small fluctuations of the wind speed are captured and the evolution of the wind field over the terrain can be tracked. As a result, short-term wind power forecasts based on lidar measurements offer the possibility to close the spatial and temporal gap to existing methods, especially for wind ramping events. Although a wind ramp is defined as a change in power generation, the cause for the ramp is a change in wind speed. This change can be measured ahead in time using a long-range lidar and thus more accurate predictions could be possible.

2. Measurement Setup

The lidar system used in VORKAST is a HALO Photonics Stream Line XR pulsed scanning lidar as proposed in [5]. The main features, which are crucial to the project objectives, are the 10 km range and the beam steering over the total hemisphere using a two-axis scanner. The footprint which is smaller than a square meter and the power requirement which is below 600 W are beneficial characteristics of the lidar and ideal for elevated, remote measurement sites. The device is depicted in Figure 1 and an overview of its characteristics is given in Table 1. In a first measurement campaign, the lidar is mounted at the top level platform 91 m above ground level of the radio tower "Lauterstein 1" at the Swabian Alps in the south of Germany. The location is chosen due to the undisturbed sight in the prevailing wind direction and the nearly motion free mounting.

Figure 1: Stream Line XR on the radio tower.

Table 1: Technical specifications of the Stream Line XR [6].

Wavelength	1.5 µm
Max. range	10.000 m
Bandwidth	$\pm 19.4\,\mathrm{m\,s^{-1}}$
Doppler resolution	$0.0384\,\mathrm{m\,s^{-1}}$
Length of range gate	60 m
Max. no. of range gates (without overlapping)	167

In Figure 2 the terrain profile of the site in prevailing wind direction (west-northwest) is shown. As can be seen, in 1.3 km distance of the lidar a 100 m met mast is located which is operated by SWE and fully equipped with meteorological sensors with the top height 78.5 m below the lidar. The terrain is complex and in 3 km to 4 km distance of the lidar an escarpment cuts through the topography. Previous measurement campaigns and the analysis of the met tower data have shown that the complex topography of the site leads to complex flow and effects such as reverse flow conditions have been detected. This must be considered in the

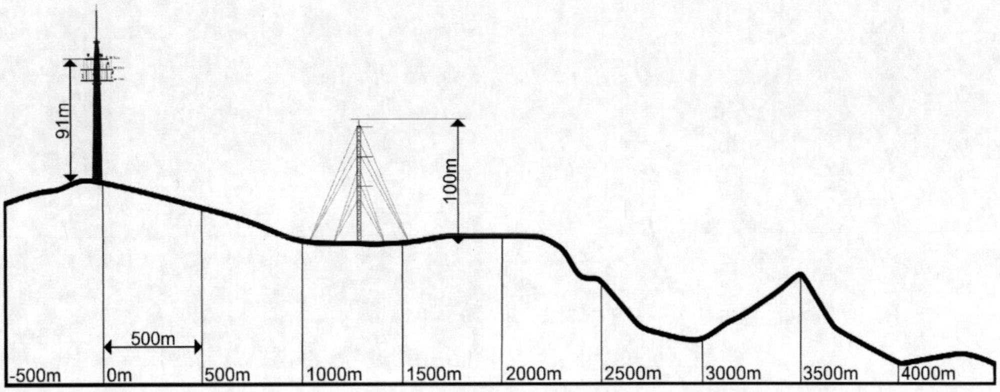

Figure 2: Terrain profile of the test site in main wind direction with radio tower and met mast. The figure is true to scale.

further development of the project. In the sector from southwest to north of the lidar there are 9 adjacent wind turbines located. From one of the turbines the nacelle anemometer and power data was made available in ten-minutes mean resolution. Simultaneous measurements from lidar and turbine were carried out for a period from July 2016 to August 2017.

3. Device alignment

Determining the device alignment as well as possible is crucial to the operation of long range lidar measurement campaigns. An accurate knowledge of the exact orientation of the laser beam in every degree-of-freedom is essential for the success of the project as a deviation of a fraction of a degree leads to a substantial error of several dozens of meters in measurements in long distances. Methods for device alignment are not well established yet, therefore a process needs to be investigated. The following steps, which are further explained in the following, are conducted:

- Determination of free sectors with a 360 degree azimuth scan
- Azimuth alignment of the beam with a hard target detection of surrounding wind turbines
- Elevation alignment by using a digital terrain model
- Testing of the repeatability of the alignment correction by scanning a single wind turbine

After the deployment of the device, a determination of free sectors is carried out. Therefore multiple high resolution full azimuth scans are executed. The analysis of the Signal-to-noise ratio (SNR) of the first five range gates gives an overview of undisturbed sectors with 1° resolution. Disturbed sectors will be avoided throughout the measurement campaign. The analysis for the site on the radio tower is shown in Figure 3. Important is the undisturbed sector from 90° to 270° which includes the prevalent wind directions from the west to northwest sector.

For the azimuth alignment of the laser beam the strong direct reflection of the beam on solid objects in the nearer field of view is used. In the analysis a hard-target is assumed when the wideband SNR in a range gate and its surrounding range gates is above 5 dB and the determined line-of-sight velocity is close to $0\,\mathrm{m\,s^{-1}}$ as seen in Figure 4. Here, the wind turbines scattered around the lidar are used as hard-targets for alignment. With their known position the calculation of the devices misalignment is possible and can be corrected during the data analysis.

IOP Conf. Series: Journal of Physics: Conf. Series **1102** (2018) 012013 doi:10.1088/1742-6596/1102/1/012013

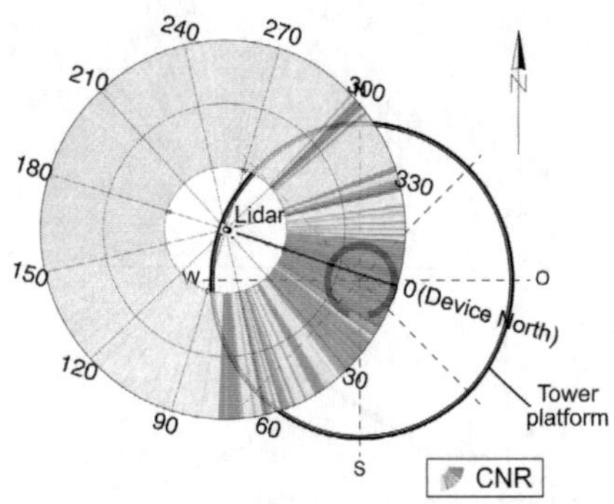

Figure 3: Topview of the radio tower and analysis of the CNR of the first 5 range gates to determine undisturbed sectors (green).

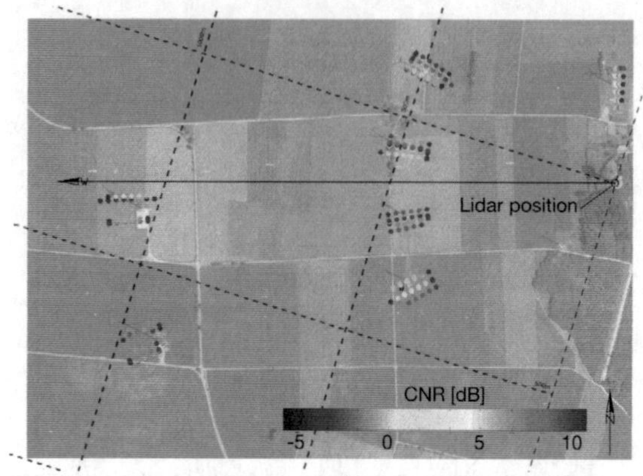

Figure 4: Top view of the test site showing the CNR values as returned from the surrounding wind turbines.

Furthermore, with the use of a digital topographic model from Shuttle Radar Topography Mission (SRTM) [7] the elevation misalignment is detected. Therefore and azimuth scan pointing towards the ground is executed and the heights above ground from the SRTM model are compared to the measured heights of the lidar.

To gain the last bit of awareness about uncertainties in beam steering, a horizontal raster scan pattern is performed over the structure of a wind turbine in 1.1 km distance. As seen in Figure 5 the structures of the turbine and also of the met mast to the left behind it are clearly visible in the lidar data. The repeated crossing of the tower shows some inaccuracies in the position repeatability of the lidar. But the maximum misalignment detected with this campaign was 0.03° which can be the result of several not explicit determined factors such as minimal displacements of the radio tower structure, an inaccuracy of the laser beam steering or beam broadening and scattering.

Figure 5: SNR signal as received from a detailed raster scanning of a wind turbine and its surroundings.

4. Filtering

Filtering of the radial velocity measurements is necessary to prepare the data for the wind field reconstruction by removing corrupted data, which could be the result of too low or too high aerosol, dust or particle concentration in the area. Different filter methods are implemented and tested, with the aim to meet the following filter requirements:

- conservative filtering with least possible data loss
- adaptability to varying environmental conditions
- high velocity for real time capability

To evaluate different filters, several data sets with varying environmental conditions are tested and the quality of the filtering is compared with regard to the requirements. In Figure 6a and 6b the timeline of the unfiltered radial velocity v_{radial} and the Carrier-to-noise-ratio (CNR) distribution for an exemplary data set are shown respectively. The data set which is shown is used to compare different filter methods and is chosen due to its variability in measurement range.

The first filter that was investigated is the CNR filter. For each measured radial velocity at any range gate the intensity of the backscattered signal is stored besides the actual velocity value. Relating the backscattered signal intensity to the sent signal leads to CNR which is used to determine the quality of the measured data. A common approach when working with radial velocity lidar, is to filter the lidar output respectively to their CNR value, removing all values with a lower CNR than a certain threshold. Many tests with various lidar data have shown that this filter works for short distances very well, since the signal is still strong. However, many radial velocity values from farther range gates that seem to be accurate, but with lower CNR are being removed. This can be seen clearly in Figure 6c and 6d where the exemplary data set was filtered with a CNR threshold of $-25\,\mathrm{dB}$. Therefore this threshold filter does not concur with the requirements of least possible data loss, since an enormous amount of accurate values is being removed. This observation leads to the necessity of developing different filters that are not, or not only based on the CNR values.

The Normal Distribution Grubbs (NDG) filter is based on a statistical test for outlier identification first implemented by Frank E. Grubbs in 1968 [8]. The idea is to sort the possible outliers into two groups. One with the observation of extreme values deviating from the rest of the data, but still representing physically explainable data. In lidar data this could be a velocity drop due to the wake of wind turbines, or a very short but strong gust. The second group of outliers deviates severely from the rest of the data as a result of an error in recording the numerical value [8]. It is the goal of the filter to detect this second group, leaving the first group untouched embedded in the data. The NDG filter does work for normally (Gaussian) distributed values, therefore an initial test for normal distribution of the raw radial velocity data is necessary. For lidar velocity measurements this usually is the case for environmental conditions with very far visibility and only few scattered corrupted values. As for the CNR filter, a threshold Th for a value that still can be plausible has to be set previously as filter input. For smaller series Grubbs defined various thresholds [8] but for this usage a threshold most suitable for any environmental condition has to be defined. In Figure 6e and 6f, the results of the filter are shown for $Th = 3\,\mathrm{m\,s^{-1}}$.

The last filter that was analyzed is the Range filter. Originally used within image processing for edge detection, to detect relatively large areas of different color intensity, this filter has been adapted to suit the needs to filter long-range lidar data. Assuming a maximum range between minimum and maximum of realistic radial velocities in windows of consecutive range gates, it is possible to detect areas of highly scattered radial velocities which are the result of erroneous measurements. The threshold value for the range filter is set in relation to the turbulence intensity of the data set. In Figure 6g and 6h results are shown for a range of

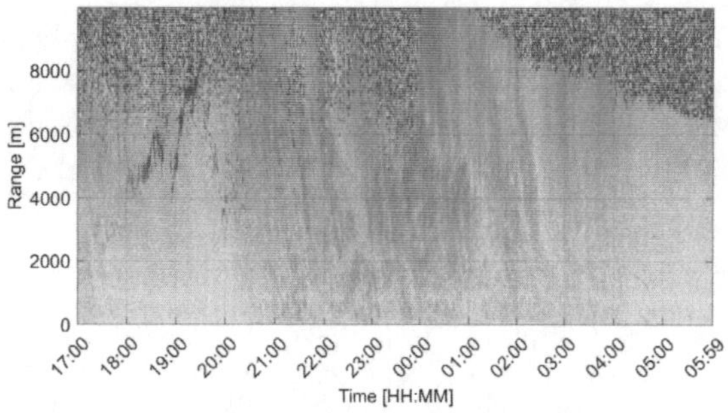

(a) V_{radial} raw data

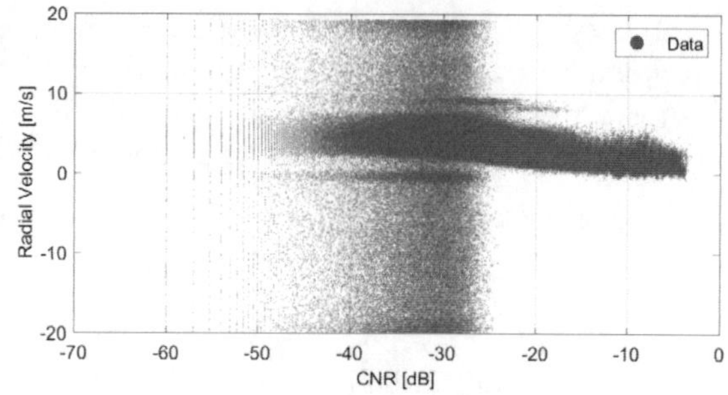

(b) CNR vs. V_{radial}

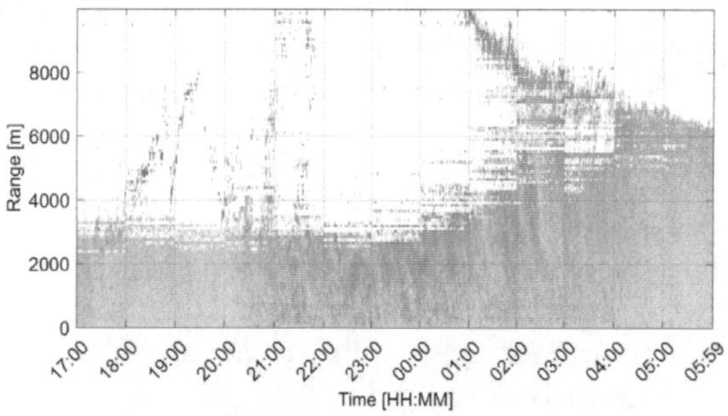

(c) V_{radial}, CNR filtered ($-25\,\mathrm{dB} < \mathrm{CNR} < 5\,\mathrm{dB}$)

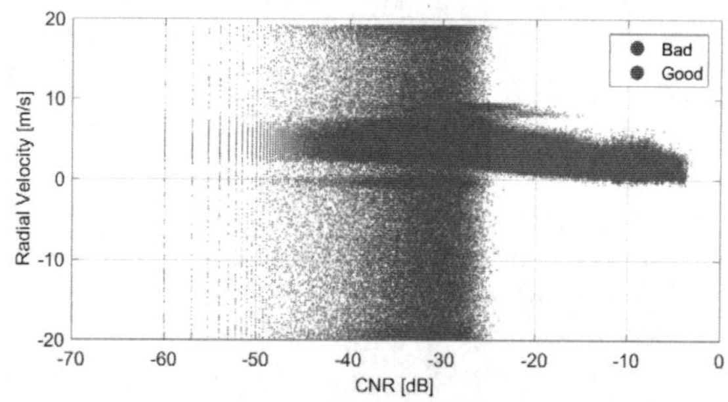

(d) CNR vs. V_{radial}, CNR filtered

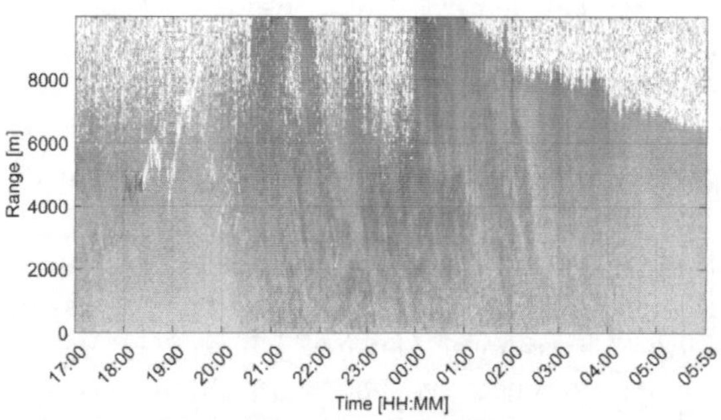

(e) V_{radial}, NDG filtered ($Th = 3\,\mathrm{m\,s^{-1}}$)

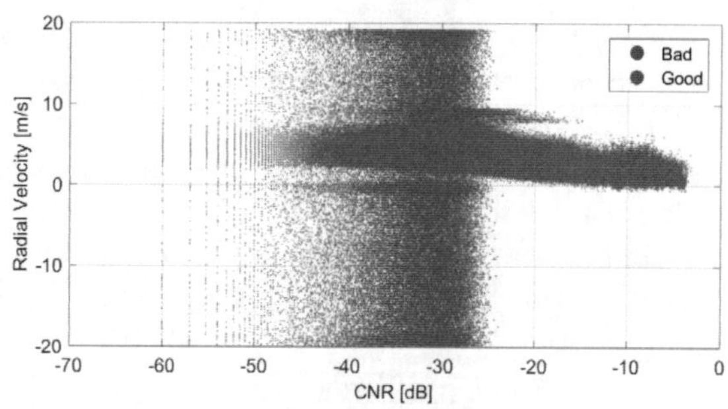

(f) CNR vs. V_{radial}, NDG filtered

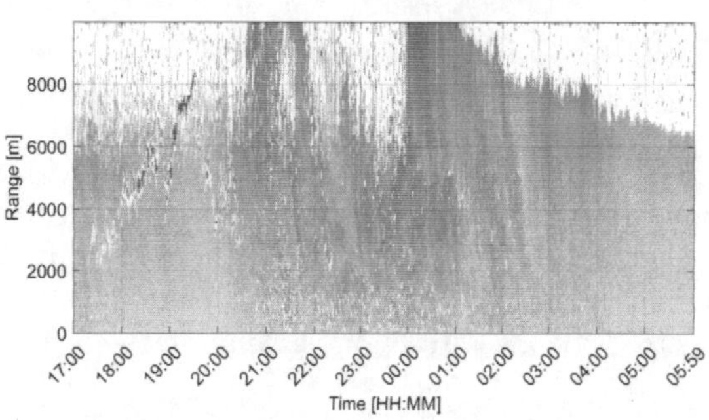

(g) V_{radial}, Range filtered (range $< 3\,\mathrm{m\,s^{-1}}$)

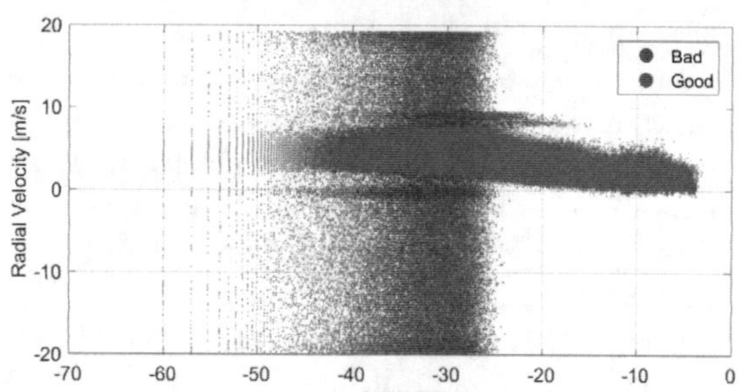

(h) CNR vs. V_{radial} , Range filtered

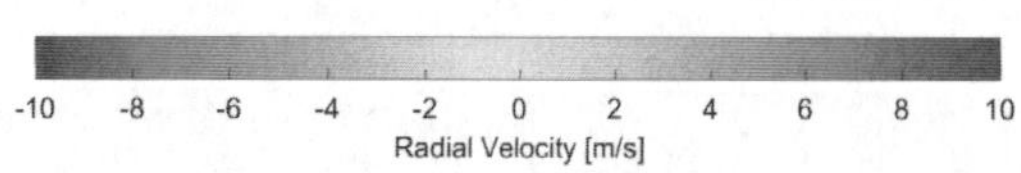

Figure 6: Results of the different filter methods for an exemplary data set. To the left the timeline of the radial velocity is shown and to the right the radial velocity is plotted against the CNR.

IOP Conf. Series: Journal of Physics: Conf. Series **1102** (2018) 012013 doi:10.1088/1742-6596/1102/1/012013

$3\,\mathrm{m\,s^{-1}}$. Less obviously good radial velocity data is removed from the data set in comparison to the standard CNR and NDG filter, which is mostly noticeable in large distance measurements with weak signal returns. Additional positive aspects of the filter are the independence from the dominant wind speed and the removal of randomly distributed radial velocities due to failure in the measurement.

The comparison of different filter methods that are set up to analyze radial velocity data, that was taken during varying environmental conditions, shows the necessity to use different filters for different conditions. This is because each filter does only perform in its best way during certain environmental conditions, and not always. Therefore key precondition to the filter process is the correct classification of the data (e.g. normal distribution) and the correct selection of the most suitable filter method. Overall, the range filter shows the most promising results since it is very robust and works for different environmental conditions. In addition, less valid radial velocity data is removed compared to the CNR and NDG filter. Therefore it is used for the following analysis.

5. Wind field reconstruction

One of the challenges when working with lidar data is to reconstruct the wind vectors at discrete points of the measured wind field correctly from the line of sight velocity data v_{los}. The basic equation that needs to be solved is given as

$$v_{los,i} = \frac{x_i}{f_i} u_i + \frac{y_i}{f_i} v_i + \frac{z_i}{f_i} w_i \tag{1}$$

where the lidar measures in point $[x_i\, y_i\, z_i]$ with a measurement distance f_i and the unknown local wind vector is $[u_i\, v_i\, w_i]$ [9]. To solve this equation, at least three measuring points are necessary and assumptions must be made. Assuming a constant and homogeneous wind field during the measurement, the previous equation can be described in matrix form for all measured radial wind speeds.

$$\underbrace{\begin{bmatrix} v_{los,1} \\ \vdots \\ v_{los,n} \end{bmatrix}}_{m} = \underbrace{\begin{bmatrix} \frac{x_1}{f_1} & \frac{y_1}{f_1} & \frac{z_1}{f_1} \\ \vdots & \vdots & \vdots \\ \frac{x_n}{f_n} & \frac{y_n}{f_n} & \frac{z_n}{f_n} \end{bmatrix}}_{A} \cdot \underbrace{\begin{bmatrix} u \\ v \\ w \end{bmatrix}}_{s} \tag{2}$$

If sufficient linear independent measuring points are available, the global wind velocity components u, v and w can be estimated using the inverse A^{-1} or the Penrose inverse A^{+} determined by the least-square method.

$$\begin{bmatrix} u \\ v \\ w \end{bmatrix} = A^{-1} \begin{bmatrix} v_{los,1} \\ \vdots \\ v_{los,n} \end{bmatrix} \tag{3}$$

The following procedure was used for the wind field reconstruction in this paper. Assuming that the vertical component w of the wind field can be neglected, a global wind field reconstruction per lidar scan was performed in a first step with Formula 3 to determine the horizontal wind vector over the entire measurement range. From the reconstructed global velocity components u_{global} and v_{global} a global wind direction was derived. In a second step, assuming that the wind direction did not change for the duration of a scan, the local horizontal wind speeds were determined by projecting the radial wind speeds $v_{los,n}$ onto the global wind direction. To validate the procedure, the reconstructed global wind direction and local wind speed are compared to met mast data (Figure 7). The data is averaged over 10 minutes and the

IOP Conf. Series: Journal of Physics: Conf. Series **1102** (2018) 012013 doi:10.1088/1742-6596/1102/1/012013

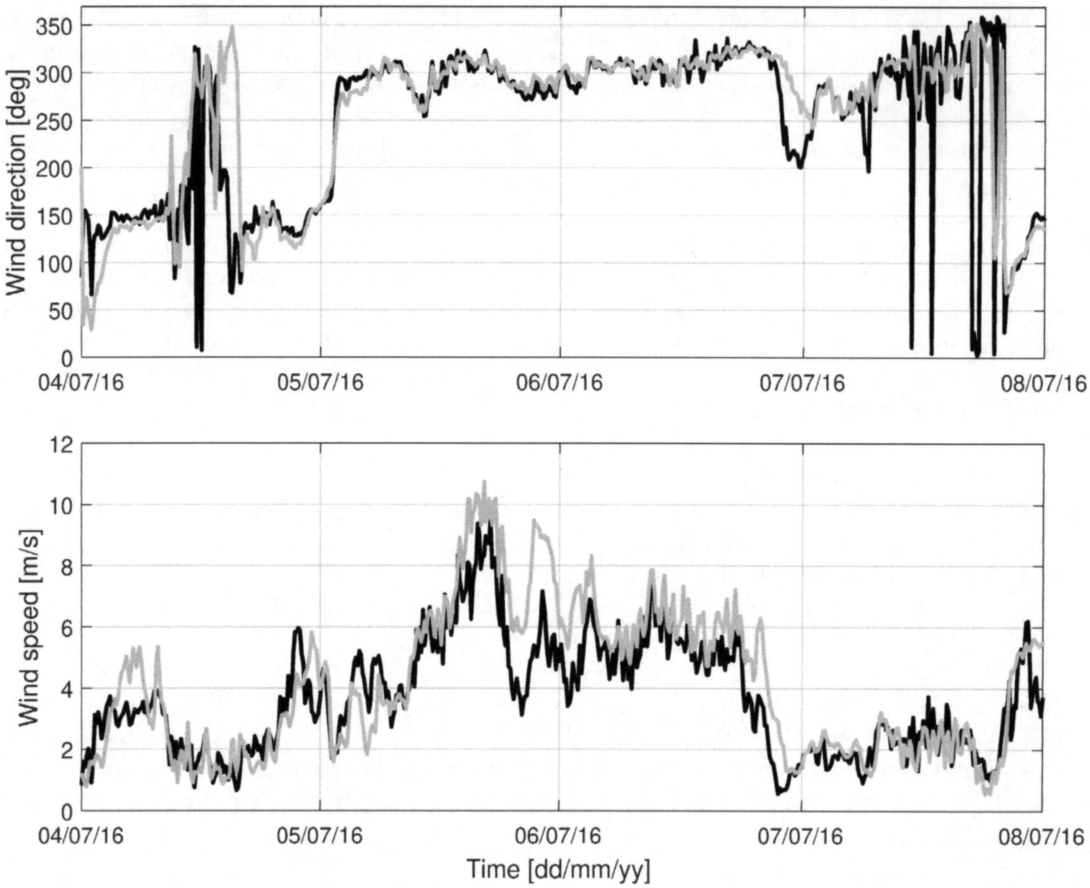

Figure 7: Comparison of 10-min averaged wind direction and horizontal wind speed data from met mast (black) and lidar (grey).

wind speed data of the range gate above the met mast is used. The data shows good accordance and wind direction changes are well captured.

6. Wind evolution

Wind evolution is a crucial factor for the success of lidar-based forecasting. Only if the wind conditions at the turbine show similarities with the wind conditions measured in the distance, a forecast can be made and expected to be meaningful. To assess this similarity, multiple pairs of reconstructed horizontal wind speed data sets from the distance and close to the turbine and the lidar unit are compared. As a tool for comparison, the coherence is calculated which quantifies the similarity between two signals in terms of frequency. The coherence is a function which can take values between 0 and 1. If the coherence is 1, two time signals are similar over the whole frequency range, if it is 0, they have no frequency in common. Consequently, measurements which are taken at two spatial separated points should result in a coherence of 1 if the wind field does not change in between and thus also the frequencies remain the same.
The coherence is applied to data sets that contain 60 min of continuous wind speed measurements from 4 km and 0.2 km distance to the lidar. For the intended comparison, the full 60 min of measurements must be from periods when the wind direction matches the lidar units measuring

orientation. If this criterion is fulfilled, the wind conditions passing the 4 km mark can be expected to pass the 0.2 km mark a few minutes later in time. Another criterion that had to be fulfilled in each data set is that at least 90 percent of valid data remained after applying the range filter. With these criterion applied, a total number of 13 data sets are available for the analysis.

Figure 8 shows the averaged coherence of distant measurements and those close to the lidar of all 13 investigated data sets on a logarithmic frequency scale. It is apparent that the coherence is only high for very low frequencies and quickly decreases to a value of 0.4 or less for any frequency higher than 0.003 Hz. A high coherence of at least 0.7 is obtained only for frequencies lower than 0.0021 Hz. This frequency corresponds through the relationship $f = 1/T$ to a period T of 476.2 s or 7 min 56 s. This means that only variations of the wind speed measured in 4 km distance with periods of above roughly 8 min are also captured in a distance close to the lidar. Smaller fluctuations in the wind speed are lost along the way. The reason for this can be found in the complexity of the terrain at the measuring site as seen in Figure 2. It appears that the variations in surface height as well as roughness have a considerable influence on the wind conditions and lead to significant changes in wind speed. Consequently, the forecast horizon is also limited to 8 minutes as changes in the wind speed with shorter periods cannot be expected to be forecast with accuracy. However, it should be noted that the evolution of the wind depends on atmospheric and environmental conditions forecast horizons of less or more than 8 minutes can well be possible.

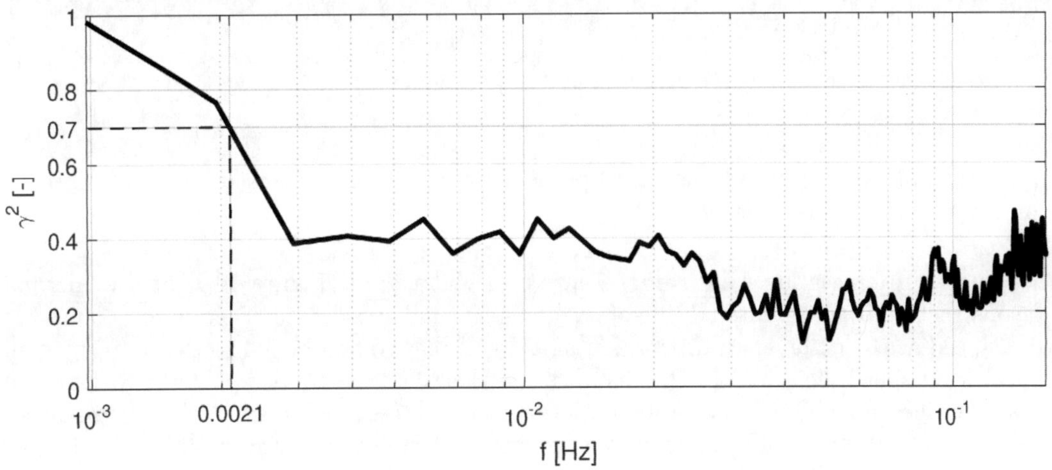

Figure 8: Averaged coherence of measurements from 4 km distance and 0.2 km in front of the lidar, derived from 13 measurement blocks of 60 min.

7. Forecasting of wind ramp events

Wind ramps are defined as a quick and strong increase or decrease of wind power generation of a wind turbine or wind farm. The increase is referred to as an upward ramp and the decrease as a downward ramp. There is no standard mathematical definition of the duration ΔT or the change in Power ΔP of a wind ramp event [10][4]. Instead, the definition usually depends on the location as well as the size of the considered wind turbine or wind park.

For this investigation, the turbine data of the reference turbine in the vicinity of the lidar measurement location is analyzed for a period from July 2016 to August 2017. The only turbine data that was available were 10-minute averaged blocks (referred to herein as 10-min data). By applying the minimum-maximum method described in [36] to the power data, several dozen upward and downward ramps are detected for a ramp duration of $\Delta t = 60\,\text{min}$ and a change in power of $\Delta P = 40\,\%$, compare Table 2. Only data from the wind direction sector compliant with the measurement direction is taken into account. An exemplary time line of generated power with marked wind ramps is given in Figure 9.

Δt [min]	ΔP [%]	Up-Ramps	Down-Ramps
60	40	35	22

Table 2: Ramp findings in wind turbine data for July 2016 - August 2017 after wind direction filtering.

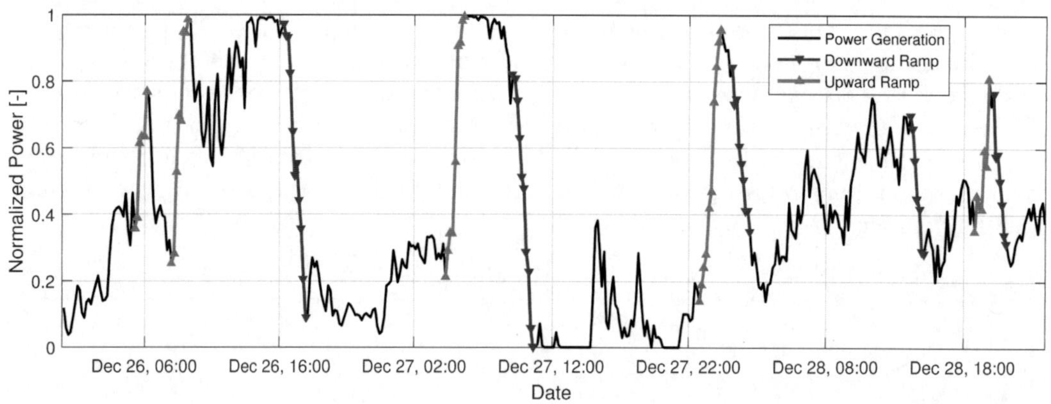

Figure 9: Exemplary time line of generated power with wind ramps marked for a time window of 60 min and a change of power of 40 %.

The process of forecasting wind ramp events using lidar data (Figure 10) starts with wind field reconstruction from the filtered line-of-sight (LOS) wind speed data (Section 5). To predict the wind speed at the turbine, a propagation model then has to be applied to the reconstructed horizontal wind speed measured in the distance. In this analysis, the simple assumption from Taylor's Hypothesis is applied. It claims that turbulent eddies are transported with the mean flow and do not change their properties but remain unchanged, or frozen [11]. Using this assumption, the time shift t_{shift} for each measured range gate i is calculated with the relationship $t_{shift,i} = \overline{x_i}/\overline{u_i}$ where $\overline{x_i}$ is the averaged distance to the range gate and $\overline{u_i}$ is the averaged horizontal wind speed. The time shift is calculated for every 10-min of measurement data. By applying the time shift of each range gate to the respective time series of horizontal wind speed data and then averaging, the forecast of the wind speed at the turbine location is achieved. It should be noted that for this analysis the forecast is carried out in 10-min blocks due to the fact that the turbine data was only available in these intervals. This means that lidar data of the last 10 minutes is used to forecast the next 10, and 10-20 minutes of wind speed.

In a last step the power curve of the wind turbine is used to forecast its power output. The power curve is derived from 10-min power and nacelle anemometer data and gives the relationship between the wind speed and the generated power. The power is forecast as a time series using the predicted wind speed and is then averaged over 10 minutes to compare it with the actual power data.

To evaluate the forecast, the root-mean-square error (RMSE) of predicted and actual power are calculated (Figure 11) for two forecasting horizons 0-10 min and 10-20 min. The lidar-based

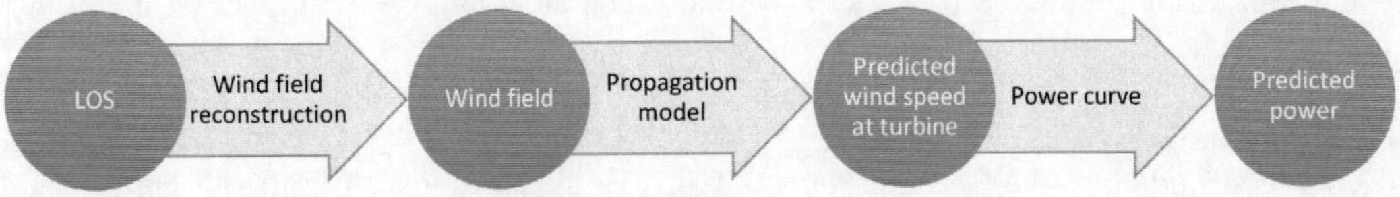

Figure 10: Forecast Process

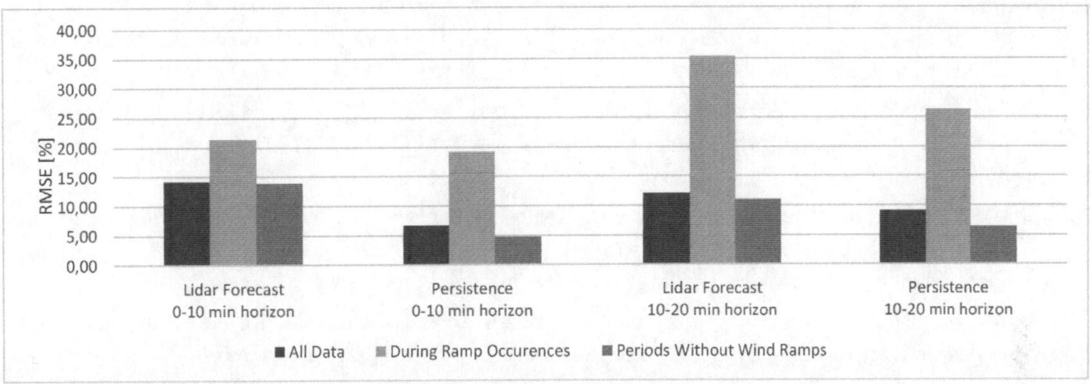

Figure 11: Comparison of errors of lidar-based and persistence forecasting for different time horizons. The error values are normalized with rated power.

forecast is compared to the persistence method, which uses the present power output value for the forecast. It is also differentiated between periods with and without ramp events. The results show that for the present set-up the lidar-based forecasts results in a higher error than persistence for both forecasting horizons. During ramp events, the error is higher for both methods but it must be noted that only few data is available. The error of the lidar-based forecast is reduced with increasing forecasting horizon (not during ramp events), whereas the persistence error increases as expected.

The reason for the high error of the lidar-based forecast is the complex terrain at the measuring site and the propagation model that was applied. As shown in Section 6, the wind speed measured in the distance evolves due to the terrain and is not very well correlated to the wind speed that reaches the turbine. The simple propagation model of Taylor does not account for those changes. Consequently an error is induced in the forecasting process when predicting the wind speed at the turbine. This error is then magnified by the cubic nature of the power curve. The use of realistic flow models instead of simple propagation could be a solution.

8. Conclusion and outlook

This paper evaluates the use of a scanning lidar for forecasting the power output of a wind turbine on a minute-scale horizon at a complex site. Results from the lidar-based forecast are generally worse than persistence in this complex flow, and contrast with promising results from other investigations in coastal regions [12]. This is most likely due to the difficulty of propagating wind speed over long distances in such conditions. During wind ramps, local roughness and buoyancy effects dominate compared to larger scale phenomena. This reduces correlation distances which in turn reduces the accuracy of the lidar-based forecast.

Short term forecasting remains extremely important for integrating wind farms into the electrical grid. There is scope for improving the results in complex terrain through a more realistic propagation model, such as meso- to microscale flow models. This will be investigated in future projects.

Acknowledgments
This work was funded by the German Federal Ministry for Economic Affairs and Energy under code number 0325740B. We thank our VORKAST project partner Center for Solar Energy and Hydrogen Research Baden-Württemberg (ZSW) for the excellent cooperation.

References

[1] 2009 The energy of the future: Fourth energy transition monitoring report summary Tech. rep. Federal Ministry for Economic Affairs and Energy (BMWi)

[2] Marquis M, Wilczak J, Ahlstrom M, Sharp J, Stern A, Smith J C and Calvert S Forecasting the wind to reach significant penetration levels of wind energy 2011 *Bulletin of the American Meteorological Society*

[3] Giebel G, , Badger J, a I M P, Louka P, Kallos G, Palomares A M, Lac C and Descombes G Short-term forecasting using advanced physical modelling the results of the anemos project. 2006 *European Wind Energy Conference, Athens, Greece*

[4] Gallego-Castillo C, Cuerva-Tejero A and Lopez-Garcia O A review on the recent history of wind power ramp forecasting 2015 *Renewable and Sustainable Energy Reviews*

[5] Pearson G, Davies F and Collier C An analysis of the performance of the ufam pulsed doppler lidar for observing the boundary layer 2009 *Journal of Atmospheric and Oceanic Technology* **26** 240–250

[6] HALO Photonics Ltd. Worcestershire, UK 2015 *HALO Photonics Stream Line XR Scanning Doppler LiDAR system - Manual*

[7] United States Geological Survey (USGS) 2000 (College Park, Maryland: Global Land Cover Facility, University of Maryland) processed Version 2.1

[8] Grubbs F E Procedures for detectiong outlying observations in samples 1969 *Defense Technical Information Center*

[9] Schlipf D, Rettenmeier A, Haizmann F, Hofsaeß M, Courtney M and Cheng P W Model based wind vector field reconstruction from lidar data 2012 *German Wind Energy Conference DEWEK 2012, Bremen, Germany*

[10] Kamath C 2010 *Proceedings of the IEEE PES Transmission and Distribution Conference and Expo*

[11] Taylor G I The spectrum of turbulence 1938 *Proceedings of the Royal Society of London*

[12] Valldecabres L, Peña A, Courtney M, von Bremen L and Kühn M Very short-term forecast of near-coastal flow using scanning lidars 2018 *Wind Energy Science*

Systems Engineering for Lidar-Assisted Control: A Sequential Approach

David Schlipf, Holger Fürst, Steffen Raach, Florian Haizmann
sowento, Stuttgart, Germany.
E-mail: Schlipf/Fuerst/Raach/Haizmann@sowento.com

Abstract. This work presents a sequential approach to explore and optimize the benefits of lidar-assisted control for wind turbines. The optimization is divided in three steps: lidar hardware, lidar data processing, and feedback controller optimization. Appropriate optimization criteria and computational efficient models are used for the intermediate steps and energy production is optimized in the last step with a full aero-elastic model to provide an estimation of lidar-assisted control without the need of a detailed cost model. The case study shows that lidar-assisted control together with an adjustment of the power level are promising to extend the life-time of wind turbines and finally increase the energy capture.

1. Introduction

Lidar-Assisted control (LAC) is a promising technology to improve the control performance of wind turbines [1]. For the industrial use, the benefits in control performance need to be translated into economic benefits. This is a complex task due to the multi-disciplinary character of the problem involving meteorology, remote sensing technology, signal processing, controls and mechanics. In general, there are two scenarios for the application of LAC. First, the retrofit scenario, where a lidar system is installed on an already designed and installed turbine. Second, the integrated design scenario, where a lidar system is fully integrated into the systems' design process of a wind turbine. In both scenarios, there are a large number of parameters impacting the economics. Here, the systems engineering initiative [2] provides methods and tools. Systems engineering has been already successfully applied to wind turbines [3] and wind farms [4]. Basically, the design process is considered to be a large optimization problem: the free parameters are used as optimization variables, constraints are defined in mathematical expressions, and economic considerations are translated into a cost function. The optimization problem is then solved automated by appropriate algorithms to find the best set of parameters. The crucial point here is to define appropriate interfaces for tools and to set up the optimization problem such that the algorithms are able to find the solution computationally efficiently.

This paper presents an initial application of systems engineering for lidar-assisted control to the retrofit scenario. It outlines how different tools can be combined to divide the process into three sub-problems which can be solved sequentially. In Section 2 the approach is detailed and in Section 3 the simulation scenario is described. Section 4 presents the results and Section 5 the conclusions.

IOP Conf. Series: Journal of Physics: Conf. Series **1102** (2018) 012014 doi:10.1088/1742-6596/1102/1/012014

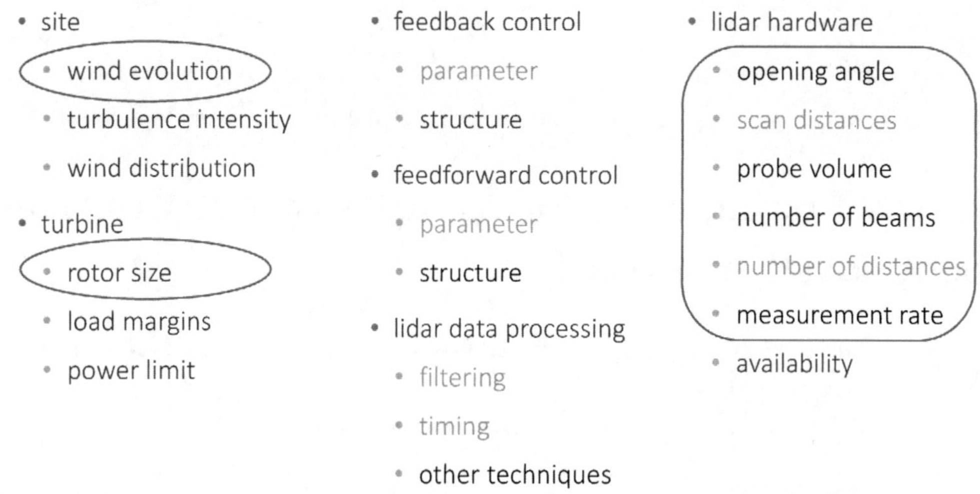

Figure 1: Cartography of parameters impacting the benefit of lidar-assisted control. Green: Free parameters in considered scenario. Blue: Parameters impacting the measurement coherence.

2. A Sequential Approach to Systems Engineering for Lidar-Assisted Control

There are several parameter which impact the benefit of LAC in a systems engineering study. They can be clustered in parameters specific for the site, turbine, feedback and feedforward control, lidar data processing, and lidar data hardware. The most important ones are listed in Figure 1. The free parameters in this study are written in green. Eventually, the main goal of LAC is the maximization of the economic benefit within all possible constraints (mostly turbine and lidar hardware). In a monolithic approach for systems engineering, the optimal set of the free parameters can be determined by solving a very large constrained optimization problem maximizing the economic benefit, see Figure 2 (left). The advantage of such an approach is that all possible interdependencies of parameters are considered. However, a monolithic approach is computationally expensive and since it is a complex problem, special care is necessary to avoid the optimization algorithm to end up in a local minimum.

In this work, a sequential approach is presented, Figure 2 (right): In a first step, the lidar hardware is optimized to provide the best measurement coherence for the rotor-effective wind speed [5, 6, 7]. This decoupling is reasonable, since a lidar hardware providing the best correlated rotor-effective wind speed is optimal for all LAC setups rejecting this disturbance. Here, the measurement coherence bandwidth, the wavenumber $k_{0.5}$ at which the coherence drops to 0.5 is maximized. When assuming isotropy, $k_{0.5}$ can be considered to be inverse proportional to the "smallest detectable eddy size" $d_{\text{eddy,min}}$, see Figure 3:

$$d_{\text{eddy,min}} = \frac{2\pi}{k_{0.5}}. \tag{1}$$

In this case, $d_{\text{eddy,min}}$ is the size of the eddies which can be still detected with a 50 % level of the coherence. Normalizing $d_{\text{eddy,min}}$ with the rotor size allows to have a measure independent of the rotor size. Detecting smaller structures than $1\,D$ is not important for the proposed feedforward control, since collective blade pitch and generator torque impact the whole rotor at once. Usually a value of $d_{\text{eddy,min}}$ close to $1\,D$ indicates a very good lidar for fatigue load reduction. Improvements above this point are still possible, since in this case structures of $1\,D$ size are captured by 50 % only.

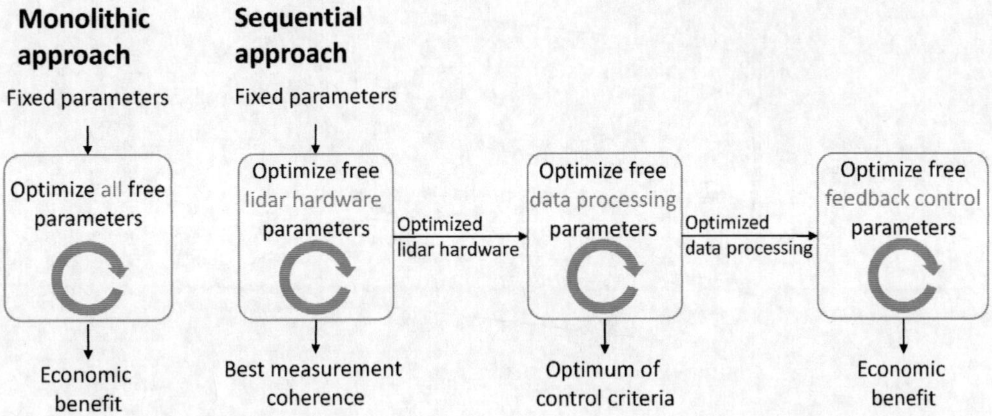

Figure 2: Monolithic (left) and sequential approach (right) to lidar-assisted control optimization.

Figure 3: "Smallest detectable eddy size": Turbulence structure still detectable by a lidar with a 50 % level of the coherence.

In a second step, the lidar data processing is optimized to make best use of the information content within the signal. Again, this decoupling is reasonable, since all LAC setups using the rotor-effective wind speed will benefit from an optimal lidar data processing. Minimum lifetime weighted tower loads are used as a criterion. The lidar data processing consists of an adaptive filter used in several field testing campaigns, which aims to filter the rotor-effective wind speed estimate based on current correlation changing with the mean wind speed and to provide the signal with a fixed preview time to the controller. Simulation can be performed with a reduced simulation model to have less computational effort, since an accuracy of the load reduction is less important for this step.

In a third and final step, the feedback control parameters are optimized to maximize the economic benefit. In the considered retrofit scenario, LAC is used for life-time-extension. In this study we focus on balancing the load reduction over tower, blades and low-speed shaft. With detailed information of the load envelops and a detailed cost model of all components, the structure could be also optimized in this step, but is out of the scope of this work. Since a accurate estimate of the load reduction is important in this step, a full aero-elastic simulation model is used.

In the following section, the simulation scenario is detailed.

IOP Conf. Series: Journal of Physics: Conf. Series **1102** (2018) 012014 doi:10.1088/1742-6596/1102/1/012014

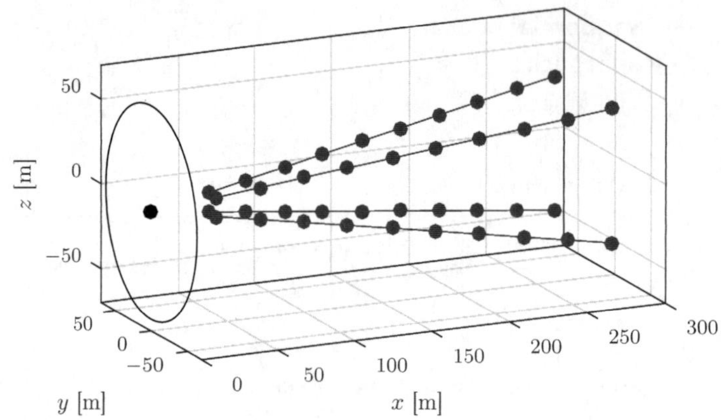

Figure 4: Used pulsed lidar scan configuration (blue). Location of lidar system and rotor (black).

3. Description of Simulation Scenario

In this section, the simulation scenario is briefly described.

3.1. Wind Turbine Model

In this simulation study, the NREL 5 MW reference onshore wind turbine with a rotor diameter of $D = 126\,\mathrm{m}$ implemented in FAST [8] and SLOW [9] is used. The FAST sFunction is integrated in MATLAB Simulink and is extended by a collective pitch actuator implemented as linear second order model with a damping factor of 0.7 and a undamped natural frequency of 1 Hz. No further changes have been done to the FAST model. The advantage of the SLOW model compared to the FAST model is that simulations are much faster (up to 1000×).

3.2. Wind Disturbance and Lidar Simulator

For this study, a 4-beam pulsed nacelle lidar is simulated, see Figure 4. The opening angle is fixed to $\arctan(0.5/2.5) = 11.3\,\mathrm{deg}$, such that the lidar can be additionally used for power performance testing (at 2.5 rotor diameter, the measurements are on a circle of the size of the rotor). The minimum distance is limited to 40 m and the measurement rate is fixed to 4 Hz. For this study, 10 measurement distances and a Gaussian volume weighting function are considered.

In this study we focus on normal operations and thus a Design Load Case (DLC) 1.2 following current standard [10] is carried out, using eleven one-hour turbulent wind fields from 4 m/s to 24 m/s based on class IA, see Appendix C.3 of [9].

Further, a tool-independent lidar simulator is used [11]. In a first step, the tool calculates the auto-spectra and cross-spectra of the rotor-effective wind speed and its lidar-estimate based on [12]. The following effects are considered: the limitation to line-of-sight wind speeds, the spatial averaging of the range weighting function, the discrete scanning, wind evolution (using a longitudinal decay of $\alpha = 0.1$ for an exponential wind evolution model [13]), and wind field reconstruction (all line-of-sight wind fields are condensed to one estimate of the rotor-effective wind speed using a dynamic wind field reconstruction method [9]). In a second step, the wind fields are reduced to the rotor-effective wind speed. Further, the corresponding lidar-estimate of the rotor-effective wind speed is calculated with the auto-spectra and cross-spectra. The preview of the calculated signal corresponds to the measurement distances and the mean wind speed. The advantage of this method is that it can be combined with the trajectory optimization, which already calculates the auto-spectra and cross-spectra of the first step, and with both simulation tools.

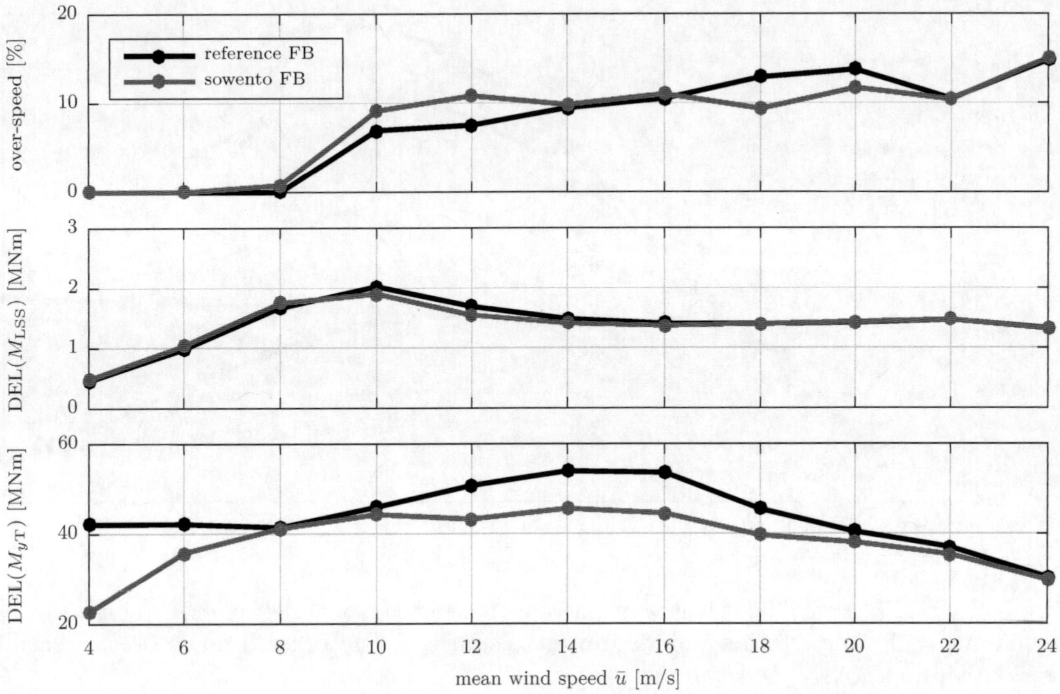

Figure 5: Comparison of life-time weighted results for feedback controllers from FAST simulations. Over-speed (top), shaft loads (center), tower loads (bottom). Reference controller (black) and used controller (blue). Each point is based on a one-hour simulation.

3.3. Baseline Feedback, Feedforward Controller and Lidar Data Processing

For the baseline controller, a modified version of the NREL 5 MW reference controller [8] is used. The modifications are:

- The reference torque controller is replaced by a PI torque controller. In region 1.5 the rotor speed set-point is 8 rpm. In region 2, the limits of the PI torque controller are adjusted such that it follows the reference torque controller. In region 2.5 the rotor speed set-point is rated rotor speed. In region 3, the upper limit aims for constant power.
- The set-point of the pitch controller in region 2.5 and of the torque controller in region 3 is adjusted by a set-point fading based on [14] providing a smooth transition.

The advantage for this study is that the sowento FB controller is much closer to industrial standards, allows straight-forward adjustment of the rated power, and enables less loads (see Figure 5): the life-time-weighted Damage Equivalent Loads (DEL) at the tower base are reduced by 12.5 %, at the blade root by 5.0 %, and at the low-speed-shaft by 2.5 %, while the maximum over-speed is at a comparable level (15 % over all wind speeds).

For the LAC, the collective pitch feedforward controller combined with an adaptive filter [9] is used. The adaptive filter is a first order low pass filter which changes the cut-of-frequency f_{cutoff} with the mean wind speed $\bar{u}$ depending on the maximum coherent wave number k_{max}. Further, the filtered signal is buffered such that it is provided to the feedforward controller with a fixed prediction time τ for all wind speeds.

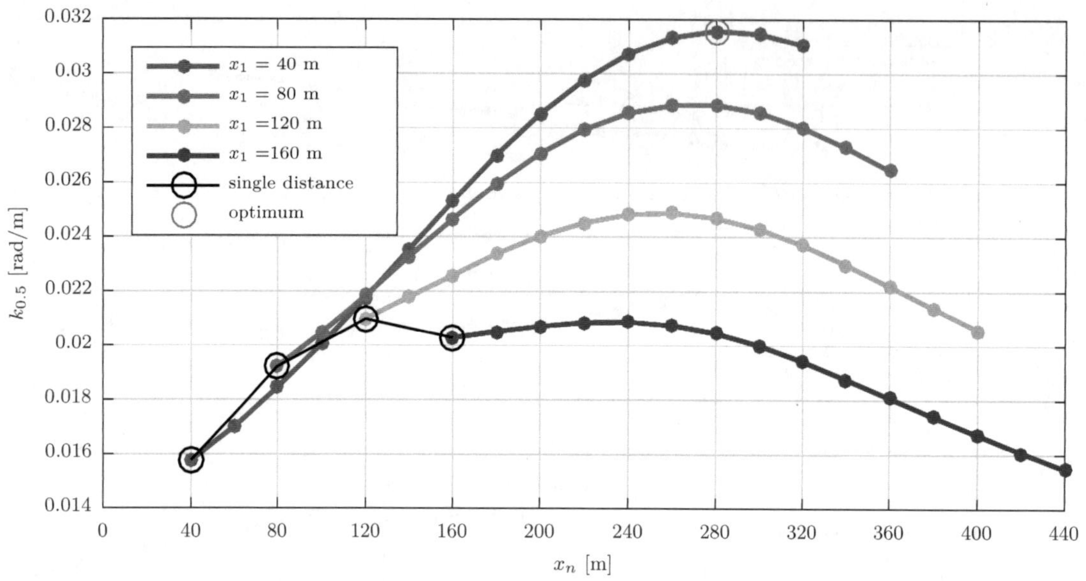

Figure 6: Optimization of lidar hardware based on coherence model: left points (in black circles) are minimum distances, other points are last distances. The other 8 distances are equally distributed, respectively. Optimum in red circle.

4. Results

For simplicity, better illustration and to avoid local minima, brute force optimization is used for all optimization steps described in the following subsections.

4.1. Optimization of Lidar Hardware

In a first step, the lidar scan configuration is optimized using a frequency based correlation model [12]. Here, the measurement coherence bandwidth is calculated for several scan configurations and the one with the highest value is considered to be optimal. Several parameters can in theory be changed. Here we focus on the location of the 10 measurement ranges, where the first one is fixed to 40 m, the last one is the free parameter of the optimization, and the rest of the measurement distances is equally distributed between the first and the last one.

Figure 6 illustrates the results: For this turbine, site, and lidar constraints, the best scan from 40 m to 280 m reaches a $k_{0.5} = 0.0316\,\text{rad/m}$, corresponding to a smallest detectable eddy size of $d_{\text{eddy,min}} = 1.58D$. At a certain distance, no further improvement can be gained, since the effect of the wind evolution is increasing with distance and after 2.5 D=315 m the measurement will be outside of the rotor disc and thus less representative.

In this case combining multiple distances (here 10) with dynamic wind field reconstruction leads to better results compared to using a single distance alone. This cannot be generalized. An individual investigation for each lidar system is necessary. However, multiple distances have additional advantages, for example for extreme event detection, since large changes can be detected earlier by distances further away, and for differentiation of wind shears and wind direction.

Additional parameters such as the scan angles, the number of beams, scan rate, measurement volume etc. can be added to the optimization, but this is out of the scope of this work. More details about optimizing lidar systems for wind turbine control applications can be found in [7].

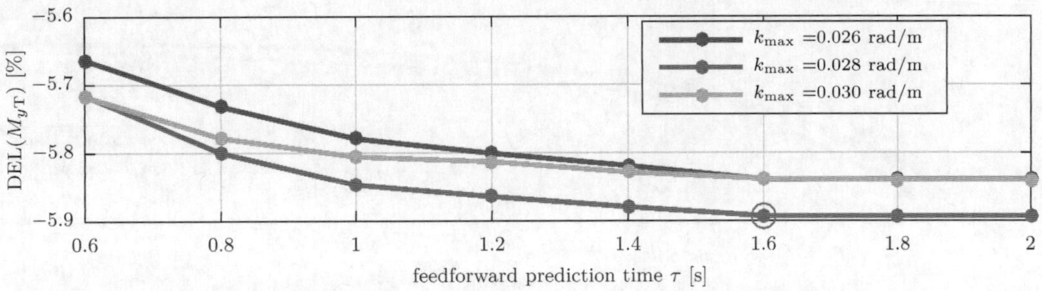

Figure 7: Optimization of lidar data processing for different prediction times τ and wave numbers $k_{\max}$ using the SLOW model: Criterion is reduction of life-time weighted tower loads by LAC compared to feedback only. Optimum is in red circle. Each point is based on a full DLC 1.2.

4.2. Optimization of Lidar Data Processing

In a second step, the optimal lidar hardware is used and the lidar data processing is optimized using DLC 1.2 simulations with the combined feedback-feedforward controller. Here, the SLOW model is used, since a full DLC 1.2 can be run in 10 s on a common PC. The free parameter are the feedforward prediction time and the maximum coherent wave number. The optimization criterion is the life-time weighted DEL of the tower base fore-aft bending moment $M_{y\mathrm{T}}$. The results are displayed in Figure 7. The optimal values are in this case a prediction time τ of 1.6 s and maximum coherent wave number $k_{\max}$ of 0.028 rad/m.

4.3. Optimization of Combined Feedback-Feedforward Controller

Finally, the combined feedback-feedforward controller is optimized. Here, the FAST model, the optimal lidar scan configuration, and the optimal lidar data processing are used. LAC reduces significantly the tower and blade loads [9] but usually does not affect much the shaft loads. Increasing the rated power is not feasible in the retrofit scenario, if the margins of these loads or the power limit are already reached. However, LAC can be combined with de-rating the turbine for life-time-extension (LTE). The main optimization goal of this step is to increase the energy production by balancing the loss by de-rating and the gain by LAC.

The LTE potential for each component i based on the load reduction of the LAC compared to the baseline controller (FB) can be calculated by [15, 16]:

$$\mathrm{LTE}(i) = \left(\frac{\mathrm{DEL}(i)_{\mathrm{LAC}}}{\mathrm{DEL}(i)_{\mathrm{FB}}}\right)^{-m}, \tag{2}$$

where $m = 4$ is the Wöhler exponent for steel components (tower and shaft) and $m = 10$ for composite components (blades). Here, the tower base fore-aft bending moment $M_{y\mathrm{T}}$, the blade root out-of-plane bending moment M_{oop1} of blade 1, and the low-speed shaft torque M_{LSS} are considered. The overall LTE potential is then given by

$$\mathrm{LTE} = \min\left\{\mathrm{LTE}(M_{y\mathrm{T}}), \mathrm{LTE}(M_{\mathrm{oop1}}), \mathrm{LTE}(M_{\mathrm{LSS}})\right\}. \tag{3}$$

Finally, the increase in energy production $\Delta\mathrm{EP}$ can be calculated by

$$\Delta\mathrm{EP} = \mathrm{AEP}_{\mathrm{LAC}} \times \mathrm{LTE} - \mathrm{AEP}_{\mathrm{FB}}, \tag{4}$$

where $\mathrm{AEP}_{\mathrm{FB}}$ and $\mathrm{AEP}_{\mathrm{LAC}}$ are the Annual Energy Production of the feedback controller and the LAC, respectively.

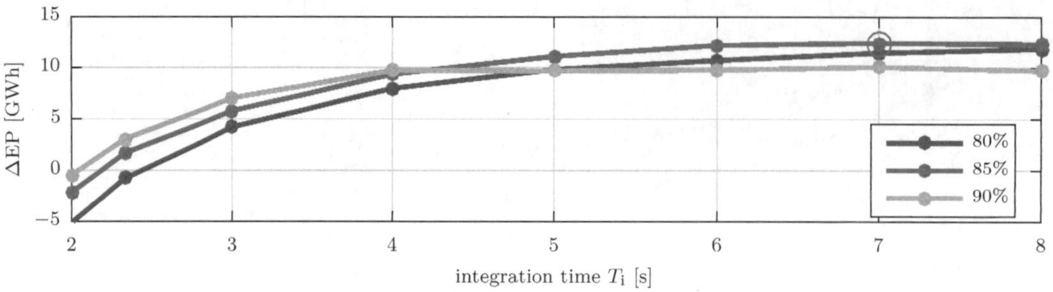

Figure 8: Optimization of de-rated feedback controller with LAC for different integration times T_i and de-rating levels: Criterion is increase of energy production. Each point is based on a full DLC 1.2. Optimum is in red circle.

The optimization of ΔEP is done by changing two parameters: the integration time T_i of the de-rated feedback controller (original value 2.33 s) and the percentage of de-rating. The gain scheduling parameter $\theta_k = 6.3$ deg and the proportional gain $k_p = 0.0188$ rad/(rad/s) are kept constant in this study for simplicity. Having the same proportional gain for feedback only operation and feedback with LAC operation is advantageous in a real application, since it allows to switch to the re-tuned controller without a step in the demanded pitch angle.

The results are displayed in Figure 8. The optimum is found for a de-rating level of 85 % and an integration time of 7 s, where the life-time weighted DEL of the tower fore-aft bending moment and the low-speed shaft torque are reduced by 11.8 % and 11.5 %, respectively (see Figure 9 and 10). The maximum over-speed can be kept below the 12 % of the baseline controller. The life-time weighted DEL for the blade root bending moment are reduced by 6.4 % (not displayed).

With Equation (2), the life-time can be increased for the tower by 65.1 %, for the blades by 93.7 %, and for the low-speed shaft by 63.1 %. An overall increase in life-time of 63.1 % (LTE = 1.631) for the entire wind turbine can be assumed with Equation 3: Instead of one year of feedback only control, the turbine can be operated 1.631 years with additional LAC. Due to the de-rating of 85 %, the AEP is first reduced from $\mathrm{AEP}_{\mathrm{FB}} = 25.8$ GWh to $\mathrm{AEP}_{\mathrm{LAC}} = 24.3$ GWh (-9.17 %). Due to the longer operation time based on the load reduction however, the increase in energy production ΔEP is 12.4 GWh using Equation (4).

Assuming roughly 40 EUR to be reasonable gain per MWh (income minus additional operation and maintenance cost), the economic benefit is 496.000 EUR minus the investment of a lidar and additional costs for the certification and control adjustments. If the LAC is used for several years, the benefits will increase with the same investment.

Since in Equation (4) several assumption are made, this number should be considered as an indication. Further, the following effects have not been included in the simulations and will impact the results in a real application:

- Data availability: If the lidar signal is not available, the benefits are not accessible. However, the effect of data availability on the benefits can be assumed to be linear, if the loss of the signal is detected and the baseline control activated.
- Interactions of blade blockage: Since nacelle-based lidar system are mounted behind the rotor, some measurements are blocked by the blades. The effect is usually small, if the measurement rate is high enough.
- Terrain complexity: In this study, no specific terrain or inflow angle has been considered.
- The fatigue loads during normal operation are often not the design driver. Reducing the fatigue loads only might not necessarily lead to a life-time extension.

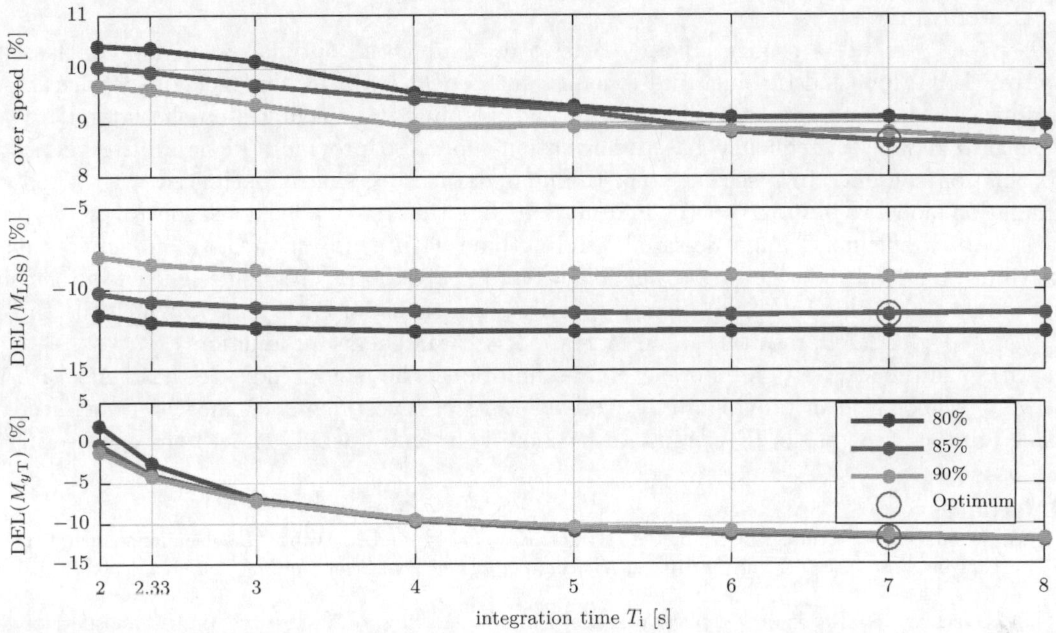

Figure 9: Details of optimization of de-rated feedback controller: Over-speed (top), reduction in shaft loads (center) and tower loads (bottom) by feedback with LAC compared to feedback only. Each point is a based on a full DLC 1.2. Optimum is in red circle and detailed in Figure 10.

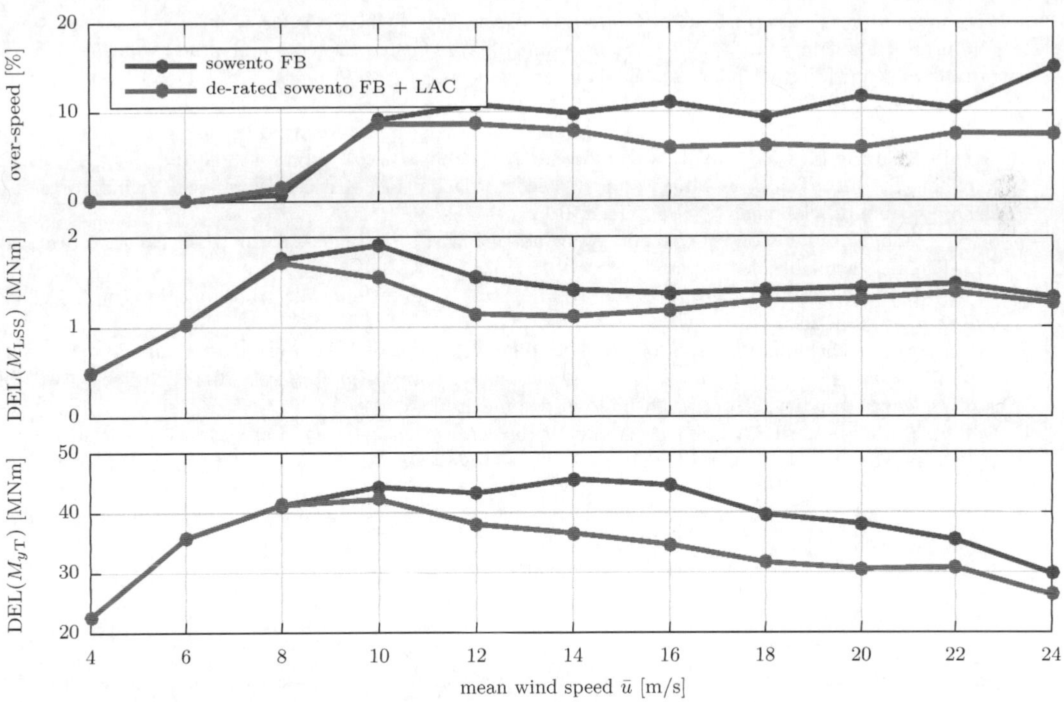

Figure 10: Comparison of life-time weighted results from FAST simulations. Over-speed (top), shaft loads (center), tower loads (bottom). Feedback only (blue) and de-rated feedback with LAC (red, optimal case of Figure 9). Each point is based on a one-hour simulation.

5. Conclusions

This paper presents a sequential approach to determine the optimal setup for lidar-assisted control (LAC) for wind turbines and to assess the benefit by using this technology. Brute force optimization is done for better illustration and simplicity. In a first step the lidar hardware is optimized with a frequency based correlation model to provide a signal with the highest information content. In a second step, the lidar data processing is optimized with a reduced simulation model to best extract the information. In a third step, a feedback controller combined with LAC is optimized for a de-rated wind turbine with a full aero-elastic model to equally distribute the load reduction on tower, blades and shaft. For a fatigue-designed wind turbine, this would yield a life-time extension of roughly 7.6 months for each year, in which the turbine is operated with LAC, resulting in an increase of 48 % in energy production.

Although the study is based on several assumptions, the results indicate that LAC with the current technology can provide an attractive return of investments. A more detailed study is planed in the scope of the IEA Wind Task 32 on lidar[1] and Task 37 on systems engineering[2].

References

[1] A. Scholbrock, P. Fleming, D. Schlipf, A. Wright, K. Johnson, and N. Wang, "Lidar-enhanced wind turbine control: Past, present, and future," in *Proceedings of the American Control Conference*, July 2016, pp. 1399–1406.

[2] K. Dykes and R. Meadows, "Applications of systems engineering to the research, design, and development of wind energy systems," NREL, Tech. Rep. NREL/TP-5000-52616, 2011.

[3] C. L. Bottasso, F. Campagnolo, and A. Croce, "Multi-disciplinary constrained optimization of wind turbines," *Multibody System Dynamics*, vol. 27, no. 1, pp. 21–53, Jan 2012.

[4] R. N. King, K. Dykes, P. Graf, and P. E. Hamlington, "Optimization of wind plant layouts using an adjoint approach," *Wind Energy Science*, vol. 2, no. 1, pp. 115–131, 2017.

[5] E. Simley and L. Y. Pao, "Reducing lidar wind speed measurement error with optimal filtering," in *Proceedings of the American Control Conference*, Washington, DC, USA, 2013.

[6] D. Schlipf, F. Haizmann, N. Cosack, T. Siebers, and P. W. Cheng, "Detection of wind evolution and lidar trajectory optimization for lidar-assisted wind turbine control," *Meteorologische Zeitschrift*, vol. 24, no. 6, pp. 565–579, 11 2015.

[7] E. Simley, H. Fürst, F. Haizmann, and D. Schlipf, "Optimizing lidars for wind turbine control applications – results from the iea wind task 32 workshop," *Remote Sensing*, vol. 10, no. 863, 2018.

[8] J. Jonkman, S. Butterfield, W. Musial, and G. Scott, "Definition of a 5-MW reference wind turbine for offshore system development," NREL, Tech. Rep. TP-500-38060, 2009.

[9] D. Schlipf, "Lidar-assisted control concepts for wind turbines," Ph.D. dissertation, University of Stuttgart, 2015. [Online]. Available: http://dx.doi.org/10.18419/opus-8796

[10] IEC 61400-1, *Wind turbines - Part 1: Design requirements*, International Electrotechnical Commission Std., Rev. third edition, 2005.

[11] S. Raach and D. Schlipf, "Cross-tool realistic lidar simulations," in *Presentation at the IEA Wind Task 32 workshop on Certification of Lidar-Assisted Control Applications*, 2018. [Online]. Available: https://www.ieawindtask32.org/ws08-internal-documents/

[12] D. Schlipf, J. Mann, and P. W. Cheng, "Model of the correlation between lidar systems and wind turbines for lidar assisted control," *Journal of Atmospheric and Oceanic Technology*, vol. 30, no. 10, pp. 2233–2240, 2013.

[13] R. A. Pielke and H. A. Panofsky, "Turbulence characteristics along several towers," *Boundary-Layer Meteorology*, vol. 1, pp. 115–130, 1970.

[14] E. A. Bossanyi, "The design of closed loop controllers for wind turbines," *Wind Energy*, vol. 3, no. 3, pp. 149–163, 2000.

[15] W. Engels, A. Marina, S. Kanev, and D. v. d. Hoek, "Condition based control," ECN Wind Energy, Tech. Rep. ECN-E–17-047, 2017.

[16] L. Ziegler, U. Smolka, N. Cosack, and M. Muskulus, "Brief communication: Structural monitoring for lifetime extension of offshore wind monopiles: can strain measurements at one level tell us everything?" *Wind Energy Science*, vol. 2, no. 2, pp. 469–476, 2017.

[1] www.ieawindtask32.org

[2] www.windbench.net/iea37

IOP Conf. Series: Journal of Physics: Conf. Series **1102** (2018) 012015 doi:10.1088/1742-6596/1102/1/012015

NAUTILUS-DTU10 MW Floating Offshore Wind Turbine at Gulf of Maine: Public numerical models of an actively ballasted semisubmersible

J Galván[1], M J Sánchez-Lara[1,2], I Mendikoa[1], G Pérez-Morán[1], V Nava[1,3] and R Rodríguez-Arias[1,2]

[1] Tecnalia Research & Innovation, Energy division, Offshore Renewable Energy area; Technology park of Bizkaia, Build. 700, Derio 48160, Basque Country, Spain
[2] Nautilus Floating Solutions S.L.; Technology park of Bizkaia, Build. 612, Derio 48160, Basque Country, Spain
[3] BCAM - Basque Centre for Applied Mathematics; Alameda Mazarredo 14, Bilbo 48009, Basque Country, Spain

josean.galvan@tecnalia.com

Abstract. This study presents two numerical multiphysics models of the *NAUTILUS-10* floating support structure mounting the *DTU10 MW Reference Wind Turbine* at Gulf of Maine site, and analyses its dynamics. With the site conditions and the FAST model of the onshore turbine as the starting point, the floating support structure: tower, floating substructure with its corresponding active ballast system and station keeping system, was designed by NAUTILUS. The numerical models were developed and the onshore *DTU wind energy controller* was tuned to avoid the resonance of the operating FOWT by TECNALIA, in the framework of H2020 LIFES50+ project. This concept and its subsystems are fully characterised throughout this paper and implemented in opensource code, FAST v8.16. Here, the mooring dynamics are solved using MoorDyn, and the hydrodynamic properties are computed using HydroDyn.Viscous effects, not captured by radiation-diffraction theory, are modelled using two different approaches: (1) through linear and quadratic additional hydrodynamic damping matrices and (2) by means of Morison elements. A set of simulations (such as, decay, wind only and broadband irregular waves tests) were carried out with system identification purposes and to analyse the differences between the two models presented. Then, a set of simulations in stochastic wind and waves were carried out to characterise the global response of the FOWT.

1. Introduction

The study summarises the state-of-the-art multiphysics numerical models of *NAUTILUS-DTU10* MW Floating Offshore Wind Turbine (FOWT) designed for Gulf of Maine (GoM) site, developed in the framework of the H2020 LIFES50+ project. A public repository [1] contains detailed description of these numerical models and the designer's technical report [2].

The FOWT is composed by the Rotor-Nacelle Assembly (RNA) of the *DTU 10 MW Reference Wind Turbine* (RWT) [3] and *NAUTILUS-10* floating support structure. The main innovation of NAUTILUS FOWT is the introduction of a Platform Trim System (PTS) which cancels /reduces the wind-induced mean trim angle by means of a variable sea water ballast system. This smart floating

IOP Conf. Series: Journal of Physics: Conf. Series **1102** (2018) 012015 doi:10.1088/1742-6596/1102/1/012015

structure is complemented with a tuned version of the onshore *DTU wind energy controller* [4] acting on the wind turbine.

Two numerical models are described, both implemented in FAST code [5], [6], using 8.16 version. While both numerical models have been developed to account for the active ballast dependent properties, the difference between these two models relays on the hydrodynamic modelling.

2. *NAUTILUS-DTU10* MW FOWT description

NAUTILUS-DTU10 MW FOWT is composed by a 10 MW Horizontal Axis Wind Turbine (HAWT): *DTU 10 MW RWT* and the NAUTILUS support structure, designed for GoM site characteristics [7]. Inside this paper, the FOWT is divided in two main subsystems: (1) the RNA and (2) the support structure, being the latter composed by the tower, the floating substructure and the Station Keeping System (SKS).

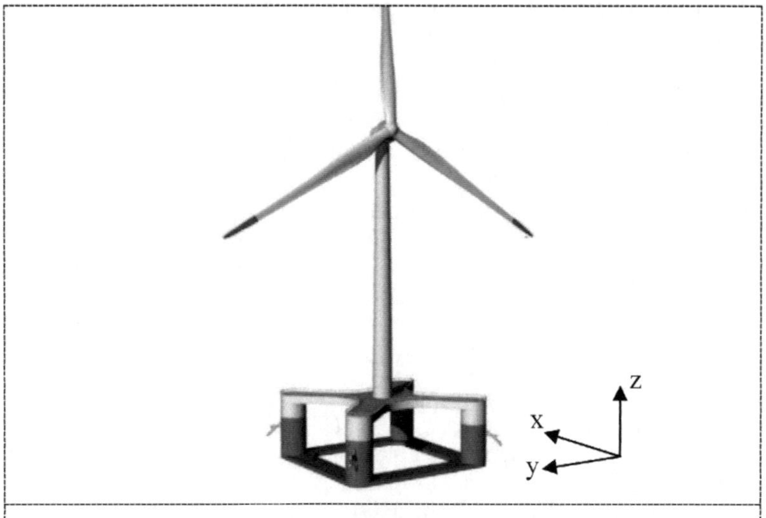

Figure 1. *NAUTILUS-DTU10* MW FOWT concept.

The *DTU 10 MW* RNA is mounted on the *NAUTILUS-10* floating substructure (Figure 1, being *NAUTILUS-DTU10* the complete FOWT system identification). The SKS designed consists of four standard catenary mooring lines anchored to the seabed of GoM. The whole system is controlled to behave smoothly by means of the wind turbine actuators and tuned WT controller, and the PTS to keep the tower vertical by flooding the tanks placed inside the columns with sea water ballast (WB) or pumping it outside the semisubmersible.

Following along this section, the properties of the FOWT and its subsystems/components are presented. This information is the basis for the development of the multiphysics numerical models.

2.1. *FOWT properties*

As it has been briefly mentioned, *NAUTILUS-DTU10* presents variable weight distributions due to the PTS that controls the active ballast to cancel the FOWT heel/trim angle during energy production. The modification of the weights distribution inside the semisubmersible carries on the modification of the weight-dependant properties of the FOWT:

- Mechanical properties: mass, centre of mass (CM) and inertia
- Draft and draft-dependant variables: hydrostatic and hydrodynamic properties; location of the hub, tower base and fairleads.

2.1.1. Mechanical properties. The mechanical properties and displacement (ρ_{SW} = 1,025 kg/m^3) determined for the empty and fully loaded active ballast tanks are shown in Table 1, while the variation of these properties as function of wind speed at hub height and wind different directions can be found in [2].

Table 1. FOWT mechanical properties considering empty and fully loaded water tanks.

FOWT property	Empty WB	Full WB	Units
Mass	8,137,093	9,337,093	[kg]
CM w.r.t. **k** CSys.	(-0.066 , 0 , 20.607)	(-0.058 , 0 , 20.113)	[m]
System roll / pitch / yaw radius of gyration	47.19 / 47.22 / 26.50	45.61 / 45.64 / 28.40	[m]
Operational draft w/ SKS	14.952	18.333	[m]
Displacement w/ SKS	8,113.06	9,280.96	[m^3]

2.1.2. Hydrostatic properties. NAUTILUS-DTU-10 MW FOWT, derived from this semisubmersible geometry, considering fully loaded WB tanks and including the SKS present a metacentric distance from keel (KM) of 34.623 m, while the centre of buoyancy is located 6.497 m above keel. Considering the FOWT CM (Table 1), the distance between CM and the metacentre (M) is then 14.881 m.

The hydrostatic stiffness terms for a floater draft of 18.333 m, horizontally positioned and without accounting for the mass restoring effects are: $C_{33} = 3.505 \cdot 10^6$ N/m, $C_{44} = 1.5373 \cdot 10^9$ N·m/rad and $C_{55} = 1.5513 \cdot 10^9$ N·m/rad.

2.1.3. Hydrodynamic properties. Due to the different approaches employed in the numerical hydrodynamic models presented in section 3, the hydrodynamic characterisation of the FOWT is presented in the mentioned section which describes both models.

2.2. FOWT subsystems/components

2.2.1. DTU 10 MW RNA. The *DTU 10 MW RWT* was designed and analysed inside LigthRotor danish project and INNWIND EU project by Vestas, DTU and the INNWIND Consortium [8]. The RNA of this wind turbine was selected as reference for the design of the floating support structures inside LIFES50+ project [9].

This RNA is designed for a wind regime IEC Class 1A and develops 10 MW of power at nominal above nominal wind speed (11.4 m/s). The three bladed upwind rotor has a diameter of 178.33 m and spins in the range from 6 to 9 rpm. Cut-in and cut-out wind speed are 4 m/s and 25 m/s, respectively. Further information about this RNA can be found in [3], [8], [9].

2.2.2. DTU wind energy controller. This controller was conceived for the *DTU 10 MW RWT* which is an onshore WT. The controller is inspired on previous DTU controllers and the modifications proposed in [10] for the *NREL 5 MW WT* controller inside UpWind EU project.

Table 2. Tuned WT controller parameters in the full load region.

	Value	Units
Proportional gain of pitch controller	0.208004	[rad/(rad/s)]
Integral gain of pitch controller	0.041415	[rad/rad]
Differential gain of pitch controller	0.0	[rad/(rad/s^2)]
Proportional power error gain	$0.4 \cdot 10^{-8}$	[rad/W]
Integral power error gain	$0.4 \cdot 10^{-8}$	[rad/(W·s)]
Linear coeff. in aerodynamic gain scheduling, KK1	5.498310	[deg]
Quadratic coeff. in aerodynamic gain scheduling, KK2	386.005941	[deg^2]
Relative speed for double nonlinear gain	1.3	[-]

The control strategies implemented in the *DTU wind energy controller* are based on the traditional onshore HAWT controllers which consider two main operating regions: (1) partial load region and (2) full load region, where the blade pitch controller acts with the main objective to maintain a constant power or torque, depending on the strategy selected.

The controller setup in the partial load region remains unmodified. In the full load region, by contrast, the controller strategy objective has been switched from a constant power output to a constant torque; and to avoid the negative damping induced motion amplifications, the controller parameters have been tuned (Table 2).

2.2.3. NAUTILUS-10 tower. At earliest stage of the design, the original onshore tower of the *DTU 10 MW RWT* was analysed installed on the moored semisubmersible. As it was expected, that onshore design falled inside the rotor 3P excitation due to the stiffness reduction of the floating substructure compared to rigid ground; therefore *NAUTILUS-10* tower was conceived to dynamically behave as a stiff/stiff tower once installed in the FOWT.

It is a conical single piece tower design of linearly varying thickness. The tower base is connected to the floating substructure by a transition piece at tower-platform interface. This component is embedded into the main deck of the floating platform and thus, it is considered part of the hull. The tower base presents an outer diameter of 10.5 m, meanwhile the tower top conserves the 5.5 m outer diameter to fulfil RNA requirements [3]. Tower base and top thickness are 0.040 m and 0.037 m, respectively. The tower has a total length of 107 m. once the FOWT is installed and considering the fully filled ballast tanks, the tower base is located 7.667 m above MSL. Note that this distance will vary with the WB levels in the tanks.

The tower is made of steel S-355-J2H. Material density has been augmented to 8,500 kg/m^3 to account for the mass of "realistic" tower sections flanges, secondary structures, etc., as it was done in other public WT tower models [11]. The total tower mass is 879,381 kg and its CM is located 47.241 m above tower base. Distributed mechanical and structural properties of the tower can be found in [2].

NAUTILUS-10 tower installed onshore and on the free-floating and moored floater, supporting the RNA mass presents the natural frequencies shown in Table 3. These results have been obtained resolving the eigenproblem of the undamped system where the RNA is modelled as a point mass.

Table 3. *NAUTILUS-10* tower undamped natural frequencies considering different boundary conditions and 1,200 tonnes of WB.

Modeshape	Onshore	Free-floating	Moored	Units
1st Side-to-Side	0.3930	0.5435	0.5407	[Hz]
1st Fore-Aft	0.3970	0.5508	0.5478	[Hz]
1st Twist	1.5125	1.5180	1.5179	[Hz]
2nd Side-to-Side (coupled with 2nd twist)	1.9493	2.0058	2.0042	[Hz]
2nd Fore-Aft	2.2372	2.2979	2.2962	[Hz]

As it can be seen the first Fore-Aft (FA) and Side-to-Side (SS) natural frequencies of the tower for the moored configuration are between the rotor 3P and 6P frequencies, corresponding to a stiff-stiff tower design.

2.2.4. NAUTILUS-10 floating substructure. NAUTILUS-10 semisubmersible type platform is an upscaled (and modified) version of the concept developed for the *NREL 5 MW WT [12]* by TECNALIA for NAUTILUS Floating Solutions S.L. The main modifications that are relevant for this study are: (1) the central heave-plate is removed and (2) the column height h_C, and therefore the freeboard, has been reduced.

NAUTILUS is a symmetric semisubmersible floating platform (Figure 1), whose **hull** is composed by four columns connected at keel plane by a square-shaped ring pontoon and by an X-shaped main deck at columns top. The transition piece to connect the tower is embedded in the main deck centre. All the structure is compartmented to keep the structure water tight. The structure is made of structural steel S-275 J2 and S355. The conceptual design of *NAUTILUS-10* semisubmersible to support *DTU 10 MW* RNA installed on the described tower at Gulf of Maine site, presents the dimensions and operational draft shown in Table 4.

Table 4. Geometrical properties of *NAUTILUS-10* concept components and floater draft.

***NAUTILUS-10* design variable**	Value	Units
Transition piece height, h_{TP}	3.00	[kg]
Transition piece diameter, d_{TP}	10.50	[m]
Main deck width, w_{MD}	10.50	[kg·m^2]
Main deck height, h_{MD}	3.00	[kg·m^2]
Distance between columns, L_C	54.75	[kg·m^2]
Column diameter, d_C	10.50	[m]
Column height, h_C (from keel to deck top)	26.00	[m]
Pontoon width, w_P	10.50	[m]
Pontoon height, h_P	1.50	[m]
Equilibrium draft (including RNA, tower and SKS)	18.33	[m]

FOWT stability is increased by means of ballasting the ring pontoon with 3,885 tonnes of concrete. The concrete ballast (CB) reaches a height of 0.71 m w.r.t. to the keel line of the floater.

The columns, which are compartmented above pontoon height, store movable seawater ballast (WB) in the lowest compartment with up to 300 tonnes of active ballast per column. This removable seawater ballast mass is modified under control of the Platform Trim System (PTS).

Following, the mechanical properties of *NAUTILUS-10* semisubmersible considering hull, passive ballast mass and (1) empty and (2) full loaded seawater ballast tanks are shown in Table 5. Further information about the mechanical properties of *NAUTILUS-10* can be found in [2].

Table 5. Mechanical properties of the semisubmersible including CB and considering empty and fully loaded WB tanks.

Floating substructure property	Empty WB	Full WB	Units
Mass	6,581,000	7,781,337	[kg]
CM w.r.t. **k** CSys.	(0 , 0 , 4.204)	(0 , 0 , 4.050)	[m]
System roll / pitch / yaw radius of gyration	24.41 / 24.41 / 29.26	24.91 / 24.91 / 30.95	[m]

As it can be derived from this table, the WB increases the floater roll and pitch inertia in 23% and 32% in the case of yaw one, which effectively is translated in higher periods. The platform mass, by its side, is increased in 18% and a slightly reduction (<4%) of the vertical position of the CM is also achieved when the full WB mass (1,200 tonnes) is considered.

2.2.5. NAUTILUS-10 Platform Trim System. NAUTILUS-DTU10 MW FOWT cancels/reduces the mean wind-induced trimming and heeling by means of a Platform Trim System (PTS), thus restoring the platform to an optimal position (vertical tower) from the WT performance point of view.

The PTS has three sensors that are responsible of measuring the FOWT roll and pitch amplitude and the draft.

The control system actuator is the pumping system. Opposite to other active ballast systems oriented to floating wind turbines which do not modify the draft [13], [14]; this system pumps sea water independently into/out to each of the individual columns. The controller acts on the valves opening and water pumps to vary the WB level. Up to 300 tonnes of sea water ballast could be taken off in approximately 30 min using two independent flow paths with redundant pumping capability.

As it can be glimpsed, when the WT is operating, the WB distribution will be driven by FOWT mean heel/trim amplitudes which are tightly related to thrust force magnitude and direction which at the same time depends on the next parameters:

- Wind speed.
- Effective wind direction, which depends on the mean wind flow direction and the nacelle yaw.

Although the sea state (SS) also influences the FOWT rotational displacements, its contribution to the total mean amplitude has been found to be minimal when the WT is operating around rated wind speed. Thus, for *NAUTILUS-DTU10* MW FOWT the PTS has been simplified to a wind speed dependant WB distribution. Considering the wind aligned with global X-axis, the WB mass in the upwind columns tanks will be constant (300 tons) and the downwind tanks will pump out WB (300 tonnes present for 0 m/s wind speed) until the FOWT heel angle is cancelled (see Figure 2 and [2] for further information).

Due to the variable draft characteristic of *NAUTILUS-10* floating support structure, the hub height above the mean sea level (MSL) will vary according to the WB distribution (Figure 2).

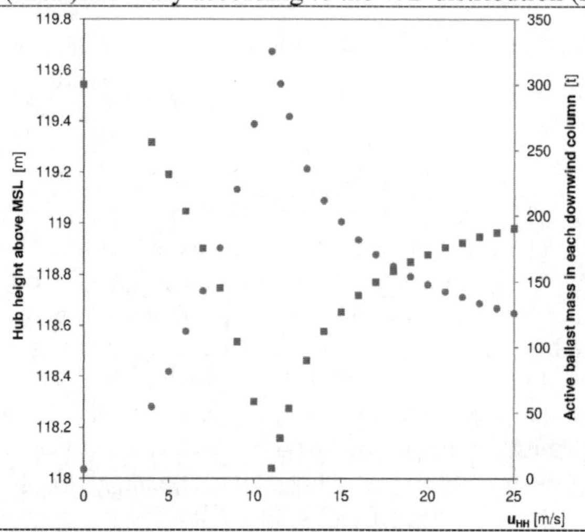

Figure 2. Hub height (blue) as function of wind speed due to water ballast (red) variation in downwind tanks.

Some additional advantages of installing a PTS in a FOWT are pointed below:

- More flexibility to vary draft which could prevent from wave slamming over the deck under non-operational extreme conditions. Regarding inspection and maintenance, the control over draft can simplify these two tasks on the floating platform.
- Under damaged condition (i.e., due to a collision), the floating platform disposes of additional resources to pump out the sea water.

2.2.6. NAUTILUS-10 station keeping system. The aim of the SKS is to avoid the FOWT from drifting and therefore to maintain the power cable within a sufficient stress range. For this purpose, the floater is moored with four catenary lines (Figure 1).

Gulf of Maine seabed presents 4 m of dense sand [7] which enables the use of conventional drag anchors to keep lines secured to seabed. The design water depth was set to 130 m for LIFES50+ EU project, meanwhile water density of 1,025 kg/m^3 was considered. The seabed is assumed to be horizontal and flat.

Fairleads are located 12 m above the floater's keel line which let them 6.333 m below MSL when the FOWT is in its undisplaced position (draft of 18.333 m), and 43.96 m, in XY-plane, from platform centreline. Drag anchors are fixed to seabed at 793.50 m from fairlead position and therefore to 837.46 m from *NAUTILUS-10* centreline. All mooring lines have the same unstretched length of 833.24 m.

The mooring lines selected are studless chains of 97 mm of bar nominal diameter. The chain distributed mass is 188.16 kg/m and the extensional (axial) stiffness is 803.5 MN. The SKS pretension derived from this layout is then 406,300 N.

3. Multiphysics modelling of *NAUTILUS-DTU10* MW FOWT

This section is fully dedicated to the description of the numerical multiphysics models that have been implemented to capture the FOWT aero-hydro-servo-elastic behaviour through time domain coupled simulations. FAST 8.16 32-bit version code [15] has been employed to build these models. Following, this section is structured according to the physics that are part of the problem.

3.1. Multibody dynamics

FAST v8.16 presents a modular framework which is articulated around the multibody dynamics model that represents the FOWT. This model is the responsible of solving the nonlinear equations of motion of the system and of feeding with the calculated displacements, velocities and accelerations the multiphysics modules that determine the loads to be applied to the multibody model. The multibody formulation is combined with a modal superposition approach to represent the flexible elements of the FOWT (blades and tower). The FOWT model implemented accounts for a total of 23 DOFs.

3.2. Elastodynamics

The drivetrain, blades and tower of the FOWT are considered flexible inside the models presented here. While the drivetrain is represented by a single degree of freedom (DOF) representing its torsion, the tower and each blade consider 4 and 3 DOFs, respectively. The tower modeshapes included in the modal superposition approach implemented in ElastoDyn are: the 1st and 2nd fore-aft and side-to-side. The tower is meshed with a total of 30 equally spaced nodes.

Taking into account the geometry, the structural and aerodynamic properties of the prebended blade of *DTU 10 MW RWT* [3], a high fidelity model like that implemented in BeamDyn [16], [17] would be advisable. Nevertheless, the blade model distributed inside the LIFES50+ project was simplified to a straight blade model [9]. This simplification was considered to overcome the numerical efficiency problems detected in BeamDyn of FAST v8.12, initially employed in LIFES50+ project.

Then, the blade model considered here accounts for 1st and 2nd flapwise modeshapes and considers just the 1st edgewise. A total of 50 structural elements are used in the discretisation of this component.

3.3. Aerodynamics

As in the case of elastodynamics, the blade properties have been adapted, so the rotor shows very similar aeroelastic performance when using the high fidelity or the modal superposition approach to calculate elastodynamics. The blade aerodynamic mesh defined in [9] is composed by 37 nodes.

Rotor aerodynamics have been modelled using an engineering-level model based on the modified Blade Element Momentum theory that AeroDyn v14.05 implements [18]–[20]. Note that nor the tower influence on the flow around the blade nor the wind loading on the tower and floater have been considered in the results presented in this paper. The reason for this important simplification is that the *DTU 10 MW RWT* aerodynamics have not been defined for AeroDyn v15 code [9], the first release including these capabilities in the analysis of floating support structures [21].

3.4. Servodynamics

3.4.1. WT controller. This controller is distributed under a 32 bit Dynamic Library Link (DLL) package, in the style of Garrad Hassan's BLADED WT software which can communicate with ServoDyn module of FAST [22]. These files and the tuned input files can be download from [1].

3.4.2. Platform trim system. On the other hand, the floating platform control system dynamics are not considered along the coupled dynamic simulations due to its large response period (as in the case of nacelle yaw controller for onshore WTs). The sea water ballast distribution is determined prior to the simulation and it remains constant along all the simulation.

Although the inclusion of the dynamics of PTS opens the study to combined strategies between the WT and floater's controllers, this topic will not be covered in the present paper.

3.5. Hydrodynamics

As result of the comparison of the hydrodynamic loads due to two different sea states (NSS and ESS) in [2], it was concluded that for the case of *NAUTILUS-DTU10* MW FOWT the Potential Flow (PF) solution obtains satisfactory results under NSS, but by contrast under ESS, the viscous effects should not be neglected. Same conclusions were obtained in [23] for the OC4 DeepCwind semisubmersible.

Following the previous recommendations, two hydrodynamics models have been implemented in HydroDyn for comparison purposes, identified as PF+AHD (potential flow + additional hydrodynamic damping) and PF+ME (potential flow + Morison elements), differing on the modelling of turbulent effects. While additional hydrodynamics models in [2] are based on a variable submerged volume, both numerical hydrodynamics models presented here consider a constant submerged volume.

3.5.1. Hydrodynamics: Potential flow solution. For the studied geometry, calculations have been carried out using a higher order panel method with the maximum panel size considered is 2 m, with a frequency resolution of 0.02 rad/s, considering no RNA eccentricity, under fully loaded WB condition and hull keel plane horizontal due to PTS.

For the calculation of the mean drift force, FOWT mechanical properties (Table 1) and the linearized mooring stiffness matrix K_M is also given to the solver, while no additional damping was included at this stage (due to the absence of reliable damping values), provided that results were considered acceptable.

3.5.2. Hydrodynamics: Viscosity through additional hydrodynamic damping vs Morison elements. In the PF+AHD model, AHD values (Table 6) were determined through an upscaling process of *NAUTILUS-NREL5* MW concept, which had been experimentally tested at two different scales [12], [25], asdescribed in [2].

Table 6. Linear and quadratic AHD values of *NAUTILUS-DTU10* MW FOWT.

	Linear AHD terms		Quadratic AHD terms	
DOF	Value	Units	Value	Units
Surge	0	[N·s/m]	1,100,985	[N·s^2/m^2]
Sway	0	[N·s/m]	827,308	[N·s^2/m^2]
Heave	335,479	[N·s/m]	5,637,998	[N·s^2/m^2]
Roll	211.97·10^6	[N·m·s/rad]	38,515.2·10^6	[N·m·s^2/rad^2]
Pitch	222.17·10^6	[N·m·s/rad]	41,617.9·10^6	[N·m·s^2/rad^2]
Yaw	22.56·10^6	[N·m·s/rad]	7,066.5·10^6	[N·m·s^2/rad^2]

The drag terms of Morison's equation (ME) can be computed by combination with the strip theory. The mesh defined to calculate the submerged volume and described in [2] is used here.

Table 7 collects the drag coefficients and dimensions of the members that the platform has been decomposed into. These values have been fitted to account for the inter-member interactions and show a hydrodynamic behaviour comparable to that obtained with the PF+AHD model, which accounts for the interaction between the submerged members under waves (but not current) conditions.

Table 7. Morison elements and drag coefficients of *NAUTILUS-10* semisubmersible.
$^*C_{D,a}$ is only applied at the lower end of the member.

	Modelling geometry		Drag coefficients	
Member	d [m]	L [m]	$C_{D,a}$* [-]	$C_{D,t}$ [-]
Pontoon(s)	4.478	44.250	-	2.0000
Ring pontoon rounded corner(s)	4.617	1.500	1.0483	-
Column(s) connection w/ pontoons	4.753	1.500	-	-
Column(s)	10.500	26.000	0.8745	0.9600
Virtual column	6.529	1.500	550.0000	-

3.6. Mooring dynamics

Meanwhile in [2] up to four different mooring dynamics codes (MAP++, MoorDyn, FEAM and OrcaFlex) are analysed and compared in time domain, coupled and decoupled from the overall multiphysics problem. Here, the SKS dynamics will be resolved using MoorDyn code [26]. The mooring line properties are determined from OrcaFlex [27] database and adapted to the theory behind MoorDyn, as shown in Table 8 where the difference between both codes can be derived.

Table 8. Mooring lines properties considered in the numerical codes.

	MoorDyn	OrcaFlex	Units
Hydrodynamic diameter in normal direction	0.097	0.097	[m]
Hydrodynamic diameter in axial direction	0.097	0.031	[m]
Diameter for line-to-seabed friction	-	0.302	[m]
Drag coefficient in normal direction, $C_{D,N}$	2.400	2.400	[-]
Drag coefficient in axial direction, $C_{D,Axial}$	0.368	1.15	[-]
Drag coefficient in normal direction, $C_{A,N}$	1.000	1.000	[-]
Added mass coefficient in axial direction, $C_{A,Axial}$	0.160	0.500	[-]
Line-to-seabed friction coefficient, $C_{D,N}$	-	0.0	[-]
Structural damping ratio, ξ_{Line}	1.0	1.0	[%]

4. Tests and Design Driving Load Cases: Definition and results

4.1. Decay tests

These tests dedicated to identifying the system properties are performed in calm water without and with the presence of uniform wind (4, 8, 11.4, 18 and 44 m/s) parallel to global X-axis (Figure 1). This set of tests provide the designers wide information about eigenfrequencies of the platform, linear and quadratic damping and linearity effects. The considered initial offsets from equilibrium position are 10 m, 2 m and 8 degrees for surge, heave and pitch, respectively.

The free decay tests without wind (Figure 3, top) reveal that a very good match between the hydrodynamic models proposed is achieved, finding slight differences in the first cycles of surge test.

When steady wind is present and the rotor is operating, there are three main factors that will modify the floater periods: (1) the SKS stiffness at equilibrium, (2) rotor introduced aerodynamic damping and (3) the PTS driven WB distribution to adapt to wind characteristics. The variation of the periods due to SKS stiffness can be clearly observed in surge decay tests; because of the lower contribution this parameter to heave and pitch motion (for the case of this semisubmersible), the variation of the periods due to mass variation and aerodynamic damping are more noticeable in heave and pitch tests.

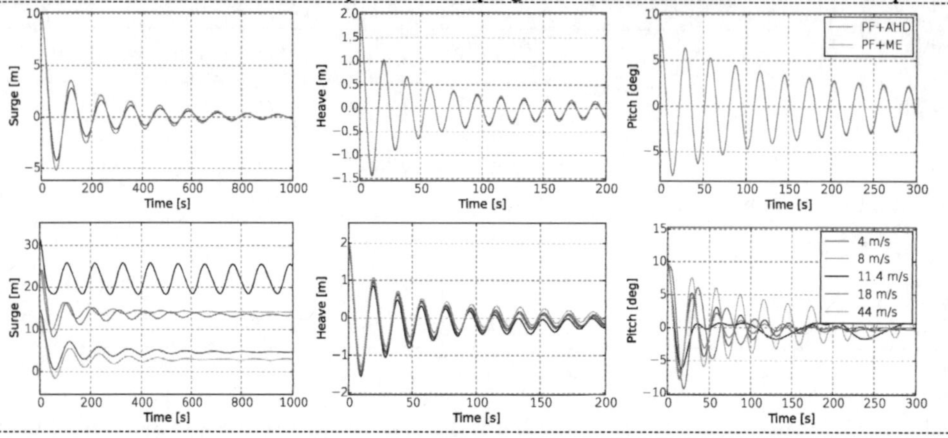

Figure 3. Decay tests of *NAUTILUS-DTU10* FOWT w/o wind (top, comparison of hydrodynamic models) and operating w/ steady wind (bottom, PF+AHD).

As it can be seen in Figure 3, surge and pitch tests operating at nominal wind speed do not completely decay and some resonance phenomena can be observed in surge motion. This resonance is mainly due to the WT controller setup, which was tuned considering a stepped wind case without ocean loading (section 4.3) showing a stable behaviour. It should be noted that this self-excitation reduces/disappears when irregular waves are present, due to the randomness of the excitation frequency. Nominal wind in still water in not a load case (LC) expected in GoM. The reproduction of this instability is very sensitive to the initial blades pitch condition.

4.2. *Wind tests*

Uniform wind type with increasing speed (1 m/s steps of 300 s) confined between the rotor cut-in to cut-out limits is considered here to assess the FOWT dynamics behaviour under operation. To ensure the dynamic stability of the system, the inclusion of waves is neglected and thus, significant values of hydrodynamic damping too. Additionally, the response has been calculated for WB distributions that correspond to 4 and 11.4 m/s wind speed.

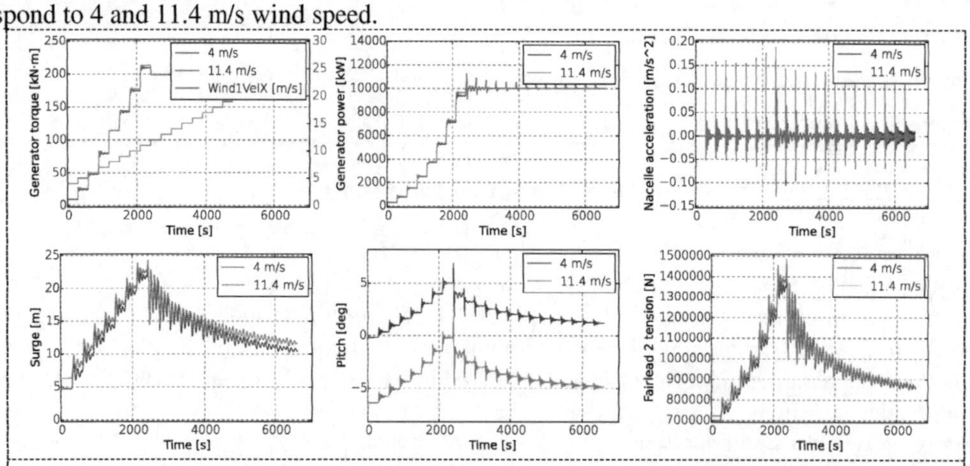

Figure 4. FOWT performance under stepped wind.

In the case of FOWTs the wind range affecting the full-load region of the controller is of special interest (rotor thrust decreasing region). The constant torque strategy is clearly shown in Figure 4. The controller tuning will also depend on the maximum allowable generated power, which already is unknown. Nacelle accelerations show a slightly lower maximum and more stable response under the

11.4 m/s WB distribution. Non-zero mean pitch offsets indicate a suboptimal setup and /or distribution of the ballast, fact visible through the wind speed ranges different to that considered (4 m/s and 11.4 m/s) at the beginning of the simulation. This effect is partially induced by the static character of the simulation, but it should be noticed that the response time of the proposed PTS (≈0.15 deg/min) will be able to answer to such speed steps that are maintained constant during 5 min. The relation between surge motion and fairlead tension can be observed in the results above.

4.3. Broadband irregular waves tests

Broadband spectrum waves tests are very useful to characterise the hydrodynamic behaviour as the system is excited in a wide range of frequencies. The energy of the waves is distributed here in a white noise pattern from 4.5 to 18.2 s for two different wave heights (2 and 4 m) and headings (-15 and 90 deg) considering two hours simulation length (excluding transients).

The results are presented in Figure 5 under the form of power spectral density (PSD) of the most relevant DOFs of the floater. No significant differences can be observed in the low frequency translations (surge and sway) independently from wave height. PF+ME calculates a more damped yaw response, opposite to heave motion where larger energy content is present at heave natural frequency. Both models obtain nearly identical roll and pitch response, independently of wave height.

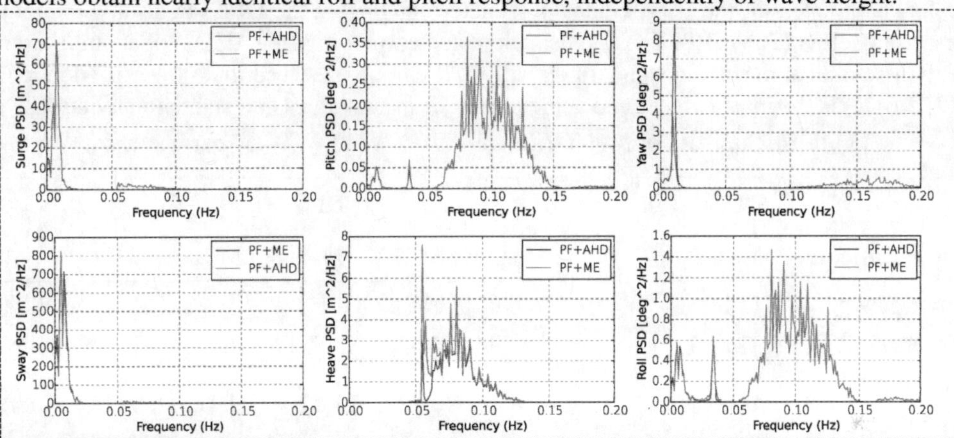

Figure 5. Comparison of the hydrodynamic models under white noise tests. $H_S = 2$ m, $\alpha_{Wave} = -15$ deg (top) and $H_S = 4$ m, $\alpha_{Wave} = 90$ deg (bottom).

4.4. Design driving load cases: Turbulent wind and irregular waves at Gulf of Maine

Finally, closing the analysis section of this paper, the spectrum-compatible irregular waves and turbulent wind expected at Gulf of Maine are analysed. A reduced set of Design Load Cases (DLCs), coherent with IEC standards, has been selected for this paper (Table 9), which indeed are part of the *NAUTILUS-DTU10* MW FOWT Design Driving Load Cases (DDLCs) set.

Table 9. Turbulent wind ($\alpha_{Wind} = -15$ deg) and irregular waves DLCs.

DLC Id.	U_{HH} [m/s]	I_{Turb} [%]	Shear [-]	Wave direction [deg]	H_S [m]	T_P [s]
DLC 1.6_1/2/6	8.0/11.4/18.0	8.3/8.4/8.7	0.14	-15	7.7/7.7/10.9	12.4/12.4/15.0
DLC 1.6_7	11.4	8.4	0.14	90	7.7	12.4
DLC 5.1	11.4	12.8	0.14	0	1.6	25
DLC 6.1_2a	44.0	11.0	0.11	-15	10.9	15.0

A fault condition DLC has been included: DLC 5.1. This LC considers the loss of the electrical grid while operating at rated wind speed and thus, a rapid reduction of the mechanical torque (and thrust consequently) will occur to due to blade pitching to avoid rotor over-acceleration. This is a DDLCs for the *NAUTILUS-10* PTS and the unphysical wave of 25 s period has been chosen to excite heave and roll/pitch motion. This DLCS should be taken into consideration in all the actively ballasted floating substructures; as the WB distribution for this wind condition will augment the maximum pitch amplitude generated during the transients.

The maximum pitch rotation remains above -15 degrees after the loss of grid (Figure 6) and the response decays until equilibrium. Due to the reduction of thrust, FOWT excursion is reduced while the nacelle accelerations induced by the transients remains below 0.6 m/s^2.

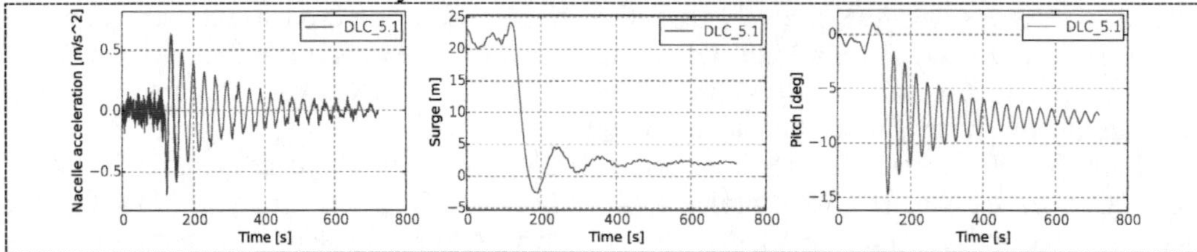

Figure 6. FOWT response operating at nominal wind speed and with loss of grid.

Some key parameters of the FOWT behaviour determined from the DLCs dedicated to study ULS are summarised below (Table 10). The results, calculated employing PF+AHD model, are presented in terms of their statistical mean and standard deviation. It can be observed that power production under SSS (DLC 1.6_X) is the most exigent and achieves larger mean values and standard deviations.

Table 10. Statistics of *NAUTILUS-DTU10* MW FOWT key parameters.

	DLC 1.6_X			DLC 6.1_2a			
Output parameter	Max	Mean	STD	Max	Mean	STD	Units
Generator power	11,860	9,996	882	-	-	-	[kW]
Blade root My	37,670	25,324	4,630	5,821	868	1,585	[kN·m]
Combined RNA acceleration	1.99	0.40	0.22	1.51	0.41	0.20	[m/s^2]
Combined tower base moment	329,892	138,956	59,300	183,367	51,446	26,282	[kN·m]
Floater excursion	27.92	21.50	2.87	10.49	5.52	1.68	[m]
Combined floater heel/trim	6.90	1.44	0.99	3.02	0.91	0.45	[deg]
Mooring line tension	2,511	1,558	213	1,025	741	69	[kN]

5. Conclusions

The inclusion of the active ballast control system in *NAUTILUS-10* semisubmersible brings some challenges to the numerical modelling of this concept. While most of these challenges increase the time during the FOWT design and validation processes, some simplifications that can be applied to this concept to reduce the demand of these resources have also been mentioned.

To validate the concept against a reduced set of DDLCs, two numerical models have been presented. When both are applicable, PF+AHD is 30% faster than PF+ME model, which is particularly convenient during the first design loops of the concept. Nevertheless, when implemented thorugh FAST's HydroDyn module, PF+AHD cannot model the current induced loading/damping.

The numerical results show that controller dynamics and hydrodynamic loading/damping play an important role in the dynamic stability of the FOWT and demonstrate the satisfactory behaviour of *NAUTILUS-DTU10* MW Floating Offshore Wind Turbine.

After the first design iteration presented here, future work will focus on the WT controller and communication with the PTS and the calibration of the hydrodynamic models against experimental tests. The inclusion of the submodels to capture tower and floater aerodynamics will be of special interest too, linked to the development of the BeamDyn model of the blades.

In order to make this information public to the research community and support the development of Floating Offshore Wind technology, the FOWT designer's information and the numerical models developed and presented here can be found at: https://www.researchgate.net/project/NAUTILUS-DTU10-MW-Floating-Offshore-Wind-Turbine-at-Gulf-of-Maine.

Acknowledgments
The research leading to these results has received funding from the European Union Horizon2020 programme under the agreement H2020-LCE-2014-1-640741, LIFES50+ project.

References

[1] Tecnalia R&I; Offshore Renewable Energy area, 'The NAUTILUS-DTU10 MW Floating Offshore Wind Turbine Project site', 03-Apr-2018. [Online]. Available: https://www.researchgate.net/project/NAUTILUS-DTU10-MW-Floating-Offshore-Wind-Turbine-at-Gulf-of-Maine. [Accessed: 04-Apr-2018].
[2] J. Galvan *et al.*, 'Definition and Analysis of NAUTILUS-DTU10 MW Floating Offshore Wind Turbine at Gulf of Maine: Experiments at Sintef Ocean & PoliMi', Tecnalia R&I, Derio, Basque Country, Spain, Public use under permission TRI-ORE-PUB-001\2017, Dec. 2017.
[3] C. Bak *et al.*, 'Description of the DTU 10 MW reference wind turbine', DTU Wind Energy, DTU Wind Energy Frederiksborgvej 399 4000 Roskilde, Denmark, I–0092, Jul. 2013.
[4] M. H. Hansen and L. C. Henriksen, 'Basic DTU Wind Energy controller', DTU, Department of Wind Energy, DTU, Denmark, E-0028, Jan. 2013.
[5] B. Jonkman and J. Jonkman, 'Readme file for FAST v8.16.00a-bjj', 26-Jul-2016. [Online]. Available: https://wind.nrel.gov/nwtc/docs/README_FAST8.pdf. [Accessed: 16-Oct-2014].
[6] J. M. Jonkman and M. L. Buhl Jr, *FAST user's guide*. Golden, CO, USA: National Renewable Energy Laboratory, 2005.
[7] A. Krieger, G. K. V. Ramachandran, L. Vita, P. Gómez Alonso, J. Berque, and G. Aguirre, 'D7.2 Design basis', DNV GL; IBERDROLA Ingeniería y Construcción; Tecnalia R&I, Public LIFES50+ D7.2, Nov. 2015.
[8] P.K. Chaviaropoulos, D. Chortis, and D. Lekou, 'Definition of the reference Wind Turbine - Analysis of Rotor Design Parameters', CRES, Public INNWIND D1.2.1, May 3013.
[9] M. Borg, M. Mirzaei, and H. Brendmose, 'D1.2 Wind turbine models for the design', DTU, Public LIFES50+ D1.2, Dec. 2015.
[10] E. Bossanyi, 'Controller for 5 MW reference turbine', Garrad Hassan & Partners Limited, UpWind report 11593/BR/04, Jul. 2009.
[11] J. Jonkman, S. Butterfield, W. Musial, and G. Scott, 'Definition of a 5-MW reference wind turbine for offshore system development', National Renewable Energy Laboratory, Golden, Colorado, U.S.A., NREL/TP-500-38060, 2009.
[12] Nava, V., Aguirre, G., Galvan, J., Sanchez-Lara, M., Mendikoa, I., and Perez-Moran, G., 'Experimental studies on the hydrodynamic behavior of a semi-submersible offshore wind platform', in *Proceedings of 1st International Conference on Renewable Energies Offshore*, Lisbon, Portugal, 2014, p. 7.
[13] I. Couchman and R. Bowyer, 'Coordinated control of a floating wind turbine', US 2015/0147174 A1, 28-May-2015.
[14] D. Roddier and C. Cermelli, 'Floating Wind Turbine platform with ballast control and mooring system', US 9,139,266 B2, 22-Sep-2015.
[15] NREL, 'FAST v8 | NWTC Information Portal', *NREL: NWTC Information Portal*, Jul-2016. [Online]. Available: https://nwtc.nrel.gov/FAST8. [Accessed: 29-Nov-2016].

[16] Q. Wang, M. Sprague, J. Jonkman, and N. Johnson, 'BeamDyn: A high-fidelity wind turbine blade solver in the FAST modular framework', presented at the SciTech 2015, Florida, USA, 2015.
[17] *BeamDyn User's Guide*. National Renewable Energy Laboratory Golden, CO, 2015.
[18] D.J. Laino and A.C. Hansen, *AeroDyn user's guide, version 12.50*. 2002.
[19] P. J. Moriarty and A. C. Hansen, *AeroDyn theory manual*. National Renewable Energy Laboratory Golden, Colorado, USA, 2005.
[20] B. Jonkman, *Using the NewTower feature of AeroDyn 14*. NREL.
[21] J. Jonkman, G.J. Hayman, B. Jonkman, and R.R. Damiani, *AeroDyn v15 user's guide and theory manual*. 2015.
[22] J. Jonkman, 'Overview of the ServoDyn control & electrical-drive module', presented at the NREL Wind turbine modelling workshop, Bergen, Norway, 11-Sep-2014.
[23] A. Robertson *et al.*, 'Offshore Code Comparison Collaboration Continuation within IEA Wind Task 30: Phase II results regarding a floating semisubmersible wind system', in *ASME 2014 33rd International Conference on Ocean, Offshore and Arctic Engineering*, 2014, p. V09BT09A012–V09BT09A012.
[24] K. Müller, R. Faerron-Guzmán, M. Borg, and A. Manjock, 'D7.7 Identification of critical environmental conditions and design load cases', STUTT, Public D7.7, Jul. 2018.
[25] V. Ayllon *et al.*, 'Análisis experimental de una plataforma flotante para la extracción de energía eólica en aguas profundas', IHC - Instituto de Hidráulica ambiental de la universidad de Cantabria, Cantabria, Memoria final de ensayos, Nov. 2014.
[26] M. Hall, *MoorDyn User's Guide*. University of Maine, 2014.
[27] Orcina Ltd., *OrcaFlex 10 user manual*. Orcina Ltd., 2016.

Assessing the mechanical stresses of dynamic cables for floating offshore wind applications

D G Young[1], C Ng[1], S Oterkus[2], Q Li[3] and L Johanning[4]

[1] Offshore Renewable Energy Catapult, Blyth, UK
[2] University of Strathclyde, Naval Architecture, Ocean and Marine Engineering Glasgow, UK
[3] University of Edinburgh, School of Engineering, Edinburgh, UK
[4] University of Exeter, Offshore Renewable Energy , Exeter, UK

E-mail: david.young@ore.catapult.org.uk

Abstract. Offshore wind farms are progressing further offshore and into deeper waters, presenting the need for new substructures, including floating offshore wind turbines. These floating turbines will require dynamic cables to run through the water column, exposing them to the dynamic loadings of the marine environment. This paper presents a tool which models the stresses across a dynamic cable cross section's insulation layers when attached to a floating wind platform. Differing wave, wind and current flow conditions are applied and their impact on the stress distributions of the dynamic cable's insulation layers are presented. Finally from these stress histories, accumulated fatigue damage of the insulation is calculated and presented. The outcome of this can be used to estimate fatigue damage of a cable components cross section at any point along the cable length, and aid in cable installation configuration decisions.

1. Introduction

Offshore wind farms are progressing to having larger turbines installed as a cost effective means to reduce the cost of energy. These larger wind farms are moving further from shore and into deeper water depths. This move is presenting the need for new substructures including floating platforms for offshore turbines. Floating platform wind turbines are envisioned to be installed in water depths ranging from 50m to 200m, with the Carbon Trust estimating up to 90MW of floating wind farms are to be installed in UK waters by the end of 2018 [1].

Floating platforms will introduce new challenges for the umbilical, or array, cables that hang from the base of the platform to the seabed. This cable installation arrangement will expose the cables to the dynamic forces of the environment they would be installed in, including the actions of the waves, current flows and movement of the platform itself in response to wind/turbine interactions. Cables exposed to dynamic environmental loadings will experience dynamic mechanical stresses across their cross sections and along their length. This is a new operating consideration as traditional fixed bottom platforms have allowed cables to operate statically.

The causes of cable failures during normal operation, and excluding third party interference, can be approximately grouped into mechanical damage, water ingress and overheating which all lead to electrical failure or breakdown of the insulation layers [2]. A renewable energy insurance provider [3] highlights that most cable failure claims are related to incorrect installation, incorrect operation or electrical faults. As dynamic cables will be exposed to dynamic environments and the corresponding

IOP Conf. Series: Journal of Physics: Conf. Series **1102** (2018) 012016 doi:10.1088/1742-6596/1102/1/012016

force loadings, they are at higher risk of cable failure with more expected to fail during normal service lifetime.

The impact of dynamic loadings on the fatigue characteristics of dynamic cables has been investigated by [4], [5], [6] while mechanical fatigue due to environmental loadings is shown to be a concern as a propagating mechanism for water treeing in the cables insulation layers [7], [8]. The impact of differing cable bending and axial stiffness on cable fatigue damage was investigated by [9]. The axial and bending stiffness of a cable affects the cables motions in response to the dynamic environment, where having a larger stiffness reduces the cable motion. The inter-wire friction between helically wound armouring layers is found to be a determining factor in cable bending stiffness [10] while [11] investigated the impact of armouring radius reduction effects on cable axial stiffness when under tension. Similar findings are reported in [12] when comparing calculated cable axial stiffness which account for radius reduction or not.

This paper presents an assessment of the marine environment at a chosen site and presents a model which determines how the mechanical stresses across the cable cross section distribute when installed in this environment. A fatigue assessment of the cable insulation layers is completed and conclusions are drawn from the accumulated damage in differing dynamic environment loadings.

The approach presented begins with an assessment of the environment to determine the wave, wind and current loadings on the cable as detailed in section 2. An introduction to the software used and the created model is given in section 3. Section 4 provides details on how the damage is determined from the created model. Results from the model and calculated damage are presented in section 5, while section 6 gives a discussion of these results. Section 7 draws conclusions from the work, and finally section 8 highlights ongoing work.

2. Environmental Loading Assessment

Wave and current flow data from sites off the north east coast of England, as shown in Figure 1, was obtained and analysed. Wave data was obtained from a wave buoy placed at a site in 90m water depth, while current data was obtained from a 56m water depth site by an acoustic wave and current profiler, AWAC. Wind speed data correlated to wave height was obtained from two locations based in the central North Sea from the NEXT study, an extension of the North European Storm Study, NESS [13]. Briefly, NESS was initiated to produce a hind cast database of winds, waves and currents for the North European continental shelf.

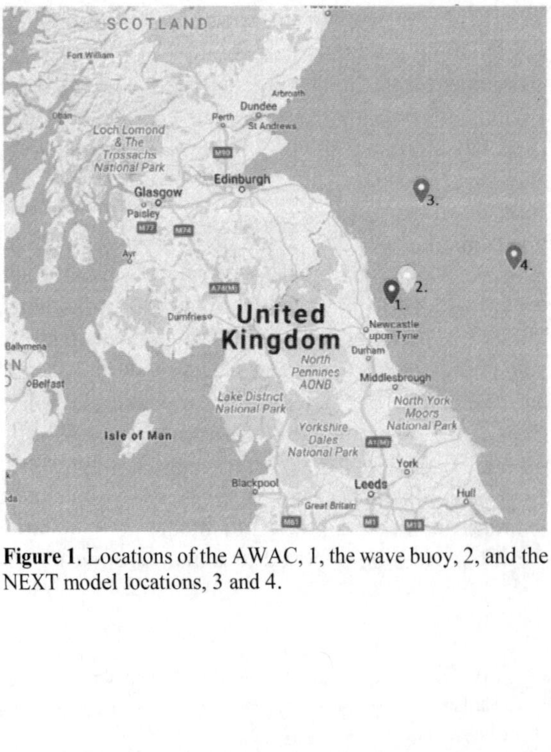

Figure 1. Locations of the AWAC, 1, the wave buoy, 2, and the NEXT model locations, 3 and 4.

The locations of the wave buoy, AWAC and NEXT study sites were determined to be within the same coarse grid of the central North Sea. Therefore it was assumed the environment exhibited at each site would be similar, meaning the wave environmental data captured at the wave buoy site would be the same wave data that would be captured at the AWAC site or the NEXT sites, and vice versa. The water depths at the sites were differing, so the captured current flow speed data was fitted to the power law velocity profile using equation 1 [14] , assuming a shear profile due to seabed roughness. This was then extrapolated, allowing for current velocity calculations at other water depths:

$$\frac{U_z}{U_{max}} = (\frac{Z}{H})^{1/n} \quad (1)$$

Where U_z is the water velocity in ms^{-1} at point Z, U_{max} is the maximum water velocity in ms^{-1}. Z is the vertical position from the seabed in m, H is the full tidal depth in m and n is the power law coefficient.

To use wind data near the wave data site it was necessary to assume that the wind generated the waves and thus the wave height was directly related to the wind speed. The NEXT data sites present wave height binned with a corresponding wind speed 10m above sea level. This data was used to create a best fit curve that could be used to back calculate a corresponding wind speed for the wave heights recorded by the wave buoy. For this reason it was also assumed that the wind and waves were coming from the same general directions.

This model includes a tension legged floating wind platform incorporating the NREL 5MW turbine. This turbine has a hub height of 90m, therefore by using the equations laid out in the guidelines for the certification of offshore wind turbines [15] the wind speed at the hub height was calculated using:

$$V_z = V_{hub}(\frac{h}{h_{hub}})^{\alpha} \quad (2)$$

Where V_z is the wind speed at height h in ms^{-1}, h is the height about still water line in m, h_{hub} is the hub height above still water line in m, and α is the power law exponent, [15] assumes that α is equal to 0.14 for all wind speeds.

3. Dynamic Cable Model

To understand the stresses that the loading environment would impose upon a dynamic cables cross section, the whole floating platform system needs to be analysed initially as a global model. The output from the global model can then be used to simulate the stress distributions and time history for a local mechanical model of the cable cross section.

Details of this model can be found in [16] however for clarity a condensed version of the essentials will be provided here.

3.1. Global Dynamic Model

This model was created using the proprietary marine dynamics software OrcaFlex by Orcina. OrcaFlex was used to estimate the loadings on the dynamic cables installation length in response to the actions of the waves, currents, wind and the movements of the floating platform itself. For this work the cable was split into 0.5m lengths and the software employs a lumped mass approach to solve the cables dynamic behaviour.

Required inputs for the dynamic model include the environmental loadings and properties of the cable, including its geometry, length, axial and bending stiffness's.

The MIT/NREL reference 5MW tension leg platform [17] is included in this work and its movements modelled. This is a platform of 18m waterline depth and almost 50m draft. It has four 27m long pontoons, each having two mooring lines attached to its end. Fastlink is a FAST-OrcaFlex interface package that couples the software OrcaFlex with FAST. The hydrodynamics of the cable are modelled using OrcaFlex. The NREL 5MW reference turbine, tower, floating platform and structural dynamics, aerodynamics control and electrical drive dynamics are modelled using FAST. This allows for the coupling of the effects of the wind on the turbine with the effects of the waves and currents on the platform.

3.1.1. Dynamic Cable. The cable modelled for this work is based on 280m length of a JDR dynamic cable [18] used in previous oil and gas applications. Figure 1 gives some of the cables mechanical parameters. The cable axial stiffness was calculated using works laid out in [11], and the bending stiffness using [10].

Table 1. Mechanical Parameter of the Dynamic Cable.

Mechanical Properties	
Outer Diameter	154mm
Outer Sheath Material	Extruded Polyethylene
Cable Armouring	Galvanised Steel Wire
Minimum Bend Radius	2000mm
Safe Working Load	250kN
Diameter to Weight Ratio	4.27m^2/Te

3.2. Mechanical Model

The mechanical model is used to determine the stress time histories and distributions across the cables cross section. This local model uses the outputs of the global dynamic model to allow the user to determine the stresses acting on individual cable components, highlighting areas of greatest mechanical stress. Plane and shear stresses are calculated for each of the cables six degrees of freedom as well as combined into a von Misses stress.

3.2.1. Finite Element Analysis. A finite element analysis, FEA, of the cable cross section was constructed using ANSYS Mechanical software. This proved an effective method to determine the interactions between different components of the dynamic cables cross section. Similar works on dynamic umbilical cables [19], [20] both used an alternative FEA software called UFlex2D. This is a custom built software for umbilical cross sections however ANSYS was used here for its coupling abilities with OrcaFlex.

3.3. Model Assumptions

To balance accuracy and computational expense some assumptions were made during this analysis. The cable stiffness was assumed to come from the cable armouring layers. The armouring layers were simplified to thin walled cylinders for the mechanical model, similar to [9]. This assumption follows the thought that the armouring layers are always in stick mode rather than slip mode so that individual armouring wires do not slide over one another during operation. The cable component materials were assumed to be linear elastic and that they were glued together in the FEA, meaning they could not penetrate another components surface.

4. Fatigue Damage Calculations

An output from the mechanical model are the stress histories of the components of the dynamic cables cross section. A stress based approach to fatigue was used as the stresses exerted upon the cross section are lower than the design yield stresses of the cable component materials.

Accumulated fatigue damage of the dynamic cable insulation layers are calculated using the Palmgren-Miner linear damage rule, equation 3, and the corresponding S-N curve for that material.

$$D = \sum_{i=1}^{k} \frac{n_i}{N_i} \tag{3}$$

Where for k different stress levels and the average number of cycles to failure at the ith stress, S_i, is N_i and n_i is the number of cycles are accumulated stress S_i. No fatigue limits were set for this initial work as compression and tension are both assumed to contribute to fatigue. The S-N curve used for the XLPE insulation layers was obtained from [8].

5. Results

5.1. Environmental loading Assessment

Using the data from [13] the relationship between the significant wave height and the wind speed is plotted in Figure 2. This plot allows for the determining of the wind speed for the wave heights recorded by the buoy. Figure 3 shows a binned scatter plot of the significant heights and zero crossing period of the waves recorded by the buoy.

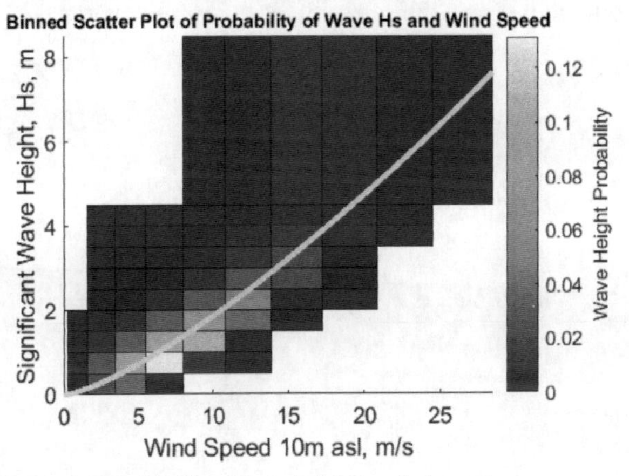

Figure 2. Relationship of wave height to wind speed.

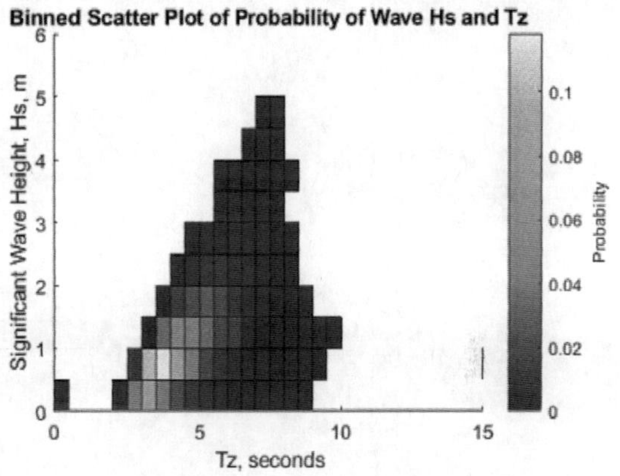

Figure 3. Binned scatter plot of significant wave height and zero crossing period.

Using the recorded wave heights from Figure 3, and equation (2), the wind speeds the turbine is exposed to are presented in table 2.

Table 2. Corresponding wave heights and wind speed.

Significant Wave Height(m)	Wind Speed 10m asl (ms^{-1})
0.25	2.2
0.75	5.1
1.25	7.4
1.75	9.5
2.25	11.5
2.75	13.4
3.25	15.1
3.75	16.8
4.25	18.5
4.75	20.1

From these results, the largest wind speed the turbine would face 10m asl is 20.1ms^{-1}, roughly corresponding to gale force winds and a Beaufort Force of 8 on the Beaufort scale.

The wave directions were also recorded by the wave buoy and are presented in a wave direction rose shown in Figure 4. As the NESS model sites did not provide much information to back calculate the wind directions, the winds and waves are assumed to come from the same direction for this work.

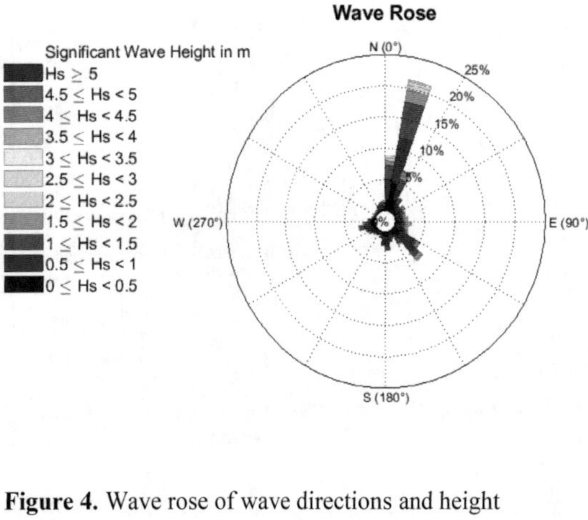

Figure 4. Wave rose of wave directions and height

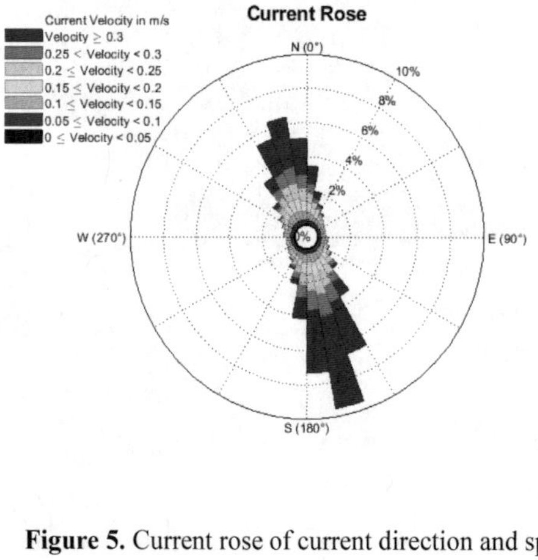

Figure 5. Current rose of current direction and speed

From figure 4 it can be seen that the wave direction is predominantly from the North, so for the model the waves and therefore wind are assumed to be coming from one direction only.

The data recorded by the AWAC was used to determine the current magnitude and direction and is presented in Figure 5. From this it can be seen that the current direction is approximately bi-directional, traversing up and down the North East of England coastline. The average current speed for each recorded water depth was calculated and extrapolated to a 200m water depth with the curve plotted in Figure 6. This curve was compared to the theoretical current profile calculated using equation (1).

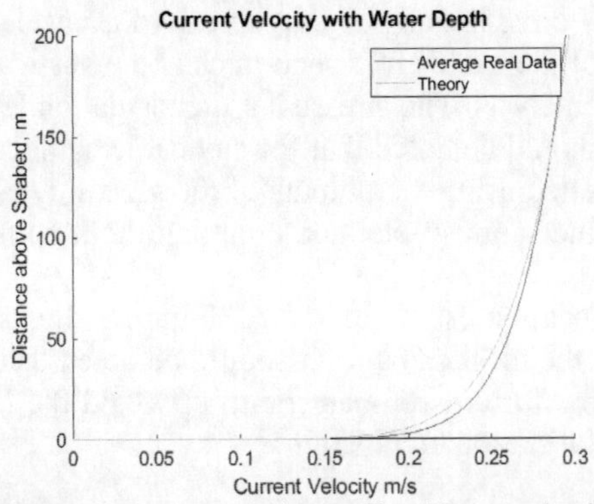

Figure 6. Current velocity profile with water depth

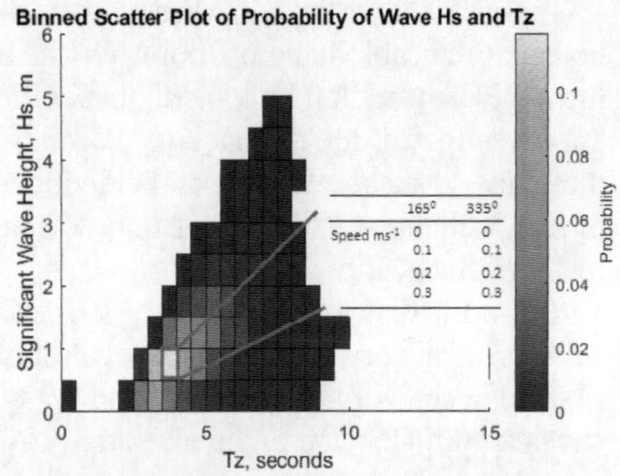

Figure 7. Subdivisions of wave loadings by current speed and direction

Figure 7 illustrates how each wave loading scenario was subdivided by current speed and directions. This plot was generated based on the assumption that the current flow is bi-directional and accelerates and decelerates from 0.3ms^{-1} in one direction to 0.3ms^{-1} in the other, similar to results found in [21]. From the calculated results the following environmental loading conditions were used for the model.

Table 3. Environmental Loadings

Environmental Loadings	
Wave height (m)	0.25 – 4.75
Wave period (s)	0.25 – 9.75
Wave direction (^{0})	15
Wind Speed (ms^{-1})	2.2 – 20.1
Wind Direction (^{0})	15
Current Speed ms^{-1}	0-0.3
Current Direction (^{0})	165 and 335

5.2. Dynamic Cable Model Output

Similar to works presented in [16] and [4], the hang off point of the cable is an area of concern for fatigue. Further investigation of the stress distributions and accumulated damage at this point specifically shall be given in these works, however it should be noted that any point along the length of the cable can be chosen.

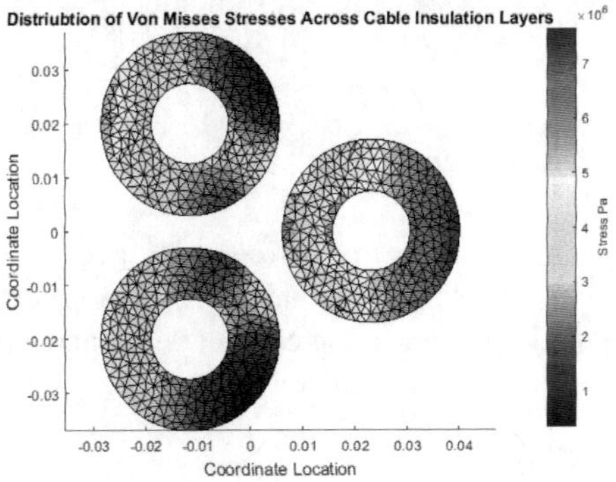

Figure 8. Stress distributions at hang off point.

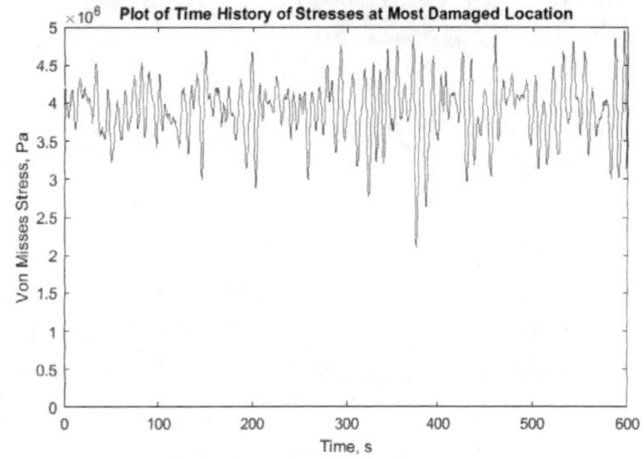

Figure 9. Stress time history at most damaged location.

IOP Conf. Series: Journal of Physics: Conf. Series **1102** (2018) 012016 doi:10.1088/1742-6596/1102/1/012016

Figure 8 shows the von Misses stress distributions of the cross linked polyethylene, XLPE, insulation layers at the cable hang off point. While any component of the cable cross section can be investigated, the XLPE layer shall be looked at specifically for further analysis. The stresses of the insulation layers are investigated due to concerns of water tree growth. It should be noted that for any cable component the plane or shear stresses can be deduced. Figure 8 is plotted using the outputs of the global dynamic model including, effective tension, x-bend moment, y-bend moment etc. and coupling them with the local mechanical model.

From the distribution, the stress time history of each location can be extracted. Figure 9 shows the stress time history of the XLPE location that experienced the most damage. It should be noted that the first 30 seconds of each simulation was discarded to allow for environment ramp up avoid the large stress amplitudes due to the platform moving from an initial stationary position.

5.3. Fatigue Damage

For each location the time history can be extracted, and the accumulated fatigue damage can be calculated using equation (3). This allows for the distribution of fatigue damage across the cable insulation layers to be plotted, shown in figure 10.

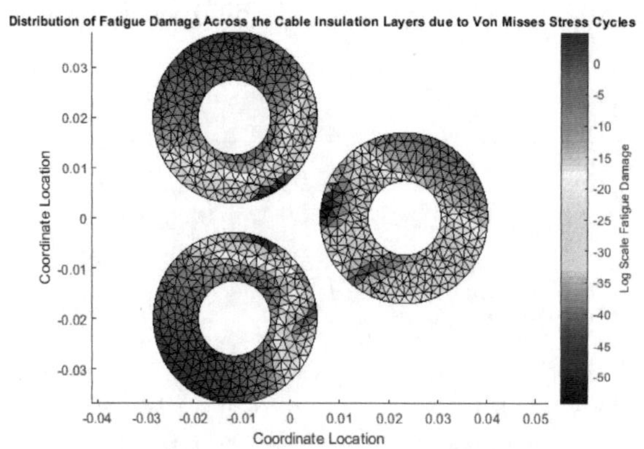

Figure 10. Log scale distribution of fatigue damage.

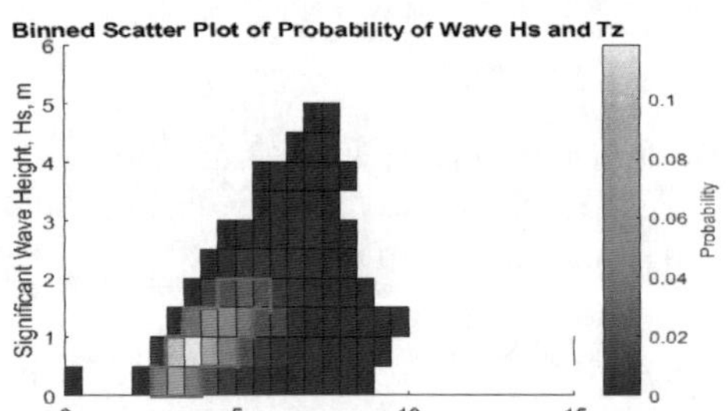

Figure 11. Wave spectrum distribution with greatest probability conditions highlighted.

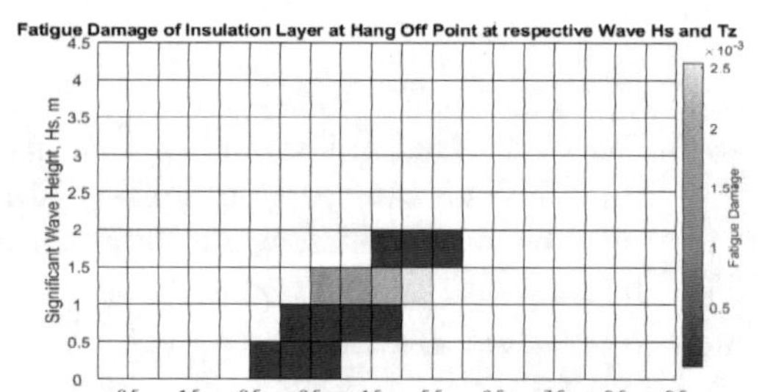

Figure 12. Calculated fatigue damage for the highlighted wave spectrum conditions.

Figure 11 shows the previously calculated wave spectrum, but with the most frequently occurring waves circled. Figure 12 shows the damage calculated for each of these wave conditions. It should be noted that these total damages incorporate the damages calculated for each of the current velocities and directions as depicted in Figure 7, but does not account for the occurrence of each wave type.

6. Discussion

The presented work can be used to give an estimation of the fatigue damage of any component of a dynamic cable's cross section. This work focused on the von Misses stress distribution, σ_{vm}, of the dynamic cable's insulation layers and assumed that in these preliminary works that both compression and tension contributed to damage within the cable insulation layers. Work was completed to deduce the impact of varying wave heights, periods and current velocities on the mechanical fatigue damage of the insulation layers of a dynamic cable. It was found that greater wave heights resulted in a greater level of fatigue damage. This could be contributed to greater wave heights causing a greater heave of the floating platform, therefore increasing the tension on the dynamic cable.

It was found that a longer wave period increases the fatigue damage calculated. This is contrary to the thought that a greater wave frequency would inhibit the platform from returning to its undisturbed position, keeping the cable in a state of stress. While wave celerity does increase with wave period, the orbital, or horizontal velocities, of the water particles decrease with wave period. These velocities can determine the force of the impact of the waves on the floating platform using Morrison's equation, where the force in the x-direction on a body in unsteady flow with velocity *U(t)* is:

$$F_x(t) = \rho C_m V\dot{U} + \frac{1}{2}\rho C_d A\,|U|U \tag{4}$$

Where V is the body volume in m^3, ρ is the mass density of the fluid in kgm^{-3}, $\dot{U}$ is the flow acceleration in ms^{-2}, A is the body reference area in m^2, C_d and C_m are the drag and momentum coefficients respectively.

Using equation 4, it can be quickly deduced that a reduction in velocity would result in a reduced force exerted on the platform from the waves, thus reducing the displacement of the platform. From figure 12 the wave periods ranging from 3.75s to 5.25s were investigated to determine the displacements of the floating platform, shown in table 4.

Table 4. Displacement of floating platform at varying wave period and wave height 1.25m

Wave Period Tz (s)	Average X displacement (m)	Average Y displacement (m)	Average Z displacement (m)	Damage
3.75	3.34	-0.081	-0.037	0.0013
4.25	3.34	-0.080	-0.037	0.0013
4.75	3.35	-0.080	-0.037	0.0019
5.25	3.34	-0.08	-0.037	0.0024

From table 4 it is seen that at varying wave periods the average platform displacement minimally changed, if at all, however the damage of the dynamic cable insulation layers increased. This increase in damage can perhaps be explained by a greater frequency of platform movement. If a closer look is taken at the stress time histories of the most damaged insulation locations, it can be seen that a wave period of 5.25s compared to a period of 3.75s, produces a stress history with more stress amplitudes, evident in the jagged plot for the 5.25s wave period shown in figure 13. This greater stress frequency can perhaps be due to the period of the waves matching the eigen period of the system, however this has not been investigated in any detail.

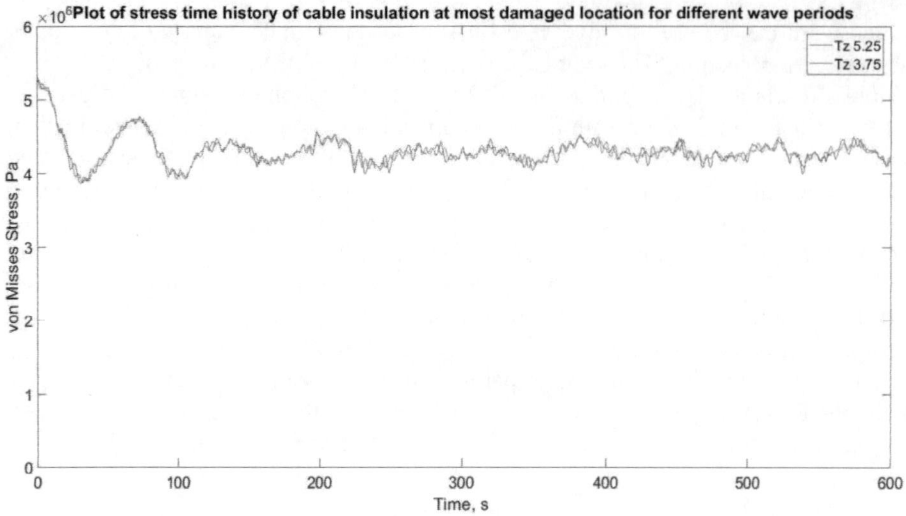

Figure 13. Stress time histories of the cable insulation at differing wave periods

It is important to note that the magnitude of the fatigue damage of the insulation layers is small. This is to be expected as the armouring layers surrounding the insulation layers are protecting them from damage. However when the extreme cases were simulated for a H_s of 4.75m and a Tz of 7.75s, the fatigue damage calculated was much larger than 1, indicating a cable failure, even of the insulation layers. Unfortunately this result was omitted from figure twelve due to its fatigue value being much larger than the others. As figure 12 shows the fatigue damages scaled to one another, this distorts the plot and results in the details of the smaller fatigue differences being lost. Furthermore as this work is the initial steps in analysing water treeing within the insulation layers, extreme wave conditions would cause a cable failure outside of that of water treeing.

When considering the impact of varying current speeds and directions it was found that when the current is travelling up the coast in Figure 1 (direction 165^0) the damage is greater than when it is travelling in the opposite direction, 335^0. It can be seen that when the current speeds are at their greatest in the tidal cycle, this also corresponds to a greater level of cable insulation fatigue damage. This output from the model could be used in the future to aid in cable installation layout and orientation decisions to reduce cable fatigue.

While this work presented von Misses stress distributions, the model allows the user to look at individual plane or shear stresses a component experiences. When these individual stresses are looked at for a particular environmental loading, it can be see that the largest contributing stress is σ_{zz} as seen in table 5.

Table 5. Magnitude of insulation fatigue damage at most damaged insulation location due to corresponding stress type.

Stress Type	Fatigue Damage
σ_{xx}	4.53e-41
σ_{yy}	5.28e-50
σ_{zz}	5.15e-06
σ_{xy}	2.74e-46
σ_{yz}	4.65e-25
σ_{xz}	3.41e-30
σ_{vm}	9.00e-06

In this case, this highlights that it is cable bending that contributes to insulation fatigue at the hang off point, rather than cable tension.

Unless in extreme storm conditions, the fatigue damage of the cable insulation calculated was very small. This could be a weakness in the model being under conservative. However it is important to note that the cable modelled was a dynamic cable used in historic oil and gas applications, and would have a much larger cable axial and bending stiffness than one that would be used for floating offshore wind applications. A further weakness of the model is that for the fatigue damage on the insulation layers, only one S-N curve could be found in previous works. To give a greater credibility to the results a larger number of S-N curves would have been desirable.

7. Conclusions

This work presents a method to determine mechanical fatigue damage across a dynamic cable's cross section, specifically the insulation layers, when exposed to a dynamic marine environment. It can be used as a tool to compare the insulation fatigue damage at differing lengths along the cables length. Individual components of the cable cross section can be investigated, giving way to a more comprehensive fatigue assessment of dynamic cables.

It was found that increasing wave height increases cable insulation fatigue damage in deep water applications. Increasing current velocities also increases cable insulation fatigue damage. Current direction plays a role in cable insulation fatigue damage, aiding in installation arrangement decisions. Finally, the works suggests that it is the stresses due to cable bending, as opposed to tension that contribute most to a dynamic cable's insulation fatigue damage.

8. Future Work

The next stages of this work is to use the most damaged cable insulation locations as initiation points of insulation water treeing. Using the stress cycles calculated within these works is hoped to allow for the modelling of cracks within the insulation as propagating mechanisms of water treeing.

The electrical stresses across the insulation layers are then to be modelled. It is envisioned this work serves at the first step in the development of a coupled mechanical and electrical fatigue model of the cable insulation which considers both mechanical stresses and electrical stresses due to the cable being energized.

9. Acknowledgements

The authors would like to thank the Energy Technology Institute and the Research Council Energy Programme for funding this research as part of the IDCORE programme (grant EP/J500847), and the Offshore Renewable Energy Catapult for providing technical assistance and equipment. A special thanks to Orcina for providing OrcaFlex software and to JDR cables for allowing presentation of cable data.

References

1. Carbon Trust. Floating Wind Joint Industry Project: Policy & Regulatory Appraisal [Internet]. 2017 [cited 2018 Mar 14]. Available from: https://www.carbontrust.com/resources/reports/technology/floating-wind-policy/

2. Marazzato H, Barber K, Jansen M, Graeme B. Cable Condition Monitoring to Improve Reliability. Olex Australia; 2004.

3. GCube. Down to the Wire: An Insurance Buyer's Guide to Subsea Cabling Incidents, June 2016 Report. GCube Insurance Services; 2016.

4. Thies PR, Johanning L, Smith GH. Assessing mechanical loading regimes and fatigue life of marine power cables in marine energy applications. Proc Inst Mech Eng Part O J Risk Reliab. 2011 Oct;1748006X11413533.

5. Buitrago J, Swearingen SF, Ahmad S, Popelar CF. Fatigue, Creep and Electrical Performance of Subsea Power Cable. In American Society of Mechanical Engineers; 2013 [cited 2018 May 23]. p. V003T03A028-V003T03A028. Available from: http://proceedings.asmedigitalcollection.asme.org/proceeding.aspx?articleid=1786236

6. Thies PR, Johanning L, Dobral C. Parametric Sensitivity Study of Submarine Power Cable Design for Marine Renewable Energy Applications. 2017 Jun 25;V03BT02A010.

7. Ildstad E, Bardsen H, Faremo H, Knutsen B. Influence of mechanical stress and frequency on water treeing in XLPE cable insulation. In: , Conference Record of the 1990 IEEE International Symposium on Electrical Insulation, 1990. 1990. p. 165–8.

8. Wang Z, Marcolongo P, Lemberg JA, Panganiban B, Evans JW, Ritchie RO, et al. Mechanical fatigue as a mechanism of water tree propagation in TR-XLPE. IEEE Trans Dielectr Electr Insul. 2012 Feb;19(1):321–30.

9. Yang S-H, Ringsberg JW, Johnson E. Parametric study of the dynamic motions and mechanical characteristics of power cables for wave energy converters. J Mar Sci Technol. 2017;

10. M. Raoof, Davies TJ. Determination of the bending stiffness for a spiral strand. J Strain Anal Eng Des. 2004;39(1).

11. Chen X, Fu S, Song L, Zhong Q, Huang X. Stress analysis model for un-bonded umbilical cables. Ocean Syst Eng. 2013;3(2):97–122.

12. Ekeberg KI, Ottesen T, Aarstein J, Saevik S, Ye N, Igland R. Predicting, Measuring and Implementing Friction and Bending Stresses in Dynamic Umbilical Design. In Offshore Technology Conference; 2006 [cited 2018 Mar 15]. Available from: https://www.onepetro.org/conference-paper/OTC-17986-MS

13. Fugro GEOS, Health & Safety Executive. Offshore Technology Report, "Wind and wave frequency distributions for sites around the British Isles." 2001.

14. Lewis M, Neill SP, Robins P, Hashemi MR, Ward S. Characteristics of the velocity profile at tidal-stream energy sites. Renew Energy. 2017 Dec 1;114:258–72.

15. Germanischer Lloyd Rules and Guidelines, Industrial Services. Guideline for the Certification of Wind Turbines,. 2005.

16. Young D, Ng C, Oterkus S, Li Q, Johanning L. Predicting failure of dynamic cables for floating offshore wind. In: 3rd International Conference on Renewable Energies Offshore. 2018.

17. Jonkman B. FAST-OrcaFlex Interface | NWTC Information Portal [Internet]. 2016 [cited 2018 Mar 15]. Available from: https://nwtc.nrel.gov/OrcaFlexInterface

18. Featherstone J. Dynamic Subsea Power Cables for Floating Wind & 66kV Subsea Power Cables, 6th Annual Advanced Submarine Power Cable & Interconnection Forum, Berlin. 2017 Jun.

19. Silva D, Balena R, Lisbôa R. Methodology for Thermoplastic Umbilical Cross Section Analysis. 2012 Jul 1;407–11.

20. Gaidai O, Ye N, Jin J, Reid D, Mainçon P. Fatigue Analysis Methods of Dynamic Umbilicals. In International Society of Offshore and Polar Engineers; 2015 [cited 2018 May 15]. Available from: https://www.onepetro.org/conference-paper/ISOPE-I-15-595

21. Pheonix A. Development of a tidal flow for optimisation of tidal turbine arrays. Access to Research at NUI Galway; 2018.

Directional fatigue accumulation in wind turbine steel towers

RMM Slot[1,2], J Schwarte[3], L Svenningsen[1], JD Sørensen[2], and ML Thøgersen[1]

[1]EMD International A/S, Niels Jernes Vej 10, 9220 Aalborg, Denmark
[2]Dep. Of Civil Engineering, Thomas Manns Vej 23, 9220 Aalborg, Denmark
[3]Nordex Energy GmbH, Turbine Engineering, Am Vögensteich 23, 18057 Rostock, Germany

E-mail: rmms@civil.aau.dk

Abstract. Wind turbines are subject to fatigue loads during their entire lifetime of 20-25 years. A main source of the fatigue loads is the turbulence, which varies with direction due to the surrounding terrain and wake effects inside wind farms. A common approach to assess wind turbine fatigue loads is to simulate the structural response based on a site-specific wind climate, described in the IEC 61400-1 standard. To reduce the amount of needed simulations the standard introduces an "effective turbulence" approximation that integrates directional variation of turbulence, resulting in an omnidirectional value. This method implicitly assumes that all wind turbine components face the wind directly, which is a conservative simplification for components below the yaw bearing.

Using wind measurements from almost one hundred international sites, we show how this simplification leads to over-predictions of tower fatigue loads of up to 23% compared to directional fatigue accumulation. Three simplified models are developed to approximate the directional fatigue damage using various levels of information ranging from only the wind rose to full sector wise simulations. The first two recommended models may be used as proxies to decide if sector wise simulations are feasible, and the last model accurately predicts the full directional fatigue damage. The simplified models can contribute to a reduced material consumption of wind turbine towers, thereby reducing the cost of wind energy.

1. Introduction

Wind turbines experience fatigue loads during their lifetime of 20-25 years due to the fluctuating excitation from the turbulent wind field. An important task for engineers is therefore to verify that a given wind turbine design can withstand the site-specific fatigue loads in the entire design lifetime. For this purpose the design standard IEC 61400-1 ed. 3 [1] describes how a site-specific wind climate can be evaluated in terms of the windspeed distribution, windspeed standard deviation (turbulence), vertical windspeed variation (wind shear), air density, and flow inclination. Together, these parameters form the basis for simulating structural loads using aero-elastic codes, to check whether the site-specific fatigue loads are within the design loads of the considered wind turbine.

It is computationally expensive to simulate wind turbine loads, which makes it important to ensure that a site-specific wind climate is specified such that no excessive simulations are required. In the IEC standard[1] [1] this is addressed by adopting the "effective turbulence" approximation, first introduced by

[1] The design standard IEC 61400-1 ed. 3 is referred to as "IEC standard" in the remainder of the paper.

Frandsen [2], which integrates out the directional variation of the turbulence. The accuracy of the effective turbulence is beyond the scope of this paper but has been investigated in e.g. [3] using wind data from two existing wind farms, which showed good agreement between fatigue loads obtained by sector-wise and omnidirectional simulations. This was further confirmed in [4] where it was shown that combining the effective turbulence with bi-linear material fatigue strength models leads to accurate predictions of fatigue loads. However, neither [3] nor [4] investigated the consequence of neglecting directional variation of turbulence, which directly implies that fatigue damage is accumulated at the same point on a given component, regardless of the varying wind direction. This was shortly discussed in [5] stating that the rule of thumb in the industry was that neglecting directions overpredicted tower fatigue damage with 10%. As potential consequence wind turbine towers may have been over-designed for more than a decade, thereby increasing the cost of wind energy.

The aim of this work is to quantify the consequence of not taking wind direction into account when accumulating fatigue damage in a steel wind turbine tower. A modern multi-megawatt turbine is considered and wind measurements from 99 international sites are used to define real characteristic site-specific wind climates. The large number of sites is subsequently used to develop three simple models that approximate the directional fatigue damage based on different levels of information ranging from only the wind rose to full directional simulations. Together, these models form the basis for improved decision support when it is unclear if sector-wise simulations are worth the increased investment, compared to the expected material reduction by doing directional fatigue accumulation in the tower.

2. Wind measurements and site-specific wind climates

The site-specific wind climates used in this study are based on high quality 10 min. measurements of mean windspeed, mean wind direction, turbulence, and wind shear from 99 international sites. At all sites the measurements span exactly one year to account for seasonal variation and the surrounding terrain represent varying types of orography with and without nearby forest. Neither air density nor flow inclination has been measured and for consistency these parameters are fixed at all sites as 1.225kg/m^3 and 0.0°, respectively. This is a simplification but these two parameters have insignificant influence on wind turbine tower fatigue loads [6,7]. This setup was also used in [6] where a more detailed description of the sites and the measurements is presented.

To determine the characteristic site-specific wind climates according to the IEC standard [1], the wind measurements are grouped into windspeed bins of 1 m/s and 12 sectors (θ) covering 30° each. The characteristic turbulence ($\sigma_{U,c}$) is determined as the 90% quantile according to Eq. (1) using the windspeed and sector-wise turbulence mean value ($\sigma_{U,\mu}$) and standard deviation ($\sigma_{U,\sigma}$).

$$\sigma_{U,c}(U,\theta) = \sigma_{U,\mu}(U,\theta) + 1.28\sigma_{U,\sigma}(U,\theta) \tag{1}$$

The characteristic wind shear (α_c) is derived from the wind measurements as the mean value, and the windspeed distribution and sector frequencies are taken directly from the measurements.

To reflect that most modern turbines are installed in wind farms, artificial wake added turbulence has been considered. Neighbouring wind turbines are assumed located 5 rotor diameters (RD) up and downwind in the main wind direction, and 3 RD perpendicular to that. The wake added turbulence is modelled by Eq. (2) according to [1], where C_T is the thrust coefficient. Note that the wakes are assumed to cover an entire sector independent of the distance to the assumed neighbouring turbine. This simplification is not unrealistic as the main purpose of introducing the wakes is to have a directional variation of the turbulence that reflects modern utility turbines.

$$\sigma_{U,T}(U,\theta) = \sqrt{\frac{U^2}{\left(1.5 + \frac{0.8RD}{\sqrt{C_T(U)}}\right)} + \sigma_{U,c}(U,\theta)^2} \tag{2}$$

Finally, the characteristic turbulence values including wake contributions are used to calculate the damage equivalent effective turbulence according to Eq. (3), taken from [1], where m is the "Wöhler exponent" related to material fatigue strength.

$$\sigma_{U,\mathrm{eff}}(U,m) = \left(\int_0^{2\pi} \sigma_U^m(U,\theta) f_\theta(\theta|U) d\theta \right)^{1/m} \tag{3}$$

In table 1 the characteristic wind climate parameters are summarized.

Table 1. Characteristic wind climate parameters used for site-specific fatigue assessment.

Description	Measured	Characteristic value	Notation
Wind direction	Yes	Sector-wise frequency in 30° bins	f_θ
Windspeed	Yes	Windspeed frequency in 1m/s bins	f_U
Turbulence	Yes*	Windspeed and sector dependent 90% quantile including wake contribution	$\sigma_{U,\mathrm{T}}$
Wind shear	Yes	Mean value	α_c
Air density	No	Fixed value (1.225 kg/m^3)	ρ_c
Flow inclination	No	Fixed value (0.0°)	φ_c

**only the ambient turbulence is measured.*

3. Wind turbine simulation and fatigue assessment method

Fatigue damage is assessed by aero-elastic simulations of the 5MW reference wind turbine by NREL [8] during normal operation (design load case 1.2 in [1]). The turbine is simulated using FAST [9] and the turbulent wind fields are generated by TurbSIM [10], using 20 seeds for each combination of characteristic wind climate parameters.

The material fatigue strength of the steel tower is modelled by a linear SN-curve with a Wöhler exponent of 4, which relates a given stress range to the number of cycles to failure. To accumulate fatigue damage from varying stress ranges, linear summation by Miner's rule [1] is performed based on Rainflow counting [11] of the output response from the simulations. This allows the fatigue damage to be expressed in terms of a "damage equivalent load" (DEL) as defined in Eq. (4), where N_{eq} is an equivalent number of cycles set to 10^7 and n_i is the number of cycles with moment range ΔM_i.

$$\mathrm{DEL} = \left(\frac{1}{N_{\mathrm{eq}}} \sum_i n_i \Delta M_i^m \right)^{1/m} \tag{4}$$

The combined equivalent fatigue load of the tower (F_{eq}) from different windspeeds and directions can be combined according to Eq. (5), where U_{in} and U_{out} represent the cut-in and cut-out windspeed of the turbine, respectively.

$$F_{\mathrm{eq}}(f_\theta, f_U, U, \sigma_U, \alpha, \rho, \varphi, m) = \left(\int_0^{2\pi} f_\theta(\theta) \int_{U_{\mathrm{in}}}^{U_{\mathrm{out}}} f_U(U|\theta) \mathrm{DEL}(U, \sigma_U, \alpha, \rho, \varphi)^m \, d\theta dU \right)^{1/m} \tag{5}$$

IOP Conf. Series: Journal of Physics: Conf. Series **1102** (2018) 012017 doi:10.1088/1742-6596/1102/1/012017

4. Quantification of the consequence of neglecting direction

This section quantifies the consequence of not explicitly accounting for wind direction in fatigue accumulation of wind turbine towers. First, the method used to fully account for direction directly in the time domain is presented followed by a short discussion of the results.

To assess the directional fatigue damage the tower cross-section is divided into 36 equidistant points as shown in figure 1. For the sake of simplicity, the cross-section is assumed to be circular symmetric, hence, no door is considered. This is not expected to influence the conclusion of this paper, as the effect of the door is handled by local re-enforcements whereas the scope of this paper is the entire tower.

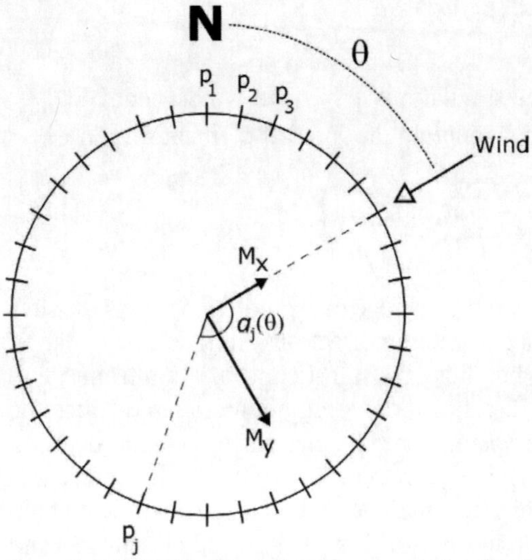

Figure 1: Sketch of the tower bottom cross-section (no door is considered).

For each simulated time step (t_s) the projected moment at each of the 36 points is evaluated by Eq. (6) where M_x is the tower side-side moment, M_y is the tower fore-aft moment, M_{p_j} is the projected moment at point p_j, and a_j is the angle between the wind direction and point p_j.

$$M_{\mathrm{p}_j}(t_s, \theta) = M_x(t_s)\sin\left(a_j(\theta)\right) + M_y(t_s)\cos\left(a_j(\theta)\right) \tag{6}$$

Based on the projected moment time series the fatigue damage in the tower is evaluated at each point ($F_{\mathrm{eq,p}_j}$) as outlined in Eq. (7).

$$F_{\mathrm{eq,p}_j}\left(f_\theta, f_U, U, \sigma_{U,\mathrm{T}}, \alpha_c, \rho_c, \varphi_c, m, \theta\right) = \left(\int_0^{2\pi} f_\theta(\theta) \int_{U_{\mathrm{in}}}^{U_{\mathrm{out}}} f_U(U|\theta)\mathrm{DEL}_{\mathrm{p}_j}\left(U, \sigma_{U,\mathrm{T}}, \alpha_c, \rho_c, \varphi_c, \theta\right)^m \, d\theta dU\right)^{1/m} \tag{7}$$

The point-wise damage equivalent load for each wind climate combination ($\mathrm{DEL}_{\mathrm{p}_j}$) is given by Eq. (8) where $\Delta M_{\mathrm{p}_j,i}$ are the projected point-wise moment ranges.

$$\mathrm{DEL}_{\mathrm{p}_j}(U, \sigma_{U,\mathrm{T}}, \alpha_c, \rho_c, \varphi_c, \theta) = \left(\frac{1}{N_{\mathrm{eq}}}\sum_i n_i \Delta M_{\mathrm{p}_j,i}^m\right)^{1/m} \tag{8}$$

The critical point with the maximum combined fatigue damage ($F_{\mathrm{eq,p_{max}}}$) is then assumed representative for the tower design, and the consequence of not taking direction into account is quantified as the fatigue damage ratio defined by Eq. (9).

$$\Delta F_{\text{eq,eff}} = \frac{F_{\text{eq,eff,}y}(f_U, U, \sigma_{U,\text{eff}}, \alpha_c, \rho_c, \varphi_c, m)}{F_{\text{eq,p}_{\max}}(f_\theta, f_U, U, \sigma_{U,\text{T}}, \alpha_c, \rho_c, \varphi_c, m, \theta)} \tag{9}$$

Where $F_{\text{eq,eff,}y}$ is the site-specific tower fore-aft fatigue damage found by omnidirectional simulations, see Eq. (10).

$$\begin{aligned} &\text{F}_{\text{eq,eff,}y}(f_U, U, \sigma_{U,\text{eff}}, \alpha_c, \rho_c, \varphi_c, m) \\ &= \left(\int_{U_{\text{in}}}^{U_{\text{out}}} f_U(U)\text{DEL}_{\text{eff,}y}\big(U, \sigma_{U,\text{eff}}, \alpha_c, \rho_c, \varphi_c\big)^m \, d\theta dU \right)^{1/m} \end{aligned} \tag{10}$$

The damage equivalent load from the effective turbulence simulations at a given windspeed, $\text{DEL}_{\text{eff,}y}$, is defined by Eq. (11) where $\Delta M_{\text{eff,}y,i}$ is found by Rainflow counting the fore-aft moment response.

$$\text{DEL}_{\text{eff,}y}\big(U, \sigma_{U,\text{eff}}, \alpha_c, \rho_c, \varphi_c\big) = \left(\frac{1}{N_{\text{eq}}} \sum_i n_i \Delta M_{\text{eff,}y,i}^m \right)^{1/m} \tag{11}$$

Figure 2 shows the results of Eq. (9) across all 99 sites, where the average overprediction across the sites is 14% with a maximum of up to 23%. This documents that neglecting directional fatigue accumulation leads to unnecessary steel consumption at most real sites, but there is a significant variation from site to site. Since full sector-wise calculations require at least 12 times more simulations this motivates the need for simple approximate methods to assess the site specific overprediction, to provide decision support whether this increased computational cost is justified. Furthermore, the full method presented here requires postprocessing of the results from the aero-elastic simulations directly in the time domain (besides cycle counting), thereby introducing an extra step in the workflow from wind climate to wind turbine loads. To alleviate this effort, an accurate method is proposed that use traditional fore-aft and side-side fatigue loads from the sector-wise simulations, without losing any significant accuracy compared to the full method presented in this section.

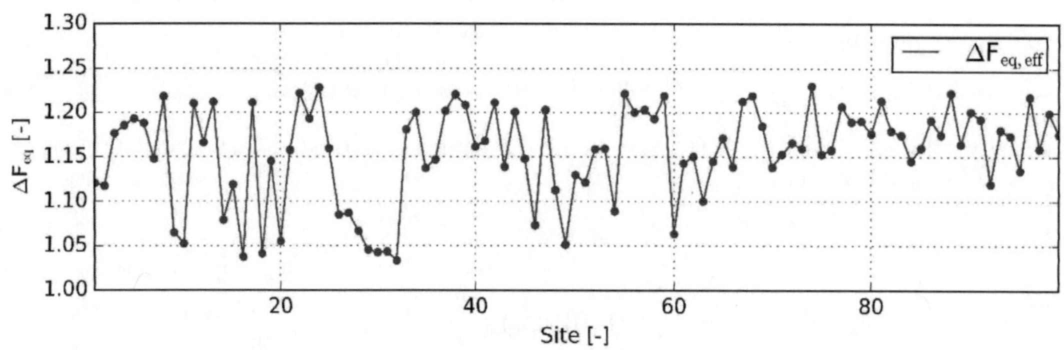

Figure 2: Overestimation of damage equivalent loads across all sites by doing omnidirectional fatigue assessment compared to explicitly accounting for direction.

5. Simplified framework to account for directional dependence of tower loads

In this section three models are developed to approximate the results of the full directional approach described in the previous section. Each model from one to three requires increased information and computational investment ranging from using the wind rose only to using sector-wise simulations.

5.1. Model 1: Spread of wind rose

A straightforward explanation for the large deviation of the overprediction of fatigue loads seen in figure 2 is the site-specific wind roses. When the wind is more unidirectional the overprediction is less significant. This is illustrated in figure 3 where $F_{\mathrm{eq,eff}}$ is plotted as function of the site-specific max frequency of the wind across the 12 sectors ($f_{\theta,\max}$) for all 99 sites. Note that $f_{\theta,\max}$ is a very simple metric to describe the spread of the wind rose and it contains no information of the windspeed or turbulence distribution in the sectors which are both governing factors for the fatigue damage. This results in significant scatter in the datapoints, but the tendency that lower overpredictions follow with higher $f_{\theta,\max}$ is clear.

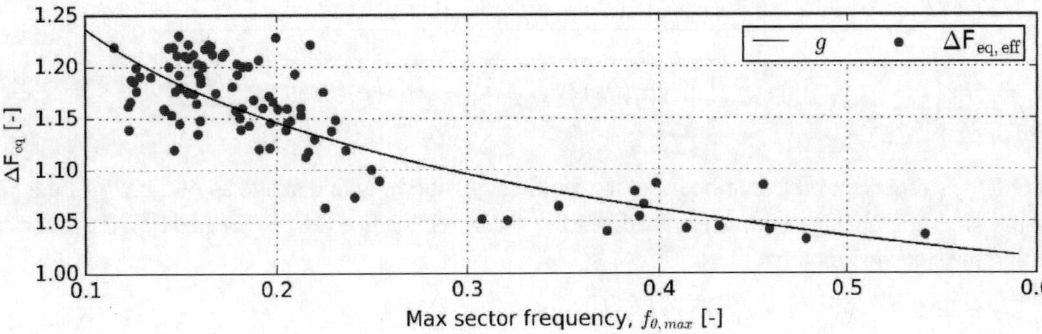

Figure 3: Overprediction of fatigue damage by not accounting for direction shown as function of the frequency of wind in the main direction.

By fitting a power function, g, to the observations a very simple tool for decision support is provided to predict if sector-wise simulations are feasible. The accuracy of this method is quantified by comparison to the full method as defined in Eq. (12), where the fitted function is given by Eq. (13).

$$\Delta F_{\mathrm{eq,M1}} = \frac{g^{-1}(f_{\theta,\max}) \cdot F_{\mathrm{eq,eff},y}(f_U, U, \sigma_{U,\mathrm{eff}}, \alpha_c, \rho_c, \varphi_c)}{F_{\mathrm{eq,p_{max}}}(f_\theta, f_U, U, \sigma_{U,\mathrm{T}}, \alpha_c, \rho_c, \varphi_c, \theta)} \tag{12}$$

$$g(f_{\theta,\max}) = 0.96 \cdot f_{\theta,\max}^{0.11} \tag{13}$$

5.2. Model 2: Omnidirectional simulations

To approximate $F_{\mathrm{eq,p_{max}}}$ directly based on fatigue load estimates from omnidirectional simulations, the simplified method described by Eq. (14) is proposed, where $f_\theta(\theta_k)$ is the frequency of wind measurements in sector k, a_j is the angle between sector k and point p_j, and $F_{\mathrm{eq,eff},x}$ describes the side-side bending fatigue damage from the omnidirectional simulations (estimated by the same procedure as $F_{\mathrm{eq,eff},y}$).

$$F_{\mathrm{eq,M2}}(f_\theta, f_U, U, \sigma_{U,\mathrm{eff}}, \alpha_c, \rho_c, \varphi_c, \theta) = \max_{\mathrm{p}_j}\left(\sqrt[m]{\sum_{k=1}^{12} f_\theta(\theta_k) \cdot \left|\sin\left(a_j(\theta_k)\right) F_{\mathrm{eq,eff},x} + \cos\left(a_j(\theta_k)\right) F_{\mathrm{eq,eff},y}\right|^m}\right) \tag{14}$$

To quantify the accuracy of this method the predictions are compared to the full method as defined in Eq. (15).

$$\Delta F_{\mathrm{eq,M2}} = \frac{F_{\mathrm{eq,M2}}(f_\theta, f_U, U, \sigma_{U,\mathrm{eff}}, \alpha_c, \rho_c, \varphi_c, \theta)}{F_{\mathrm{eq,p_{max}}}(f_\theta, f_U, U, \sigma_{U,\mathrm{T}}, \alpha_c, \rho_c, \varphi_c, \theta)} \tag{15}$$

5.3. Model 3: Sector-wise simulations

This last method is intended to accurately approximate the fatigue loads of the full method without explicitly accounting for the fore-aft and side-side moment interaction in the time domain. It is initially based on the assumption that the tower fore-aft and side-side moment accumulate fatigue damage fully independent, in which case the full method of assessing the fatigue damage boils down to the fatigue damage combination in Eq. (16).

$$F_{\text{eq,indep}}(f_\theta, f_U, U, \sigma_{U,\text{T}}, \alpha_c, \rho_c, \varphi_c, \theta) = \max_{\text{p}_\text{j}}\left(\sqrt[m]{\sum_{k=1}^{12} f_\theta(\theta_k)\cdot\left(\sqrt[m]{\left|\sin\left(a_j(\theta_k)\right)F_{\text{eq},x,k}\right|^m + \left|\cos\left(a_j(\theta_k)\right)F_{\text{eq},y,k}\right|^m}\right)^m}\right) \tag{16}$$

Where $F_{\text{eq},x,k}$ represent the combined sector specific side-side fatigue damage as outlined in Eq. (17), with the side-side damage equivalent load ($\text{DEL}_{x,k}$) defined by Eq. (18), where $\Delta M^m_{x,k,i}$ is found by counting the side-side moment response.

$$F_{\text{eq},x,k}(f_U, U, \sigma_{U,\text{T}}, \alpha_c, \rho_c, \varphi_c, \theta_k) = \left(\int_{U_\text{in}}^{U_\text{out}} f_U(U|\theta_k)\text{DEL}_{x,k}\left(U, \sigma_{U,\text{T}}, \alpha_c, \rho_c, \varphi_c\right)^m d\theta dU\right)^{1/m} \tag{17}$$

$$\text{DEL}_{x,k}(U, \sigma_{U,\text{T}}, \alpha_c, \rho_c, \varphi_c) = \left(\frac{1}{N_\text{eq}}\sum_i n_i \Delta M^m_{x,k,i}\right)^{1/m} \tag{18}$$

Equations (17) and (18) describe the sector-wise side-side fatigue damage, but the procedure is the same for the fore-aft fatigue damage ($F_{\text{eq},y,k}$).

By checking the accuracy of Eq. (16) it was found that the assumption of independence leads to a slight non-conservative bias compared to the full method of ~1%. To calibrate the final version of model 3 a site-specific offset is therefore introduced by weighting the side-side loads slightly higher in the load combination by using a power of 2 instead of m as shown in Eq. (19).

$$F_{\text{eq,M3}}(f_\theta, f_U, U, \sigma_{U,\text{T}}, \alpha_c, \rho_c, \varphi_c, \theta) = \max_{\text{p}_j}\left(\sqrt[m]{\sum_{k=1}^{12} f_\theta(\theta_k)\cdot\left(\sqrt[2]{\left|\sin\left(a_j(\theta_k)\right)F_{\text{eq},x,k}\right|^2 + \left|\cos\left(a_j(\theta_k)\right)F_{\text{eq},y,k}\right|^2}\right)^m}\right) \tag{19}$$

The accuracy of the calibrated model 3 is quantified by comparison to the full method as defined in Eq. (20)

$$\Delta F_{\text{eq,M3}} = \frac{F_{\text{eq,M3}}(f_\theta, f_U, U, \sigma_{U,\text{T}}, \alpha_c, \rho_c, \varphi_c, \theta)}{F_{\text{eq,p}_\text{max}}(f_\theta, f_U, U, \sigma_{U,\text{T}}, \alpha_c, \rho_c, \varphi_c, \theta)} \tag{20}$$

5.4. Summary of the developed models

In Table 2 an overview of the three simplified methods is presented. For comparison the computational time is specified in a relative sense (Relative cost), as the actual computational time is highly dependent on the aero-elastic code and specific turbine that is considered.

Table 2. Simplified methods to account for directional fatigue damage variation.

Simplified approach	Accuracy assessment*	Relative cost	Short description
Model 1 - Eq. (12)	$\Delta F_{\text{eq,M1}}$	0	Proxy to assess the overprediction with no information of actual loads.
Model 2 - Eq. (15)	$\Delta F_{\text{eq,M2}}$	1/12	Proxy to predict the fatigue damage of the full model by omnidirectional simulations.
Model 3 - Eq. (19)	$\Delta F_{\text{eq,M3}}$	1	Accurate model to evaluate the fatigue damage of the full model without projecting moment time series.

**The accuracy of all methods is evaluated with respect to the same fatigue damage obtained by the full model for comparison.*

Note that model 3 requires the same amount of simulations as the full method making it very suitable for e.g. response surface methods to predict the fatigue loads based on pre-run simulations [12,13].

6. Results and recommendations

The results of applying the three developed models across all 99 sites are shown in figure 5 and summarized in Table 3. For comparison the results of using the omnidirectional effective turbulence are also presented, which show that the developed model results are more accurate at all sites. Care should be taken though, as both model 1 and 2 result in non-conservative fatigue assessments at some of the sites. Based on the observations the following recommendations are considered by the authors:

***Model 1** – Spread of wind rose*:
Using information of the wind rose only, this model provides an initial estimate of the consequence of neglecting direction when accumulating fatigue damage. A core benefit of this model is that it requires no expert knowledge of wind turbine simulation. The accuracy is comparable to the second method, but it cannot be used to predict fatigue loads directly. Instead, it may be used either to decide if omnidirectional simulations should be skipped entirely, or in conjunction with the results from the second method to assess whether directional simulations are feasible at the specific turbine location.

***Model 2** – Omnidirectional simulations:*
Using omnidirectional fatigue loads an estimate of the actual directional fatigue damage is obtained, but since the method may lead to non-conservative assessments it is only recommended as a proxy to decide if sector-wise simulations are justifiable. In contrast to the first approach, a clear benefit of this method is that it allows the designing engineer to compare the expected material savings in the fatigue limit state with the margins from other design load cases. The importance of this is illustrated by the following simple example: The initial simulations show that the ultimate limit state governs the tower design, hence, any expected reduction in material consumption in the fatigue limit state is irrelevant.

***Model 3** – Sector-wise simulations:*
This last model may be used directly to predict fatigue loads within 1% of the full method across all analysed sites. The computational cost is the same as for the full method, but it alleviates the designing engineers for the extra step in the workflow from wind climate to wind turbine loads of computing projected moments.

Table 3. Summarized results of all simplified methods.

Simplified approach	**ΔF_{eq} minimum [-]**	**ΔF_{eq} Maximum [-]**	**ΔF_{eq} Average [-]**	**ΔF_{eq} Std.dev. [-]**
Effective turbulence	1.033	1.230	1.156	0.053
Model 1	0.940	1.076	1.004	0.027
Model 2	0.938	1.097	1.013	0.029
Model 3	0.995	1.001	0.999	0.001

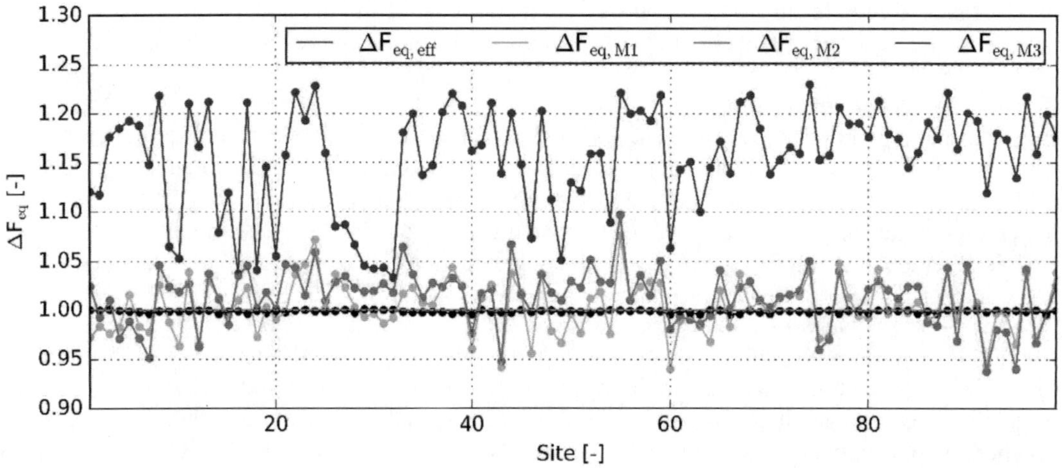

Figure 4: Results of analysing all 99 sites using the effective turbulence approximation (blue) and the three simplified models (yellow, red, and black) compared to the full directional approach.

7. Conclusions

Fatigue assessment of wind turbine components is commonly based on simulations using omnidirectional effective turbulence to limit computational time. This approach implicitly assumes that fatigue damage accumulates independently of the wind direction, which is a simplification for all components below the yaw bearing. Using wind measurements from 99 real sites this assumption has been shown to overestimate DELs by an average of 14% for wind turbine steel towers, compared to explicitly accounting for direction in the fatigue accumulation.

Utilizing the large number of available sites three simplified models have been developed to approximate the full sector-wise model. Model 1 is based only on wind measurements which makes it accessible to wind and site engineers without expert knowledge of wind turbine simulation. The method can predict the actual overprediction within 8% of the full model. Model 2 is based on results of omnidirectional simulations, making it possible to predict the actual fatigue loads of the full model within 10%. Both models 1 and 2 show promising results, but for some sites they are non-conservative, hence, they should only be used as proxies to decide if full sector-wise simulation are justified at the considered site and turbine location. The third method accurately predicts the directional fatigue damage using sector wise simulations, but without explicitly accounting for tower fore-aft and side-side bending in the time domain. This method captures the fatigue loads of the full model within 1% without any bias and is regarded as accurate as the time-based calculation.

The models and results that have been presented in this work may contribute to an overall reduction in the steel consumption of wind turbine towers, thereby reducing the cost of wind energy and the environmental impact of wind turbines when the entire life-cycle is considered.

IOP Conf. Series: Journal of Physics: Conf. Series **1102** (2018) 012017 doi:10.1088/1742-6596/1102/1/012017

Acknowledgement

The authors wish to thank the data providers; KNMI, ICDC, CliSAP/Klima-Campus, University of Hamburg, DTU, Vattenfall, and VENTUS INGENIERÍA. The presented work is part of the project "From wind climate to wind turbine loads – Efficient and accurate decision support and risk analysis" cofounded between Innovation Fund Denmark, EMD International A/S, and Aalborg University. Their financial support is greatly appreciated.

References

[1] IEC. 2010 International Standard IEC 61400-1 ed. 3, "Wind Turbines - Part 1 Design Requirements".

[2] Frandsen S T 2007 *Turbulence and turbulence- generated structural loading in wind turbine clusters* (Technical University of Denmark)

[3] Henriksen S S, Malcolm D J and Thomson J 2012 Effective Turbulence in Wind Turbine Site Suitability Assessment *European Wind Energy Conference and Exhibition 2012, EWEC 2012. 2.* pp 959–66

[4] Sørensen J D, Frandsen S and Tarp-Johansen N J 2008 Effective turbulence models and fatigue reliability in wind farms *Probabilistic Eng. Mech.* **23** 531–8

[5] Veldkamp D 2006 *Chances in Wind Energy* (Delft University)

[6] Slot R M M, Svenningsen L, Sørensen J D and Thøgersen M L 2018 Importance of Shear in Site Assessment of Wind Turbine Fatigue Loads *J Sol Energ* **140** 041012

[7] Dimitrov N, Natarajan A and Kelly M 2015 Model of wind shear conditional on turbulence and its impact on wind turbine loads *Wind Energy* **18** 1917–31

[8] Jonkman J, Butterfield S, Musial W and Scott G 2009 *Definition of a 5-MW reference wind turbine for offshore system development* (NREL/TP-500-38060)

[9] Jonkman J 2015 FAST An aeroelastic computer-aided engineering (CAE) tool for horizontal axis wind turbines

[10] B. Jonkman N K 2016 TurbSim A stochastic, full-field, turbulence simulator primarialy for use with InflowWind/AeroDyn-based simulation tools

[11] ASTM 2011 ASTM No. E1049-85, "Standard Practice for Cycle Counting in Fatigue Analysis".

[12] Toft H S, Svenningsen L, Moser W, Sørensen J D and Thøgersen M L 2016 Assessment of wind turbine structural integrity using response surface methodology *Eng. Struct.* **106** 471–83

[13] Murcia J P, Réthoré P-E, Dimitrov N, Natarajan A, Sørensen J D, Graf P and Kim T 2017 Uncertainty propagation through an aeroelastic wind turbine model using polynomial surrogates *Renew. Energy* **119** 910–22

IOP Conf. Series: Journal of Physics: Conf. Series **1102** (2018) 012018 doi:10.1088/1742-6596/1102/1/012018

Gearless generator with magnets on the stator for wind turbine

V Dmitrievskii [1,2], **V Prakht** [1,2]

[1] EMACH LLC, 620100, Tveritina 17-59, Yekaterinburg, Russia

[2] Ural Federal University, Department of Electrical Engineering and Electrotechnological Systems, 620002, Mira19, Yekaterinburg, Russia

E-mail: emf2010@mail.ru

Abstract. In this work a new design of multi-pole gearless low-speed Generator with Magnets on the Stator (GMS) for wind turbine is described. Gearless GMS combines both high number of magnetic poles created by magnets and low-pole winding creating strong electromotive force. Therefore GMS has high efficiency and low operational speed. The GMS mathematical model has been developed. The low-speed gearless GMS was designed with the help of the developed mathematical model. Comparison of the designed GMS with a generator with magnets on the rotor (SG) is given. The GMS active material cost is 2.3 times less than that for the SG. The GMS mass is 1.3 times less. The GMS has higher efficiency at the rated speed. Also GMS maintain high efficiency in wide range of torques and speeds and GMS torque ripple is low.

1. Introduction

Multi-pole synchronous generators with magnets on the rotor (SG) are commonly used in gearless wind turbine generators. However, making such SG of small outer diameter is not easy. For example, the gearless SG (350 rpm, 3141 W, 228 mm of outer diameter, model 300STK2M) with distributed winding has 24 poles on its rotor and 72 teeth on the stator (the slot number per pole and phase is $q = 1$) [1] (Figure 1a). Further increasing the number of the poles is difficult because it leads to increasing number of the stator teeth, decreasing the copper fill factor and, as a result, decreasing the efficiency.

A SG with concentrated winding can have a little higher number of poles than a SG with distributed winding. For example, gearless SG (200 rpm, 3000 W, 280 mm of outer diameter) has 30 poles on the rotor and 36 teeth on the stator (the slot number per pole and phase is $q = 0.4$) [2] (Figure 1b).

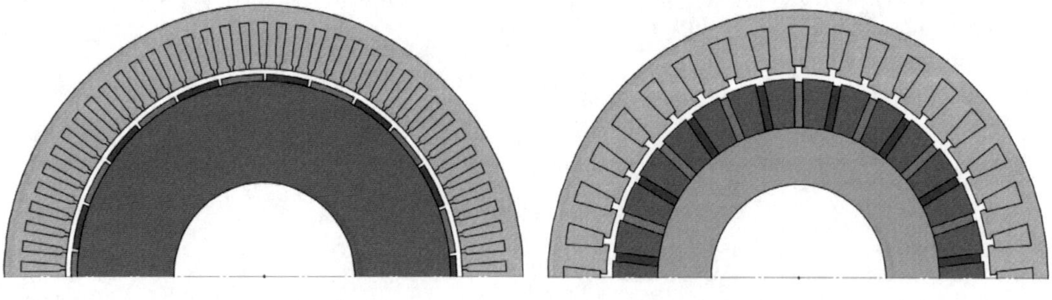

a) b)

Figure 1. a) SG design according to [1]; b) SG design according to [2].

However, significant drawbacks of the SG with concentrated winding and small fractional number q of the slots per pole and phase are high torque ripple and nonsinusoidal magnetomotive force (MMF) distribution in air gap which worsens the generator efficiency, specific power and torque.

IOP Conf. Series: Journal of Physics: Conf. Series **1102** (2018) 012018 doi:10.1088/1742-6596/1102/1/012018

These two SGs are commonly used solutions for wind generators and characterized in lessening the pole pitch when the number of poles increases. As a result, the winding MMF at the same current density decreases compared to that of the SG with lower number of poles which leads to decreasing the generator efficiency.

Generators with magnets on the stator (GMS) can be designed on the basis of Flux Reversal Machines (FRM). The first FRM was described in [3]. FRMs can be used as low-speed motors and generators because of easiness of increasing number of poles, even in small diameter of stator and without losing MMF. For example, the energy-efficient multi-pole single-phase FRM is described in [4]. Magnets on the FRM stator teeth create large number of poles (32 poles) in the small stator diameter (100 mm). The FRM stator winding is concentrated and has just 4 poles. Using the winding with small number of poles increases its MMF with the same current density. Therefore, the winding losses are decreased significantly which increases the efficiency. The comparison of the 32-pole FRM with the outer stator core diameter of 100 mm to the 16-pole synchronous machine with magnets in the rotor and with the outer stator core diameter of 110 mm is given in [4]. The FRM efficiency is much higher than that of the 16-pole synchronous machine with magnets in the rotor. However, single-phase machines such as in [3], [4] have large cogging torque and torque ripple and cannot be used in wind generators.

A vast overview of various three-phase FRM designs with magnets on the stator teeth is given in [5]. However, the windings of these machines are concentrated and with small pole pitch which results in decreasing the winding MMF, the efficiency and specific power. Also, those FRM have large stator slot openings creating a dip in magnetic field, decreasing the efficiency and the specific characteristics and increasing torque ripple. Hence, these machines are not widely used.

Multi-pole three-phase FRM with distributed winding (Figure 2a) is described in [6]. Authors of both paper [5], [6] note that FRM suits for direct driven wind power generation.

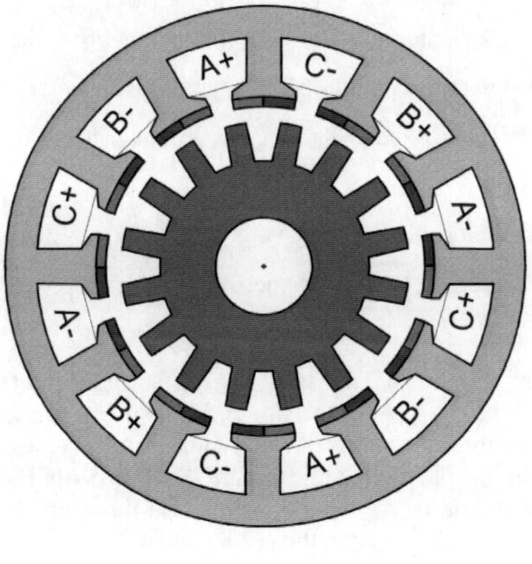

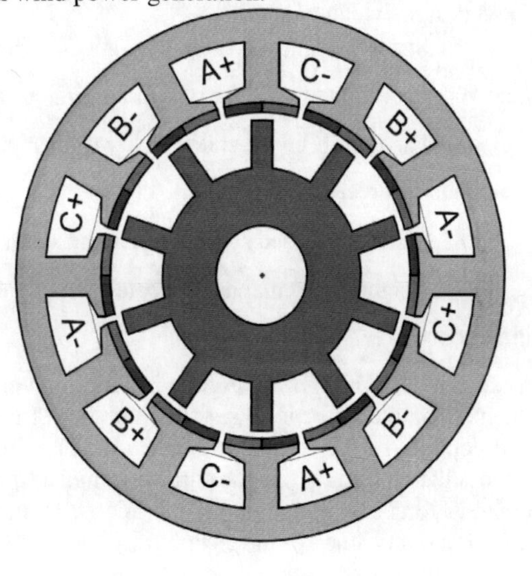

a) b)

Figure 2. a) FRM design according to [6] $z = 12$, $r = 16$, $q = 1$, $p = 2$, $m = 24$;
b) New FRM design $z = 12$, $r = 10$, $q = 1$, $p = 2$, $m = 24$.

The multi-pole FRM from [6] with the distributed winding contains (see Figure 2a):

- Stator with a twelve-tooth magnetic core and a distributed winding with four poles, where the number of pole pairs of the stator winding p is 2, and the number of poles per pole and phase q is 1.
- Each tooth of the stator has two magnets of equal size, the adjacent magnets located on one and the same stator tooth are magnetized in the opposite direction.
- The stator has open slots, which openings are approximately equal to the width of the magnet.

- Neighboring magnets located on adjacent stator teeth are magnetized identically.

In this machine, the number of teeth of the stator is 12, the number of magnets on the stator is 24, and the number of teeth of the rotor is 16, the surface of the stator teeth occupies about 2/3 (66.7%) of the inner stator surface, and 1/3 (33.3%) of the inner stator surface is not used (slots openings).

The disadvantages of this design are:

- Efficiency and specific power are not very high, because 1/3 (33.3%) of the stator surface is not used.
- Complex harmonic composition of the magnetomotive force (MMF) of the magnets due to a violation of the periodicity of their alternation by the large slot openings and the presence of additional undesirable harmonics in the MMF. Because of this, there are large torque ripple and cogging torque, acoustic noise and vibrations.

All FRMs considered in [5], [6], [7], [8] have a similar operating principle, the same faults (large moment pulsations and not very high efficiency) and are not widely used in practice.

2. Proposed new FRM design

A new FRM design with distributed winding and with small stator slot openings was described in the little-known paper [9]. However, multi-pole three-phase gearless machines based on that FRM design was not investigated in detail. In this paper, such design of the new multi-pole gearless FRM is proposed as a generator for wind turbines. Figure 2b demonstrates the simplest embodiment of such FRM design, which contains:

- Stator with a twelve-teeth magnetic core and a distributed winding with four poles, where the number of pole pairs of the stator winding p is 2, and the number of poles per pole and phase q is 1.
- Each tooth of the stator has two magnets of equal size, adjacent magnets are magnetized in the opposite direction.
- There are closed slots between the stator teeth, the opening of which are very small.
- Neighboring magnets located on adjacent stator teeth are magnetized in the opposite direction.

In the FRM shown in Figure 2b, the total number of teeth on the stator is 12, the number of magnets on the stator is 24, and the number of teeth of the rotor is 10, the slot opening is very small and depends on the thickness of the winding wire. Therefore, the surface of the stator teeth occupies almost the entire inner stator surface (almost 100%). In the FRM shown in Figure 2b, torque ripple, cogging torque, acoustic noise, vibrations, moment of inertia of the rotor is lower, and the specific power and efficiency of the electric machine is higher compared to that in the conventional FRM.

The number of teeth on the stator, the number of magnetic poles on the stator surface in the FRM with the distributed winding can be easily enhanced by increasing the number of slots per pole and phase q, which is especially important for gearless applications. According to the recommendation in [9], for each pair of the stator winding poles the number of rotor teeth is greater or less than the number of the magnets magnetized in the same direction by one. Namely, the number of rotor poles can be calculated as $r = m/2 \pm p$. For example, in the FRM shown in Figure 3, the total number of teeth on the stator is z = 48, the number of magnets on the stator is m = 96, q = 4, the number of pole pairs of the stator winding is p = 2. The number of rotor teeth is r = 96/2-2=46. The FRM (Figure 3) has distributed four-pole winding which allows obtaining high MMF at small current density which significantly decreases the winding losses and increases the efficiency of the FRM.

The electric frequency of the FRM generator is $f = r \cdot n / 60$ (measured in Hz), where n is the rotor speed measured in rpm.

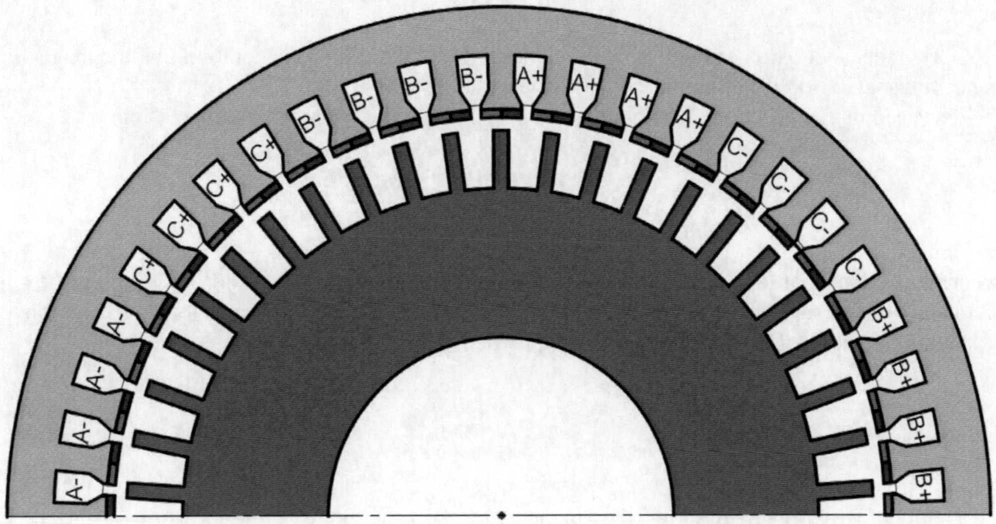

Figure 3. The proposed new FRM design with $z = 48$, $r = 46$, $q = 4$, $2p = 4$, $m = 96$.

In order to design and study the characteristics of FRM, the mathematical model based on the finite element method (FEM) has been developed.

3. Mathematical model description

Various approaches to the mathematical modeling of the single-phase FRM are given in [4], [10]. Those models are developed for supply by voltage of rectangular form. In this paper, sinusoidal form of supply currents is assumed for three-phase FRM. Such approach is wildly used for simulating other three-phase motors and generators fed with frequency converters. The model is based on solving a set of 2D magnetostatic problems. Usual gauge fixing $A_x = A_y = 0$ is used for vector potential:

$$B_x = \frac{\partial A_z}{\partial y}, \; B_y = \frac{\partial A_z}{\partial x}, \qquad (1)$$
$$\iint (\mathbf{H}\delta\mathbf{B} - J_z \delta A_z) dxdy = 0, \; J_z = \sum_k \Xi_k I_k,$$

where **H** and **B** are in-plane magnetic field and flux density. J_z is perpendicular to the plane component of the current density. Other components are equal to zero: $J_x = J_y = 0$. I_k is the current of the k-th phase, k = A, B, C; $\Xi_k(x,y)$ are the current density fields when only the k-th current I_k is nonzero and equal to 1 A. $\Xi_k(x,y)$ can be nonzero only in the stator slot areas.

In general case a set of magnetostatic boundary problems should be solved for various rotor positions. One and the same calculation area is used for all boundary problems. For doing so, the entire calculation area is divided into two parts. One part contains the rotor and the half of the air gap. Another part contains the stator and another half of the air gap. These two parts are joined with a boundary condition depending on the rotor position. As a result, rotor and stator are considered in their own reference system.

Due to the generator symmetry, the calculation area can be halved (Figure 3). Periodic condition is applied on the emerged boundary. Also, only rotor position interval of sixth of electrical period that is of 360 / 46 / 6 = 1.3 degrees should be taken into consideration.

The generator power, losses in the winding, in the steel and in the magnets, the efficiency, and the torque ripple and so on are calculated in post-processing.

Winding losses can be easily calculated using the Joule–Lenz law. The estimation of losses in stator and rotor cores is more complicated and is done in postprocessing.

IOP Conf. Series: Journal of Physics: Conf. Series **1102** (2018) 012018 doi:10.1088/1742-6596/1102/1/012018

During the calculation of the core losses it should be taken into consideration that the magnetic flux is not sinusoidal. So the harmonics contribution should be evaluated.

Averaged magnetic losses density in steel is supposed to be given by the expression:

$$p_{st} = \frac{\alpha}{\omega^2} < \left(\frac{d\mathbf{B}}{dt}\right)^2 >= \frac{\alpha B_{equ}^2}{2}, \quad (2)$$

where α is the proportionality factor determined on the basis of empirical data, $\omega = 2 \cdot \pi \cdot f$ is the angular frequency of the supply, < > is averaging over the interval of rotor position for which the boundary problems are solved, f is frequency and the equivalent flux density is

$$B_{equ} = \sqrt{2 < \left(\frac{d\mathbf{B}}{d\varphi_e}\right)^2 >} \quad (3)$$

The total derivative of ***B*** shows the change ***B***, at the same point of the medium (steel cores) with changing the rotor position (electric) angle φ_e.

Having B_{equ}, the C. Steinmetz expression was assumed for calculating the losses density:

$$p_{st} = \rho \cdot p_{50} \left(f / 50\right)^s B_{equ}^2 \quad (4)$$

where ρ is the steel density, p_{50} is losses in 1 kg of steel at 50 Hz and 1 T; s is the parameter which is taken as $s = 1.3$ for the majority of electrical steels.

The stator and rotor core losses can be found with integrating the expression (4).

Although coefficient in (4) is calculated using the steel datasheet data, the expression for α taking into consideration the steel eddy current losses is also known:

$$\alpha = \frac{\omega^2 \sigma \chi^2}{12}, \quad (5)$$

where σ is electric conductivity of considered material (steel), χ is a lamination thickness, that is a thickness of the material pieces in the direction perpendicular to the magnetic flux.

The equation (5) was used for estimating the losses in the magnets. The conductivity of the magnets is assumed to be 700 kS/m. The magnet arc length is taken for χ.

Formula (5) is derived from the assumption of uniformity of the magnetic flux density field. This is a good assumption for thin laminated steel. The magnetic field density can differ significantly along the magnet arc. However the calculations show that the magnet losses are not very high and cannot influence significantly on the FRM efficiency and its performance. Hence, accurate calculations of the eddy current losses in the magnets are not required.

4. Example of FRM design for wind generator

3 kW, 200 rpm FRM was designed on the basis of the developed mathematical model. The calculations are done for the laminates steel of M250-50A grade and for the rare-earth magnets which remanent flux density is 1.1 T and coercivity with respect to magnetization is 750 kA/m. The FRM geometry and vector potential field are shown in Figure 4. This figure demonstrates the emergence of the four-pole magnetic field. Also it can be seen that vector magnetic potential is not continuous and that of the stator is shifted with respect to that of the rotor which demonstrates the boundary condition joining both parts of calculation area.

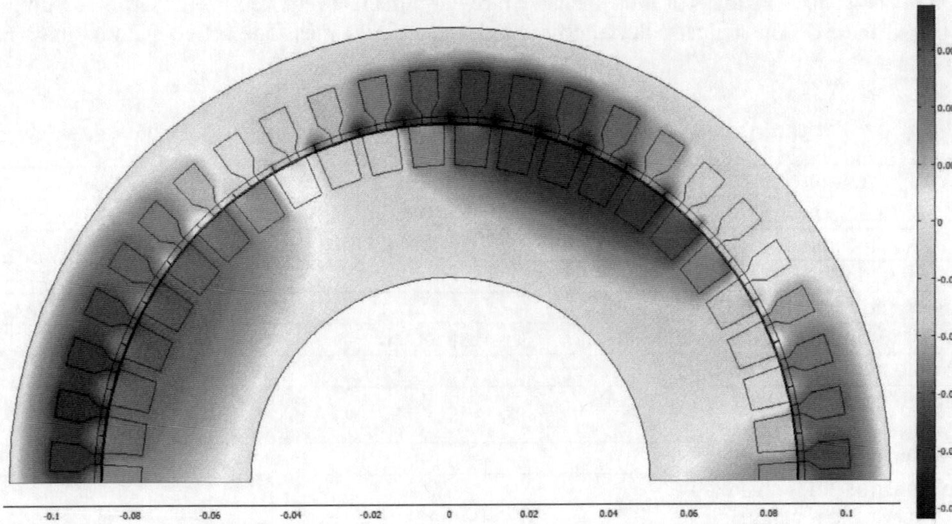

Figure 4. Geometry FRM and vector potential field (Wb/m).

Figure 5 demonstrates the flux density in FRM. In the most parts of the machine flux density does not exceeds 1T.

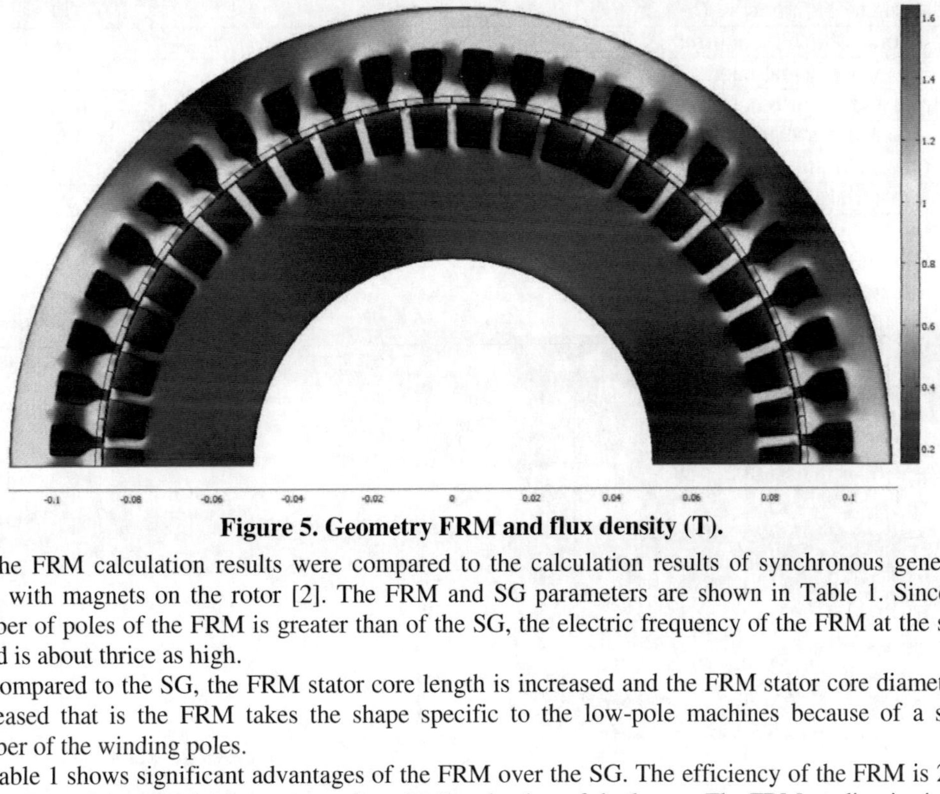

Figure 5. Geometry FRM and flux density (T).

The FRM calculation results were compared to the calculation results of synchronous generator (SG) with magnets on the rotor [2]. The FRM and SG parameters are shown in Table 1. Since the number of poles of the FRM is greater than of the SG, the electric frequency of the FRM at the same speed is about thrice as high.

Compared to the SG, the FRM stator core length is increased and the FRM stator core diameter is decreased that is the FRM takes the shape specific to the low-pole machines because of a small number of the winding poles.

Table 1 shows significant advantages of the FRM over the SG. The efficiency of the FRM is 2.3% higher than of the SG which corresponds to 32 % reduction of the losses. The FRM cooling is simpler and it can run at higher ambient temperature. The core volume that is $V = \pi D^2 L / 4$ (where D is the outer stator core diameter and L is the stator core stack length) is reduced by 25.6%.

Active materials costs also were calculated. The reference prices that were assumed for this calculation are: 50 Euro/kg for PMs, 7.5 Euro/kg for copper and 3 Euro/kg for stator and rotor steel [2].

IOP Conf. Series: Journal of Physics: Conf. Series **1102** (2018) 012018 doi:10.1088/1742-6596/1102/1/012018

The table 1 shows that the mass of the expensive rare-earth magnet required for the GMS is 5 times less than that of the SG. The active material cost is reduced by 2.3 times. The active material mass is reduced by 1.3 times.

Table 1. Parameters of 3kW 200 rpm gearless generators.

Parameters	FRM	SG [2]
Rated power, kW	3	3
Rated speed, rpm	200	200
Rated efficiency, %	94.6	92.3
Rated frequency, Hz	153	50
Winding type	distributed	concentrated
Number of winding poles, *p*	4	30
Number of stator teeth per pole and per phase, *q*	4	0.4
Number of stator teeth	48	36
Number of poles created by permanent magnets	96	30
Magnet type	rare-earth	rare-earth
Remanence of magnets, T	1.2	1.2
Stator outer core diameter (*D*), mm	222	280
Stack length of stator core (*L*), mm	237	200
Core volume (*V*), litres	2.9	3.9
Copper mass, kg	5.3	9.1
Stator and Rotor Steel mass, kg	38.6	44.7
Magnet mass, kg	1.0	5.6
Total active material mass, kg	45.0	59.4
Active materials estimated cost, Euro	206	480

The FRM efficiency map is shown in the Figure 6a. The FRM has high efficiency in wide range of torques and speeds.

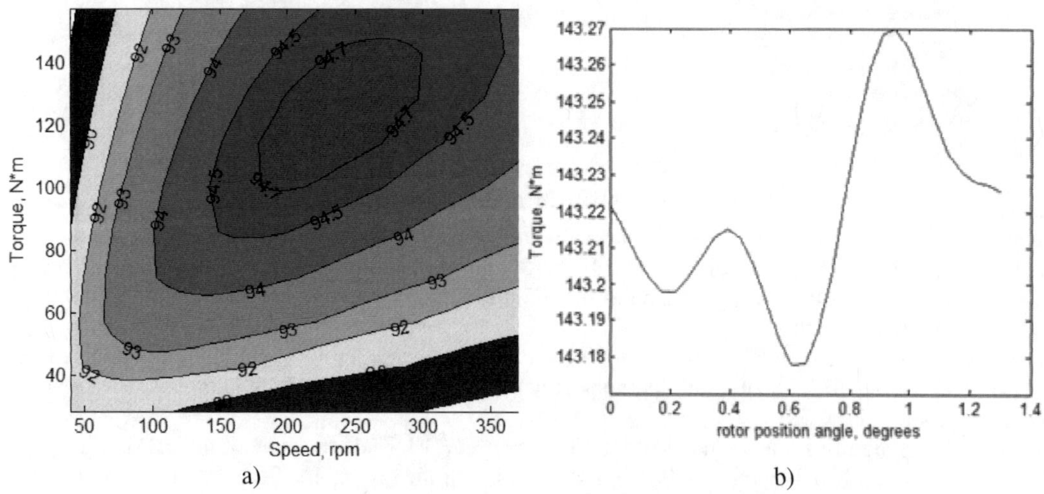

Figure 6. a) FRM efficiency map; b) FRM torque ripple.

Figure 6b shows the torque ripple at the rated load. The peak to peak value of the torque ripple is less than 0.1 % of the torque mean value. The torque ripple is extremely small although it was not even chosen as optimization parameter.

Since the 3kW SG was taken for the comparison, the 3kW GMS (FRM) was designed. Generators with higher rated power will have higher dimensions which make it possible to increase the number of poles. Therefore, a greater increase of the FRM rated power density compared to that of a traditional SG is expected in this case. This question requires more detailed investigations and will be considered in future work.

5. Conclusions

The paper describes a new GMS (Flux Reversal Machine) design developed for wind turbine application.

The main advantages of the gearless low-speed GMS compared to SG: 1) Making low-speed multi-pole generator with 80 poles on the stator and more in small diameter is easy; 2) Mass of the expensive rare-earth magnets is 5 times less than that in analogues with magnets on the rotor; 3) The toothed rotor is simple and reliable because its active part is made only of the laminated steel; 4) The active material cost twice as low as that of the analogue with magnets on the rotor; 5) The GMS has higher efficiency than analogues; 6) The torque ripple is extremely small although it was not even chosen as optimization parameter.

References

[1] Alternators Alxiion STK for Wind Turbines, PM Brushless Alternators for Direct Drive, ALXION Automatique & Productique, Catalogue, Edition February 2011 – Revision 2.1

[2] Pop A, Jurca F, Oprea C, Chirca M, Breban S, Radulescu M, Axial-flux vs. radial-flux permanent-magnet synchronous generators for micro- wind turbine application, *Proc. of the European Conference on Power Electronics and Applications, EPE 2013 (Lille, France)*

[3] Deodhar R, Andersson S, Boldea I, Miller T, The flux-reversal machine: a new brushless doubly-salient permanent-magnet machine, 1997 *IEEE Transactions On Industry Application* **33** 4

[4] Prakht V, Dmitrievskii V, Kazakbaev V, Sarapulov S, Steady-state model of a single-phase flux reversal motor, *Proc. of the 58th Annual Scientific Conference of Riga Technical University, RTUCON 2017 (Riga, Latvia)*

[5] Gao Y, Li D, Qu R, Li J, Design Procedure of Flux Reversal Permanent Magnet Machines, 2017 *IEEE Transactions On Industry Application* **35** 5

[6] More D, Fernandes B, Power density improvement of three phase flux reversal machine with distributed winding, 2009 *IET Electric Power Applications* **4** 2

[7] Pellegrino G, Gerada C, Modeling of Flux Reversal Machines for direct drive applications, *Proc. of the 2011 14th European Conference on Power Electronics and Applications (Birmingham, UK)*

[8] More D, Kalluru H, Fernandes B, Outer Rotor Flux Reversal Machine for Rooftop Wind Generator, *Proc. of the 2008 IEEE Industry Applications Society Annual Meeting, (Edmonton, AB, Canada)*

[9] Dmitrievskii V.A, Prakht V.A., Various configurations of electric machines with magnets on the stator teeth, *The journal of scientific publications of postgraduate and doctoral students, 12 (66), 2011.*

[10] Dmitrievskii V, Prakht V, Pozdeev A, Klimarev V, Mikhalitsyn A, Angular Grinder with New Flux Reversal Motor, *Proc. of the 18th International Conference on Electrical Machines and Systems, ICEMS 2015 (Pattaya, Thailand)*

IOP Conf. Series: Journal of Physics: Conf. Series **1102** (2018) 012019 doi:10.1088/1742-6596/1102/1/012019

Optimal yaw strategy for optimized power and load in various wake situations

Albert M. Urbán[1], Torben J. Larsen[1], Gunner Chr. Larsen [1], Dominique P. Held[1,2], Ebba Dellwik[1], David Verelst[1]

[1] Department of Wind Energy, Technical University of Denmark, Frederiksborgvej 399, 4000 Roskilde, Denmark.

[2] Windar Photonics A/S, Helgeshøj Alle 16, 2630 Taastrup, Denmark

Abstract. The interaction between nearby wind turbines in a wind farm modifies the power and loads compared to their stand-alone values. The increased turbulence intensity and the modified turbulence structure at the downstream turbines creates higher fatigue loading, which can be mitigated by wind farm and/or wind turbine control. To alleviate loads and maximize power possible strategies such as wake steering, where the turbine is yawed to redirect the wake such that it does not impinge the downstream turbine, have been studied. The work presented here focuses on situations where the wake is nevertheless affecting the downstream turbine, and more specifically how high loads can be avoided by yawing the wake-affected turbine. The analysis is conducted on a 2.3 MW machine, and the flow field is simulated using the Dynamic Wake Meandering model. The study investigates the impact on power and loads for different longitudinal interspacing and turbulence intensities. Optimal yaw strategies are defined for above rated regions where no power loss occurs. The potential load alleviation for different load sensors are studied, but the presentation is focussed on the blade root flapwise fatigue loading. For full wake at 3D interspacing and turbulence intensity of 5 %, around 35 % of load reduction on the 1 Hz Damage Equivalent Loads can be achieved at high wind speeds. Smaller reductions are achieved for higher atmospheric turbulence; the analogue case with 15 % turbulence intensity shows 17 % potential alleviation. The alleviation on the wind turbine lifetime is also calculated and compared for different turbulence intensities and mean wind speeds. Small reductions are achieved for sites with low mean wind speed and high turbulence intensity, but high reductions, of around 19 %, are accomplished in low turbulence intensity with high mean wind speed.

1. Introduction

The competitiveness on the wind energy market has pushed the wind turbine industry to develop larger rotors, which yield higher power output. The control system of the turbine is also evolving with the inclusion, for example, of lidar systems. The additional information provided by such sensors has the potential to be used by modern control systems to optimize power and/or reduce the wind turbine loading.

Even though the final objective is the same, maximizing power while minimizing loads, different approaches have been presented in the literature. On a single wind turbine, strategies such as individual pitch control for load reduction is studied by [1], where it is demonstrated that a significant reduction in loading can be achieved using relatively straight forward control strategies. Several modern control strategies, like model predictive control, uses information of the wind field upstream of the turbine and a turbine model to optimize the controller operation set point. The capability of using lidar systems in load reduction and energy yield is shown in [2].

The controller becomes even more complex, when interaction with nearby wind turbines needs to be taken into account. During the operation of a wind farm, full and partial wake situations occur frequently. Wind farm control has been investigated in the last years. In [3], the benefits of wake steering to increase power production and reduce loading on wind farm level are presented.

Yaw control is another interesting option to maximize power and decrease loads. Several studies have addressed the importance of wind turbine alignment in below rated region to optimize power capture. The study in [4] presented the case of a 2.3 MW offshore turbine, where an error on the static yaw calibration leads to a loss of around 0.2% in AEP. Load reduction can also be achieved by active misalignment of an isolated machine. In [5] the load alleviation potential, when vertical shear is present, was studied. The same phenomena is investigated by [6] and compared with field measurements. It was shown that by misaligning turbines in above-rated regime, blade loading can be reduced while maintaining power production. However, the loads on the non-rotating components was increased.

This paper focuses on finding an optimal yaw strategy of a downstream wind turbine under the influence of different wake conditions. First, the methodology is explained describing the setup, characterizing the downstream flow and defining the optimal yaw control strategy. Then, the impact of yawing under different wake condition and its effects on the load level is discussed.

2. Methodology

2.1. Simulation Setup

Power production as well as selected load sensors on the wind turbine are obtained by aero-servo-elastic simulations computed with HAWC2 [7]. The simulated wind turbine is a collective pitch regulated, variable speed turbine with a rotor diameter of 90 m and a rated power of 2.3 MW at 12 m/s. It is using DTU's open-source controller.

Two turbines are used in the setup: the first turbine is used to generate the wake, while the power production and loads are computed on the second turbine, see left panel of figure 1. The unsteady inflow conditions are defined on the upstream turbine. Thus, the downstream turbine feels a disturbance of the original inflow due to the wake. It is important to remark that wake steering by the upstream turbine is not part of this analysis.

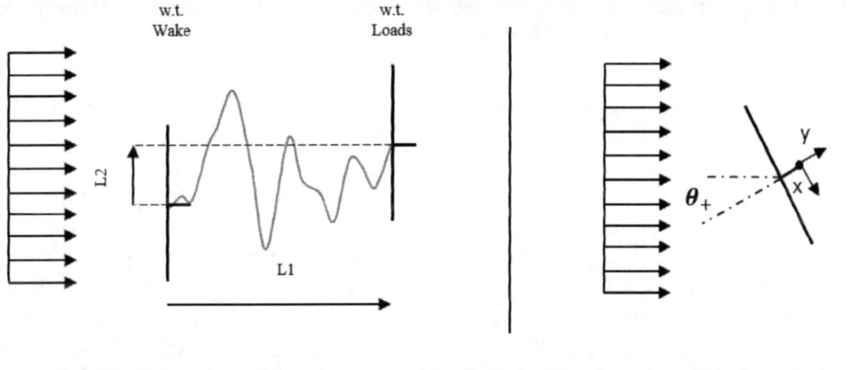

Figure 1: *Left*: Simulation setup with the upstream turbine that is emitting the wake and the downstream turbine where the yaw strategy is applied and the loads are calculated. The red line indicated the movement of the wake. *Right*: Illustration of the turbine yaw convention.

Figure 1 presents the wind turbine configuration, where the left turbine (upstream) is used to generate the wake observed by the right turbine (downstream). L1 corresponds to the longitudinal, along wind speed direction, and L2 relates to the transversal direction. A positive value of L2 corresponds to a displacement of the wake origin to the left as seen from the second turbine facing the mean wind speed direction. The yaw misalignment angle is defined such that a positive angle corresponds to the wind reaching the turbine from the right seen from the turbine, see figure 1.

Normal production, according to IEC 61400-1, is run for a range of yaw misalignments of the downstream wind turbine, where the power and the Damage Equivalent Loads (DELs) of selected channels are computed. In order to cover the variability of the optimal yaw strategy for a wide range of sites, the simulated cases need to cover a large number of scenarios. The investigated scenarios include mean wind speeds ranging from 4m/s to 25m/s in 1m/s steps and turbulences intensities from 5%, to 25% in 5% steps. Six turbulence seeds are used. Further, different discrete yaw scenarios were used, covering a misalignment angle $\theta = \pm 30°$ with 2.5° steps. A power shear law is used with an exponent of 0.2.

2.2. Wake modelling

The upstream wake is simulated using the DWM (Dynamic Wake Meandering) model [8,9], which has been successfully compared to full-scale measurements from the Egmond aan Zee wind farm [10]. The simulated wake deficit and the wake-added turbulence as function of wind speed is shown in figure 2. In the left panels different ambient turbulence levels are studied, while the right panels show the influence of different turbine spacing.

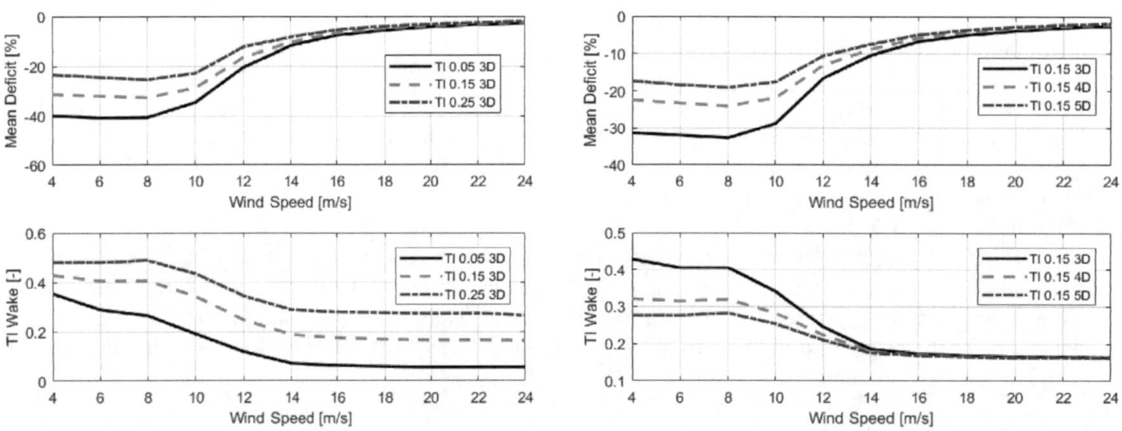

Figure 2: Wake deficit and wake-added turbulence intensity as function of wind speed for different ambient turbulence intensities (*left*) and different turbine spacing (*right*).

The presented values are extracted at the downstream rotor position using the observed wind speed at the rotor plane with a discretization of 1x1 meter and the mean wind speed value over 6 seeds for each wind speed is extracted. The presented results are aligned with the results shown in [8]. The mean inflow deficit is larger for the low turbulence intensities due to the modest mixing of the wake with the undisturbed wind field (figure 2, upper left). The wake recovery is also a function of the distance to the upstream turbine (figure 2, upper right), where the further the distance to the upstream turbine, the lower is the wind speed deficit observed by the wake-affected turbine. The turbulence intensity difference is dependent on both the original level and the wake distance. For identical atmospheric turbulence, the turbulence intensity difference felt by the downstream turbine (fig 2, bottom right) decreases with downstream distance.

IOP Conf. Series: Journal of Physics: Conf. Series **1102** (2018) 012019 doi:10.1088/1742-6596/1102/1/012019

An optimal yaw strategy, especially in partial wake situations, depends on the condition that the simulated wake correctly resembles real wake conditions. Thus, it is important to ensure that the wake is meandering in the simulations. HAWC2 provides the wake deficit position resulting from the meandering process. To illustrate the wake meandering effect, the wake center position as function of time was tracked (figure 3). A Cartesian coordinate system with origin at the rotor centre of the downstream turbine is used.

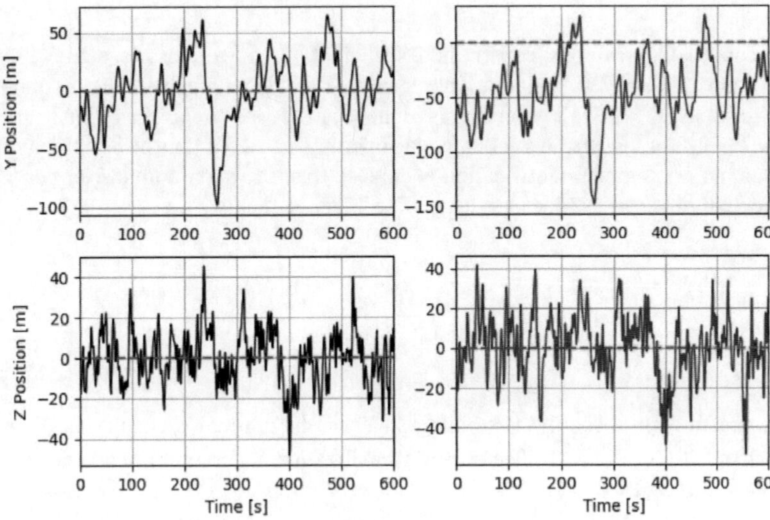

Figure 3 : Wake center position of a full-wake situation (*left*) and a half-wake situation (*right*).

Figure 3 shows a visualization of the wake meander process associated with a full wake case, where the upstream turbine is aligned with the downstream (left figure) and a half wake case (right figure), where the upstream turbine is displaced half diameter (45m) from origin. In both plots, the red line represents the rotor centre of the downstream turbine. Regarding the full wake case, both displacements in y and z have a mean around 0, which implies that the wake meanders symmetrically on the rotor plane. On the other hand, the half wake case presents a negative mean displacement, which corresponds to the half wake. This has an implication on loads and power production, since there is an asymmetry of the mean inflow wind speed on each side of the rotor.

2.3. Yaw control workflow

The implemented method aims to: (1) define a yaw strategy as function of the atmospheric condition and wake position; and (2) to compare the load level to a non-yawed turbine. The output of the workflow results in a yaw strategy for each of the load channels and its associated load alleviation on the lifetime damage of the turbine compared to the non-yawed case. The figures shown in this chapter correspond to a full wake case at 3D and an ambient turbulence intensity of 15%. The mean wind speed is described in terms of a Weibull distribution with 8 m/s mean wind speed and shape factor of 2. The figures are presented to facilitate the explanation of the method.

The data is binned in wind speed and yaw direction, and the annual energy production is obtained. The wind speed frequency values are obtained using a truncated two parameters Weibull distribution. Then, the lifetime damage equivalent loads are calculated using:

$$L_{eq} = \left(\frac{N_{life} \sum_{i=1}^{N} f \, S_{eq}{}^{m}}{N_{eq}} \right)^{m}, \tag{1}$$

where L_{eq} is the lifetime DEL, S_{eq} the mean binned 1 Hz DEL, N_{life} the number of seconds in 20 years lifetime, $N_{eq} = 1e7$ the reference equivalent load and m is the Wöhler exponent. For blades $m = 10$ and for the tower and the shaft $m = 4$.

The optimal yaw control algorithm is constrained such that no power loss is achieved. This is a simplification of a more generalized problem where, in case of the existence of a cost-model, power could be sacrificed, if the value of load reduction is higher than the loss in power production. Thus, the first step is to ensure that the optimal yaw controller operates in maximum power for a given wind speed. The maximum power regime identification is done so that, the maximum power per wind speed is found and subsequently the other values are normalized against that.

$$\Delta P(\theta, v) = \frac{P(\theta, v)}{\max(P(\theta, v))}, \tag{2}$$

where v is the free stream wind speed in the simulation. Thus, the optimal yaw angle θ_{opt} can only be found in regimes where $\Delta P(\theta_n, V_0) = 1$. The power normalization and maximum power regime identification are shown in figure 4.

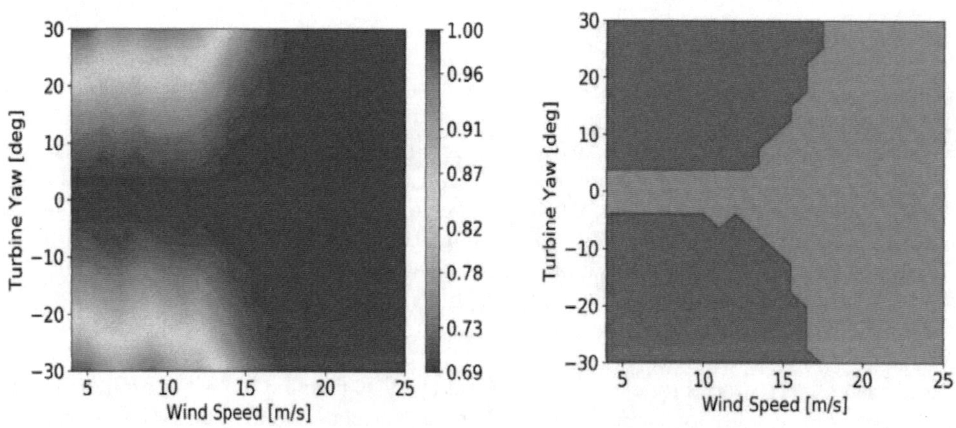

Figure 4: Full wake at 3D from origin and turbulence intensity of 15%. Left: power normalization to the aligned case; Right: region identification for optimal yaw controller.

On the left, it is possible to observe the normalized power as function of wind speed and yaw misalignment. In above-rated wind speed regions large yaw misalignments are possible without power loss. On the right, the regime identification is done based on condition of no power loss.

The 1Hz DEL of blade flapwise bending moment is used to design the yaw optimal controller. In this case, the variation of fatigue loading is normalized against the aligned case $\theta = 0°$:

$$\Delta S(\theta, v) = \frac{S_{eq}(\theta, v)}{S_{eq}(0°, v)}. \tag{3}$$

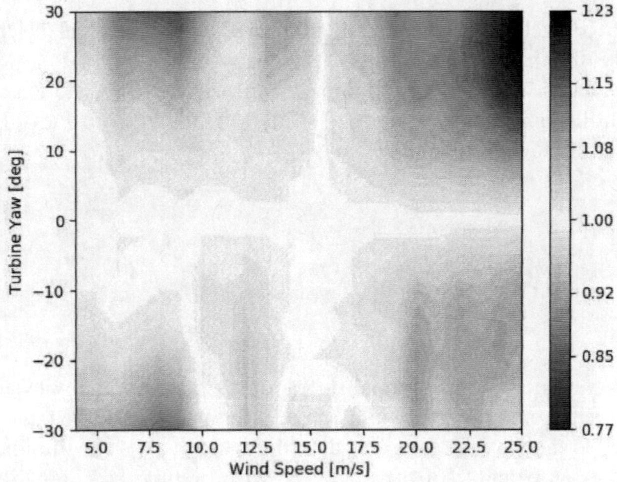

Figure 5: Blade flapwise 1Hz DEL normalized with non-yawed case with wake at 3D from origin and turbulence intensity of 15%.

Figure 5 shows the flapwise blade root 1 Hz DEL as a function of wind speed and turbine yaw. It is observed that the increase of fatigue damage ≅20% for positive angles at high wind speed, while a reduction around ≅ 17% is achieved for negative angles, below rated power. The superposition of the constraint regimes based on power and the fatigue variation provides the information required to define the optimal yaw strategy. The optimal angle, conditioned on mean wind speed, is defined in equation 4:

$$\theta_{opt}(v) = \arg\min_{\theta}\big(\Delta S(\theta, v)\big) \text{ subject to } \Delta P = 1. \tag{4}$$

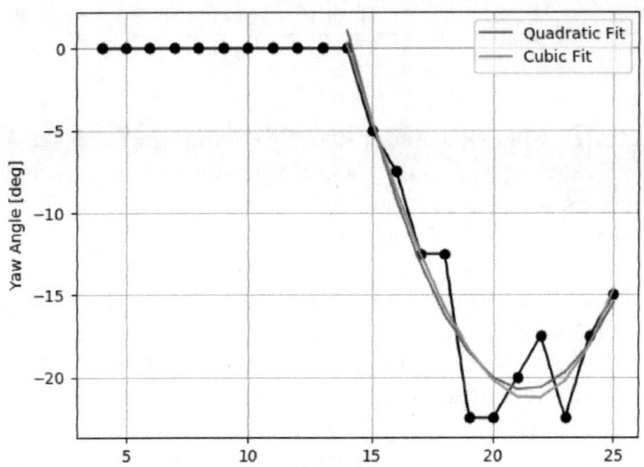

Figure 6 : Optimal yaw angle fit based on blade flapwise bending momment at full wake at 3D from origin and turbulence intensity of 15%.

The result of the blade flapwise yaw strategy can be observed in figure 6. The used discretization, 1 m/s for wind speed and 2.5° on yaw, is enough to provide a rough visualization on the yaw angle, but it is however found too course to define a strategy, which could be implemented in a control system. The discrete optimal angle distribution is used to fit a quadratic and cubic expression.

The fit is evaluated for each wind speed, and compared to the original discrete controller fatigue reduction. In this case, the mean squared error, MSE, between the original yaw angle and the fitted, is not relevant for this analysis. Instead, it is important to evaluate the difference on the 1 Hz damage equivalent. The controller is considered correctly fitted, if the error on the 1 Hz damage equivalent load for all wind speeds is smaller than 0.5 %.

3. Results

3.1. Load reduction

The lifetime DEL is presented for the normal operation case, where the wind turbine is aligned with the wind, see table 1. Different upstream wind turbine spacing and position cases are presented and the values are normalized to the un-waked case to determine the increase of lifetime loading due to wake (ΔW). Table 1 presents an example of the blade flapwise lifetime DEL for different upstream wake positions.

Table 1: Lifetime DEL increase comparison for ambient turbulence intensity 15 % and Weibull parameter A=8 m/s k=2.

Channel	Case	$\Delta W[\%]$
Blade Flapwise	Reference 0.15	100
	3D 0.15	110.5
	4D 0.15	109.1
	5D 0.15	107.7
	3D 0.15 +0.5D	117.6
	3D 0.15 -0.5D	110.1

The lifetime DEL is increased with decreased distance to the upwind turbine. An asymmetry can be found in the partial wake cases - positive displacement on the perpendicular direction of the wake deficit presents higher loading than negative, which can be explained by the tilt angle combined with the wake deficit and shear.

Once the increment of loading due to the wake is presented, the potential fatigue DEL reduction is evaluated for several load channels. Here the blade flapwise bending moment (BF), the blade edgewise bending moment (BE), the tower top fore-aft bending moment (TT Fa) and the tower top side-side bending moment (TT Ss) are compared. An example for v = 18 m/s, TI=15% is shown in figure 7.

IOP Conf. Series: Journal of Physics: Conf. Series **1102** (2018) 012019 doi:10.1088/1742-6596/1102/1/012019

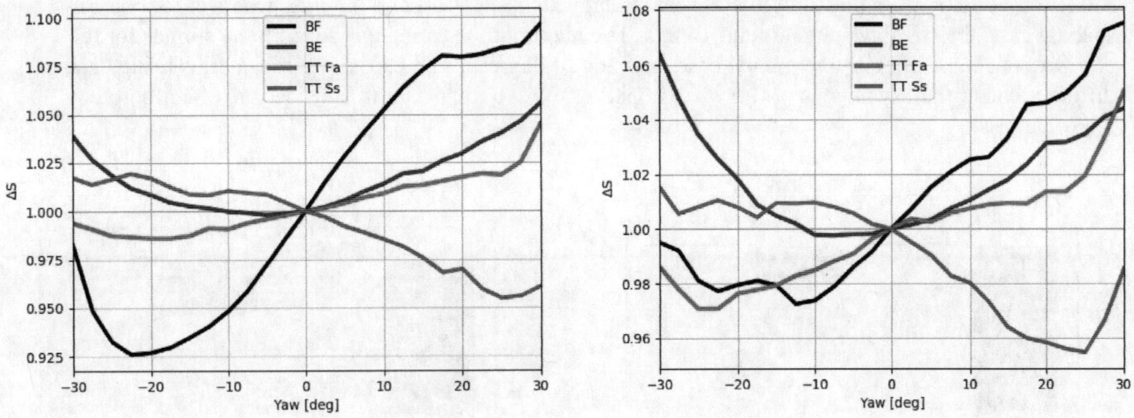

Figure 7: Load variation as function of yaw misalignment for a non-wake case (left) and a wake at 3D (right) at an TI of 15% at mean upstream wind speed of 18 m/s .

It can be seen that the BF is reduced in both cases when negative yaw misalignments are applied. For the non-wake case reductions around 7.5% are achieved at a misalignment of 22°. The reduction is lower in a waked case. There is a small increase for TT Ss and BE towards negative angles.

In general, lower TI will yield a higher load alleviation because the loads are dominated by the shear. For the lowest simulated TI = 5% at 14 m/s a 35% reduction was achieved at high wind speed and large yaw misalignment.

3.2. Optimal yaw strategy

The wind turbine yaw controller was defined as the setting that resulted in minimized blade flapwise fatigue loads. Four different yaw strategies corresponding to different turbine positions are presented in figure 8 using the same background atmospheric turbulence intensity of 15%.

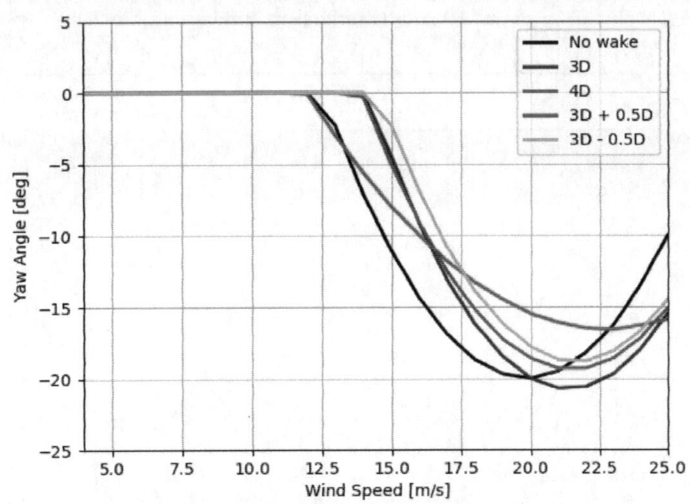

Figure 8: Yaw control strategy comparison for different wake positions for ambient turbulence intensity at 15%.

For all upstream turbine positions the optimal yaw angle is negative. The yaw controller starts acting at different wind speeds depending on where equation 4 is fulfilled. Thus, for the case without a wake (black curve) it is possible to define a yaw strategy that starts around 12 m/s, whereas for the 3D spacing case, the strategy starts around 14 m/s. The maximum yaw angle is found to be similar for the case without wake, the 3D and the 4D case, around -20 degrees. The partial wake scenarios results are different due to the asymmetry of the wake impact. It is also important to consider the effect on other load channels.

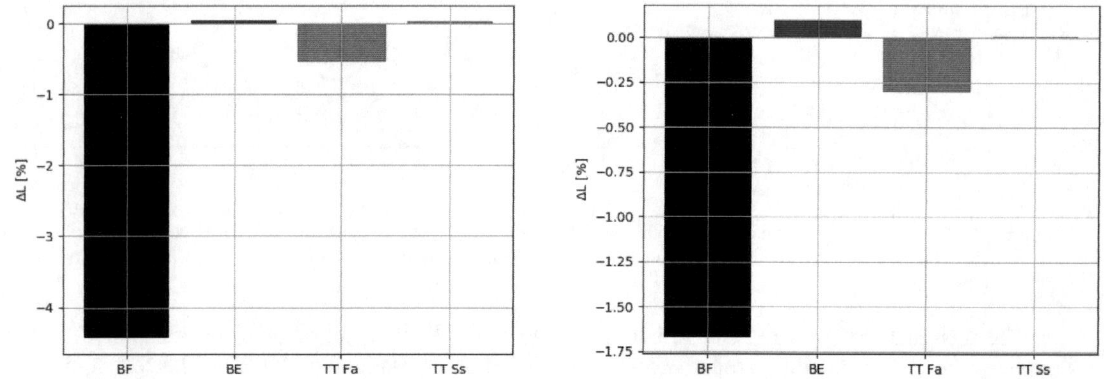

Figure 9: Lifetime DEL using the optimal yaw controller for a non-wake case (left) and a waked case (right) at TI of 15%, Weibull parameter $A = 8$ m/s and $k = 2$

In figure 9 the lifetime DEL reduction is presented using the optimal yaw strategy. It is shown that BF loads are alleviated. The reduction is smaller for wake cases because the wake add additional loading, which cannot be reduced by yawing, see table 1. The reduction achieved ranges from -4.5 % for the free inflow condition case to -1.7 % for the full wake case. Slight increases for BE and TT Ss can be observed. The results are dependent on the used Weibull distribution, and the discussion follows in the next section.

The loading and optimal yaw control strategy is depending on TI. The optimal controller for different turbulence intensities is shown in figure 10 and its reduction appear in table 2.

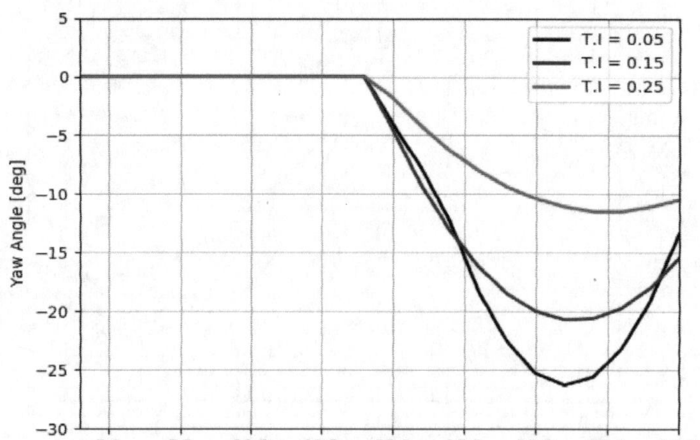

Figure 10: Yaw control strategy comparison for full wake at 3D and different ambient turbulence intensities.

Table 2: Lifetime DEL and reduction comparison for full wake at 3D and different ambient turbulence intensities.

Channel	Case	Lifetime DEL comparison [%]	ΔL using $\theta_{opt}(v)$ [%]
Blade Flapwise	3D 0.25	100 %	-1.4 %
	3D 0.15	69.5 %	-1.7 %
	3D 0.05	32.4 %	-7.8 %

The optimal yaw control strategies achieve higher peak values the lower the turbulence intensity is. The lifetime reduction value is higher, 7.8 %, for the lower turbulence intensity.

3.3. Sensitivity to Weibull distribution

The yaw controller performance is evaluated, in this paper, by comparing the lifetime DEL reduction to its value in normal operation. Another possibility is to compare the 1 Hz fatigue loading for the operating wind speeds. This procedure is discarded, since it requires a large number of information to evaluate the controller performance if more than one inflow condition is compared.

By using the lifetime load value, the wind speed probability density function needs to be included, as defined in equation 1. The yaw controller is independent on the wind probability, since it minimizes the 1 Hz damage equivalent load as function of wind speed. However, the reduction results provided in table 1 and 2 depends on the probability distribution. The Weibull distribution is changed to analyze the load reduction sensitivity to the applied Weibull distribution. The shape factor remains constant, k = 2, for the three Weibull intensity parameter choices, where the mean wind speed is changed to 6 and 12, respectively. The reduction is calculated, considering the lifetime value for each Weibull distribution, and presented in tables 3 and 4.

Table 3: Lifetime reduction comparison for ambient turbulence intensity 15 % and Weibull distributions with k =2 and A = 6, 8 and 12 m/s.

Channel	Case	ΔL using $\theta_{opt}(v)$, A= 6 [%]	ΔL using $\theta_{opt}(v)$, A= 8 [%]	ΔL using $\theta_{opt}(v)$, A= 12 [%]
Blade Flapwise	Reference 0.15	-1.12 %	-4.5 %	-7.75 %
	3D 0.15	-0.3 %	-1.7 %	-5.8 %
	4D 0.15	-0.3 %	-2.0 %	-6.6 %
	5D 0.15	-0.2 %	-2.1 %	-6.7 %
	3D 0.15 +0.5D	-1.0 %	-1.9 %	-5.6 %
	3D 0.15 -0.5D	-0.1 %	-1.2 %	-5.1 %

Table 4: Lifetime reduction comparison for full wake at 3D and different ambient turbulence intensities and Weibull distribution with k =2 and A = 6, 8 and 12 m/s.

Channel	Case	ΔL using $\theta_{opt}(v)$, A= 6 [%]	ΔL using $\theta_{opt}(v)$, A= 8 [%]	ΔL using $\theta_{opt}(v)$, A= 12 [%]
Blade Flapwise	3D 0.25	-0.1 %	-1.4 %	-2.2 %
	3D 0.15	-0.3 %	-1.7 %	-5.8 %
	3D 0.05	-1.9 %	-7.8 %	-19.5 %

For Weibull distributions with predominance of lower wind speeds, A = 6 m /s, the reductions obtained are lower than for the distributions with higher mean wind speed. The designed controller operates mainly in above-rated conditions, guarantying power production, thus not reducing fatigue loading in the partial load regime.

4. Conclusion and future work

The fatigue load potential reduction is investigated in this study and a simple method for defining the optimal yaw strategy for a wind turbine affected by wake from an upstream turbine. The defined strategies can be implemented in the wind turbine control system where, for example, lidars can be used to detect the presence of the wake in the approaching flow as demonstrated by [11,12].

In previous studies the accuracy of inflow measurements by nacelle lidar systems have been investigated. Intercomparisons with sonic anemometers showed very good correlation and no biases of the mean wind speeds. Similarly, turbulence intensity measurements can be achieved through the calculation of average Doppler spectra. The ability to detect wakes is shown in [12]. There, the spectral width of the Doppler spectra is used to determine small-scale turbulence levels. Comparing these levels to ones measured during wake-free situations, it is possible to identify instances where a wake is impinging the inflow.

The simulations carried out with HAWC2 for the aero-servo-elastic model and DWM for the wake modelling has been conducted on a 2.3 MW machine. The defined yaw controller is constrained to achieve maximum power, thus, operating in the above-rated condition where the inclusion of yaw misalignment does not alter the rated power. The optimal yaw controller could be applied more broadly in below rated conditions, to further alleviate the high wake loads, for example, by applying a cost model.

The flapwise fatigue loading is presented in this paper, showing load alleviation towards negative yaw angles. The optimal yaw strategy presented in figure 8 and figure 10 depends on the inflow conditions. For the same atmospheric turbulence intensities, the controller presents a similar pattern but shifted towards the highest wind speed for low interspacing. An exception is found in one of the partial wake cases, 3D and ½ D lateral displacement. The wake asymmetry and the effect on tilt angle will be investigated in future papers. For the same spacing, but different turbulence intensities, the lower the turbulence intensity, the higher the value of the optimal yaw controller.

Low turbulence intensity sites have a larger potential for load alleviation. For example, a 35 % load reduction in the 1 Hz flapwise DEL is observed for high wind speeds at 3D full wake at 5 % turbulence intensity. The higher the turbulence intensity is, the smaller the potential load reduction is. The lifetime damage equivalent load is computed for three different mean wind speed sites, concluding that the highest potential load alleviation is found in cases with low turbulence intensity and high mean wind speed. The presented values do not take into account the operating frequency of the various wake conditions. Turbine and site specific reductions are found dependent on the park layout, machine and wind distribution.

The sensitivity of shear is not included in this paper even though the load reduction is depending on it, since it produces a once-per-revolution variation. In future work, one dimension will be added to the analysis to study the importance of the combined effect of shear and wake.

The investigation of the optimal yaw controller at partial wake should be extended to cover more wake positions than half-wake and full-wake. Also, the half-wake situations in this paper are not guaranteed to only affect half of the rotor, since the dynamic wake meandering model adds a random component to the position of the wake.

5. Acknowledgements

This research was supported by the EUDP funding agency through the project Lidar detection of wakes for wind turbine optimization.

6. References

[1] E. A. Bossanyi, "Individual blade pitch control for load reduction," *Journal of Wind Energy* 2003, **6**: 119–128.

[2] D. Schlipf, "Lidar-Assisted Control Concepts for Wind Turbines," PhD-thesis, Faculty of Aerospace Engineering and Geodesy of the University of Stuttgart (2015).

[3] P. Flemming et. al, "A simulation study demonstrating the importance of large-scale trailing vortices in wake steering," Wind Energy Science 2018, **3**: 243-255

[4] K. A. Kragh and M. H. Hansen, "Potential of power gain with improved yaw alignment," Wind Energy 2015; **18**:979-989

[5] K. A. Kragh and M. H. Hansen, "Load alleviation of wind turbines by yaw misalignment," Wind Energy 2014; **17**:971-982

[6] R. Damiani et.al, "Assessment of Wind Turbine Component Loads Under Yaw-Offset Conditions," Wind Energy Science 2017; **3**: 173–189

[7] T. J. Larsen and A. M. Hansen, "How 2 HAWC2, the user's manual," Tech. Rep., 2007.

[8] G. C. Larsen et. al., "Wake meandering: a pragmatic approach," Wind Energy 2008; **11**(4): 377–395

[9] H. Aa. Madsen et. al., "Calibration and validation of the dynamic wake meandering model for implementation in an aeroelastic code," Journal of Solar Energy Engineering 2010; **132**(4): 041014

[10] T. J. Larsen et. al. "Validation of the dynamic wake meander model for loads and power production in the Egmond aan Zee wind farm" Wind Energy 2013; **16**:605-624

[11] T G Herges et. al. "Detailed analysis of a waked turbine using a high resolution scanning lidar," The Science of Making Torque from Wind, J. Phys.: Conf. Ser. 2018; **1037**: 072009

[12] D. P. Held et al. "Wake detection in the turbine inflow using nacelle lidars," in review

IOP Conf. Series: Journal of Physics: Conf. Series **1102** (2018) 012020 doi:10.1088/1742-6596/1102/1/012020

Flight Phase Control Strategies for Airborne Wind Energy Systems

John Warnock, David McMillan, Samuel Tabor

University of Strathclyde, Electronic & Electrical Engineering Department, Glasgow, UK

KPS, Glasgow, UK

john.warnock@Strath.ac.uk

Abstract. Traditional Danish concept wind turbines face many constraints when upscaling in order to access higher wind speeds, such as size, mechanical loading and weight. It is possible that some of these constraints could be circumvented through use of airborne wind energy systems (AWES).

With research into AWES becoming more prominent, the topic of launching and landing the system must be analysed in detail. Currently several concepts are being pursued with differing launch and land technologies. For all systems it is likely that minimising launch and land cycles will be a key objective due to increased energy costs and hardware risk in these phases.

This research focuses on a cross-wind ground-based generation system and discusses the problem of the launch and land policy with regards to the wind speed at operational height. The paper also discusses the use of an airborne powered loiter phase and a grounded waiting phase. A key consideration when analysing this problem is wind speed measurement uncertainty (including the degree of temporal averaging) and how to integrate this uncertainty into any launch & land policy.

The present research concerns cost-benefit analysis with respect to generated and consumed energy cost functions for each flight phase. It is found that for any given AWES there will be an optimum airborne loiter time after which a system should be landed. This avoids landings due to short-duration low wind periods.

This research will be followed up by further analysis of additional cost functions such as reliability and failure aspects associated with each of the above phases. Further research will also consider the impact of short term forecasting of various accuracy levels on the optimal control policy and performance of AWES.

1. Introduction

Increasing energy demands have led to a sharp increase in the number of renewable energy technologies being utilised worldwide. Wind power is perhaps the most prominent of these.

As the horizontal axis wind turbine installed capacity has grown, the size of these turbines has increased to reduce overall costs and access higher wind speeds at great altitude. Due to lower material usage, airborne wind energy systems, or AWES, could offer further cost reductions and access to still higher wind speeds.

AWES have been a topic of discussion in the field of energy generation for many years. Early studies and research in this area was published as far back as 1980 by Miles Loyd [1] in which the author outlines the concept of wind power generation using kites. Despite this early research it is only recently that airborne wind research has seen investment as a viable means of renewable energy generation.

IOP Conf. Series: Journal of Physics: Conf. Series **1102** (2018) 012020 doi:10.1088/1742-6596/1102/1/012020

Early work in the field was carried out by a dedicated team at TU Delft in the Netherlands. The initial focus of this research was into the concept of the Laddermill, a tiered kite-based AWES [2 - 4]. Through this research a number of branches of AWES technology have been investigated such as multiple kite systems, systems with numerous kites connected to one tether and kites connected to branches of the tether [5]. Other research by the team have investigated the control of the tethering system, using inelastic tethers to control the altitude and flight path of the kite [6, 7].

A functional prototype of the system developed by TU Delft took flight in 2010, the 20kW demonstrator is a pumping kite power system with an inflatable wing. This system, like the system used by KPS, is a ground generating AWES [8, 9]. Several research teams and companies have also begun to develop prototypes, such as Makani, a company acquired by Google in 2013, who have successfully constructed a 600kW cross-wind flight demonstrator [10]. The Makani demonstrator utilises generators fixed to the aircraft to generate energy which is transferred down the tether, this is known as "Fly-Gen" AWES as described by Cherubini et al. in their 2015 paper "Airborne Wind Energy Systems: A review of the technologies" [11].

Other technologies are also being investigated such as ground generating cross-wind flight AWES which can take several forms. Generally, a ground generating system operates via the aircraft or kite pulling on a tether which is attached to a generator on the ground [11, 12]. However, this does require a phase in which energy is consumed to recover the tether.

The co-authors of this paper, KPS, are currently developing a cross-wind ground generating device [13]. The device uses rotors for take-off and landing, and these rotors may also be used for sustained flight in low-wind conditions. The energy requirements of these flight phases, as well as generation phase energy production, have been calculated and are used as indicative values in the following analysis.

2. Flight Phase Policy

An AWES will exist in one of several flight phases. In the model used to conduct this research, they are considered to be Grounded, Launch, Generate, Loiter and Land. The transition between states is visualised in figure 2:

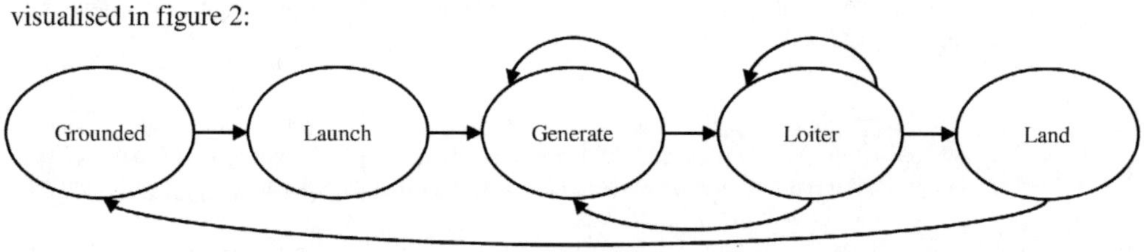

Figure 1: States and Transitions between states of a ground generating airborne wind energy system

In the model the transition between states is dictated by wind speed and time. If a system is grounded it should only transition to a launch state if the wind speed surpasses a predetermined minimum wind speed (cut-in speed). Similarly, if a system is in the generate phase and the wind speed drops below the cut-in speed then it will enter a loiter phase. Launch and land phases will always transition to generate or grounded respectively.

The transition from the loiter phase to either loiter, generate or land is determined by wind speed and time. Landing will be initiated if the system has been in a loiter phase for a predetermined length of time with the wind speed below cut-in. If the wind speed rises above cut-in while the system is loitering, generation will resume.

The loiter phase is utilised to add hysteresis into the launch/land decision, thus avoiding an excessive number of launch and land cycles when the wind speed fluctuates around cut-in wind speed. A loiter time should be established, preferably an optimal loiter time which is discussed later in this paper, that is the maximum duration the system will remain in the loiter phase with wind speed below minimum cut-in speed.

Through review of airborne wind energy literature there appears to be no research in the area of flight phase control policy, so it is difficult to establish whether all the presently considered flight phases are implemented across all ground generating systems.

The following sections consider the specific example of optimising the loiter time of the system. This research aims to establish the utility of airborne loitering of an AWES, and the effect of system parameters on the optimal loitering time.

2.1. Loiter Time

In order to optimise the energy output from an AWES using such a policy, the maximum time allocated to loitering must be established. Consider the following:

$$\frac{E_{Land} + E_{Launch}}{E_{Loiter}} = 1 \quad [1]$$

$$\frac{E_{Land} + E_{Launch}}{P_{Loiter}} = t_{Loiter} \quad [2]$$

Where P is the power and E the energy used when landing / launching / loitering and t is the time spent landing / launching / loitering.

However, many AWES will require an intermission between land and launch phases, e.g. for charging batteries or resetting the system. In this scenario it is possible that without a loiter capability a system could risk losing substantial generation time due to a premature landing. This leads to the conclusion that the optimal loiter time must also account for the time lost whilst landing.

2.2. Airborne Loiter Policy

The optimum loiter time will depend on the relative energy costs and yields of the flight phases. If an energy-intensive VTOL launch and land method is implemented by KPS, the energy cost of launching and landing the system is significantly higher than the cost of loitering.

Considering only the costs associated with loiter, launch and land – or generally the parasitic costs – would suggest that there would be little benefit to implementing a loiter policy due to the increasing parasitic costs with increased loiter times. However, this does not account for the increased energy yield that could be determined due the implementation of an effective loiter policy.

Any loiter policy would benefit from a reliable wind speed prediction tool being implemented into the system. This could be a physical prediction tool such as a forward scanning LiDAR that would provide wind speed readings to inform decision making. However, the contribution to capital cost of such a system might be detrimental in some markets.

Another method which is being investigated by the authors is the implementation of a prediction model based on historical data. This system could also have some drawbacks due to the reliability of the wind speed data being predicted. The feasibility and impact of such methods are currently being evaluated.

3. Data

Analysis was carried out using the historical wind speed data from two different sites. The wind speed data for these sites was acquired through various means, including LiDAR and met mast data extrapolated to operational heights using wind shear data.

The results of any analysis related to optimal loiter times will be dependent on the time resolution of the input data. The resolution of the data is variable, however must be large enough in order for the AWES to react to any wind speed changes without incorrectly changing phase. In this study historical wind speed data has been used to emulate the wind speed, as this has been taken from a number of sources such as LiDAR and met mast data, the time resolution is also dependent on the nature of the data provided by the sources.

4. Analysis

The annual energy yield for each of the sites was calculated for a range of loiter times, 0 -60 minutes, and was offset against the 'parasitic' energy (the energy lost to loiter, launch and land.)

IOP Conf. Series: Journal of Physics: Conf. Series **1102** (2018) 012020 doi:10.1088/1742-6596/1102/1/012020

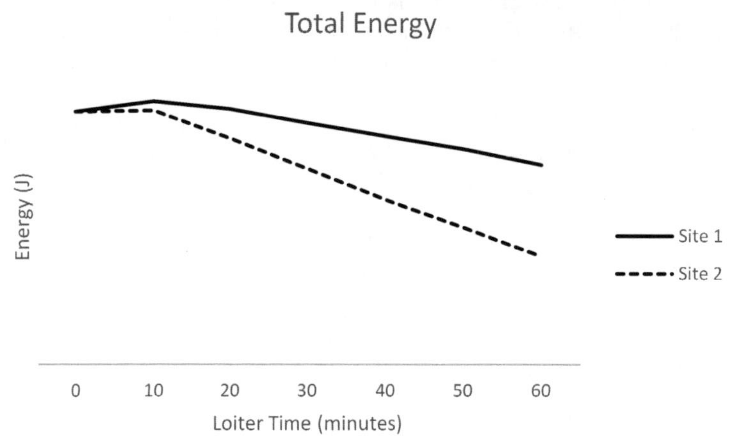

Figure 2: Total Energy Output

Figure 3 shows that given the parameters used in the calculations for both sites 1 and 2 the peak loiter time occurs at 10 minutes after which the total energy output for the system over 12 months diminishes. To further understand the results shown in figure 3 it is important to evaluate both the energy yield and parasitic energy for each of the sites.

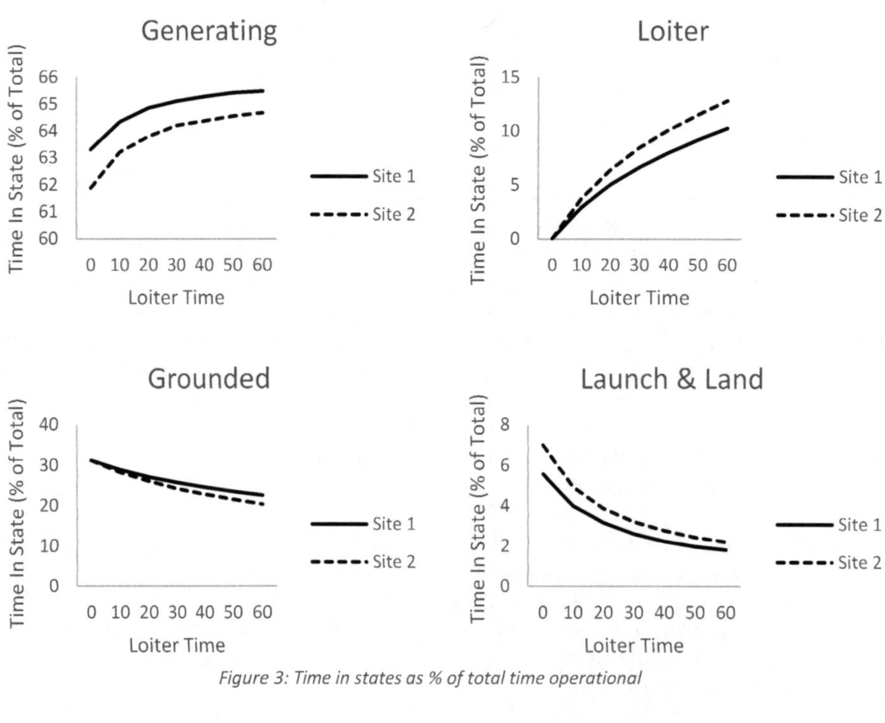

Figure 3: Time in states as % of total time operational

The overall energy yield increases with the loiter time. As expected as the loiter time increases less time is spent grounded and more time in the air primed to capture energy as the wind speed exceeds the cut in wind speed. Similarly, for each of the sites the parasitic energy increases as the loiter time increases. This is due to the cost of loiter exceeding the cost of landing the system.

Note also the effect of loiter time on relative time in each of the five states. From figure 4 it can be seen that in each of the sites the predetermined loiter time has a direct impact on the amount of time spent in each of the phases. As expected as the loiter time increases as does the time spent in the air (generate and loiter phases) and the time spent on the ground (grounded, launch and land phases) decreases. This could be essential when considering the relative reliability implications of these conditions.

5. Discussion

The research presented in figure 3 shows the total energy output by the system in two sites. In this graph it appears that the optimal loiter time for both of the sites is 10 minutes. These results could prove to be specific to the technology being researched throughout this paper and could be altered by several factors, particularly the energy costs of the flight phases and the system power curve.

Perhaps most importantly the results will vary depending on the phase length allocated to the system. This will have a significant impact upon both parasitic energy and energy yield. Currently this research assumes 10 minute phases due to the wide availability of 10-minute datasets.

Currently launch and land phases are allocated a full ten-minute phase to complete, whereas in practice this would take significantly less time. This results in a significant loss in energy yield, as the ten-minute launch phase is lost generation time. Similarly, loiter phases are also allocated ten minute phases.

The information presented in figure 4 highlights that the behaviour of the system is as expected when the loiter time is increased, with the time in the air increasing as loiter time increases and the time on the ground decreasing.

To improve the modelling process, and to further evaluate the utility of airborne loitering, several prediction methods are being investigated and implemented. This is explained further in the future work section of this paper.

6. Conclusion

This paper has set out to investigate the airborne loiter policy that will be implemented by ground generating airborne wind energy systems. Equations have been presented to theoretically calculate the maximum amount of time that should be spent loitering by an AWES.

Through analysis of historical wind speed data at several sites the flight phase control strategy was evaluated. Whilst an optimal loiter time appeared to have been established the reasons that this may change for other systems in different environments have been highlighted.

Investigation into the behaviour of the system given changes in the predetermined loiter time was conducted. It was confirmed that the system behaves as expected.

To optimise the loiter time of any given ground generating airborne wind energy system weather prediction tools must be implemented to allow the system to make better informed decisions about the time spent loitering or waiting on the ground.

7. Future Work

As described throughout this paper due to the infancy of this area of research there are a number of areas in which the research is ongoing.

Several parameters of the flight phase policy require investigation, such as the utility of a grounded waiting phase entered above cut-in wind speed before launch is initiated. Estimation of realistic costs other than energy costs, such as load cycles of launch/land components and costs of increased risk during these phases, will be key to determining true optima of policy parameters.

A further key study in advancing this research is that of implementation of a prediction tool into the model. This would allow a number of advancements for decision making compared to the current model, that considers only the current wind speed. There are many methods being investigated in order to implement this tool, autoregression being the primary method being explored currently whereby estimations are extrapolated from reverse autoregressive fits.

In future work, the flight phase policy decision will be formalised in a Markov Decision Chain framework.

IOP Conf. Series: Journal of Physics: Conf. Series **1102** (2018) 012020 doi:10.1088/1742-6596/1102/1/012020

References

[1] Loyd, M. “Crosswind Kite Power.” Journal of Energy. Vol 4. No 3. Pages 106-111. June 1980.

[2] J.P. Meijaard, W.J. Ockels, A.L. Schwab: "Modelling of the dynamic behaviour of a laddermill, a novel concept to exploit wind energy". 3rd International Symposium on Cable Dynamics, Trondheim, Norway, pp.229-234, 16-18 August 1999.

[3] W.J. Ockels: "Laddermill, a novel concept to exploit the energy in the airspace". Aircraft Design, Vol. 4, No. 2-3, pp. 81-97, 2001.

[4] W.J. Ockels, B. Lansdorp, J. Breukels, G. Spierenburg: "The Laddermill: work in progress". European Wind Energy Conference & Exhibition, London, UK, 22-25 November 2004.

[5] A.R. Podgaets, W.J. Ockels: "Flight control and stability of a multiple kites tethered system." Renewable Energy 2006, Makuhari Messe, Chiba, Japan, 9-13 October 2006.

[6] P. Williams, B. Lansdorp, W.J. Ockels: "Modeling and Control of a Kite on a Variable Length Flexible Inelastic Tether". AIAA 2007-6705, AIAA Modelling and Simulation Technologies Conference and Exhibit, Hilton Head, SC, USA, 20-23 August 2007.

[7] P. Williams, B. Lansdorp, W.J. Ockels: “Optimal Crosswind Towing and Power Generation with Tethered Kites”. AIAA Journal of Guidance, Control and Dynamics, Vol. 31, No. 1, 2008.

[8] P. S Manoj: "Airborne wind energy technology - Potential alternative for clean energy". Atlantis Magazine, Volume 27, Issue 1, 2016.

[9] C. Grete: “The Economic Potential of Kite Power”. Leonardo Times – Journal of the Society of Aerospace Engineering Students 'Leonardo da Vinci', No. 4, pp. 10-11, 2014.

[10] https://x.company/makani/technology/

[11] Cherubini, A. Papini, A. Vertechy, R. Fontana, M. “Airborne Wind Energy Systems: A review of the technologies.” Renewable and Sustainable Energy Reviews. Vol 51. Pages 1461-1476. 2015.

[12] Fagiano, L. Milanese, M. “Airborne Wind Energy L: an overview.” “American Control Conference. 2012.

[13] http://www.kps.energy/technology/

Observability Analysis for Horizontal Axis Wind Turbines using Empirical Gramian Matrices

Bastian Ritter[1,2], **Edwin Mora**[2‡], **Axel Schild**[3]

[1] Industrial Science GmbH, Alexanderstr. 25, 64283 Darmstadt, DE
[2] Technische Universität Darmstadt, Dept. of Control Systems and Mechatronics, DE
[3] IAV GmbH, Rockwellstraße 16, 38518 Gifhorn, DE
‡ now with the Dept. of Energy Information Networks and Systems (same university)

E-mail: bastian.ritter@industrial-science.de

Abstract. This contribution deals with the state observability and parameter identifiability analysis for nonlinear wind turbine control (WTC) systems based on empirical observability Gramian (EOG) matrices. The concepts of observability and Gramian matrices are introduced to investigate the inverse condition number (ICN) and the singular value decomposition (SVD) of the EOG matrices and to evaluate different sensor configurations with respect to their degree of observability. The obtained results are then reviewed for practical plausibility over state estimates, provided by sigma-point Kalman filters (SPKF), in order to relate the observability measures directly to the expectable estimation error. The investigation reveals indeed a correlation between the employed measures for individual state observability and the estimates produced by the nonlinear filters. The joint assessment of ICN, SVD and SPKF results is found as a strong tool when aiming at a holistic and practically relevant observability discussion.

1. Introduction

Wind energy plays an important role in the transition from fossil to renewable energy resources. It is available free-of-charge and almost unlimited. Today, horizontal axis wind turbines (HAWT) with three rotor blades are the most promising concept to harvest this large potential. In order to gain more electrical power from the wind, the dimensions have continuously grown over the last thirty years with the main objective to lower the levelized costs of energy (LCoE) [1]. Due to still strong competition from the conventional energy sectors, the LCoE must drop further for wind energy to prevail. One possibility to do so, is to improve energy yield and component life-time by advanced model-based state-space controllers [2, 3, 4]. These require the complete state information and also the knowledge of relevant model parameters to outperform standard industrial control schemes (cf. [5]).

Therefore, previous work has already focused in detail on the state estimation problem for wind turbines [6] and how to tackle practical issues like real-time capabilities and to explore benefits of advanced sensor configurations [7]. Moreover, noise adaptive filter algorithms have been discussed to cope with time-varying noise statistics and to prevent poor estimation performance [8].

On the contrary, this contribution puts emphasis on the analysis of state observability and parameter identifiability which is conducted using the empirical Gramian approach [9, 10]. For that purpose, the remainder of this paper is structured as follows: Section 2 will briefly explain the used design model and the realistic simulation environment. Afterwards, Sec. 3 elaborates on the definitions of observability, the theorems and the criteria applied. Sec. 4 will condense the relevant outcomes of the conducted research and present illustrative simulation results. Finally, Sec. 5 will sum up with the conclusions and provide a glimpse on future work.

2. Control-oriented modeling and wind turbine simulation

Control-oriented modeling requires a significant level of abstraction of the real system. Abstraction means that only the relevant dynamics are considered for the problem to be solved. Inevitably, this deliberate simplification introduces modeling errors which the practitioner must be aware of. Control-oriented models are widely used for control design, simulation purposes, state estimation and also for observability analysis. This section introduces the simplified HAWT design model and the realistic simulation model used as reference.

2.1. Design model equations and parameter set

The design models are an indispensable part of internal model-based control schemes or state estimators. The goal-oriented choice of model granularity and system order has quite an importance for real-time applications due to computational requirements.

The nonlinear system considered in this publication is derived from the conservation of angular and linear momentum which indicates that mainly the mechanical dynamics are decisive for WTC. Applied to the wind turbine's drive-train, this yields

$$\ddot{\varphi}_{\mathrm{g}} = \frac{\varrho}{2}\frac{\pi R^3}{\Theta_{\mathrm{r}}+\Theta_{\mathrm{g}} i_{\mathrm{gb}}^2} C_{\mathrm{M}}(\lambda,\beta)\left(v_{\mathrm{w}} - \dot{x}_{\mathrm{T}}\right)^2 - \frac{i_{\mathrm{gb}}}{\Theta_{\mathrm{r}}+\Theta_{\mathrm{g}} i_{\mathrm{gb}}^2} M_{\mathrm{g}} \quad \text{with} \quad \lambda = \frac{\dot{\varphi}_{\mathrm{g}} R}{v_{\mathrm{w}} - \dot{x}_{\mathrm{T}}} \tag{1a}$$

where λ is the so called tip-speed-ratio. The second dynamic equation is obtained from the conservation of linear momentum for the nacelle in axial direction. It reads

$$\ddot{x}_{\mathrm{T}} = \frac{\varrho}{2}\frac{\pi R^2}{m_{\mathrm{T}}} C_{\mathrm{T}}(\lambda,\beta)\left(v_{\mathrm{w}} - \dot{x}_{\mathrm{T}}\right)^2 - 2\zeta_{\mathrm{T}x}\omega_0 \dot{x}_{\mathrm{T}} - \omega_0^2 x_{\mathrm{T}} \quad . \tag{1b}$$

Additionally, the instantaneous rotor power P_{a} and the electrical generator power P_{g} denote

$$P_{\mathrm{a}} = \frac{\varrho}{2}\pi R^2 C_{\mathrm{P}}(\lambda,\beta)\left(v_{\mathrm{w}} - \dot{x}_{\mathrm{T}}\right)^3 \tag{2}$$

$$P_{\mathrm{g}} = \eta_{\mathrm{g}} i_{\mathrm{gb}} \dot{\varphi}_{\mathrm{g}} M_{\mathrm{g}} \tag{3}$$

where $C_{\mathrm{P}} = \lambda\, C_{\mathrm{M}}$ is the aerodynamic power coefficient. Fig. 1 shows the illustrative contour plots of the aerodynamic characteristics for the widely used 5-MW reference turbine [11] of the NREL (National Renewable Energy Laboratory). Moreover, a complete parameter set for the same turbine is provided in Tab. 1.

Besides the system dynamics, also the available sensors must be described in an output model. Speaking of the state-of-the-art sensor instrumentation, there is first the precise generator speed measurement $y_1 = \frac{60}{2\pi} i_{\mathrm{gb}} \dot{\varphi}_{\mathrm{g}}$, then the nacelle acceleration $y_2 = \ddot{x}_{\mathrm{T}}$ provided by an inertial measurement unit and the generator power measurement $y_3 = P_{\mathrm{g}}$. The mechanical system states are the generator angular speed $x_1 = \dot{\varphi}_{\mathrm{g}}$, the nacelle velocity $x_2 = \dot{x}_{\mathrm{T}}$ and position $x_3 = x_{\mathrm{T}}$. The main disturbance input is the rotor effective wind speed $d_1 = v_{\mathrm{w}}$. As unknown and/or time-varying parameters are considered the tower eigenfrequency $p_1 = \omega_0$ and the air mass density $p_2 = \varrho$. These states, disturbance inputs and parameters are the ones to be estimated/identified in practise. The control inputs are the generator torque $u_1 = M_{\mathrm{g}}$ and the collective blade pitch angle $u_2 = \beta$.

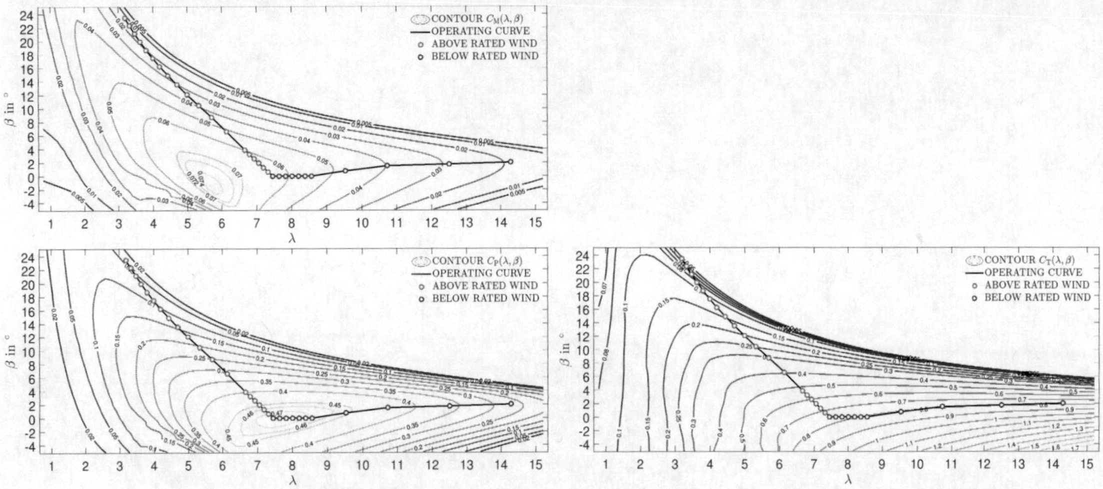

Figure 1. Aerodynamic coefficients for rotor torque $C_{\mathrm{M}}(\lambda, \beta)$, rotor power $C_{\mathrm{P}}(\lambda, \beta)$ and rotor thrust $C_{\mathrm{T}}(\lambda, \beta)$ of the NREL5MW reference turbine

With the above definitions, the nonlinear wind turbine system can be brought into a generalized state-space representation that denotes as follows:

$$\dot{\boldsymbol{x}} = \boldsymbol{f}(\boldsymbol{x}, \boldsymbol{u}, \boldsymbol{d}, \boldsymbol{p}) \tag{4a}$$

$$\boldsymbol{y} = \boldsymbol{h}(\boldsymbol{x}, \boldsymbol{u}, \boldsymbol{d}, \boldsymbol{p}) \quad . \tag{4b}$$

Therein, the state vector $\boldsymbol{x}$, the control inputs $\boldsymbol{u}$, the disturbance inputs $\boldsymbol{d}$ and the unknown model parameters $\boldsymbol{p}$ are introduced. The output vector $\boldsymbol{y}$ contains all measurement variables considered for state estimation. The detailed sensor configurations used for observability analysis are presented in Sec. 4.1.

2.2. Realistic simulation environment

In order to test new control schemes and/or observers under preferably realistic conditions, a simulator is needed to evaluate the performance prior to application in the field. In this contribution, the FAST8 [12] simulator as well-known and freely available reference is deployed as detailed aero-elastic simulation model, considering the following degrees of freedom: Tower fore-aft and side-side motion, blade flap and edge motion, generator and drive-train, nacelle yaw. Another simulation model is given by the design model in Eqs. (1). To evaluate the effects of modeling errors due to deliberate simplification, comparative simulation results are shown in Fig. 2. Both simulation models have been run simultaneously with the same wind scenario

Table 1. Relevant parameters of NREL's 5 MW turbine

$\zeta_{\mathrm{T}x}$	tower/nacelle modal damping	0.01	-
η_{g}	generator efficiency	0.944	-
Θ_{r}	rotor inertia (low speed side)	3.56×10^7	kg m^2
Θ_{g}	generator high-speed inertia	534	kg m^2
ϱ	air mass density	1.225	kg/m^3
ω_0	tower/nacelle eigenfrequency	2.06	rad/s
C_{P}^*	optimum power coefficient	0.47	-
i_{gb}	drive-train gear-box ratio	97	-
m_{T}	equivalent tower top mass	450×10^3	kg
R	blade tip radius	63	m

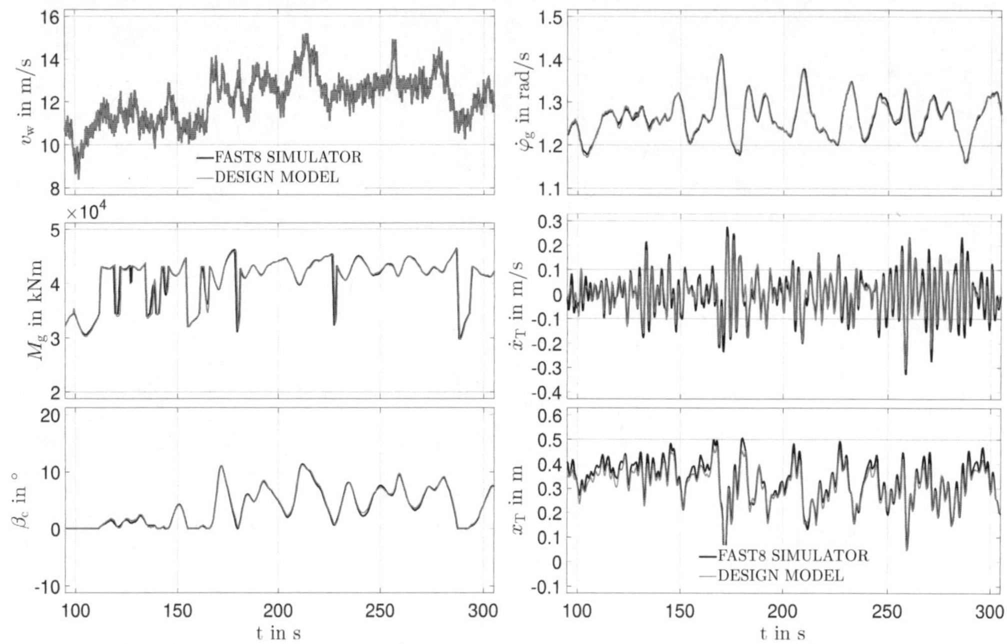

Figure 2. Comparison of design model with FAST8 for an illustrative wind scenario

and the same standard industrial controller type. As can be seen from the plots, the deviations between both simulators for the rotor and nacelle dynamics are negligible.

3. Observability definition, measures and problems

The observability analysis deals with the investigation of a given dynamic system whether it is possible to reconstruct the system states $\boldsymbol{x}(t)$ only from the knowledge of control inputs $\boldsymbol{u}(t)$ and system outputs $\boldsymbol{y}(t)$. Such an analysis is one fundamental step prior to estimator design. For linear systems, the Gramian matrix and the Kalman criterion are well-known concepts to perform this evaluation [13, 14, 15]. However, these are not applicable for nonlinear systems without a preceding linearization step.

The analytical investigation for nonlinear control systems is often more demanding, in particular for higher number of states, cf. [16, 17, 18]. Sometimes, it might be a valid approach for low-order systems, as briefly highlighted for wind speed estimation in [7]. Due to this complexity of the analytical procedure, a simplified but still practical approach is desirable.

3.1. Observability definitions

Before measures to assess observability are introduced, one needs to define what is actually meant by saying that a dynamic system is observable. A generally accepted definition of observability reads as follows [18, 19]:

Definition 3.1 (System Observability) *A system is called fully (state) observable if its initial state $\boldsymbol{x}_0 = \boldsymbol{x}(t_0)$ can be uniquely determined over a finite interval $[t_0, t_e]$ with given input $\boldsymbol{u}_{[t_0,t_e]}$ and output trajectories $\boldsymbol{y}_{[t_0,t_e]}$. Otherwise, the system is said to be unobservable.*

Though, some systems may be unobservable but single states are not (which depends on the chosen measurement configuration). Thus, besides Def. 3.1 the definition of observability for single states is useful to distinguish between unobservable and observable states:

Definition 3.2 (State Observability) *A dynamic state $x_i(t)$ with $i \in [1, 2, \ldots, n]$, where n is the number of states, is called observable if and only if its initial value $x_i(t_0)$ can be exactly* *determined over a finite time interval $t \in [t_0, t_e]$ solely from the input trajectories $\boldsymbol{u}_{[t_0,t_e]}$ and measurement outputs $\boldsymbol{y}_{[t_0,t_e]}$.*

Hence, if the single state of interest is not observable for a given output configuration, then either the model of the system or the sensor configuration must be reformulated.

For nonlinear systems, the following definition of local observability is given in [18]:

Definition 3.3 (Local Observability of Nonlinear Systems) *A nonlinear system with control is given by*

$$\begin{aligned} \dot{\boldsymbol{x}} &= \boldsymbol{f}(\boldsymbol{x}, \boldsymbol{u}) \quad \textit{with} \quad \boldsymbol{x}(t_0) = \boldsymbol{x}_0 \\ \boldsymbol{y} &= \boldsymbol{h}(\boldsymbol{x}, \boldsymbol{u}) \end{aligned}$$

where $\boldsymbol{x} \in D_{\boldsymbol{x}} \subseteq \mathbb{R}^n$ and $\boldsymbol{u} \in C_{\boldsymbol{u}} \subseteq C^{n-1}$ be defined and $\boldsymbol{y} \in \mathbb{R}^q$ holds. If every initial state $\boldsymbol{x}_0 \in D_{\boldsymbol{x}}$ in the neighbourhood U of a point $\boldsymbol{x}_p \in D_{\boldsymbol{x}}$, with $U = \{\boldsymbol{x}_0 \in \mathbb{R}^n : ||\boldsymbol{x}_0 - \boldsymbol{x}_p|| < \epsilon\}$, can be reconstructed uniquely from the knowledge of $\boldsymbol{u}$ and $\boldsymbol{y}$ for a time interval $[t_0, t_e < \infty]$ for all $\boldsymbol{u} \in C_{\boldsymbol{u}}$, then the system is locally observable for all $\boldsymbol{x}_p \in D_{\boldsymbol{x}}$.

Thus, according to Def. 3.3 it is possible that the observability of a nonlinear system depends on the input signal and its excitation characteristics.

3.2. Observability Gramian matrices

The Gramian matrices (GM) are used here as practical criteria to assess observability. The analytical approach is described initially, followed by an introduction to the empirical observability Gramians (EOG). Consider at first the following theorem:

Theorem 3.1 (Linear Gramian Observability Matrix) *The linear time-variant (LTV) dynamic system*

$$\begin{aligned} \dot{\boldsymbol{x}} &= \mathbf{A}(t)\boldsymbol{x} + \mathbf{B}(t)\boldsymbol{u} \quad \textit{with} \quad \boldsymbol{x}(t_0) = \boldsymbol{x}_0 \\ \boldsymbol{y} &= \mathbf{C}(t)\boldsymbol{x} + \mathbf{D}(t)\boldsymbol{u} \end{aligned}$$

is said to be locally observable over a finite time interval $t \in [t_0, t_e]$ if and only if the analytical observability Gramian matrix

$$\mathcal{G}_{\mathrm{O}}^{\mathrm{lin}} = \int_{t_0}^{t_e} \mathbf{\Phi}^{\mathrm{T}}(t, t_0)\mathbf{C}^{\mathrm{T}}(t)\mathbf{C}(t)\mathbf{\Phi}(t, t_0)\mathrm{d}t \tag{5}$$

is invertible. Otherwise, the system is said to be unobservable.

Therein, $\mathbf{\Phi}(t, t_0)$ represents the state transition matrix as discussed in [14]. Generally speaking, the Gramian matrix $\mathcal{G}_{\mathrm{O}}$ represents an energy-based measure for observability since it characterizes the generalized transfer of energy

$$E_{\mathrm{meas}} = \int_{t_0}^{t_e} \boldsymbol{y}^{\mathrm{T}}(t)\, \boldsymbol{y}(t)\mathrm{d}t = \boldsymbol{x}_0^{\mathrm{T}}\, \mathcal{G}_{\mathrm{O}}\, \boldsymbol{x}_0 \tag{6}$$

from the initial state $\boldsymbol{x}_0$ to the measurement outputs $\boldsymbol{y}(t)$ [14, 15, 20]. Hence, the GM influences directly which states contribute to the total measurement energy and which not. States that do not contribute significantly to the measurement energy are poorly observable. A drawback of these linear observability GM is their limitation to linear (or at least linearizable) systems.

Fortunately, a more favourable aspect is that the analytical approach from Theo. 3.1 has been extended to nonlinear systems [9, 21, 10]. Moreover, Krener & Ide [17] and Powel & Morgansen [22] proposed the EOG approach as a tool to assess the (local) observability of nonlinear systems with control. This concept avoids the linearization procedure and builds upon nonlinear system simulation only. As an introduction to empirical Gramians, consider the following theorem:

Theorem 3.2 (Empirical Gramian Observability Matrix) *The nonlinear controlled system described by*

$$\dot{\boldsymbol{x}} = \boldsymbol{f}(\boldsymbol{x}, \boldsymbol{u}) \quad \textit{with} \quad \boldsymbol{x}(t_0) = \boldsymbol{x}_0 \tag{7a}$$

$$\boldsymbol{y} = \boldsymbol{h}(\boldsymbol{x}, \boldsymbol{u}) \tag{7b}$$

is said to be locally observable over a finite time interval $t \in [t_0, t_\mathrm{e}]$ *if and only if the empirical observability Gramian matrix*

$$\mathcal{G}_\mathrm{O} = \sum_{l=1}^{r} \sum_{m=1}^{s} \frac{1}{rsc_m^2} \int_{t_0}^{t_\mathrm{e}} \mathbf{T}_l \boldsymbol{\Psi}_{lm}(t) \mathbf{T}_l^\mathrm{T} \mathrm{d}t \tag{8}$$

has full matrix rank and is therefore invertible. Otherwise, the system is denoted as (locally) unobservable and not all states are uniquely observable at the same time.

Generally speaking, the EOG from Eq. (8) can be interpreted as the sum of output covariance matrices computed for a set of different initial conditions [21]. The reason is that the components ij of the correlation matrices $\boldsymbol{\Psi}_{lm} \in \mathbb{R}^{n \times n}$ are computed by

$$\Psi_{lmij}(t) = \big(\boldsymbol{y}_{lmi}(t) - \boldsymbol{y}_\mathrm{N}(t)\big)^\mathrm{T} \big(\boldsymbol{y}_{lmj}(t) - \boldsymbol{y}_\mathrm{N}(t)\big) \tag{9}$$

where $\boldsymbol{y}_{lmi}(t)$ is the solution of (7) to an initial condition (perturbed state)

$$\boldsymbol{x}_{lmi,0} = c_m \mathbf{T}_l \, \mathbf{e}_i + \boldsymbol{x}_{\mathrm{N},0} \tag{10}$$

and an input trajectory $\boldsymbol{u}_\mathrm{N}(t)$. $\boldsymbol{y}_\mathrm{N}(t)$ represents the nominal output and $\boldsymbol{x}_{\mathrm{N},0} = \boldsymbol{x}_\mathrm{N}(t = t_0)$ the initial state of the nominal state trajectory. To determine the set of perturbed states in Eq. (10), a set $\mathcal{T}$ of unitary rotation matrices, a set $\mathcal{M}$ of different perturbation sizes (weights), and a set $\mathcal{E}$ of basis vectors are defined as follows:

$$\mathcal{T} = \{\mathbf{T}_1, ..., \mathbf{T}_r : \mathbf{T}_l \in \mathbb{R}^{n \times n},\ \mathbf{T}_l^\mathrm{T} \mathbf{T}_l = \mathbf{I},\ \ l = 1, ..., r\} \tag{11a}$$

$$\mathcal{M} = \{c_1, ..., c_s \ : c_m \in \mathbb{R},\ c_m > 0, \qquad m = 1, ..., s\} \tag{11b}$$

$$\mathcal{E} = \{\mathbf{e}_1, ..., \mathbf{e}_n \ : \mathbf{e}_i \in \mathbb{R}^n, \qquad i = 1, ..., n\} \tag{11c}$$

With this sets, a region in the state space around the point of interest (including distance and state perturbation direction) can be determined suitably. The above definition of the empirical Gramians Eq. (8) reduces to Eq. (5) in case of linear systems [9, 21].

3.3. Observability measures based on empirical Gramian matrices

Once the EOG matrix $\mathcal{G}_\mathrm{O}$ is computed for a given time interval, the observability study is then conducted by analysing the rank, the condition number [17] and the singular values σ_i as well as the singular vectors $\boldsymbol{v}_i$. In particular, the inverse condition number (ICN) defined by

$$\mathrm{ICN}(\mathcal{G}_\mathrm{O}) = \mathrm{cond}^{-1}(\mathcal{G}_\mathrm{O}) = \frac{\sigma_n(\mathcal{G}_\mathrm{O})}{\sigma_1(\mathcal{G}_\mathrm{O})} \in [0, 1] \tag{12}$$

is employed as a measure for the degree of observability of the system, whereby a value closer to 1 indicates a higher observability. σ_n is the smallest singular value and σ_1 the largest one.

Taking Eq. (6) as starting point, the energy transfer from the state to the output can be expressed in terms of the singular value decomposition [23] of the EOG as follows:

$$E_\mathrm{meas} = \boldsymbol{x}_0^\mathrm{T} \, \mathcal{G}_\mathrm{O} \, \boldsymbol{x}_0 = \boldsymbol{x}_0^\mathrm{T} \, \mathbf{U} \boldsymbol{\Sigma} \mathbf{V}^\mathrm{T} \, \boldsymbol{x}_0. \tag{13}$$

IOP Conf. Series: Journal of Physics: Conf. Series **1102** (2018) 012021 doi:10.1088/1742-6596/1102/1/012021

Thus, each singular value in $\boldsymbol{\Sigma}$ weights the contribution of each state x_i to the energy function according to observability directions contained in the unitary matrices $\mathbf{U}$ or $\mathbf{V}$. Since the singular vectors are in general not unit vectors, a direct assignment of each singular value to a state is not readily possible. Though, if only the largest component of each singular vector is assumed to determine the observability direction in state space (and the other ones are omitted), this gives a simplified metric to assess the state observability based on the EOG's singular values.

4. Simulation results

This section presents the results obtained for an illustrative application example that is relevant for investigation of state-feedback based closed-loop WTC.

4.1. Test scenario and sensor configurations

As realistic test scenario a turbulent wind field with an average wind speed of 12 m/s is chosen as representative scenario, shown earlier in Fig. 2. The turbulent wind scenario is also necessary to be able to investigate the identifiability of critical model parameters since a persistent excitation is necessary. The FAST8 simulation data is used for control inputs $\boldsymbol{u}(t)$ and measurement outputs $\boldsymbol{y}(t)$ and the simplified design model is used as internal model for prediction of state perturbation. In order to use the above model (1) for empirical observability and identifiability analysis, the augmented state vector is defined as:

$$\boldsymbol{x}_\mathrm{a} = \begin{bmatrix} x_1 \\ x_2 \\ x_3 \\ x_4 \\ x_5 \\ x_6 \end{bmatrix} = \begin{bmatrix} \dot{\varphi}_\mathrm{g} \\ \dot{x}_\mathrm{T} \\ x_\mathrm{T} \\ v_\mathrm{w} \\ \omega_0 \\ \varrho \end{bmatrix} = \begin{bmatrix} \boldsymbol{x} \\ \boldsymbol{d} \\ \boldsymbol{p} \end{bmatrix} . \tag{14}$$

With this trick, it is possible to investigate state and disturbance observability as well as parameter identifiability with EOG approach simultaneously. The following sensor configurations $\boldsymbol{y}_{(j)}$ with $j \in [1, 2, ..., N_\mathrm{sc}]$ are considered for investigation:

$$\boldsymbol{y}_{(1)} = \begin{bmatrix} x_1 \; x_2 \; x_3 \; x_4 \; x_5 \; x_6 \end{bmatrix}^\mathrm{T} \qquad \boldsymbol{y}_{(4)} = \begin{bmatrix} x_1 \; x_2 \; x_3 \end{bmatrix}^\mathrm{T} \qquad \boldsymbol{y}_{(7)} = \begin{bmatrix} x_1 \; x_2 \end{bmatrix}^\mathrm{T}$$
$$\boldsymbol{y}_{(2)} = \begin{bmatrix} x_1 \; x_2 \; x_5 \; x_6 \end{bmatrix}^\mathrm{T} \qquad \boldsymbol{y}_{(5)} = \begin{bmatrix} x_1 \; x_2 \; x_4 \end{bmatrix}^\mathrm{T} \qquad \boldsymbol{y}_{(8)} = \begin{bmatrix} x_1 \; x_6 \end{bmatrix}^\mathrm{T}$$
$$\boldsymbol{y}_{(3)} = \begin{bmatrix} x_1 \; x_2 \; x_3 \; x_4 \end{bmatrix}^\mathrm{T} \qquad \boldsymbol{y}_{(6)} = \begin{bmatrix} x_1 \; x_2 \; x_6 \end{bmatrix}^\mathrm{T} \qquad \boldsymbol{y}_{(9)} = \begin{bmatrix} x_1 \end{bmatrix}^\mathrm{T}$$

The generator speed $\dot{\varphi}_\mathrm{g} = x_1$ is assumed to be known in any configuration because of today's redundant high precision measurement. Provided that all states are in principle measurable/known, the best possible configuration for state estimation is $\boldsymbol{y}_{(1)}$. This is obviously rarely the case. $\boldsymbol{y}_{(9)}$ represents the least observable sensor configuration. Hence, these two configurations determine the expectable range of the observability results. Configuration $\boldsymbol{y}_{(7)}$ is assumed to be today's standard and $\boldsymbol{y}_{(5)}$ also incorporates accurate wind information from lidar (not from wind anemometer). Using $\boldsymbol{y}_{(2)}$, the model parameters are compared to $\boldsymbol{y}_{(7)}$ known in addition.

4.2. Assessment of the EOG

For the investigation, a relatively short time window $T_\mathrm{eog} = 0.5\,\mathrm{s}$, with $N = 10$ samples and a typical sample time of $T_\mathrm{s} = 50\,\mathrm{ms}$, has been chosen. Therewith, it is possible to consider fluctuating wind speeds despite modelling them as constant disturbance in the augmented

IOP Conf. Series: Journal of Physics: Conf. Series **1102** (2018) 012021 doi:10.1088/1742-6596/1102/1/012021

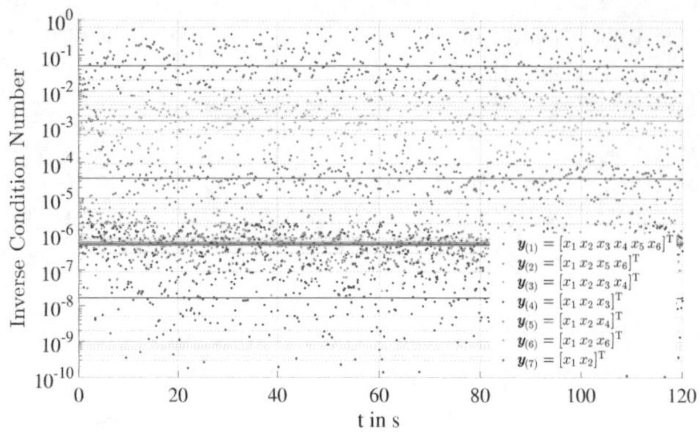

Figure 3. Observability assessment of EOG using the inverse condition number for 240 small time segments $T_{\text{eog}} = 0.5\,\text{s}$ considering the different measurement configurations (the results for $\boldsymbol{y}_{(8)}$ and $\boldsymbol{y}_{(9)}$ are excluded since EOG matrices are rank deficient for both)

system. For each sensor configuration the EOG has been evaluated for the wind trajectory of Fig. 2, what is shown in Fig. 3. The solid lines represent the median values of the ICNs for the studied sensor configurations.

First of all: If x_2 is not measured, the EOG is always rank deficient with ICNs $\ll 10^{-16}$. Thus, the complete state vector is not observable. Despite the aerodynamic coupling through the torque Eq. (1a), the measurement of x_1 contains obviously just insufficient information on the nacelle dynamics. Hence, including x_2 into the sensor configuration provides substantial new information which correlates with an increased measurement energy, cf. Eq. (6). This improves the system's observability such that all ICNs are well above 10^{-10}. $\boldsymbol{y}_{(7)}$ represents the configurations with the lowest ICN. Improvement is achieved with every additional sensors incorporated. Adding information about tower position, wind speed, or tower eigenfrequency results in similar increase of observability. Moreover, the output $\boldsymbol{y}_{(3)}$, which gathers complete tower state information combined with knowledge of the wind speed x_4, is advantageous for simultaneous identification of both, air density and tower eigenfrequency.

Unfortunately, there are two shortcomings when analyzing the ICN alone. First, it is not always clear which states are responsible for the reduced or increased observability level. And secondly, it is not clear what the actual ICN means in terms of estimation error (see Sec. 4.3).

In order to assess the state observability, the singular values and singular vectors of the EOG must be analyzed in greater detail. The singular vectors constitute the observability directions in state space and thus provide information on the single states. To assign singular values to states, the maximum (absolute) component of the singular vector $\boldsymbol{v}_i$ is used to determine the state x_j that is primarily influenced by the corresponding singular value σ_i. This constitutes a strong simplification since smaller components are omitted. However, looking at Fig. 4 illustrates that each median singular value corresponds to only one state variable which shows that this assignment works in principle (though roughly simplified) for observable sensor configurations.

In Fig. 4, the assigned singular values for $\boldsymbol{y}_{(1)}$ are almost homogeneous as expected. In contrast, the remaining sensor configurations alternate the distribution of the singular values depending on the sensor information provided. Especially the observability of wind speed, tower eigenfrequency and air density is affected to a notable extent. Looking at the wind speed x_4, a decrease of four magnitude orders is observed if no measurement information exists. Excluding also the air density from the sensors reduces the singular value by another two magnitude orders. On the other hand, the singular value associated to the nacelle position x_3 varies only slightly.

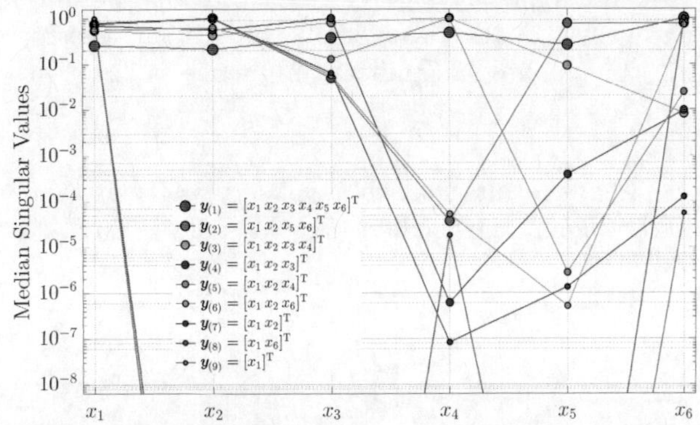

Figure 4. Median singular values associated to individual state x_i for different measurement configurations (values below 10^{-8} are $\ll 10^{-16}$, thus excluded)

4.3. State estimation with SPKF

To check plausibility of the above observability measures, a numerical investigation for the state estimation accuracy has been conducted (Fig. 5). A cubature Kalman filter [24] has been chosen as representative SPKF realization. Model errors and erronous initial state are introduced in order to consider their influence on the state estimates and to obtain meaningful results.

As expected, the results are always very accurate on the first two states. Moreover, simulation results suggest that, without measuring the wind speed, the estimates of both, tower eigenfrequency and air density, may tend to diverge (cf. Fig. 5 right). On the other hand, if both model parameters x_5 and x_6 are assumed to be known then state estimation is very accurate

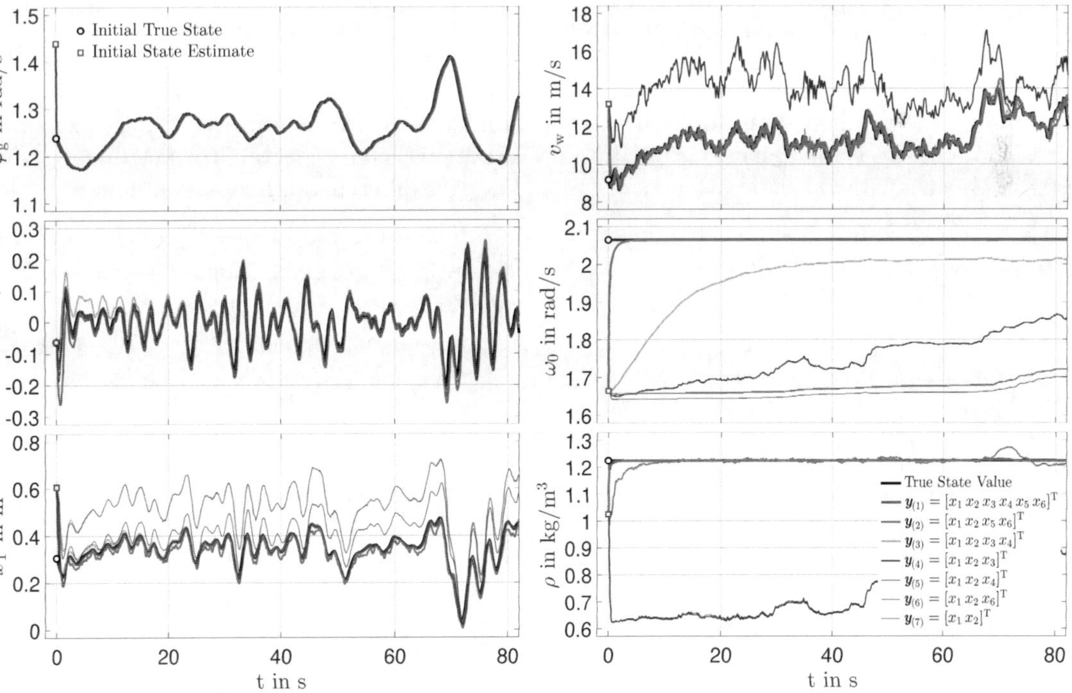

Figure 5. Comparison of state estimates from SPKF algorithm for different sensor configurations using FAST8 data including recovery from significant initial estimation error

for the first four states (which emphasizes the relevance of a well-parametrized internal model). Contrary, an erroneous air density x_6 affects adversely the estimation error of the nacelle position x_3 and the wind speed x_4, thus having a significant influence.

4.4. Discussion of results

The final comparison of the above results shall relate observability measure to estimation quality. The left plot in Fig. 6 shows again the median singular values (MSV) from Fig. 4, excluding

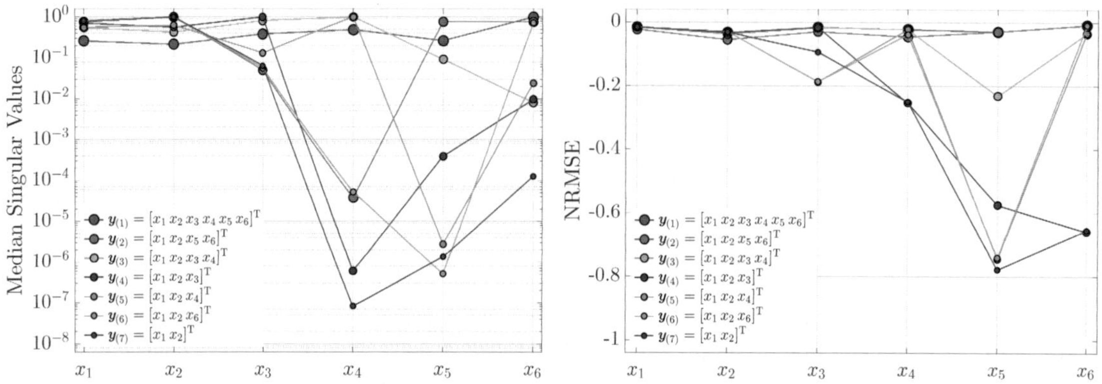

Figure 6. Median singular values obtained for different sensor configurations (left plot) and the computed NRMSE for each state estimate x_i (right plot)

configurations $\boldsymbol{y}_{(8)}$ and $\boldsymbol{y}_{(9)}$. The right plot compares the normalized root mean-squared errors (NRMSE) obtained for the SPKF results in order to evaluate the filter performance. After computation of RMSE, the vector

$$\tilde{\boldsymbol{x}} = -\begin{bmatrix}0.4\,\text{rad/s} & 0.6\,\text{m/s} & 0.8\,\text{m} & 10\,\text{m/s} & 0.5\,\text{rad/s} & 0.7\,\text{kg/m}^3\end{bmatrix}^T , \qquad (15)$$

which contains the range of each state variable as negative value, has been used for normalization to compare the results graphically to the observability measures in Fig. 6. The behaviour of MSV and NRMSE is qualitatively similar. That is, the predicted observability distribution from EOG correlates with the estimation error (EE). Though, differences are observed for

- state x_4: NRMSE is similar to configurations with measured wind speed $\boldsymbol{y}_{(1)}$, $\boldsymbol{y}_{(3)}$ and $\boldsymbol{y}_{(5)}$ despite a MSV of 10^{-4} (which seems to be sufficient for wind estimation). Other than that, a MSV smaller than 10^{-6} for $\boldsymbol{y}_{(4)}$ and $\boldsymbol{y}_{(7)}$ leads to a deterioration of EE.
- state x_6: NRMSE is similar to configurations with known air density $\boldsymbol{y}_{(1)}$, $\boldsymbol{y}_{(2)}$ and $\boldsymbol{y}_{(6)}$ despite a MSV of 10^{-2}. However, looking at $\boldsymbol{y}_{(4)}$ we can observe that the same MSV not necessarily results in the same NRMSE.

In summary, the assignment of singular values to state variables requires a simplification of the observability directions. Inevitably, this has an effect on the observability metrics obtained (and possibly produces the deviations seen and discussed above).

Nevertheless, the combined analysis of EOG and NRMSE provides extensive knowledge about the system's observability and demonstrates that the relationship between state observability and state estimation error is in general not trivial.

5. Conclusions

As concluding remark needs to be said that the above investigation shows principally a correlation between observability measures and the expectable estimation quality. In order to obtain a better picture with respect to observability of states and its actual estimation accuracy, the combined assessment of ICN, MSV and NRMSE is recommended as a strong set of analysis tools. Eventually, the observability analysis constitutes one relevant aspect for every practical realization of state estimation. It reveals the limits for the investigated estimation problems and also shows what is theoretically possible to reach.

In that context, the present contribution has introduced a promising approach based on empirical Gramian matrices to assess efficiently (local) observability metrics for wind turbine state and parameter estimation. The proposed EOG approach avoids completely the necessity of analytical or numerical linearization since it works with nonlinear system simulation only. In addition, its practical implementation is comparatively straightforward to realize, enabling assessment, even for several sensor configurations and wind scenarios, in a short time.

Furthermore, a tangible example has been discussed where an observation problem for simultaneous evaluation of states, parameters and disturbance inputs has been explored. The EOG approach has been reviewed for plausibility by evaluating the estimation results from a generic SPKF implementation against the chosen observability measures. Future work will point out the usability of this approach for higher order systems including rotor blade dynamics and effects of wind shear. Finally, the EOG approach will be reviewed as part of an online evaluation strategy to detect (at an early stage) situations with low observability.

References

[1] Bossanyi E, Savini B, Iribas M, Hau M, Fischer B, Schlipf D, van Engelen T, Rossetti M and Carcangiu C E 2012 *Wind Energy* **15** 119–145
[2] Schlipf D, Schlipf D J and Kühn M 2013 *Wind Energy* **16** 1107–1129
[3] Kumar A A, Rainey P J and Bossanyi E A 2015 *EWEC*
[4] Gros S and Schild A 2017 *International Journal of Control* **90** 2799–2812
[5] Burton T, Jenkins N, Sharpe D and Bossanyi E 2011 *Wind Energy Handbook* 2nd ed (John Wiley and Sons)
[6] Ritter B, Schild A, Feldt M and Konigorski U 2016 *Journal of Physics: Conference Series* **753** 1–12
[7] Ritter B, Schild A and Konigorski U 2016 Making nonlinear state estimation techniques ready for use in industrial wind turbine control systems *Proceedings of Wind Europe Summit 2016*
[8] Ritter B, Mora E, Schlicht T, Schild A and Konigorski U 2018 *Journal of Physics: Conf. Series* **1037** 1–11
[9] Lall S, Marsden J E and Glavaski S 1999 Empirical model reduction of controlled nonlinear systems *Proceedings of the IFAC World Congress* pp 473–478
[10] Singh A K and Hahn J 2005 On the use of empirical gramians for controllability and observability analysis *Proceedings of the 2005, American Control Conference, 2005.* pp 140–146
[11] Jonkman J, Butterfield S, Musial W and Scott G 2009 Definition of a 5-mw reference wind turbine for offshore system development Tech. Rep. NREL/TP-500-38060 Feb. 2009 National Renewable Energy Laboratory
[12] Jonkman J and Jonkman B 2015 Nwtc information portal (fast v8). https://nwtc.nrel.gov/fast8
[13] Gilbert E G 1963 *SIAM Control* **2** 128–151
[14] Brammer K and Siffling G 1989 *Kalman-Bucy Filters* (Artech House, Inc.)
[15] Chen C T 1999 *Linear System Theory and Design* (Oxford University Press)
[16] Hermann R and Krener A J 1977 *IEEE Trans. on Automatic Control* **22** 728–740
[17] Krener A J and Ide K 2009 Measures of unobservability *Proceedings of the 48h IEEE Conference on Decision and Control (CDC) held jointly with 2009 28th Chinese Control Conference*
[18] Adamy J 2014 *Nichtlineare Systeme und Regelungen* (Springer Vieweg)
[19] Skogestad S and Postlethwaite I 2001 *Multivariable Feedback Control. Analysis and Design* 1st ed (John Wiley & Sons)
[20] Glotzbach T, Crasta N and Ament C 2014 Observability analyses and trajectory planning for tracking of an underwater robot using empirical gramians *19th World Congress - The International Federation of Automatic Control*
[21] Hahn J, Edgar T F and Marquardt W 2003 *Process Control* **13** 115–127
[22] Powel N D and Morgansen K A 2015 Empirical observability gramian rank condition for weak observability of nonlinear systems with control *54th IEEE Conference on Decision and Control (CDC)* pp 6342–6348
[23] Strang G 2006 *Linear Algebra and Its Applications* 4th ed (Cengage Learning) ISBN 0-03-010567-6
[24] Arasaratnam I and Haykin S 2011 *Automatica* **47** 2245–2250

Research sized wind turbine blade modal tests: comparison of the impact excitation with shaker excitation.

M. M. Luczak[1], B. Peeters[2], S. Manzato[2], E. Di Lorenzo[2], K. Reck-Nielsen[3], P. Berring[1], P. U. Haselbach[1], K. Branner[1]

[1]Technical University of Denmark, Department of Wind Energy,
Frederiksborgvej 399, 4000 Roskilde, Denmark
[2]Siemens Industry Software NV, RTD Test Division
Interleuvenlaan 68, 3001 Heverlee, Belgium
[3] CEKO Sensors ApS
Diplomvej 381, DK-2800 Kgs. Lyngby, Denmark

mluz@dtu.dk

Abstract. Modern wind turbine blades are being tested for certification purposes in accordance to the IEC-64100 standard. Part 23 of the norm details the requirements for the full scale structural testing of rotor blades. As a minimum, it requires measurement of the first and second flap wise and first edge wise natural frequencies. It lists damping and mode shapes as other blade properties which may be of interest and optionally measured. The paper presents the modal model parameters estimation based on the experimental modal analysis. In two tests performed, the input force has been introduced through impact hammer and two electrodynamic shakers excitation. Several first modes had been identified for both excitation methods, including first torsional mode of the investigated blade. Results of the modal tests can be used to (a) provide more detailed information about the structural dynamics characteristics of the blade and (b) improve the design by adjusting the dynamic properties of the blade to some desired condition.

Keywords:
experimental modal analysis, wind turbine blade, structural dynamics identification,

1. Introduction

Wind turbine blades certification [1] demands full scale structural tests for determining the blade properties. It comprises mainly of the static tests to verify the structural strength and fatigue tests in order to ensure the designed lifetime of 20-25 years.

In the presented research, the experimental based estimation of the modal model parameters is performed. In addition to the natural frequencies values, the corresponding mode shapes and damping ratios has been estimated [2]. Investigated blade requires appropriate excitation for the adequate identification of the modes and mode shapes [3,4]. Provision of the sufficient energy for the vibration excitation requires a dedicated test setup and installation of the often heavy and stiff mounts for the load attachment. In turn, the measured system comprises of the object of investigation and the additional instrumentation. The mass, stiffness and damping of the added equipment modify the measured characteristics of the tested blade. In the presented paper the two different excitation methods were investigated to assess the trade off between test simplicity and results accuracy. The values have been

IOP Conf. Series: Journal of Physics: Conf. Series **1102** (2018) 012022 doi:10.1088/1742-6596/1102/1/012022

compared to the results of the theoretical modal analysis performed with the use of the high fidelity finite element method (FEM) model.

2. Research methodology and approach

The object of investigation is 14,3 [m] long wind turbine blade made of the Glass Fiber Reinforced Plastics. Multiple modal tests have been performed to identify the most reliable and time efficient method to identify modes for such a structure.

2.1. Test setup

The investigated blade has been supported by two elastic cords to provide free-free boundary conditions as presented in Figure 1.
To identify the best experimental setup, the measurement campaign was designed to excite the structure with both modal hammer and shakers. To optimally excite the blade, it was decided to excite the structure both in the edgewise and flapwise directions, and at different locations so that the energy could be best spread throughout the structure. Consequently, the two shakers were positioned one at approximately 4.5 m from the blade root at the suction side surface of the blade exciting the flapwise direction , and one at 8 m from the root on the leading edge exciting the edgewise mode, similarly to the approach adopted in [6]. Impacts with the hammer were applied at the same locations. The hardware selected for this test included a pair of electrodynamic shakers rated at 100N with amplifiers and a modal sledge hammer weighting 1kg with a soft tip to ensure a good excitation between 3 and 150 Hz. Both connections between the shakers and the blade were instrumented with a load cell to accurately measure the applied force; for impact testing, the force sensor embedded in the hammer was used.

Finally, to be able to cover the entire surface of the blade, a dense grid of 120 measurement points has been defined. In order to reduce the mass loading from the transducers [7] the set of 15 available triaxial accelerometers has been distributed over fifteen equidistant cross sections ranging from the root of the blade to the tip. The same measurement has been repeated 8 times by roving the available accelerometers over different airfoil locations.

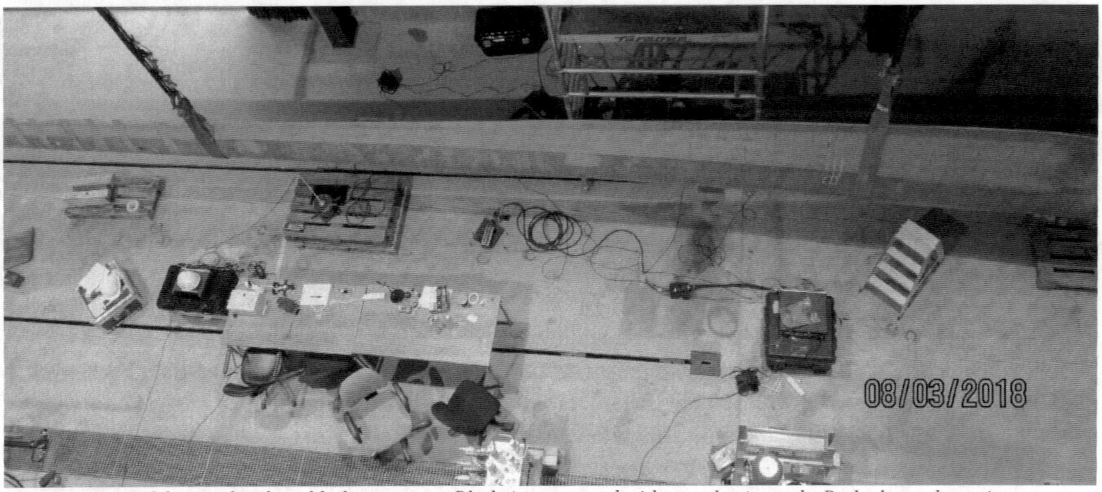

Figure 1 Top view of the wind turbine blade test setup. Blade is supported with two elastic cords. Both electrodynamic shakers are attached to the blade. Sensors are connected to the two data acquisition modules.

2.1.1. Optical accelerometer has been used for the purpose of the comparison of the metal free transducer with the traditional piezoelectric accelerometer.
All-optical sensors are ideal for applications involving harsh environments, distributed networks and remote sensing. For the vibration acceleration measurement all optical, frequency modulated, Micro-

Electro-Mechanical Systems (MEMS) sensor has been used. The sensor is a silicon microchip with integrated optical components. Spectrogram recorded during the measurement is presented on the Figure 2. Application of the electrical and optical sensors has provided useful insights about the feasibility of the optical technology application for the wind turbine blade applications.

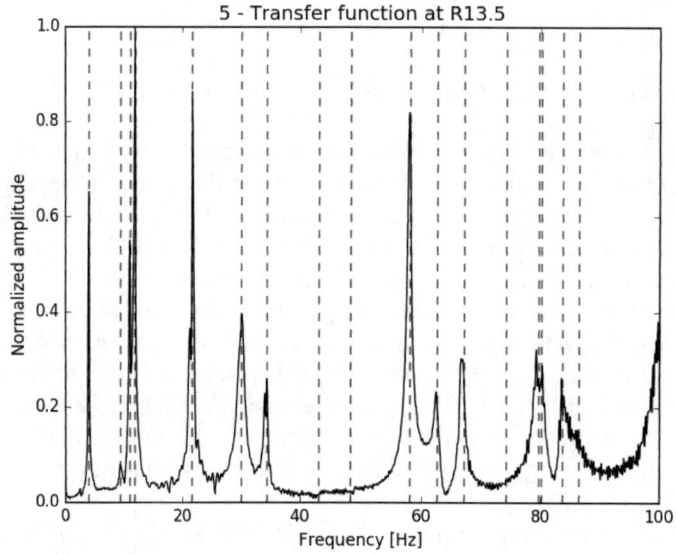

Figure 2 Frequency Transfer Function from optical accelerometer. Blue dashed lines denote the spectral lines identified from the test with the use of the electrical sensors.

3. Measurement and simulation results

3.1. Comparison of shaker and hammer FRF measurements

When performing a roving test, repeatability in the excitation is a key aspect to ensure all data can be processed together and global results derived. Shaker excitation has a clear advantage over hammer excitation, as the force profile is defined by the user; on the contrary, when doing impact testing, the ability of the user in applying always the same impact plays a crucial role. Figure 3 shows the Driving Point FRFs measured during the 8 runs with hammer and shaker excitation in both direction. When looking at the results in the edgewise direction, we can observe very repeatable FRFs regardless of the excitation technique. Some extra variability however is visible in the Impact testing results, in particular at higher frequencies and with a generally higher noise level. However, the flapwise results show very poor repeatability with shaker excitation. This is due to the fact that in this direction the shaker was often detaching from the blade because of the former pendulum motion and the necessity of using glue to avoid damaging the blade with fixed mechanical connection. In this case, consistently exciting with the hammer was much simpler.

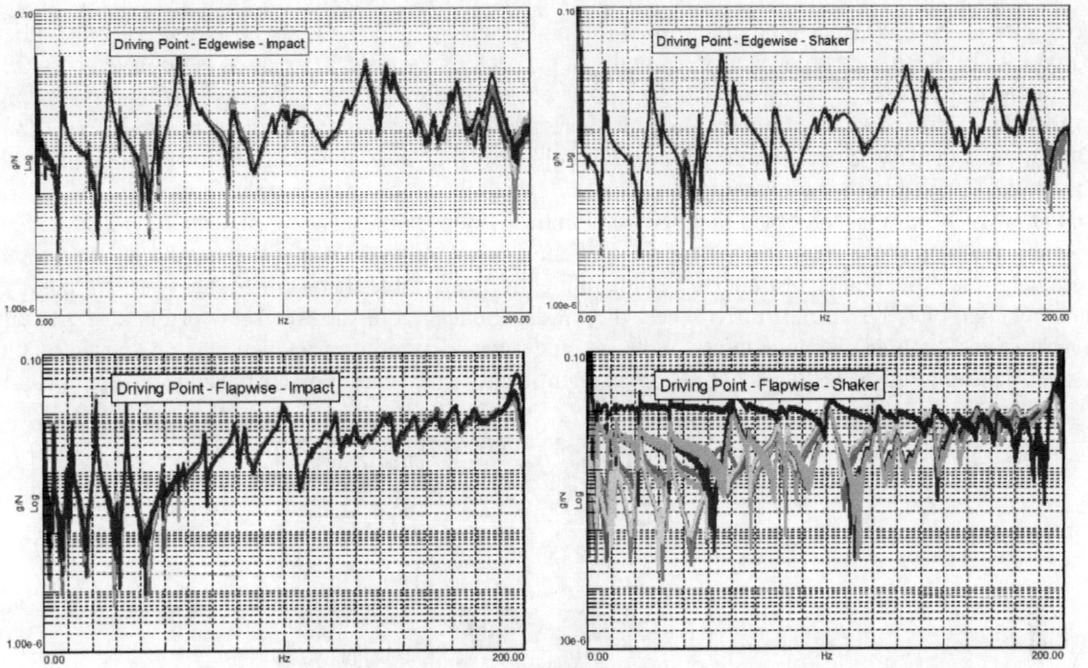

Figure 3 Repeatability analysis: Driving Point FRF over the 8 measurement runs for Hammer (left) and Shaker (rigth) excitation in both Edgewise (Top) and Flapwise (Bottom) direction

3.2. Modal analysis results

As explained in the previous section, both hammer and shaker testing for FRF measurements have advantages and disadvantages. As the ultimate goal of this analysis is to characterize the modal response of the turbine blade, the FRFs collected will be processed using Simcenter Testlab Modal Analysis to derive the modal parameters. Because of the repeatability achieved over the different tests, it is acceptable to merge all FRFs and process them in one go. The correlations of the natural frequencies (left) and corresponding damping ratios (right) are presented in Figure 4.

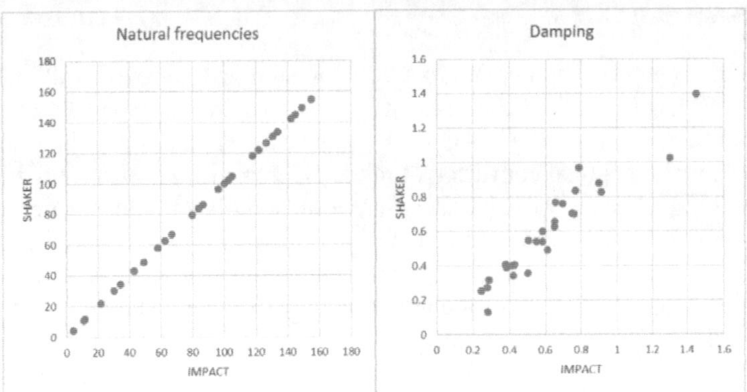

Figure 4 Correlation plot of the Natural Frequencies, Modal damping coefficients estimated from impact and shaker tests.

Natural frequencies are accurately and consistently identified in the two cases, but some differences can be observed in frequencies and damping. Damping estimates are a bit more spread, but this can be attributed to the difference in noise between shaker and impact testing: to save time, only 5 averages where collected when using the hammer, against the 50 used with shaker, thus causing noise levels to be higher in the former. Damping is known to be affected by noise and this is what causes its higher variability between the two excitations. Finally, a generally very good correlation is observed between the mode shapes, except two cases. The first poorly correlated mode is the one at 43 Hz, corresponding to the blade 1st torsion. This is also a mode where the difference in damping is highest (1.3 vs 1 %). Overall, this mode was not optimally excited in neither of the cases, thus causing low confidence with the modes derived both from hammer and shaker testing. The other non-correlated mode is the one at 104 Hz, which shows a combined flapwise-torsion behaviour. In this case, by analysing the data and by visually animating the shape, it was clear that the impact estimate was not accurate as the modal parameter identification algorithm had trouble in identifying a stable shape due to the poor FRF quality. However, despite these small differences, the modal models from the two sets correlate very well: overall, the slightly higher uncertainty on the estimates derived from impact testing are balanced by the higher flexibility of the testing setup, which is significantly more complex when using shaker.

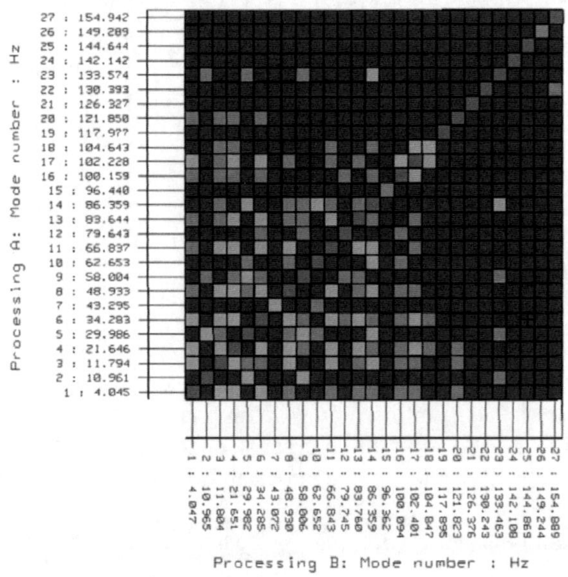

Figure 5 Modal Assurance Criterion matrix for the modal vectors estimated from impact and shaker tests.

Modal Assurance Criterion has been applied to assess the similarity of two mode shapes. For identical mode shapes from different excitations the MAC will have a value of one or 100% as show in Figure 5. For modes which are very different, the MAC value is close to zero, as shown in Figure 5. MAC values close to 100% on the main diagonal confirm that modal vectors from impact and shaker excitation are nearly identical. Low values of the off-main diagonal terms confirm the modes are uniquely observed. Through the experimental campaign overall 120 measurement points were measured in three directions. Modal model estimation has been implemented through the multi run modal analysis approach. For each of the data set corresponding with the particular sensor station the estimation of the modal parameters has been performed. It resulted with the collection of 15 partial modal models which were merged into one global modal model of the overall structure. To compensate the variability between the recorded sets which might occur due to the mass loading and the boundary conditions change, the complete mode shapes were calculated using Driving Point scaling as presented in Figure 6. Next to the experimental activity the numerical simulations presented in Figure 7 has been performed and the results has been compared to the measurement results.

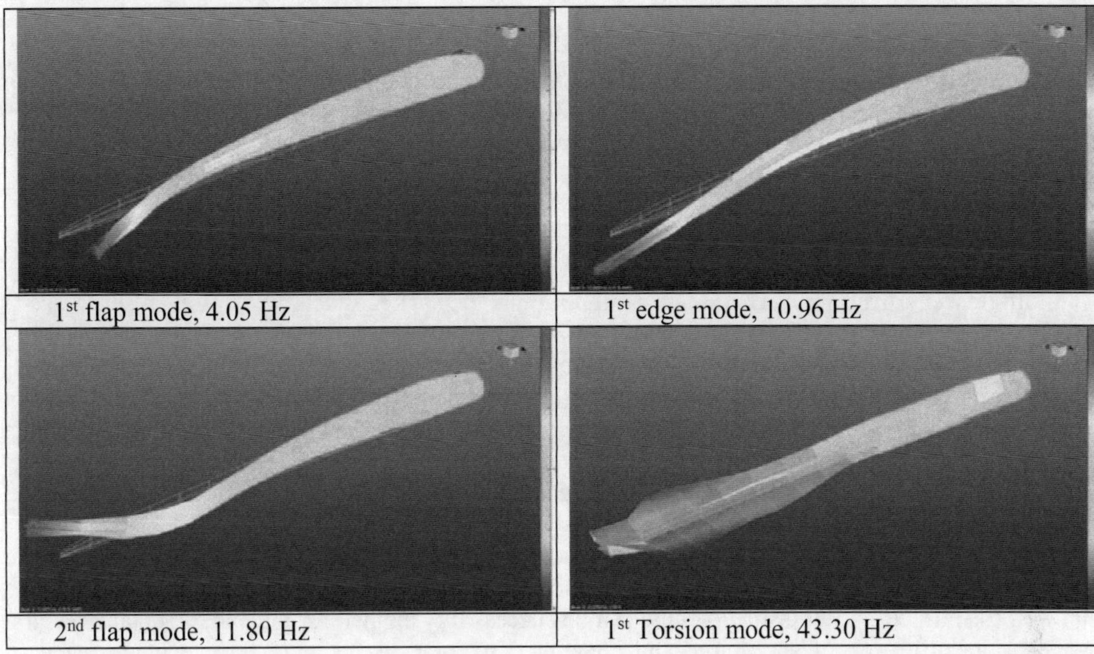

Figure 6 Experimental modal analysis results with the shaker excitation applied.

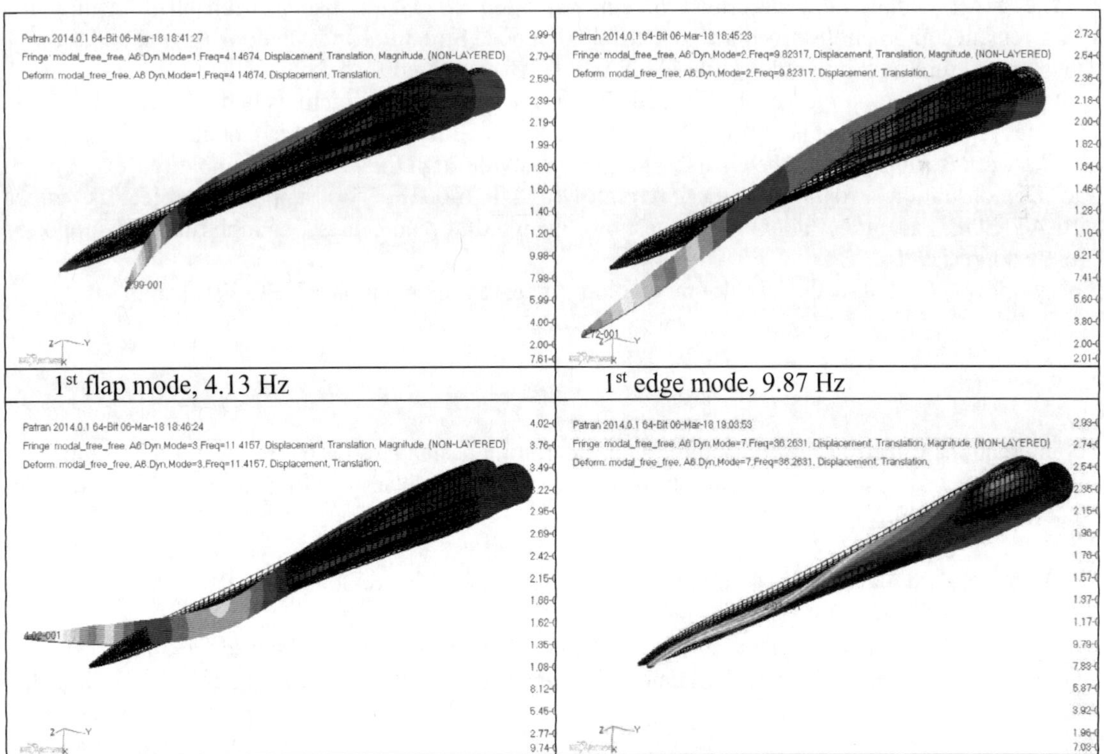

Figure 7 Theoretical modal analysis results. Calculations performed using FEA model

4. Conclusions

Presented research has been focused on assessment of the different excitation techniques applied for experimental modal testing of the same blade. Modal model parameter values estimation has been conducted based on the experimental datasets collected within both tests. In order to exclude the influence of the other test setup parameters (boundary conditions, number and location of the measurement points) these settings were kept invariant.

Impact testing in this study represents the Single Output and Multiple Input (SIMO) testing while application of wo Electrodynamic shakers falls into Multiple Input Multiple Output (MIMO) method. Main difference is that in the second test configurations the shaker attachment has been set to excite both directions, flapwise and edgewise, at the same time. This is clearly the advantage of the shaker excitation. Next to it shakers are driven by a signal generator which offers different excitation signals ranging from purely harmonic through the swept and stepped sine up to random. Shaker excitation ensures the excitation force repeatability to the extent not available from the hammer testing. Impacting the structure with the modal hammer has low repeatability in terms of the force level. Important aspect of the compared tests is the duration of the tests. Impacting by modal hammer and exciting the structure with the two electrodynamic shakers took a comparable overall testing time.

Important conclusion from the comparison of the two experimental approaches is that for the investigated blade the applied excitation method has little influence on the obtained results. It proves that well established impact testing method can be successfully applied for the experimental structural dynamics identification of such large and complex structures like a wind turbine blade made of composite material.

5. Acknowledgements

The experimental work described herein has been conducted using mechanical testing and measurement equipment from Villum Center for Advanced Structural and Material Testing (CASMaT). The support from Villum Fonden (Award ref. 00007293) is gratefully acknowledged.

This work was supported by the Danish Centre for Composite Structures and Materials for Wind Turbines (DCCSM), Grant no. 09-067212 from the Danish Strategic Research Council.

The work is supported by the Danish Energy Agency through the Energy Technology Development and Demonstration Program (EUDP), Grant No. 64016-0023. The supported project is named "BLATIGUE: Fast and efficient fatigue test of large wind turbine blades", and the financial support is greatly appreciated.

This work was funded by the VLAIO Innovation Mandate project number HBC.2016.0235.

6. References

[1] International Electrotechnical Commission. International standard iec 61400-23 Wind turbine generator systems – part 23: full-scale structural testing of rotor blades. Wind Turbines-Part 1: Design Requirements 2014.

[2] Avitabile P. Modal testing : a practitioner's guide. : John Wiley & Sons Ltd, 2018.

[3] Griffith DT, Carne TG. Experimental modal analysis of 9-meter research-sized wind turbine blades. In: Anonymous Structural Dynamics and Renewable Energy, Volume 1: Springer; 2011, p. 1-14.

[4] Griffith DT, Smith G, Casias M, Reese S, Simmermacher TW. Modal Testing of the TX-100 Wind Turbine Blade. Sandia National Laboratories Technical Report, Report# SAND2005-6454 2006.

[5] Larsen GC, Hansen MH, Baumgart A, Carlén I. Modal analysis of wind turbine blades. 2002;Risø–R–1181(EN).

[6] Luczak M, Manzato S, Peeters B, Branner K, Berring P, Kahsin M. Dynamic investigation of twist-bend coupling in a wind turbine blade. Journal of Theoretical and Applied Mechanics 2011;49:765-89.

[7] Heylen W, Lammens S, Sas P. Modal Analysis Theory and Testing. 2nd ed. Leuven: Katholieke Universiteit Leuven, Departement Werktuigkunde, 1998.

IOP Conf. Series: Journal of Physics: Conf. Series **1102** (2018) 012023 doi:10.1088/1742-6596/1102/1/012023

Beyond OC5 – Further advances in floating wind turbine modelling using Bladed

A Beardsell, A Alexandre, B Child, R Harries, D McCowen

DNV GL, Linear Park, Avon Street, Temple Quay, Bristol, BS2 0PS, UK
alec.beardsell@dnvgl.com

Abstract. Floating wind turbines offer the potential to harness the considerable wind energy resource located at deep water offshore locations. Accurate numerical modelling of floating wind turbines is of utmost importance given its prominence in the design process. DNV GL's aero-hydro-servo-elastic modelling package 'Bladed' was one of the tools validated as part of the Offshore Code Comparison Collaboration Continuation with Correlation (OC5) project [1] and gave favourable results alongside the other participants. However, in common with all of the participants using a Boundary Element Method hydrodynamic approach, the majority of the loads were under-predicted (whilst over-predicted by Morison-only approaches). In particular, the low frequency excitation was not fully captured in the Bladed numerical simulations. This paper describes the implementation and results of new simulations in which additional modelling features are implemented in the DNV GL Bladed OC5 model. The results show that including instantaneous hydrostatics and Froude-Krylov forcing produces a better match with experimental spectral energy density for platform pitch and surge motion, which in turns leads to improvements in tower base shear force predictions.

1. Introduction

Accurate modelling of the complex coupled dynamics of floating wind turbines is a challenging task. The OC5 project [1] validated a wide range of aero- and hydro-elastic modelling tools against measured tank test data and although reasonable results for simulated fatigue and ultimate loads were obtained, the results also demonstrated that further advances in modelling are required to accurately characterise floating wind turbine performance. In particular, for wave-only cases, a 'fairly consistent underestimation of wave-excitation forces outside the linear wave-excitation region' [1] was reported for tower and upwind mooring line loads, particularly at the low frequencies associated with excitation of the platform pitch (~30s) and surge (~100s) natural frequencies. Uncertainties associated with the test conditions are likely to be part of the reason for this underprediction, and detailed work has been undertaken to quantify such uncertainties [2]. Another source of the underprediction could be modelling deficiencies. DNV GL have therefore implemented several modelling improvements in Bladed with the aim of achieving a closer match to the experimental measurements. This paper describes some key results.

2. Overview of tank test measurement arrangements in OC5 project

For the OC5 project, a 1/50th scale floating wind turbine on a semi-submersible platform was subjected to an extensive measurement campaign, conducted by the DeepCwind consortium at the MARIN offshore wind basin [3]. The system was tested under Froude-scaled wind and wave loads. Note that all input and output variable values in this paper are given as equivalent full-scale quantities, using Froude scaling. Tests performed ranged from simple static displacement and free-decay tests to complex operating conditions with irregular sea states and dynamic winds. For a detailed description of the OC5 model properties and the tests performed, the reader is referred to Sections 2 and 3 of the OC5 report [1].

3. Bladed for floating wind turbine modelling

Bladed is a commercially available aero-hydro-servo-elastic multibody wind turbine modelling package, capable of fully coupled floating wind turbine simulation. Bladed has widespread use amongst wind turbine manufacturers, certification bodies, consultancies and universities around the world and so its performance in simulating floating wind turbines is of interest to a significant audience.

3.1. Bladed modelling features already implemented in OC5 study

The original submitted Bladed results for the OC5 project included the following features [1], whilst replicating measured environmental conditions:

- Time domain implementation of Boundary Element Method hydrodynamics using convolution integrals.
- Additional Morison hydrodynamics and global linear damping for better drag representation
- Second order wave excitation loads
- Dynamic mooring line model

3.2. Additional Bladed modelling features implemented for this work.

For the new simulations presented in this paper, additional features were implemented.

Model updates:

- Correction to the modelled overall 1st tower fore-aft and side-side frequencies. This adjustment was required to account for the additional flexibility identified in the model's tower base load cell. The adjustment was achieved by reducing the modelled tower Young's modulus value.

Hydrostatics and hydrodynamics updates:

- Small adjustments to the additional linear damping values for the floating structure, applied at the free surface on the tower centre line. The values were adjusted to achieve a better match to the free decay calibration simulations.
- Discretisation of floater geometry into a flat panel mesh of the whole platform, enabling the following enhanced hydrodynamic approaches to be implemented:
    - Instantaneous hydrostatics calculated based on the exact position and orientation of the floating platform at each time step
    - Instantaneous Froude-Krylov correction again based on exact position and orientation of the floating platform at each time step

  Note that for the present results set, the free surface is taken to be the mean water level. There is also the option in Bladed to use the incident wave free surface but this was not used in the present analysis.

Bladed incorporates a default quasi-linear hydrodynamics formulation based on boundary element method (BEM), potential flow solver derived properties. The force calculation is an augmented, multi-body arrangement of the well-known Cummins equation

Bladed also includes more advanced hydrodynamics options to allow the Froude-Krylov component of the wave exciting force and the hydrostatic force to be computed directly using the instantaneous wetted profile of a body. In this case, the Froude-Krylov pressure, p at any given point in the flow field may be calculated for each wave frequency constituent ω as:

$$p(\omega, t) = -\rho A\, Re[j\omega \Phi_{FK} e^{j\omega t}] \qquad [1]$$

where t is time, ρ is water density, A is wave amplitude, j is the imaginary number, Φ_{FK} is the complex velocity potential amplitude:

$$\Phi_{FK}(\omega) = -\frac{g}{\omega}\frac{\cosh k(z+h)}{\cosh kh} j e^{j(k(\boldsymbol{d}.\boldsymbol{x}))} \quad [2]$$

in which g is acceleration due to gravity, h is the water depth, k is wavenumber and $\boldsymbol{d}$ is a vector representing the horizontal wave direction. The pressure on each panel is integrated over the wetted body profile, S to directly yield the applied force and moment:

$$\begin{matrix}\boldsymbol{F_{FK}} \\ \boldsymbol{M_{FK}}\end{matrix}(\boldsymbol{\omega}, \boldsymbol{t}) = \iint_S p \begin{pmatrix}\boldsymbol{n} \\ \boldsymbol{r} \times \boldsymbol{n}\end{pmatrix} dS \quad [3]$$

where $\boldsymbol{n}$ is a unit vector pointing in to the body and r is a vector describing the position of a panel with elemental area dS.

The hydrostatic pressure is integrated in a similar manner and is given by:

$$p(t) = \rho g z(t) \quad [4]$$

where z denotes the vertical panel displacement from the reference height. Bladed allows the wetted body profile to be defined either as the instantaneous body surface below the mean waterline or the instantaneous body profile below the instantaneous incident water surface elevation, neglecting diffraction and radiation effects. The latter option introduces a greater level of nonlinearity.

More information about these methods can be found in [4] and [5].

It should be noted that some of the nonlinearities arising from the use of the instantaneous hydrostatics and Froude-Krylov forces are theoretically already accounted for in the second order excitation force which is also included in the model. For this work, exactly the same second order flow solver data was used both with and without the new features and so double accounting is present. Furthermore, accounting for some but not all nonlinearities when modelling a dynamic system can lead to greater errors than not including any since the model becomes theoretically inconsistent. However, it is the aim of this work primarily to assess the influence of including a more detailed description of some phenomena on the results, given that the above mentioned potential shortcomings may in practice have a negligible effect on results.

4. Methodology

The study focusses on two load cases from the original OC5 study. These are both 3-hour simulations with unidirectional irregular seas (JONSWAP spectrum), and no wind:

LC3.3 – Operational wave : Hs = 7.1m ; Tp = 12.1s ; γ = 2.2

LC3.4 - Design Wave : Hs = 10.5m ; Tp = 14.3s ; γ = 3.0

Wave only load cases were selected in order to highlight the effect of the implemented changes on the hydrodynamic modelling fidelity. By removing the influence of wind, uncertainties associated with the reproduction of characteristic scaled wind conditions, and the model-scale turbine's response to those conditions, was eliminated.

The data channels for analysis were selected to best illustrate the effects of the modelling changes on hydrodynamic modelling accuracy. Available channels included platform motion, structural loads and mooring line loads (fairlead tension). From these, global platform motion was identified as the primary metric for comparison of hydrostatic and hydrodynamic modelling capability in wave only load cases, as uncertainties associated with the as-built model-scale tower definition can be considered to have only a minor influence. In addition, tower base shear force was also analysed, bearing in mind the aforementioned uncertainties.

The experimental results presented below were generated directly from the measured OC5 LC33 and LC34 3-hour time series.

Output channels were processed into peak exceedance plots, autospectra and damage equivalent loads (DELs). Damage equivalent loads were calculated based on the same process as the original OC5 results (1Hz reference frequency, Wohler coefficient = 5, Goodman adjustment).

5. Results

Three versions of load cases LC33 and LC34 are compared:

1. Original 3-hour experimental data
2. Originally submitted Bladed OC5 simulation
3. New updated Bladed simulation with additional modelling features

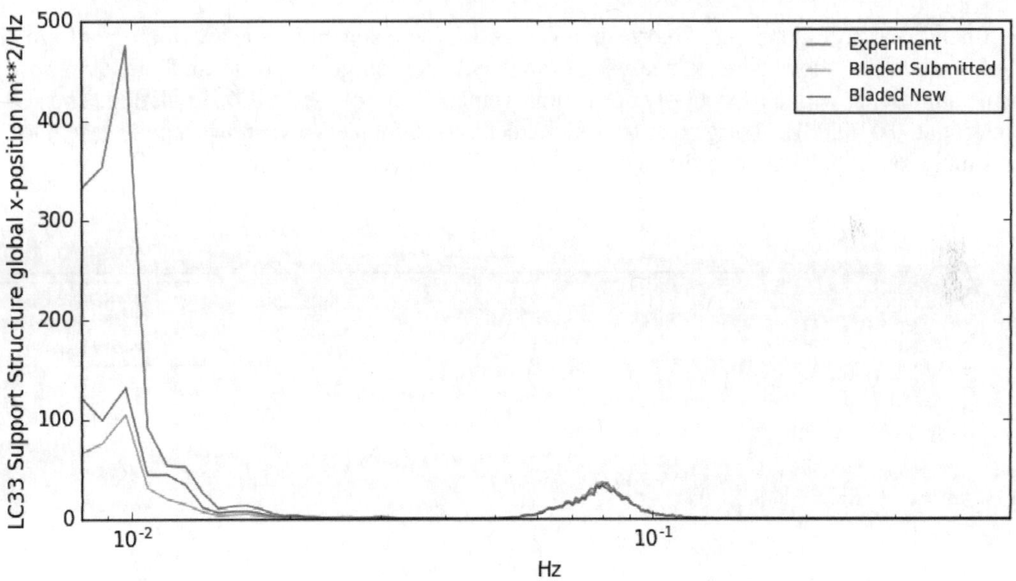

FIGURE 1: LC33 Global x- position (surge) spectral energy density

IOP Conf. Series: Journal of Physics: Conf. Series **1102** (2018) 012023 doi:10.1088/1742-6596/1102/1/012023

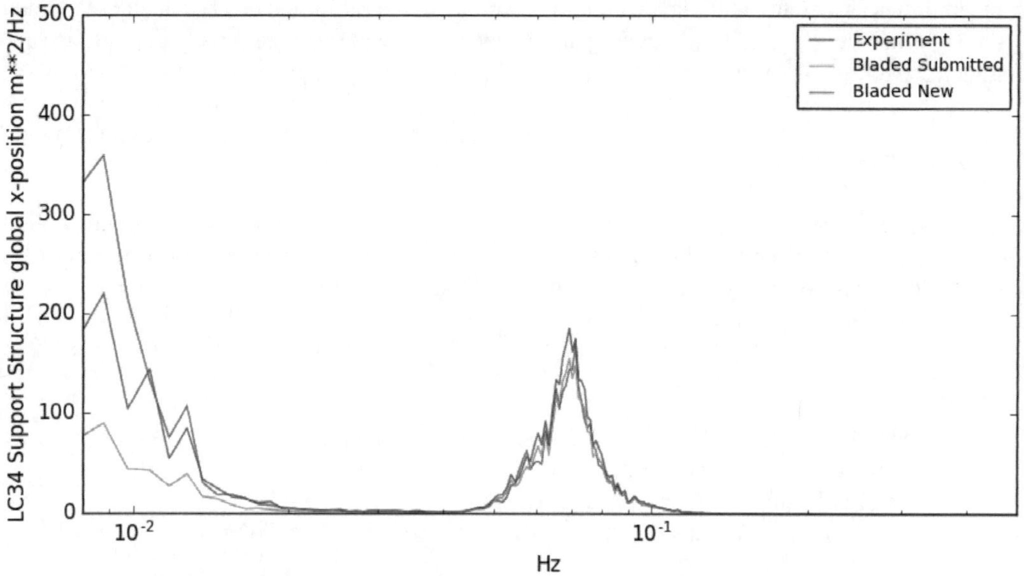

FIGURE 2: LC34 Global x- position (surge) spectral energy density

The left-hand peak in Figures 1 & 2 corresponds to the surge resonant frequency of the platform. The updated simulations show clear increases in low frequency surge response and are closer to the experimental results, with a particularly clear improvement seen for the LC3.4 simulation. The second peak at around 0.07-0.08Hz (surge response at incident wave frequencies) shows a good match for the LC3.3 simulation, but a slight overprediction for the LC3.4 extreme wave case.

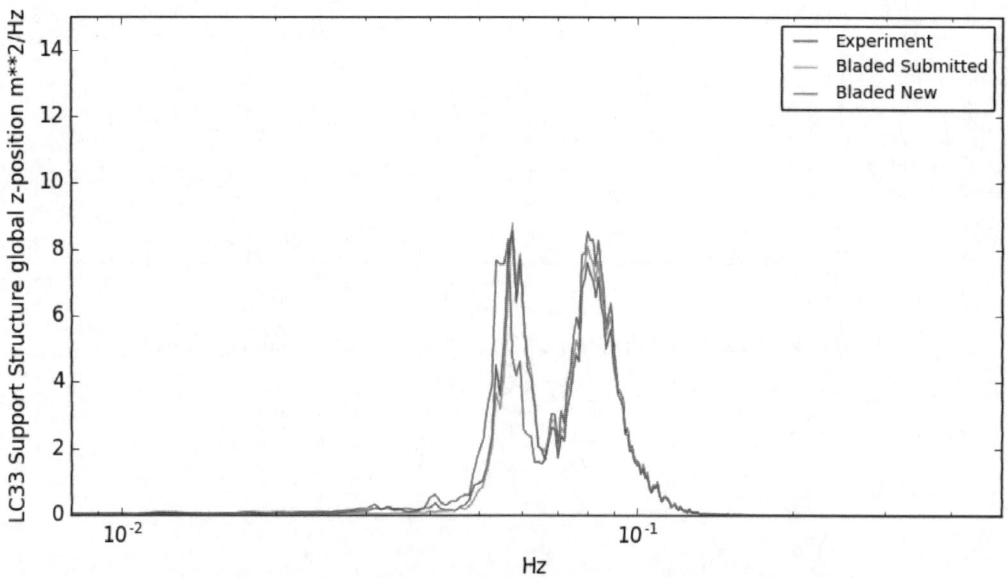

FIGURE 3 : LC33 Global z- position (heave) spectral energy density

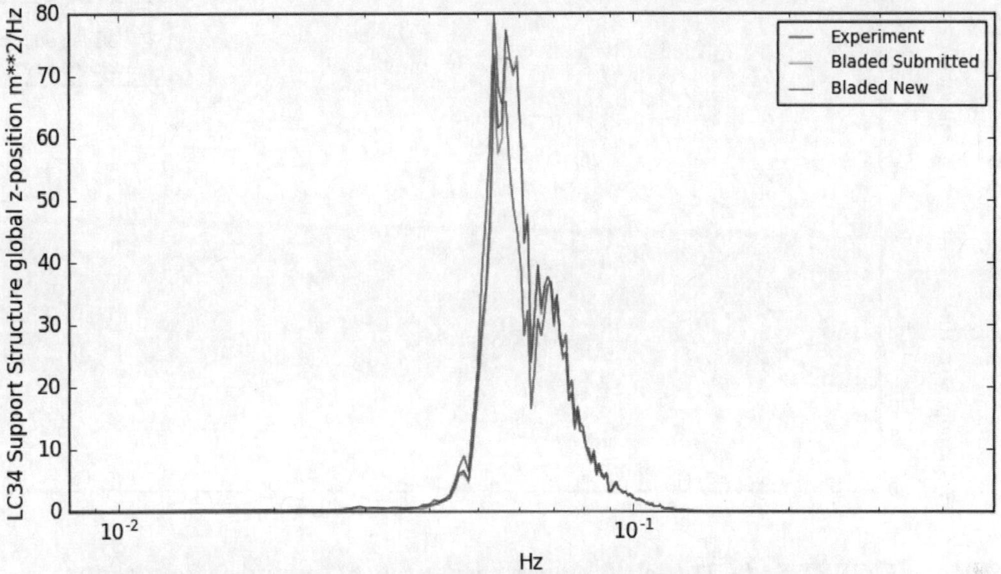

FIGURE 4 : LC34 Global z- position (heave) spectral energy density

Figures 3 & 4 show a close match between the original and updated Bladed simulations for global heave motion spectral density. Both signals are also in good agreement with experimental results. The expected difference in amplitude response between the two cases is clear, especially for the lower frequency peak corresponding to the platform heave natural frequency at 0.05 – 0.06Hz. The good match shows that heave motion is not greatly affected by instantaneous hydrostatics / Froude-Krylov forces, as expected given the floater geometry. This can be explained by the fact that the horizontal projection of the mean wetted surface is very close to that of the instantaneous position of the platform, assuming small pitch motions.

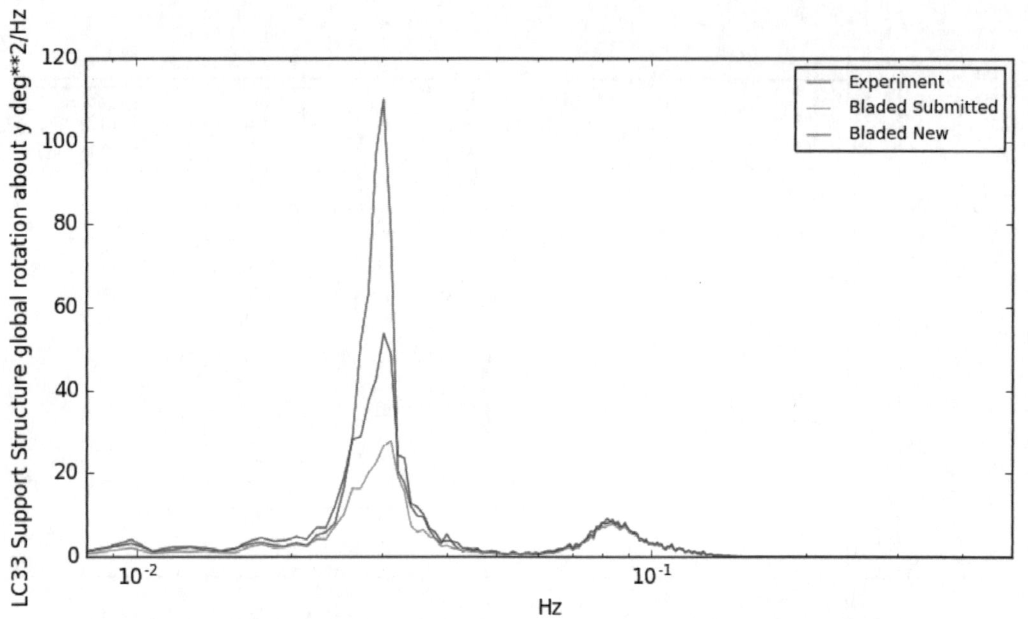

FIGURE 5 : LC33 Global rotation about y (pitch) spectral energy density

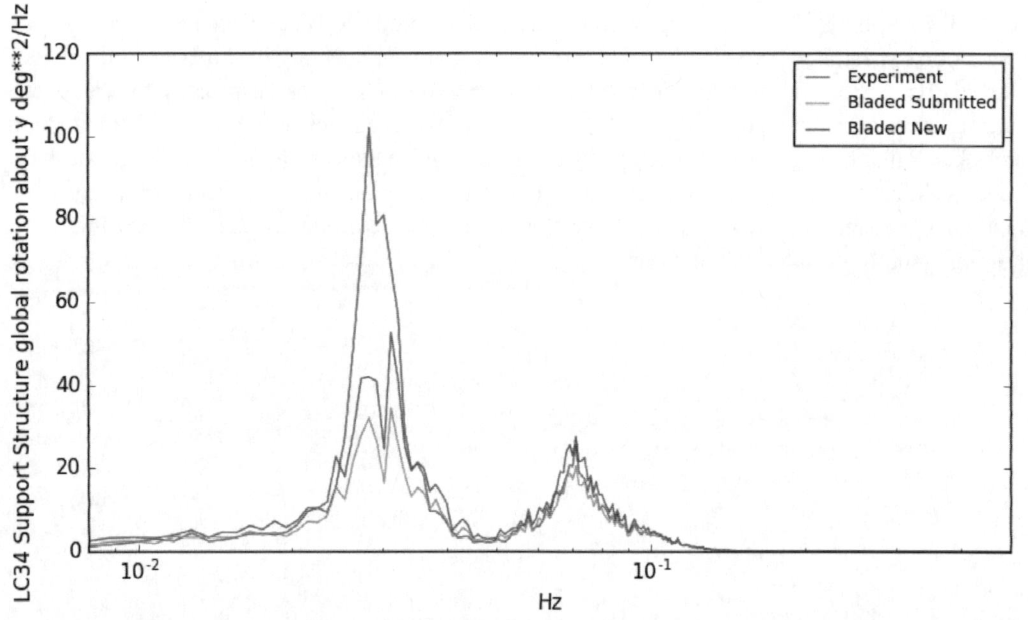

FIGURE 6 : LC34 Global rotation about y (pitch) spectral energy density

Figures 5 & 6 are very similar. The large peak on the left of both figures corresponds to the platform pitch resonant frequency, and shows the significant underprediction of pitch resonance in the original results. The second peak corresponds to the response at the incident wave energy spectrum frequencies. In both figures, the updated Bladed simulations display improved pitch excitation although the LC3.4 simulation slightly overestimates the first order pitch response.

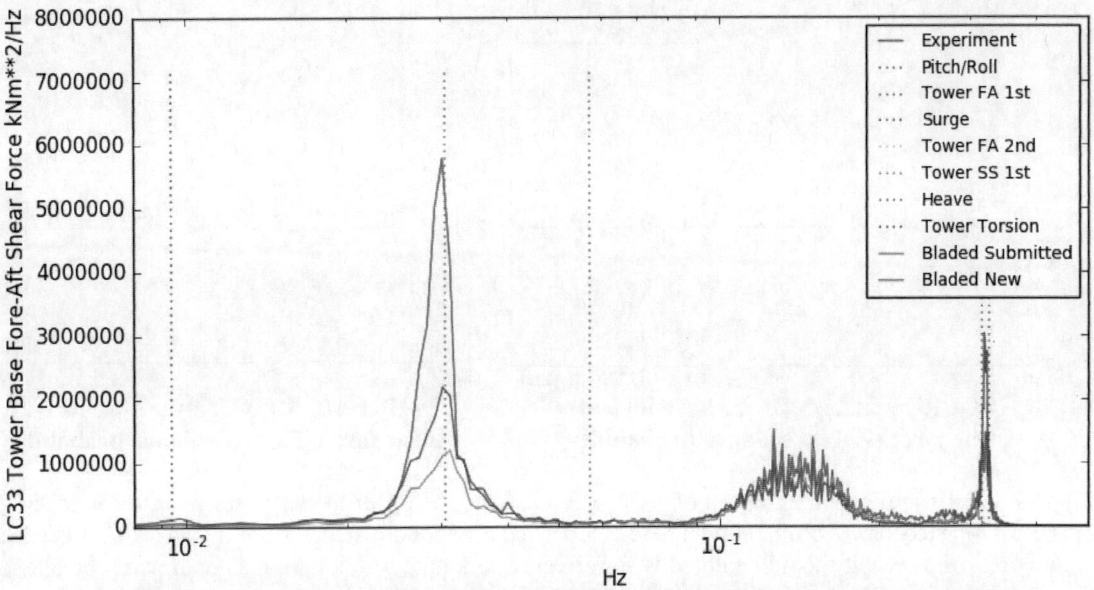

FIGURE 7 : LC33 Tower base fore-aft shear force (Fx) spectral energy density

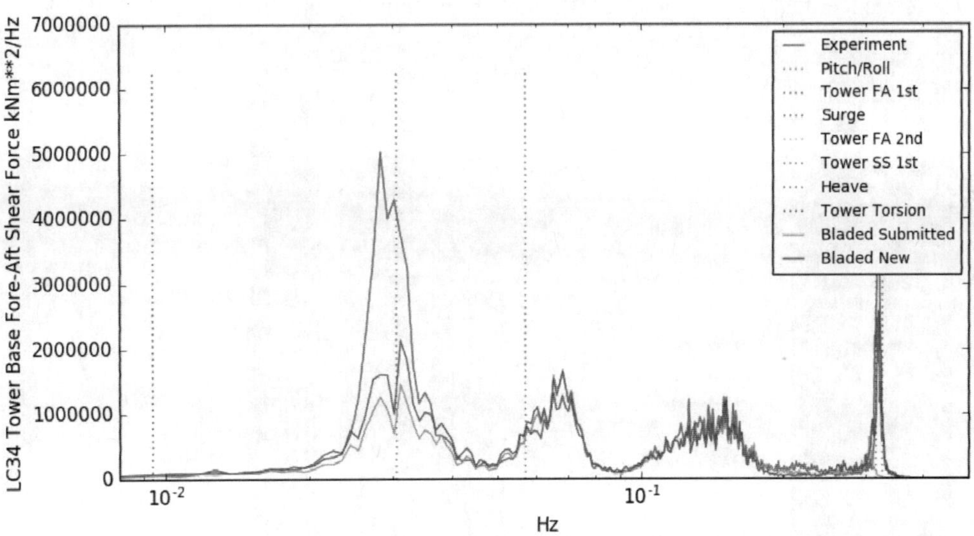

FIGURE 8 : LC34 Tower base fore-aft shear force (Fx) spectral energy density

Tower base shear force is driven largely by pitch and surge motion, and Figures 7 & 8 reflect the improved spectral response of platform pitch shown in Figures 5 & 6. The additional peak in the LC34 plot corresponding to Tp=14.3 (0.07Hz) is clearly visible. The sharp peak on the right of both plots corresponds to the tower frequency, with the updated results showing an overestimate in loads. However, uncertainty in tower damping in the constructed physical model reduces the confidence with which this tower frequency response comparison can be made.

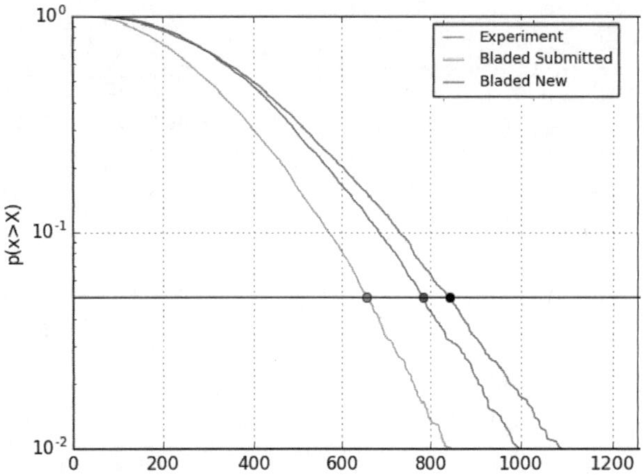

FIGURE 9 : LC33 Tower base fore-aft shear force (Fx) exceedance probability

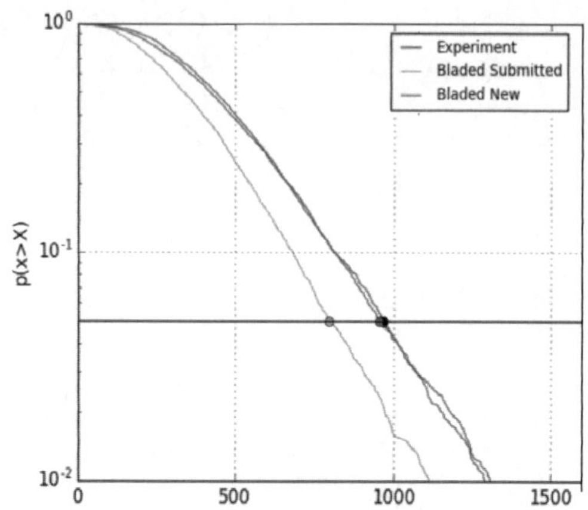

FIGURE 10 : LC34 Tower base fore-aft shear force (Fx) exceedance probability

Figures 9 & 10 show good agreement between Bladed and the experimental results for tower base Fx (shear force) exceedance probability. There is a particularly good match for the updated LC34 Bladed simulation. However, it should be noted with reference to Figure 7 & 8 that this could partly be due to the fact that there is both some over-prediction and some under-prediction of large loads at different frequencies.

Damage equivalent loads (DELs) were calculated based on the same process as for the original OC5 results (1Hz reference frequency, Wohler coefficient = 5, Goodman adjustment).

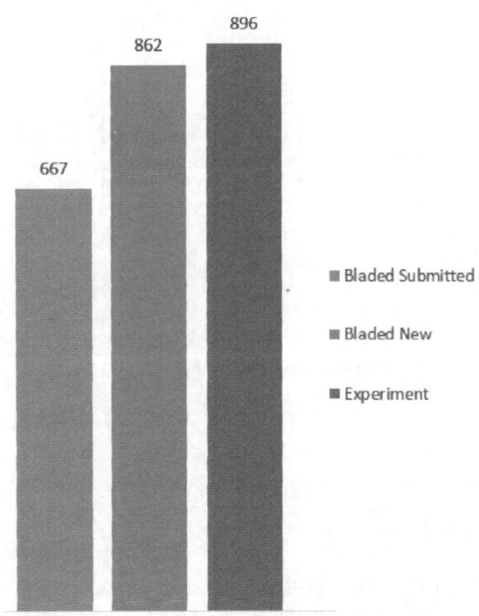

Figure 11 : LC33 Tower base shear force (Fx) DELs (kN)

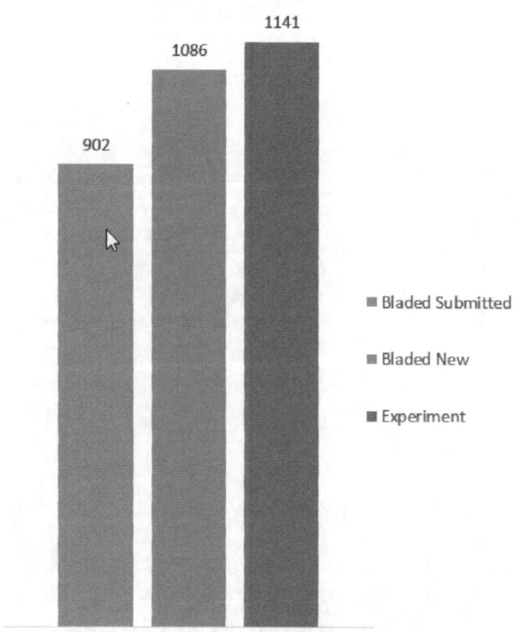

Figure 12 : LC34 Tower base shear force (Fx) DELs (kN)

Figures 11 & 12 show an improved correspondence with the experimental results for tower base Fx DELs for the updated Bladed simulations.

6. Conclusions

Inclusion of instantaneous Froude-Krylov and hydrostatics provides a clear improvement in modelling low frequency pitch and surge dynamics of the floating platform. Prediction of tower base shear force, a signal driven strongly by pitch, also shows a better match to the measured tank test data. This can be understood in terms of the instantaneous hydrostatics and Froude-Krylov force more accurately capturing the nonlinear transfer of incident wave energy to other frequencies, such as the platform surge and pitch resonant frequencies.

The instantaneous hydrostatics correction has not had a significant impact on simulated heave motion. However, given the geometry of the DeepCWind floating structure, this result was expected, and agreement with experiment was already strong. For structures in which platform motion leads to a more nonlinear variation in vertical hydrostatic forces (for example structures including vertical members of varying diameter near the free surface), instantaneous hydrostatic correction may yet prove to be significant. For the present results, the free water surface is taken to be the mean water level. A possible avenue for further work would be to repeat the analysis using the incident wave free surface.

The change to the simulated tower structural definition has improved the match to the experimental tower frequency, however the updated Bladed simulations now overestimate the tower base shear force response at the tower frequency. Uncertainties in the mass distribution and structural damping values for the tower mean that it is not possible to draw definitive conclusions from this, however.

These results suggest that improvements in floating wind turbine modelling fidelity can be achieved by implementation of the enhanced modelling approaches described in this paper. It is noted that the use of a suitable surface discretization mesh is required to implement the modelling improvements considered in this paper.

Modelling improvements such as these could help reduce risk in the design process, enabling more optimised and cost effective designs to be realised. This in turn would lead to a reduction in the cost of energy and would increase the viability of floating wind as a commercial technology.

References

[1] Robertson A, *et al.* 2017 OC5 Project Phase II: Validation of global loads of the DeepCwind floating semisubmersible wind turbine *Energy Procedia* **137(2017)** 38-57
[2] Robertson A *et al.* 2018 Assessment of Experimental Uncertainty for a Floating Wind Semisubmersible under Hydrodynamic Loading *Preprint* ASME 2018 37th International Conference on Ocean, Offshore and Arctic Engineering Conference Paper NREL/CP-5000-70877
[3] Goupee A, Fowler M, Kimball R, Helder J, Ridder E, 2014 Additional Wind/Wave Basin Testing of the DeepCwind Semi-submersible with a Performance-Matched Wind Turbine. *Presented at the Ocean, Offshore and Arctic Engineering Conference,* June 2014.
[4] DNV GL various authors, 2016, Bladed Theory Manual Version 4.8
[5] Child B, 2017 A Methodology for Identification and Simulation of Extreme Design Load Cases for Wave Energy Converters, *Marine Energy Technology Symposium,* April/May 2018

Acknowledgements

The work described in this paper builds on the OC5 project work led by Amy Robertson at the National Renewable Energy Laboratory (NREL). The authors would therefore like to acknowledge the work of all of the OC5 project participants. The authors would also like to thank Habib Dagher and Andrew Goupee at the University of Maine for very kindly providing the measured experimental data.

IOP Conf. Series: Journal of Physics: Conf. Series **1102** (2018) 012024 doi:10.1088/1742-6596/1102/1/012024

Manufacturing cost - a critical evaluation criteria for new developments in wind turbine drivetrain technologies

F. Harzendorf, B. Azzam, R. Schelenz, G. Jacobs

Chair for Wind Power Drives (CWD), RWTH Aachen University, Campus-Boulevard 61, 52064 Aachen, Germany

E-mail: Freia.Harzendorf@cwd.rwth-aachen.de

Abstract. Innovation in the wind turbine's drivetrain are mainly motivated by the goal to have better products. To have an improved final product, it should be analysed in a holistic manner over the drivetrains lifetime. One essential lifetime phase of a wind turbines drivetrain is the manufacturing phase, which is focused in this analysis. This paper presents a way to estimate manufacturing cost of wind turbine drivetrains at an early design phase. The approach is based on literature and expert interviews as well as their abstractions. The target audience are people involved in the development of new wind turbine drivetrain concepts. To customize the approach for their requirements, it is based on a minimum number of necessary input mainly geometric component specification. This approach is modular, scalable and applicable on to future drivetrain concepts. The used approach proves to be an effective way for estimating manufacturing costs of drivetrain components. A comparative analysis against a benchmark application, a 5 MW wind turbine's drivetrain, confirms the accuracy of the presented approach. A case study dealing with two possible gearbox solutions with a power rating over 5 MW underlines the importance of taking manufacturing cost in consideration when thinking about new developments in wind turbine drivetrain technology.

1. Introduction

The results of the German onshore tenders from May 2018 with an average of 5.73 ct/kWh can be seen as an indicator for the electricity generation cost of wind onshore in Germany [1]. In 2018, the average day-ahead electricity price of 3.46 ct/kWh at the electricity exchange can be seen as the main revenue source for electricity provided by wind turbines [2]. Comparing average cost and revenue potential for one kilowatt hour electricity provided by wind turbines, strong discrepancies become visible. This indicates, that wind turbine technology is not economically viable without funding at its current state in Germany. Though it can be foreseen, that state subsidies for wind will be reduced onwards and abolished someday. This circumstance puts a lot of pressure on the entire wind turbine market. Mone et al. states, that 58 % of the investement cost for an exemplary onshore wind turbine is caused by the nacelle [3]. Reder detected, that roughly 84 % of the unplanned failures of his sample geared wind turbines are induced by nacelle components [4]. This puts the nacelle and especially the drivetrain into the focus of the analysis and forces the market to come up with new ideas and develop innovations. Decisions about introducing a new drivetrain generation have to be made in an early design stage. This stage is characterized by a high degree of uncertainty and complexity as well as an information deficit concerning the concepts characteristics. Nonetheless, it is a decision with a high influence on the company's and wind technologies success in general. To have a better product in the end it should be analyzed over the drivetrains lifetime in a holistic manner as described in [5]. One essential lifetime

phase is the manufacturing phase, which is in focus of this analysis. It is estimated that 70-80% of the overall cost of a product is committed at the stage of design and conception and hence the manufacturing cost [6]. This estimate justifies the need for gaining feedback on manufacturing costs at the design stage of a wind turbine's drivetrain to improve the technology.
This paper presents a way to estimate manufacturing cost of wind turbine drivetrains in an early design phase. It starts by presenting the object of reflection and the chosen approach in chapter 2. The following chapter 3 furthermore presents all underlying assumptions for representing the manufacturing processes of the analyzed drivetrain components. A comparative analysis against a benchmark application, a 5 MW wind turbine's drivetrain in chapter 4 confirms the accuracy of the presented approach. Finally, in chapter 5 a gearbox concept comparison for an above 5 MW application proves the importance of manufacturing cost as a critical evaluation criterion. In chapter 6 the conclusion is drawn.

2. Approach

The approach offered by this paper for estimating the manufacturing cost of a wind turbine drivetrain in an early design phase is based on a bottom up cost estimation technique. Therefore, the drivetrain components are split up into their constituent parts. The components within the scope of this paper are: gearbox, generator, main shaft, main frame and the main bearing arrangement for currently available power ratings. The manufacturing cost for the constituent parts are estimated by breaking down the steps of manufacturing each of them and estimating the effort associated with completing each step. If literature does not supply sufficient information, simplified approaches are used. Estimations of the cost associated with completing each step are mainly performed by developing scalable formulas for the machining time and the cost per machining time for each manufacturing step. Initially, the required input for performing the machining time estimation included not only component specification but also a high number of process parameters that are typically known to manufacturing facilities or an experienced machinist. Since the target audience includes designers or engineers who might be less familiar with such process parameters the number of required inputs needed to be reduced to only component specification. This was performed by an extensive literature review as well as expert interviews to either assume average default values or calculate some of the process parameters based on component specification. Process parameters, which were most of the initial inputs could be excluded by making assumptions based on literature review, interviews, industry best practices, calculations of process parameters as well as eliminating inputs due to their low influences on the final estimate. The final required inputs only include component specifications. Initially, 123 input parameters were needed, with the presented approach this number could be reduced to 61 geometric inputs.
For estimating the cost per machining time, the following factors are considered:

- amortization cost
- interest cost
- maintenance cost
- room cost
- energy cost
- oil cost
- labour cost
- tool cost.

Several literature resources provided information about its constituent parts so that an estimated cost per hour can be calculated for each machining process. By multiplying the cost per machining with the machining time for each process step, the manufacturing costs for the wind turbine drivetrain components are estimated.
It is important to note that there exist major differences between various manufacturing facilities especially with respect to the machine tools and expertise of personnel available in each facility. Several formulas used to estimate the costs are based on survey data and so they represent averages. Therefore, it can also be possible to achieve lower costs than estimated with highly trained personnel using high-performance machinery or the opposite. Results of this model should be treated as an indicator and are valid for the assumptions which are going to be presented in the following chapter.

3. Model presentation

In this chapter, the underlying assumptions for calculating the machining times for each component process step are presented. First, the assumptions for the main shaft are presented. Some of which can also be applied to shafts within the gearbox and the low speed shaft. Then the main bearings, the gearbox and the generator follow. Finally, the assumptions for the machine carrier are presented. It should be kept in mind that these models are a first attempt based on publicly available data and therefore a poor database which will be extended whenever possible.

3.1 Main shaft

The manufacturing of the main shaft is split up in the following procedures (see Figure 1): Heating the ingot to forging temperature, forging and flange upsetting, forging heat treatment, sawing, turning, and drilling of the main shaft.

Figure 1: Shaft forging process [SLS15]

The heating time is estimated based on the shafts ingot diameter, a material coefficient and the arrangement of the shafts within the oven [7]. For estimating the ingots diameter, Greger's relationship between the ingots weight and its dimensions is used [7]. According to literature, roughly 3 % of the material is lost during the forging process [8, 9]. Furthermore, a material removal of 7% of the mass of the final finished shaft during the upcoming turning and drilling processes is assumed. The cost of forging will be estimated via the material cost of the shaft under investigation. The material cost proportion of the total cost of hot steel forging can be estimated to be 50% [10] [11] [12]. Therefore, the cost of the forging process is estimated to equal the material cost of the forging. It is assumed, that the material of the main/highspeed shaft is low alloy AISI 4340 steel and the material of gear shafts is 17CrNiMo6 steel [13] [14]. After forging is performed, an annealing procedure is applied to release internal stresses in the forged workpiece [7]. For cooling down the workpiece to its annealing temperature a cooling rate of 10 °/h is assumed [7]. The holding time at the annealing temperature is estimated based on the workpiece diameter divided by 25 [7]. Based on Greger, cooling in a closed furnace at a rate of 12 °C/h until a temperature of 200 °C is assumed afterwards. After forging, there exist several alternate methods to trim the forged part to size [7]. The method assumed for adjusting the length of the forged part in this analysis is sawing. Sawing costs are estimated by the number of sawing operations, the diameter of the shaft and a specific sawing cost of 0.1 €/mm [15]. The diameter of the forged shaft is assumed to be 3.5 % greater than the diameter of the finished shaft based on the assumed material removal during turning. This assumption is used to estimate the diameter of the unfinished shaft at the sawing stage to estimate sawing cost. Next process step is turning. Assumptions for estimating the turning time are based on Sandvik Coromants best practice machining time data for turning 34CrNiMo6 steel wind turbine main shafts [13]. In order to adapt to Coromant's best practice the alloy of the steel had to be changed and is not equivalent to the assumptions for the prior process steps. It is assumed that both roughing and finishing are carried out. Tungsten carbide will be assumed to be the material of the tool used to perform the turning operation. Using standard turning process parameters by Ostwald and best practices by Sandvik Coromant, the feed and turning speed will be assumed to be 1.7 mm/revolution and 2500 mm/second for roughing, and 0.8 mm/revolution and 3500 mm/second for finishing, respectively [13] [16]. The best practice guideline by Sandvik Coromant for depths of cut of the wind turbine main shaft is 3-6 mm and 1-2 mm for rough and finish turning, respectively [13]. The depths of cut for rough and finish turning will therefore be assumed to be 4 and 1.5 mm. For limiting the number of required user input parameters of the cost estimation model, the volume of the material removed during roughing and finishing will be assumed to be 5% and 1% of the volume of the finished shaft, respectively.

3.2 Main bearings

To estimate the cost of manufacturing the main bearings, a relationship between material and manufacturing costs is utilized. Industry sources estimate, that the total cost of producing the main bearing is split into 55% of manufacturing and 45% of the material cost. Where the bearing material can be assumed to be SAE 52100 chrome steel [17].

3.3 Gearbox

The main focus of the gearbox manufacturing estimation lies on the analysis of the manufacturing process of the gears. The first step is to identify the possible state-of-the-art gear arrangements, which are planetary and spur gears. Planetary gears consist out of the gear variants sun, ring and planet gear. Spur gear arrangements consist of two spur gears. These types can be broken down into two different designs of gears: exterior and interior gear. The difference between the geometry of the interior and exterior gears imposes a need for different manufacturing methods for the two gear types. Furthermore, certain requirements need to be fulfilled for manufacturing gears for wind turbines gearboxes: a high degree of accuracy for efficiency reasons, needed geometries need to be realizable and additionally the process needs to be economically feasible. Subsequently, the methods for manufacturing these gear variants are identified based on literature review and industry interviews [18] [19]. Blanks are the initial form of the gear before any teeth can be machined [18]. The envisaged cost estimation model will address the process of preparing a blank for teeth machining. This process involves two steps, sawing the blank to roughly reach the required dimension of the gear and then turning the gear to reach the required final dimensions and the needed surface finish.

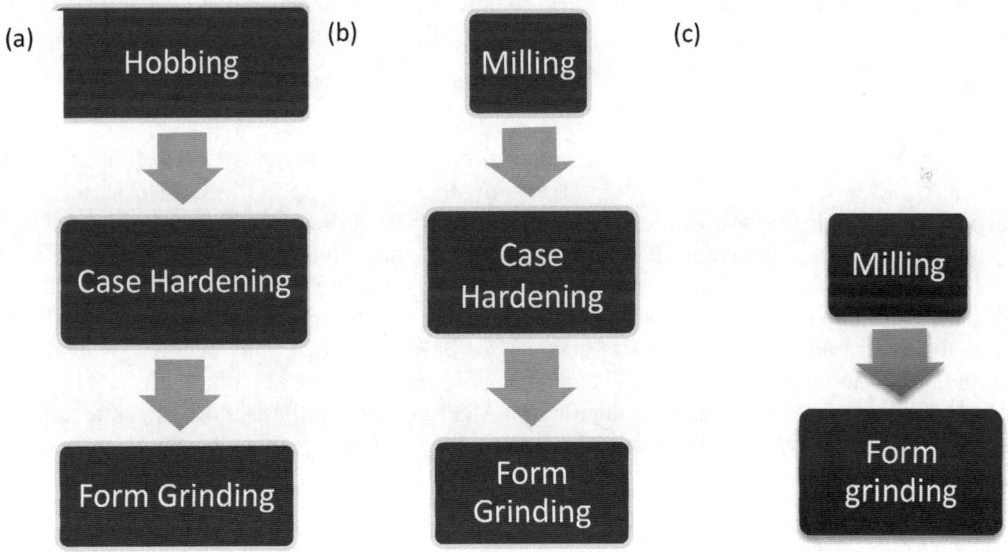

Figure 2: Sequences for manufacturing gear teeth from a blank

As for the main shaft, sawing cost is estimated based on the shafts diameter and sawing cost of 0.1 €/mm [15]. The approach for estimating turning time is similar to that for main shafts. According to expert opinion obtained through interviews, the methods for manufacturing gears can be narrowed down into three sequences. The sequence of manufacturing the planet and spur gear is a combination of hobbing, case hardening and form grinding (see Figure 2 sequence a)). For sun gears, it is a combination of milling, case hardening, and form grinding (see Figure 2 sequence b)). For ring gears, it is a combination of milling and form grinding (see Figure 2 sequence c)). Case hardening is not needed as the ring gear has to withstand less torque compared to the other gears due to the concave-convex nature of its gears meshing with the planet gears. For estimating the hobbing time, a formula of Dudley is used [18]. Process parameters are estimated based on the following assumptions: due to accuracy reasons for every 30 gear teeth one hob thread is needed [18], hob diameter is estimated based on normal module and diametrical pitch [18], hob feed per revolution and speed are estimated based on the hob diameter, the hardness of 370 HV is assumed. Milling time can be estimated based on the number of gear teeth and their face width [15]. Case hardening is a process aimed at increasing the hardness of the surface zone of a gear to increase the capability of this area to withstand the high stresses in operation. The process consists mainly of a carburization, quenching, and tempering stage. The case depth is estimated based on the normal module of the gear [18]. The carburization time can then be estimated based on the case depth and the annealing temperature of the material [18]. The material is assumed to be 17CrNiMo6. It is assumed that the tempering time has the same duration as the carburization process for the sake of simplification. Quenching is neglected due to its short duration and low energy consumption. For finishing the gears, form grinding is used as it is possible to machine the high hardness. Grinding time can be estimated based on the number of gear teeth and their face width [15]. The time of dressing the grinding wheel is interpolated for different numbers of teeth, normal module and pitch [18]. Still the gears are not the only components which need to be manufactured for a finished gearbox. For simplicity reasons the manufacturing cost of the other components are assumed to be a percentage of the gears manufacturing cost. Based on a literature review and interviews of Spieß et al. the effort for manufacturing gearbox shafts can be assumed to 50 % of the manufacturing of gears, manufacturing the housing 200 %, vendor supplied items 75 % and the assembly 160 % [15].

3.4 Generator

Unlike the case with estimating the manufacturing costs of gears, estimation methods of the costs of the different processes of manufacturing electrical generators were not available in literature at the time of writing this paper. Therefore, a simpler approach is used here. Based on the table of VDI standard 2225 sheet 2 manufacturing cost is derived from the material cost [20]. The category chosen was three-phase-generator.

3.5 Main frame

The estimation model for the main frame will rely on estimates of the mass of the main frame. From Dandong Funding Engineering Machinery Co. LTD, a manufacturer specialized in castings and subsequent machining and processing, a cast steel price of 2.57 $/kg converted in 2.1957 €/kg for complex structures and casting weights between 1 – 50 kg casting unit weight was provided [21]. However, there exist other machining processes, such as milling and pocketing, which must be performed to the main frame in order to reach the final product [13]. Therefore, a factor of 1.79 will be multiplied by the main frame cost equations to account for these processes. This factor was obtained using an online casting total cost estimator by Dandong Funding Engineering Machinery Co., LTD [21]. The material of the casting was assumed to be ductile iron ASTM A536 60-40-18, and the manufacturing processes after casting consisted of machining and painting processes to protect against corrosion, package and sea fright as Dandong is a Chinese manufacturer.

4. **Model validation**

For validating the model average prices for the different drivetrain components in consideration are derived from literature sources and compared to the results the model provides. Unfortunately, literature does not deploy information about the prices on a component level. Therefore, a benchmark is compiled from several sources. EWEA provides a split of the turbine investment cost of the different components for the MM92 [22]. MM92 is a turbine with a three-stage gearbox, a three point suspension system and a doubly fed induction generator. The Danish Megavind project estimated the average division of Capex in % of the main components and subsystems in wind turbines [23]. Unfortunately, Megavind does not state what drivetrain configuration is evaluated. WindGuard states that the main investment cost for turbines in Germany in the range of 2 to 4 MW are in between of 1,380 and 980 €/kW [24]. These assumptions lead to the cost intervals displayed in Figure 3 for a 5 MW power rating. The power level of 5 MW is chosen, as literature provides valuable information for an example drivetrain at this power range. Having a look onto Figure 3 especially for the gearbox, the investment cost can vary in a wide range. Still this figure accounts for the total investment cost for a component so takes also material cost and margin into account.

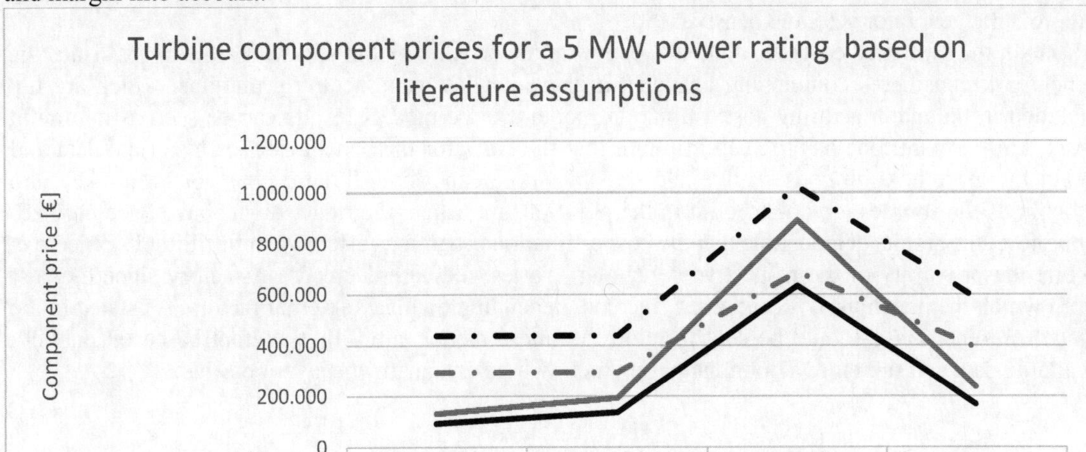

	Main shaft	Machine carrier	Gearbox	Generator
cost max EWEA 2007	131.790 €	193.200 €	890.790 €	237.360 €
cost min EWEA 2007	93.590 €	137.200 €	632.590 €	168.560 €
cost max Megavind 2012	438.000 €	438.000 €	1.022.000 €	584.000 €
cost min Megavind 2012	294.000 €	294.000 €	686.000 €	392.000 €

Figure 3: Turbine components investment cost for a 5 MW power rating based on literature assumptions

The model validation is mainly based on the information of a 5 MW high speed gearbox configuration with a four point suspension type designed by the national renewable energy laboratory (NREL) [25]. The gear specification (normal module, face width and number of teeth) from the 5 MW gearbox developed by NREL are used as an input to the model for the comparison [25]. The model calculates 105.150 € for the manufacturing of the gears of this gearbox configuration. In order to be able to compare this number to the prices in Figure 3 the gears manufacturing influence on the gearbox price needs to be assumed. Spieß states, that 14.6 % of the gearbox price is entitled to the gears manufacturing [15]. This leads to a gearbox price of 720.205 € which seems reasonable comparing it to the benchmark in Figure 3.

The 5 MW main shaft from NREL has a mass of 18,000 kg an average diameter of 1,000 mm is assumed [25]. A drawing of the main shaft was also used to indicate a length of approx. 3,500 mm. Using these specifications as input to the shafts program resulted in an estimated manufacturing cost of EUR 54.250 € and a material cost of 17.154 € leading to a total production cost of EUR 71.404. Which is much less than the literature review would estimate. Still the author assumes, that the main bearing price is included in the main shaft and needs to be added. Unfortunately, the main bearing arrangements weight is not stated in the source.

For calculating the machine carrier Fingershs cost and scaling model is used for estimating the mass of the machine carrier for a three-stage gearbox with a high speed generator [26]. To modify it to the assumptions from the NRELs 5 MW high speed gearbox configuration a rotor diameter of 126 m is assumed as Fingersh estimates the machine carrier weight based on the rotor diameter [25]. The models result state that this machine carrier leads to manufacturing cost of 111.003 € adding the material cost (1.5 €/kg cast iron [15]) leads to machine carrier investment cost of 140.659 € which is in the range of the cost calculated from the literature.

For the generator, the characteristics estimated by Sethuraman are used [27]. The 5 MW doubly fed induction generator has estimated material cost of 17.670 $ respectively 15.162 €. This leads to manufacturing costs of 35.259 € and total generator costs of 50.402 €. Compared to the number from Figure 3 the generator is to inexpensive.

All in all, the estimated costs are at the lower end of the benchmark but still within the range. Since the benchmark consideres commercial products it also includes manufacturer's margins, which are not included in the manufacturing cost estimation model. So the models results can be seen as minimum costs. Furthermore, one has to keep in mind, that the basis for the benchmark are historical data and might therefore be a bit conservative and possibly more expensive. Inflation has not been taken into account as the sources are not specific in the point of time where their data points have been enacted. Therefore, the results demonstrate that the cost estimation program is effective in fulfilling its objective. While the proximity of the respective benchmark varies, they are deemed satisfactory since the cost from which the benchmark is originated fluctuate depending on many external factors. As stated in the model presentation it should be kept in mind that these models are a first attempt based on publicly available data and therefore a poor database which will be extended whenever possible.

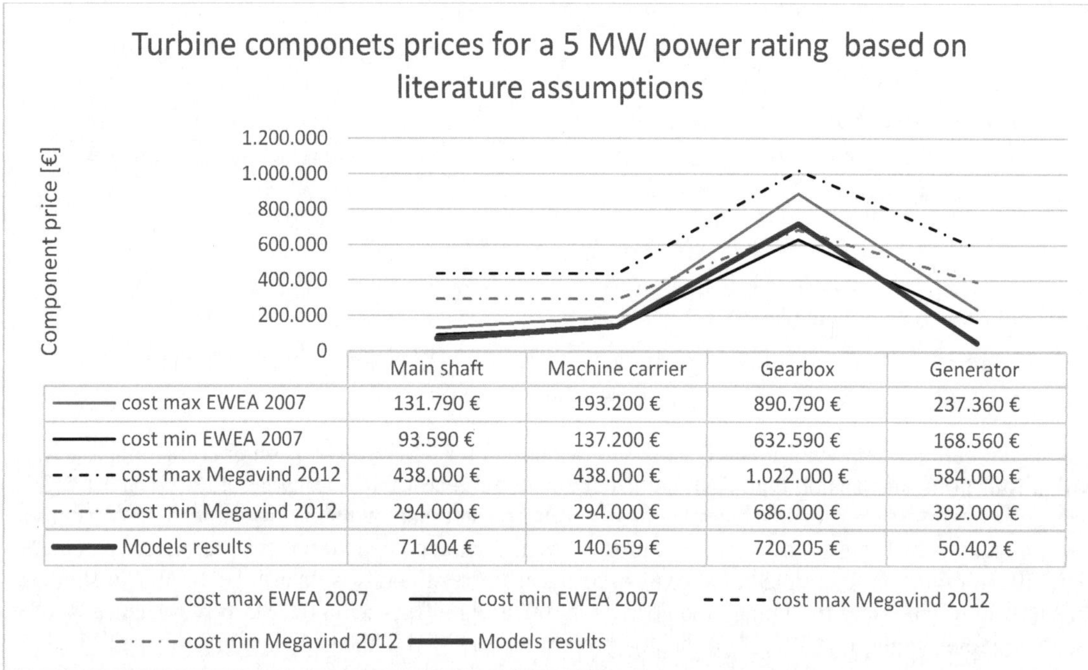

	Main shaft	Machine carrier	Gearbox	Generator
cost max EWEA 2007	131.790 €	193.200 €	890.790 €	237.360 €
cost min EWEA 2007	93.590 €	137.200 €	632.590 €	168.560 €
cost max Megavind 2012	438.000 €	438.000 €	1.022.000 €	584.000 €
cost min Megavind 2012	294.000 €	294.000 €	686.000 €	392.000 €
Models results	71.404 €	140.659 €	720.205 €	50.402 €

Figure 4: Turbine components investment cost for a 5 MW power rating based on literature assumptions compared to model results [22–24]

5. Model application

With the help of the presented model a case study for evaluating two gearbox configurations with a power rating of over five MW for manufacturing the gears is conducted. Gearbox one (I) is the 5 MW configuration provided by NREL, already used for the model validation [25]. Gearbox example two (II) has a power rating of 6.5 MW and is inspired by the gearbox used in REPower 6M [28]. Both gearboxes have an overall ratio of 96 - 97 and two planetary and one parallel stage. The 6.5 MW gearbox has four planets at the first and three at the second planetary stage [28]. The NREL gearbox has three planets at each planetary stage [25]. Figure 5 shows the calculated manufacturing cost for the gear designs with the help of the developed tool. Compared to the 105.150 € calculated for gearbox design I, design II is 30 % less expensive when taking the estimated manufacturing cost of the gears into account (73.329 €). In principle the cost distribution about the stages is comparable between these designs. For gearbox design II, the percentage share is 3 % higher for the first stage, than for design I.

Sources	(I) NREL [25]	(II) Berroth [28]
Stage 1		
Ratio	3.947	5.167
Normal module	45	25
Face width	491	560
Stage 2		
Ratio	6.167	6.522
Normal module	21	14
Face width	550	310
Stage 3		
Ratio	3.958	2.882
Normal module	14	11
Face width	360	250

Table 1: Stage characteristics for the two gearboxes under investigation [25, 28]

For explaining the differences one has to have a deeper look on the gears design. Comparing the stage ratios of these configurations it becomes visible, that for design II a higher ratio in the first planetary stage compared to gearbox design I has been chosen see Table 1. Initially, the NREL approach also had several gearbox speed ratio options (see Table 2 [25]) and speed ratio option A being similar to the version chosen for gearbox II.

Option	A	B	C	D
First stage	1:5.1944 (3p)	1:3.9549 (3p)	1:3.8322 (4p)	1:3.9502 (3p)
Second stage	1:6.2257 (3p)	1:6.1695 (3p)	1:6.2367 (3p)	1:6.1262 (4p)
Third stage	1:3.0000	1.3.9754	1:4.0586	1:4.0083
Total dry weight (x1000 kg)	53.69	48.82	43.98	47.30

Table 2: Gearbox speed ratio options for gearbox type (I) [25]

Due to the chosen weight optimization approach Nejad decided to use option B which is 9 % lighter than option A. Nevertheless, the advantage of high ratios in the first stage is, that lower torque is transferred to the subsequent stages which can therefore be designed less massive. Comparing normal module and face width of the gears of the second and third stage reflects this relation. Comparing the face width of both designs, gearbox design I is 44 % longer on the second stage and 31 % on the third stage. Gearbox design I normal module is furthermore 33% larger on the second stage and 21 % on the third stage. Keeping in mind that the geometric input for the estimation model is the number of teeth, the normal module and the face width of the respective stages, this can be an explanation for the higher cost estimation for gearbox design II.

One has to keep in mind the limitations of this simple estimation approach. Up to now process application limits due to geometry or weight are not implemented. Furthermore, the same hourly rate for all possible workpiece dimensions is used. For the future not only, the duration but also the hourly rate of the process steps needs to be adjusted to the workpiece dimensions. For the presented example manufacturing accuracy between a planetary stage of three and four planets is assumed to be the same. In reality, higher accuracy standards need to be fulfilled for planetary stages with four planets. No quantity discount due to common parts and therefore a higher utilization of the manufacturing factory is implemented. This might be interesting for evaluating high speed concepts with a standardization strategy like the RapidWind concept [29].

Still this result shows, that a weight optimization approach for the gearbox might not always lead to the best design in terms of manufacturing. It furthermore indicates, that manufacturing cost can be a critical evaluation criterion for new developments in wind turbine drivetrain technologies. Obviously there exists a trade-off between weight optimization and manufacturing efforts minimization in this design

IOP Conf. Series: Journal of Physics: Conf. Series **1102** (2018) 012024 doi:10.1088/1742-6596/1102/1/012024

decision. Further analysis should take the most important implications drivetrain weight can have on its lifetime into account e.g. material, transportation, installation and the operational phase. Comparing the weight optimization approach against the goal to have a better manufacturability will therefore lead to improved products. This is an essential goal of the authors for the future [5].

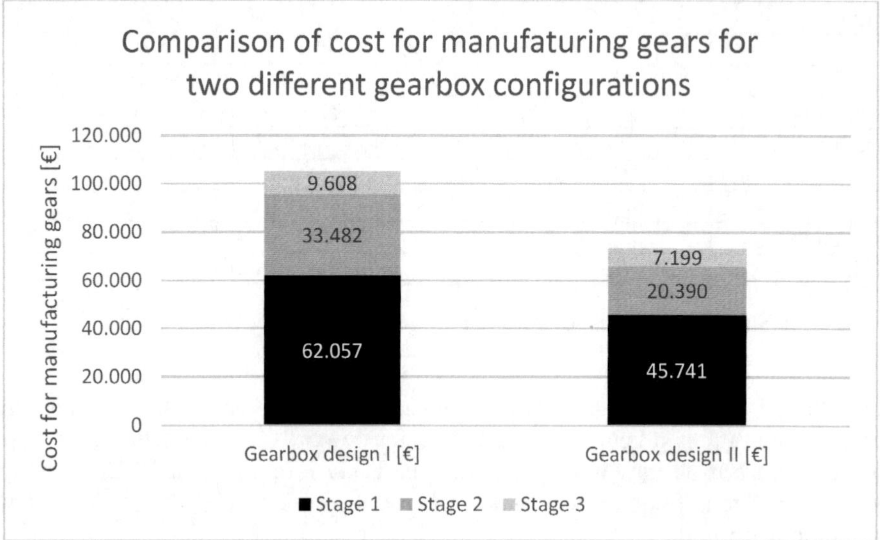

Figure 5: Comparison of cost for manufacturing gears for two different gearbox configurations [25, 28]

This case study has a strong focus on the manufacturing cost of different gearbox configurations. At the current model status, presented in Chapter 3 it can be enrolled on the main shaft in the same level of detail. For the generator, analysis on this level of detail is not possible at the moment. Still this component with its high influence on the drivetrains overall cost and its high influence on the drivetrain design especially on the gearbox aims for detailed analysis. Same treatment for the main bearing arrangement. This isn't as expensive as the generator thus industry sources estimate that 55 % of the components price account for manufacturing. This fact represents a good motivation for detailed manufacturing cost analysis especially as there exist several different suspension and bearing types.

6. Conclusion

This paper builds awareness to take manufacturing cost/implications into account at an early design stage of wind turbine drivetrains, as it is crucial when aiming for superior concepts. It furthermore offers a validated approach to estimate manufacturing costs of wind turbines drivetrain components at an early design stage phase. This way bottlenecks of current wind turbine drivetrain concepts with respect to manufacturing costs become visible paving the way for innovation and improvement of wind turbines. It furthermore offers an opportunity to evaluate the trade-off between weight optimization and manufacturability. Still, it should be kept in mind that these models are a first attempt based on publicly available data and therefore a database which will be extended whenever possible. A similar approach can also be applied to further lifecycle steps of the wind turbines drivetrain, which is planned by the authors.

7. Acknowledgment

This work was performed in the frame of the project house TESYS (Technology based energy systems analysis) of RWTH Aachen University financed by the Excellence Initiative of the German federal and state governments to promote science and research at German universities.

Article I. References

[1] Bundesnetzagentur 2018 *Beendete Ausschreibungen* https://www.bundesnetzagentur.de/DE/Sachgebiete/ElektrizitaetundGas/Unternehmen_Institutionen/Ausschreibungen/Wind_Onshore/BeendeteAusschreibungen/BeendeteAusschreibungen_node.html (accessed 28 Jun 2018)

[2] Fraunhofer ISE 2018 *Jährliche Strombörsenpreise in Deutschland* https://www.energy-charts.de/price_avg_de.htm?price=nominal&period=annual&year=2018 (accessed 28 Jun 2018)

[3] Mone C, Stehly T, Maples B and Settle E 2015 *2014 Cost of Wind Energy Review*

[4] Reder M D, Gonzalez E and Melero J J 2016 *J. Phys.: Conf. Ser.* **753** 72027

[5] Harzendorf F 2017 *Lifetime cost evaluation of different wind turbine drive train configurations (Wind Energy Science Conference)* (Kopenhagen)

[6] Hueber C, Horejsi K and Schledjewski R 2016 *Advanced Manufacturing: Polymer & Composites Science* **2** 1–13

[7] Greger M 2014 *Forging: Didactic Text*

[8] Sabroff A M 1968 *Forging Materials and Practices* (Van Nostrand Reinhold Inc.,U.S.)

[9] Bruchanow A N and Rebelski A W 1955 *Gesenkschmieden und Warmpressen* (Berlin: Verlag Technik)

[10] TUNÇ M 2003 Computerized Cost Estimation for Forging Industry THE GRADUATE SCHOOL OF NATURAL AND APPLIED SCIENCES, THE MIDDLE EAST TECHNICAL UNIVERSITY

[11] Knight W A 1992 *The international Journal for Advanced Manufacturing Technology* **7** 159–67

[12] Boothroyd G, Dewhurst P and Knight W A 2010 *Product design for manufacture and assembly (Manufacturing engineering and materials processing* vol 74*)* 3rd edn (Boca Raton, Fla.: CRC; London : Taylor & Francis [distributor])

[13] Sandvik Coromant 2011 *Best practice machining of wind power component: Tool and method advances for efficient manufacturing* (Sandvik Coromant)

[14] Coburn-Myers Fastening Systems, Inc. 2018 *Materials: Carbon Steel: Features and applications of common carbon steel alloys* (Trainer, PA, USA: Coburn-Myers Fastening Systems, Inc.)

[15] Spieß D, Wickerath J S and Strangfeld F 2014 Entwicklung neuartiger Getriebekonzepte zur Steigerung der Wirtschaftlichkeit und Effizienz von Windkraftanlagen *Projektarbeit* Werkzeugmaschinenlabor (WZL) Institute, RWTH Aachen University

[16] Ostwald P F 1992 *Engineering cost estimating* 3rd edn (Englewood Cliffs, N.J.: Prentice Hall; London : Prentice-Hall International)

[17] AST Bearings LLC 2010 *Technical Information: Bearing Materials (*ENB-04-0553*)* (Montville, NJ, USA: AST Bearings LLC)

[18] Dudley R 2012 *Dudley's handbook of practical gear design and manufacture* 2nd edn (Boca Raton, FL: CRC Press)

[19] Linke H, Börner J and Heß R (eds) 2016 *Cylindrical gears: Calculation - materials - manufacturing* (München: Hanser)

[20] VDI-Gesellschaft Produkt- und Prozessgestaltung 1998 Konstruktionsmethodik - Technisch-wirtschaftliches Konstruieren - Tabellenwerk **03.100.40; 21.020** *(*VDI 2225*)* (Düsseldorf)

[21] DANDONG FUDING ENGINEERING MACHINERY CO., LTD 2018 *Cast Iron Price Calculator* (Dandong City, Liaoning, P.R. China: DANDONG FUDING ENGINEERING MACHINERY CO., LTD)

[22] Krohn S and Morthorst, P.-E., Awerbuch S. 2009 *The Economics of Wind Energy* (EWEA)

[23] Megavind 2012 *Strategy for Wind Turbine Components and Subsystems: Report from Megavind*

[24] Luers S 2015 *Kostenstudie - Update* (Deutsche WindGuard)

[25] Nejad A R, Guo Y, Gao Z and Moan T 2016 *Wind Energ.* **19** 1089–106

[26] Fingersh L, Hand M and Laxson A 2006 *Wind Turbine Design Cost and Scaling Model: Technical Report (*NREL/TP-500-40566*)* (National Renewable Energy Laboratory)

[27] Sethuraman L and Dykes K 2017 *GeneratorSE: A Sizing Tool for Variable-Speed Wind Turbine Generators (*NREL/TP-5000-66462*)* (Golden, Colorado: National Renewable Energy Laboratory)

[28] Berroth J 2011 Design and modelling the structural components of an offshroe wind energy plant for multibody simulation *Diplomarbeit* Institut für Maschinenelemente und Maschinengestaltung, RWTH Aachen University
[29] Barenhorst F, Serowy S, Andrei C, Schelenz R, Jacobs G and Hameyer K 2016 *J. Phys.: Conf. Ser.* **753** 112001

Optimal time step of SCADA data for the power curve of wind turbine

Minh-Thang DO, Julien BERTHAUT-GERENTES
METEODYN, 33 Bd Salvador Allende - 44800 Saint-Herblain – FRANCE
info@meteodyn.com

Abstract: The SCADA system for wind turbine often logs data at 10 minutes time step, mostly due to the technical constraints. However, with the developpement in measurement and storage technology, the collection of data with higher sampling frequency is feasible. The question about an optimal time step for SCADA data is then become interesting. Our paper aims to answer this question through a statistical analysis of the power curve at different time step. In order to explain and confirm the results of this study, a spectral analysis is performed. Both approaches indicate that the time step of 10 minutes is still the best choice as it gives the best predictability of the output power from wind speed.

1. Introduction

Nowadays, in most SCADA system of wind turbine, signals are first measured at a sampling rate of 1 second; then averaged at every 10 minutes; and finally written to the database. The aggregation at 10 minutes became an industry standard [1], [2]. It provides a good balance between sampling frequency and database size. Additionally, this 10-minutes time step allows comparing and correlating with pre-construction wind data, using in the wind resource assessments [3]

With the development in storage technology, the collection of data with higher sampling frequency is feasible. Several researches are trying to analyze these high frequency data [4], [5]. However, is the time step of 10 minutes still the best choice in terms of describing the relationship between wind speed and output power of a wind turbine? This paper aims to provide a more detailed analysis of this relationship, which is represented through the power curve.

2. Approach

In this study, the data is collected by a SCADA system with a time step of 7 seconds. An algorithm of classification is used to remove anomalies data, such as: measurement failure, stop, curtailment, under and over production. Only normal data are considered.

Figure 1 shows the relationship between the wind speed and the output power of the wind turbine (green dots), each data point representing a 7s time step. This relationship could be quantified by the dispersion of the data points. Less dispersion means the relationship is better and it is easier to estimate the output power from the wind speed. High dispersion means that the estimation of output power from the wind speed is less precise. In the ideal case, each value of wind speed corresponds to one and only one value of output power, this form the power curve of the wind turbine (blue curve). The dispersion of the data points in this case is zero and we obtain a perfect relationship.

IOP Conf. Series: Journal of Physics: Conf. Series **1102** (2018) 012025 doi:10.1088/1742-6596/1102/1/012025

In this study, the fluctuation is quantified by the Root Mean Square Error between the measured value of output power and the estimated value obtained from the power curve and the measured wind speed. Note that the power curve is an estimation based on the SCADA data, through an iterative median estimation technique (the manufacture power curve is not used).

In order to explain and confirm the results of this study, a spectral analysis is performed on both signals. This approach is able to distinguish the behavior at different frequencies or time scales. It is well suited for linear phenomena, which is not the case here (non-linearity of the power-curve). By the way, the monotonic shape of the relationship (the power curve is an increasing function) allows using this analysis with caution.

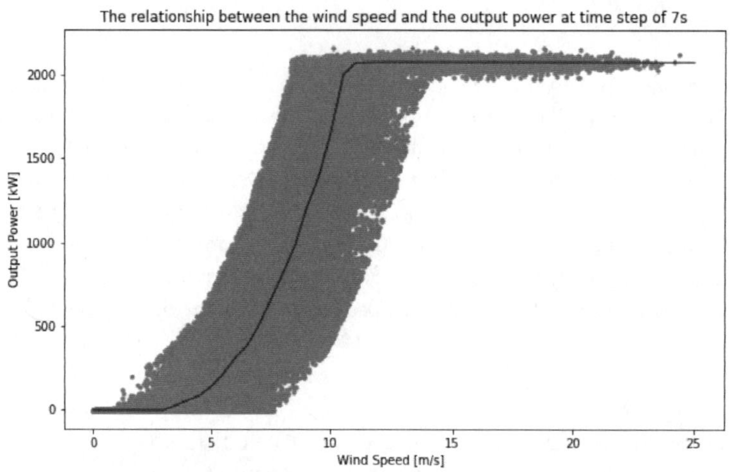

Figure 1. Scada data of wind speed and active power with a time step of 7s

3. Results and analysis

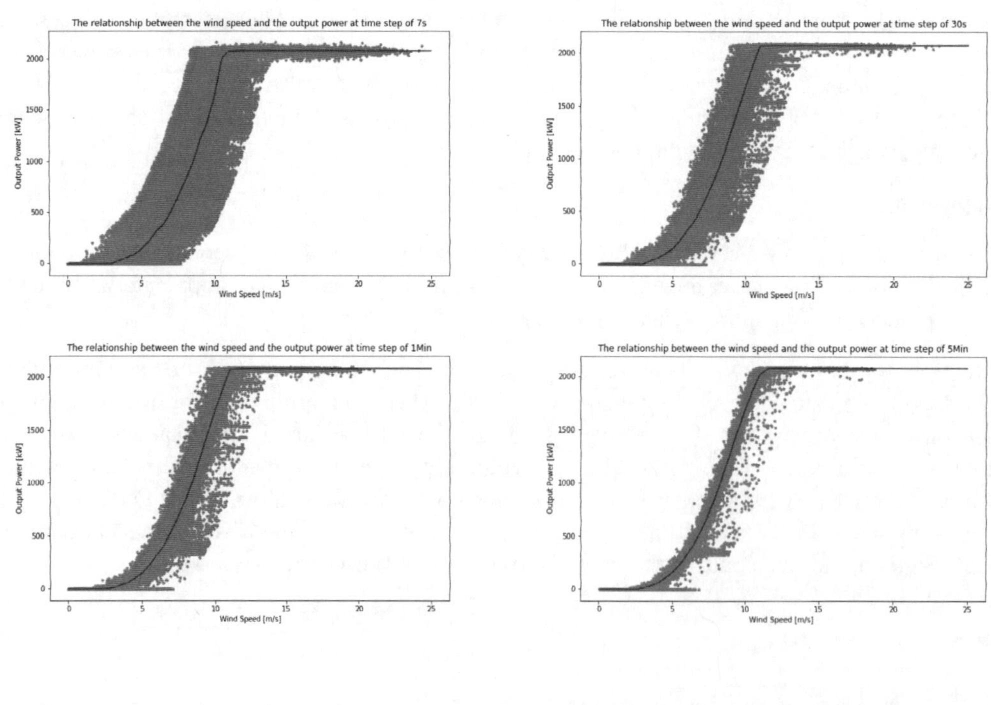

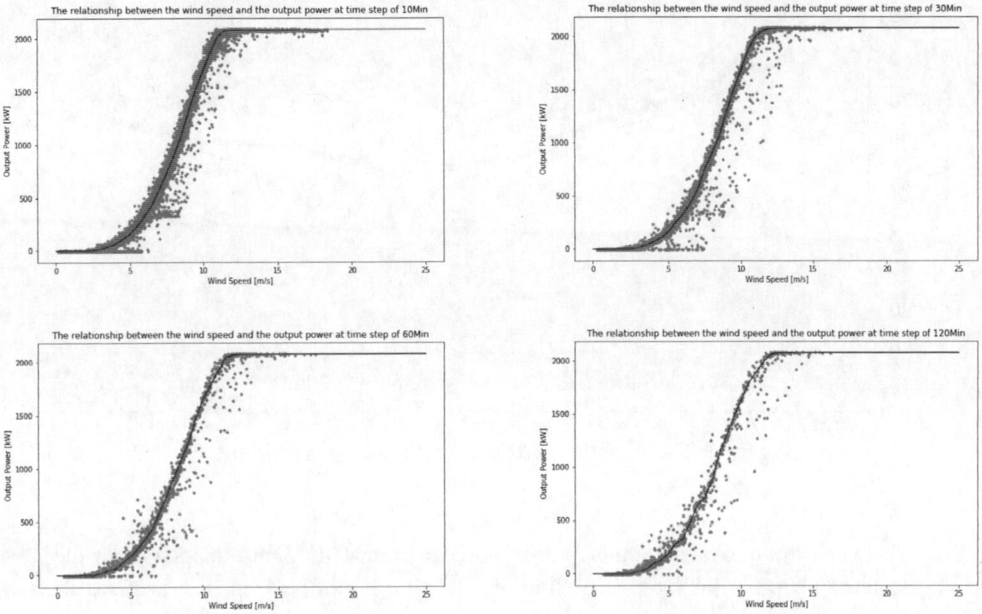

Figure 2. Scada data of wind speed and active power at different time step

In order to observe the time-dependency of the power curve, both signals (wind speed and output power) are time-averaged under different time steps. Figure 2 shows the relationship between the wind speed and the output power at the following time steps: 7 seconds, 30 seconds, 1 minute, 5 minutes, 10 minutes, 30 minutes, 60 minutes and 120 minutes. It could be observed that the data points are less dispersed at higher time step.

Wind turbine ref	7s	30s	1min	5min	10min	30min	60min	120min
F01	163	121	106	90	72	120	114	126
F02	165	122	118	105	109	117	122	121
F07	195	123	100	67	60	56	59	85
F12	181	131	120	105	99	111	96	113
F13	207	139	125	106	98	110	120	113
F19	190	145	129	114	96	130	134	150
F24	172	130	107	81	88	82	97	126
F25	159	106	90	64	60	57	63	74
AVERAGE	179	127	112	92	85	98	101	113

Table 1. RMSE of active power [kW]

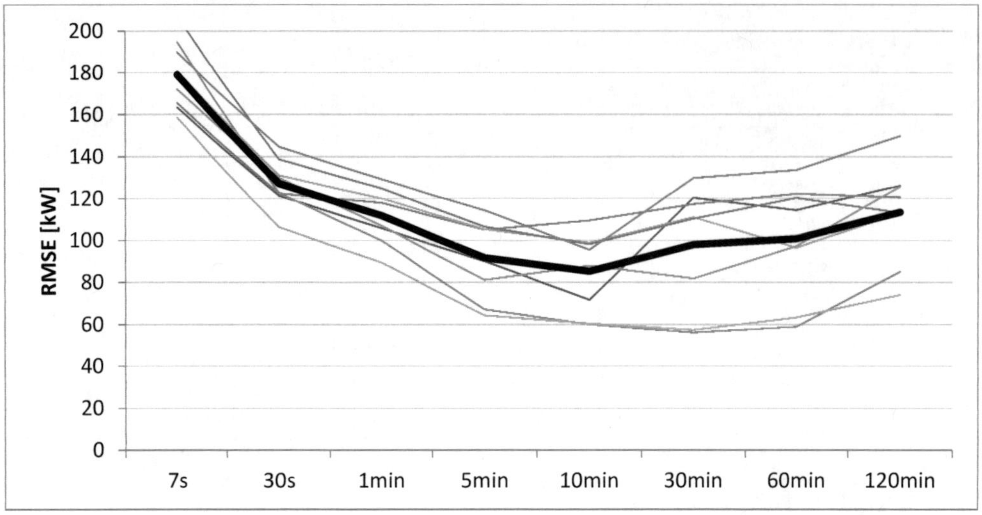

Figure 3. The variation of RMSE with each wind turbine

The RMSE between the measured output power and the estimated value obtained from the power curve is calculated for each time step (first line of Table 1). It can be seen that 7s gives the worst RMSE, while 10min seems to be the best. This process is repeated over 8 machines of a wind farm. Results are presented graphically in Figure 3.

The mean RMSE is very high at small time steps (179kW at time step of 7 seconds), but it decreases rapidly. The best mean RMSE is obtained at the time step of 10 minutes (85kW, about a half of the maximum value). Then, it slowly increases at higher time steps.

Through this analysis, it's justified that the time step of 10 minutes is the best choice in term of describing the relationship between the wind speed and the output power of wind turbine.

In the next step, we will apply the signal processing techniques with the data to understand this phenomenon [6]. The Power Spectrum Density (PSD) is computed on both signals, with a small window and large overlap to reinforce the smoothing.

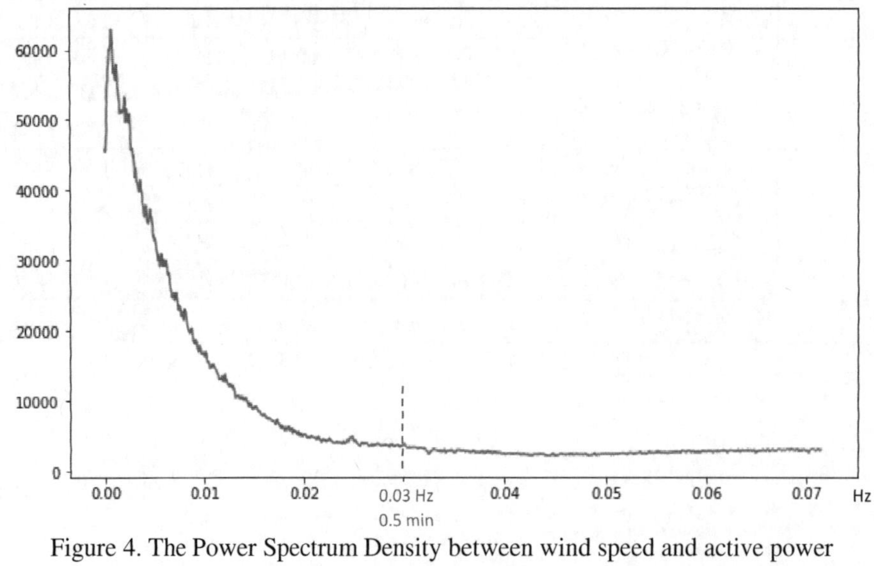

Figure 4. The Power Spectrum Density between wind speed and active power

IOP Conf. Series: Journal of Physics: Conf. Series **1102** (2018) 012025 doi:10.1088/1742-6596/1102/1/012025

Figure 4 presents the ratio between the PSD, in a linear-linear scale. It can be seen that above 0.03Hz (corresponding to phenomena repeating every half a min), this ratio is roughly constant. We suppose that this is only noise: at these frequencies, the variation of both signals is very small, and data are representing only measurement noise.

Figure 5 plots the same quantity in a log-log scale; it is similar to a Bode magnitude plot. As a first estimation, this corresponds to 1st order low pass filter tuned at 0.005 Hz (~3min). "first order" signifies that if the frequency is multiplied by x, so the response is divided by the same quantity x. In mechanics, this is typical for Mass-Damper system with no Stiffness. 3 minutes is the time-response of the whole machine inertia, including hub, high speed train and electro mechanical parts. This low-pass filter models the fact that the hub rotational speed is not able to follow the rapid fluctuations of wind speed, because of its inertia. In other words, the rapid fluctuations of wind speed (those one during less than some minutes), due to the turbulence, has no direct impact on the output power.

A more detailed exam exhibits an absolute maximum about 0.65 mHz, *i.e.* 25 min. This signifies that the power conversion is less effective for very slow variations (~2H) than for faster fluctuations (half an hour). This might be linked with the non-linearity of the power curve.

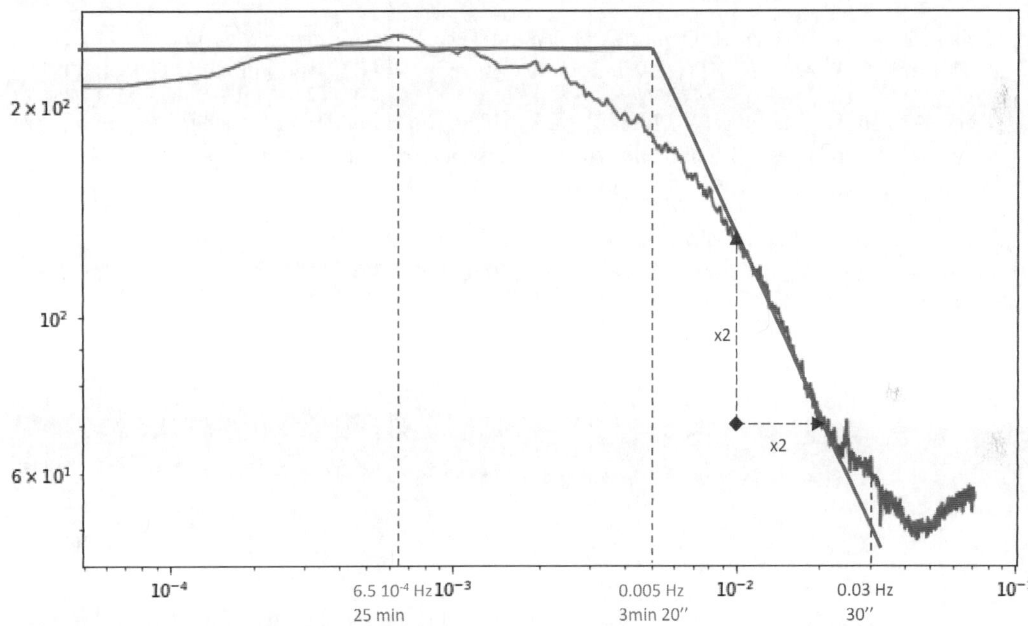

Figure 5. The Power Spectrum Density between wind speed and active power in log-log scale

4. Conclusions

This study focuses on the relationship between wind-speed and output power in wind energy, depending on the time scales.

The Bode magnitude plot shows different regions:

- for high frequencies (faster than 30 seconds), noise is dominating the spectra
- for intermediate frequencies, the generator behaves as a 0.005Hz 1st order low-pass filter:
  - response is roughly flat for fluctuations slower than ~3min
  - higher frequencies are filter (the higher the more filtered)
- for very slow fluctuation (several hours), the non-linearity of the Power curve reduced the global power generation apparent efficiency.

As a result, the 10-minute averaging standard is a good choice. At smaller time-step, the generator is not able to follow the wind speed fluctuations. At larger time step, the wind is never constant over the 1-step time span, so the power curve curvature is reduced.

References

[1]. Wang Ke-Sheng, Sharma Vishal S. and Zhang Zhen-You, “SCADA data based condition monitoring of wind turbines”, Advances in Manufacturing, vol. 2, no. 1, p. 61-69, Mar 2014.

[2]. IEC 61400-12-1, "Wind turbines – Part 12-1: power performance measurements of electricity producing wind turbines," IEC – International Electrotechnical Commission, Geneva, Switzerland, 2017.

[3]. A.J. Brand, “A downscaling method for application in wind resource assessment and wind power forecasting”, 13th International Conference on Wind Engineering, Amsterdam, The Netherlands, 11-15 July 2011

[4]. Friis Pedersen T, Wagner R, Demurtas G. “Wind Turbine Performance Measurements by Means of Dynamic Data Analysis”, DTU Wind Energy, 2016. p. 91 (DTU Wind Energy E, Vol. 0082).

[5]. Tanja A. Mücke, Matthias Wächter, Patrick Milan, Joachim Peinke, “Langevin power curve analysis for numerical WEC models with new insights on high frequency power performance”, Wind Energy 18:11, pp. 1953-1971 (2015)

[6]. Hernandez, W.; Maldonado-Correa, J.L.; Mendez, A. Frequency-domain Analysis of Performance of a Wind Turbine. Electron. Lett. 2016, 52, 221–223.

Down-regulation and individual blade control as lifetime extension enablers

Vasilis Pettas and Po Wen Cheng
Stuttgart Wind Energy at University of Stuttgart, Allmandring 5B, 70569 Stuttgart, Germany
E-mail: pettas@ifb.uni-stuttgart.de

Abstract. As more and more wind turbines are coming close to the end of their design lifetime, evaluation of end of life strategies is becoming highly relevant. Moreover, as turbine technology matures and wind farms grow larger, lifetime extension becomes a financially attractive option compared to re-powering and decommissioning. Present work suggests control strategies, namely down-regulation and individual blade control, as lifetime extension enablers. The concept of using them as retrofit control implementations is explained. Their individual and combined potential in fatigue load reduction is evaluated, along with their effect on other performance and pitch system metrics. Finally, the possible period of extension, beyond the nominal 20 years, is evaluated in an example case where the retrofit control strategy is applied after 15 years of baseline operation. The aeroelastic simulations are performed with a 10 MW reference wind turbine, according to load certification standards. Results show that the two methods complement each other in load alleviation. The pitch actuator demands are also significantly decreased when the two methods are combined.

1. Introduction

Life time extension (LTE) for existing turbines is of high interest as installed capacity increases and more projects, equipped with modern technology MW size machines, are reaching the end of their design life [1]. Furthermore, new standards [2] and guidelines [3] for LTE certification are being published showing the industrial relevance. In this context, load mitigation control strategies can be applied as retrofit in order to increase the fatigue reserve of the components while reducing the 'consumption rate' of these reserves. For this purpose, both active and passive control strategies have been researched [4], while emerging concepts like lidar assisted control [5] are also being investigated. For the purpose of life time extension, a suitable control system should be applicable with the minimum amount of changes TO the turbine components while, ideally, not affecting negatively any of the turbine's sub-systems and energy production.

Individual pitch control (IPC) [6, 7] is such a suitable method, that has been widely investigated, since most turbines already include individual pitch actuators, while relevant sensors like strain gauges for the blade root loads are becoming cheaper and more reliable. With IPC, the blades react individually to some sensor signal by pitching. Different approaches have been suggested for implementing IPC. The fundamental difference lies on whether the controlled input is on the non-rotating frame, based on coordinate transformations [8]. Thus, leading to an azimuth based control which is commonly referred to as cyclic pitching. The other approach considers the rotating frame, where a feedback (or feedforward) control law is applied on each blade's signal directly on the rotating frame, referred to as Individual Blade Control

(IBC). Moreover, different control algorithms have been investigated from PID schemes [6, 9] to more advanced or optimal controllers [10] and even feedforward approaches incorporating some preview signal [11]. The main drawback of all IPC approaches is the inevitably increased pitch actuation demand. In this work the IBC method is used and the control algorithm is based on [12].

Down-regulation of power output is a technique that has been investigated for different purposes including wind farm wake optimization [13, 14], energy curtailment [15], ancillary services [16, 17] and condition based control [18, 19]. These strategies focus mainly on accurate power production control and wake reduction. Although important for the turbine lifetime, the impact of these approaches on fatigue and extreme loads and turbine performance metrics has been less thoroughly investigated [20–23]. Down-regulation leads to the reduction of loads- mainly related to aerodynamic torque and thrust- while no new actuators or sensors are required, since it is based solely on changing the pitch and/or torque set points. In this study down-regulation is seen and investigated, purely as a load mitigation strategy for a single wind turbine in free stream conditions, leading to a machine with altered, but fixed, operational characteristics. Moreover, this perspective to down-regulation can have other secondary benefits including wake reduction, which could increase the plant performance as a whole. The obvious disadvantage of down-regulation, as a load reduction method, is the reduction in Annual Energy Production (AEP).

These two strategies can be combined for a more effective load reduction. The baseline controller regulating the operating points, by setting the collective pitch angle and generator torque based on rotor speed input, is re-tuned according to the down regulation strategy. Then IBC is tuned based on and applied on top of the new Collective Pitch Controller (CPC). This is done aiming to combine the load reductions of the system, since in bibliography it is seen that the two strategies are able to focus on different components. Moreover, the combination of them is expected to effect positively the pitch and turbine metrics too, hence this study focuses on the evaluation of these two techniques comparatively and in a combined manner.

The rest of the paper is organized as follows: In section two the retrofit control concept is explained, identifying the key aspects and suggesting an example application case. In section three, the suggested controllers for IBC and down-regulation as well as their combination are explained. Section four shows and discusses the results of full design load simulations, on fatigue loads and other performance metrics, of the two methods individually and combined. Furthermore, the life time extension potential is shown with an example case considering combined operation for 20 years. The last section summarizes the key findings in the conclusions and suggests further research on the topic.

2. Retrofit control concept for LTE

The concept of applying retrofit control for LTE is illustrated in figure 1. The new strategy is implemented after an initial period of operation with the baseline controller. The optimal switching point has to be identified based on optimization and is case specific. The benefit of retrofitting is the reduced period of operation with the new control which for the case of IPC for example would translate to reduced actuator duty cycle which can lead to O&M cost reduction compared to using it as a baseline feature. Similarly, switching to a down-regulated regime after a point will lead to reduced AEP. There, a cost function is required in order to take into account the trade-off between reduced power production and load mitigation along with the possible period of lifetime extension and identify the optimal switching strategy.

In order to evaluate the technical and financial feasibility of such concepts, multidisciplinary knowledge is required. On one hand, the Remaining Useful Lifetime (RUL) of all systems and components during operation should be known along with the estimated load and energy reductions by applying the new methods. On the other hand, specific knowledge on factors

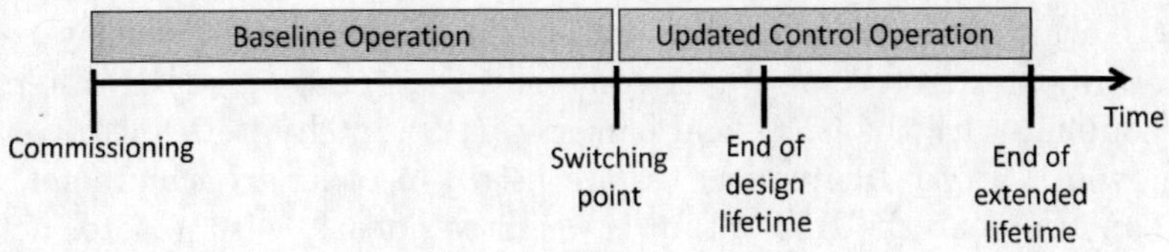

Figure 1. Concept of applying retrofit control for LTE

including local legislation, current and future electricity prices, current and updated O&M costs, and cost of implementing and certifying changes is equally important. The combination of these variables in an optimization problem, along with uncertainty quantification analysis ensuring the bankability of such a business case, can be used for a realistic assessment, as summarized in figure 2. The present study focuses only on the first part, the technical feasibility, evaluating expected fatigue load reductions and the influence on energy production and other turbine performance metrics.

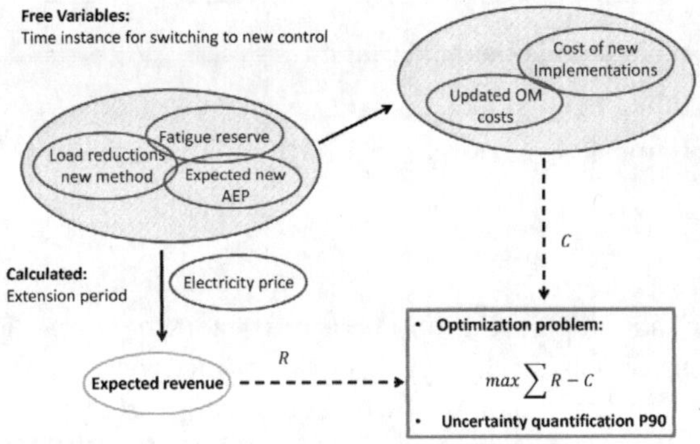

Figure 2. Evaluation of retrofit control strategies for LTE

In this context the two control methods suggested here, are implemented in an example case. The turbine operates for 15 years with the baseline controller and for the remaining 5 years of the design life time the new strategies are implemented. The resulting accumulated damage is calculated for the main turbine components and compared to the, assumed critical, damage from 20 year baseline operation. Based on these damage reserves, the possible life time extension period is calculated for each component taking into account the reduced fatigue consumption rate of the updated controller. Although this method requires a lot of assumptions and simplifications, it indicates the potential value of retrofit control in LTE applications.

3. Methodology

3.1. Individual blade control

The IPC method implemented here is referred to as individual blade control (IBC). Instead of using the traditional Coleman transformation for assigning an azimuth dependent pitch variation to the blades, with IBC three individual, decoupled controllers regulate each blade independently. The input signal is the blade root flapwise bending moment. IBC acts only in full load. This was chosen in order to avoid increasing pitch activity but also to avoid reducing energy production in partial load. Moreover, IBC is completely decoupled from the baseline PI collective pitch

IOP Conf. Series: Journal of Physics: Conf. Series **1102** (2018) 012026 doi:10.1088/1742-6596/1102/1/012026

controller (CPC), which is still responsible for regulating the low frequency bandwidth and hence the operating point (rotor speed). A combination of high pass (HP) filter for the input assuring the decoupling with the CPC, and bandstop (BS) notch filters for the output, limiting the control bandwidth up to 3P frequencies, are used. In order to achieve smooth switching between partial and full load the IBC output is weighted linearly with a factor between 0 and 1 varying linearly with the power output. The control scheme is a classical proportional feedback one. The gains are scheduled over mean pitch values in order to take into account the variation of the effect of pitch in aerodynamics in different wind speeds. A schematic representation is shown in figure 2, while more details on the controller synthesis can be found in [12].

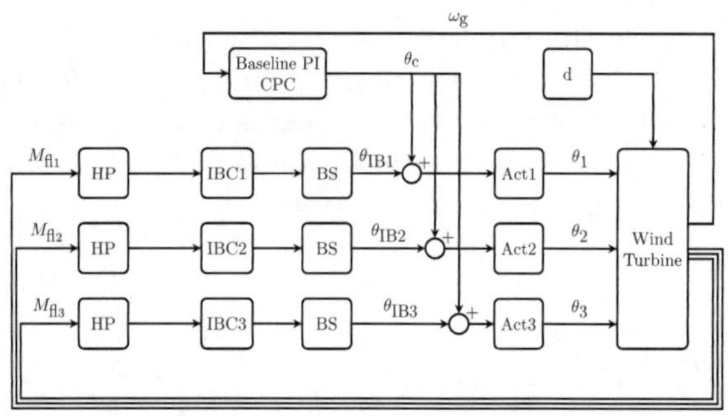

Figure 3. Schematic diagram of the individual blade controller system

3.2. Down-regulation

In bibliography it is seen that depending on the application, e.g. ancillary services to the grid or wake reduction, different objectives like optimal set point tracking or minimization of C_T are to be considered. In choosing the optimal down regulation strategy for fatigue load alleviation purposes combined with IBC specific criteria have to be considered. These are the least possible pitch actuation and the absolute load reduction especially in the loads that are not highly affected by the IBC. Hence, it was decided to implement down-regulation only in the full load region by reducing the generator torque. The partial load region was not considered in order to avoid further AEP reduction but also avoid introducing pitch activity in this region. Moreover, the generator torque reduction was preferred compared to reduction of rated rotor speed in order to avoid reducing the dominant frequencies which in turn changes the aerodynamic damping and lead to increased tower loads and pitch activity as shown also in [18] and [22]. Especially for the DTU 10 MW turbine, the first tower frequency is at 0.25 Hz while the 3P baseline value is 0.45 Hz. This shows that reducing the rated speed by 20% or more can lead to structural resonance which counteracts the load reduction objective. The down-regulation strategy explained here is also referred to as derating.

With the chosen approach, the down-regulated turbine operates with the same rated rotor speed and hence the same Tip Speed Ratio (TSR) as the baseline. The collective pitch controller is responsible for regulating the rotor speed, which due to the decreased torque demand follows a lower C_P trajectory. Figure 4 shows the C_P and C_T trajectories of the baseline and down regulated cases over the TSR-pitch surfaces. The initial jump observed close to optimal C_P value is due to the particular region 2.5 of the DTU 10 MW where the rated generator torque is reached earlier than the rated rotor speed leading to a narrow region where TSR is increased beyond the optimal value. The trajectories show the reduced thrust and aerodynamic power for the same operating points where for the same TSR the increased pitch value offsets the trajectory horizontally to the right. This effect decreases as wind speeds increase.

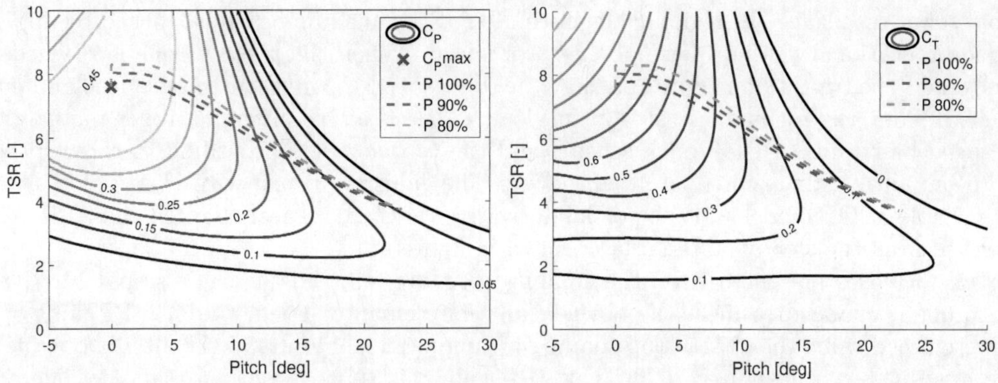

Figure 4. C_P and C_T trajectories over TSR and pitch. Wind speeds increase towards lower coefficient values

The steady state operational characteristics, as found by the non-linear simulations, are shown in figure 5. The rotor speed plot shows the reduction of rated speed from 11.4 m/s to 11.1 m/s and 10.7 m/s for 90% and 80% down-regulation. In addition, the increase of the aforementioned region 2.5 is shown here as well as in the generator torque versus generator speed plot. The pitch plot shows the increased pitch value per wind speed with the pitch value increase being lower as wind speed increases. This is also favorable for stability, since the blades operate in lower angles of attack and thus, further away from the stall region.

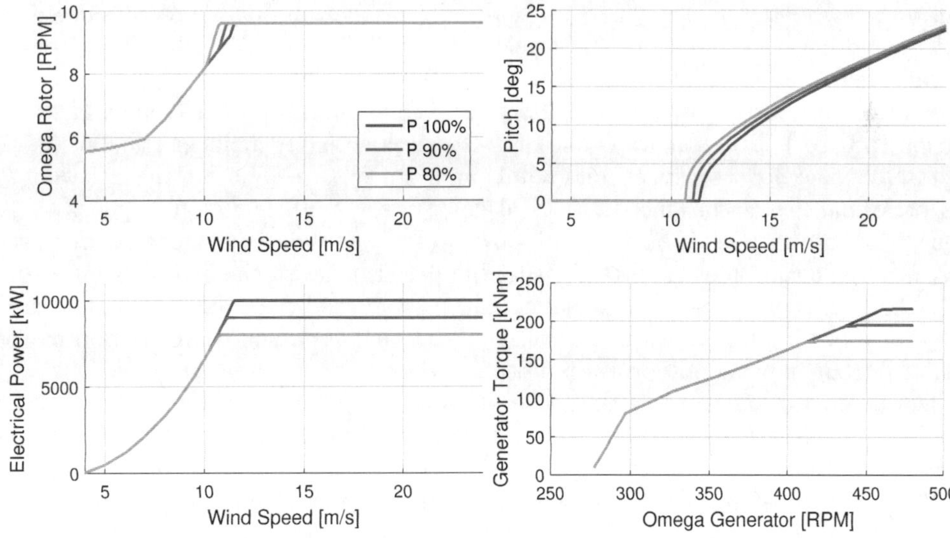

Figure 5. Steady state operational characteristics

3.3. Combining IBC and down-regulation

The baseline Variable Speed Pitch Regulated (VSPR) controller has a classical architecture. In region 1.5 the torque follows a ramp forcing the rotor speed to stay over a minimum value in order to avoid tower resonance. In region 2 the optimal TSR for maximum Cp is tracked by applying a torque proportional to the square of the rotor speed. When the rated torque is reached, the turbine enters the transition region 2.5 where the rotor speeds up until the rated speed and power is reached. Then the turbine operates in full load. The collective pitch controller regulates the rotor speed based on a PI feedback scheme and the torque varies according to rotor speed in order to output constant, equal to rated, power. The gains are tuned with closed loop shaping using a simple 1DOF model and the partial derivatives of C_p in respect to pitch and TSR. These are derived from the steady state surface shown in figure 4.

For all the cases presented here, the tuning procedure followed the same steps. Firstly, the rated torque is changed to meet the down-regulation demand. Then, the new PI gains of the CPC were tuned using the closed loop shaping technique. This results to the altered operational characteristics shown in figure 5. The new IPC gain was tuned based on black-box identified linear blade models as shown in [12] for each wind speed.

4. Results and discussion

4.1. Controller performance

The control strategies investigated include the turbine operating with IBC only, down-regulation to 80% and 90% and their combination. Results of simulations are presented with the DTU 10 MW reference turbine [24] considering DLC1.2 class IA wind conditions as indicated by the IEC standard 61-400 ed. 3 [25] using the high fidelity aeroelastic code FAST v8 [26]. The Damage Equivalent Load (DEL) calculations are done according to the Palmgren-Miner rule [27] using the rainflow algorithm for counting the load cycles. The Wohler coefficient is considered equal to 4 for all the steel components, while for the blades a value of 10 is used. For the, Weibull distribution weighted, lifetime calculations the nominal design period is considered 20 years with a reference load cycle value of 10^7.

For an overview of the whole system's response, the following loads are considered: blade root flapwise and edgewise moments and torsion (BladeFW, BladeEW, BladeTor), tower base fore-aft and side to side moments and torsion (TwrBFA, TwrBSS, TwrBTOR), tower top/yaw bearing non-rotating pitch, yaw and roll moments (TTpitch, TTyaw, TTroll) and low speed shaft torsion (LSSTor). The metrics defined for the pitch actuator include: standard deviation of pitch rate and acceleration ($\dot{\theta}$,$\ddot{\theta}$), and total pitch travel (θ). The metrics relevant to the turbine's performance are: produced electrical energy, standard deviation of generator speed (ω_g), generator torque (Gen Torque) and power (P). Initially, the results are discussed cumulatively for the total lifetime investigating the performance of the control strategies. This can also be seen as a 'per year' relative metric since same conditions and simulation cases are considered for every implementation. In the next section the possible contribution to lifetime extension as retrofit implementation is discussed.

The lifetime, or equivalently per year, relative to the baseline operation load reductions are shown in figure 6. Negative sign shows reduced loads. In general the load reductions achieved by applying only IBC are also achieved with the combined, while other loads that are out of the focus of IBC are reduced by the down-regulation. The main tradeoff is at the BladeFW load reduction. This comes from the transition region (2.5) where the lower the de-rating the larger the region becomes. This explains also why the largest reduction in BladeFW is seen when IBC is combined with P 90% where for all the other loads the combination with P 80% is more efficient. When the two methods are combined the BladeFW load reduction is less effective but still showing a reduction at a level of 10% which, as it will be shown also in the following sections, is satisfactory for LTE purposes. The blade edgewise loading, which is mainly driven by gravity and rotor speed, is practically not influenced. The blade torsional loads are decreased further with the down-regulation, since the aerodynamic pitching moment is decreased by operating in higher pitch, and the reduction is proportional to the level of the down regulation. Regarding the tower bottom, the TwrBFA DEL is marginally decreased by IBC to a level of 3%. With the combined approach the expected thrust reduction from down-regulation leads to slightly higher reductions. The TwrBSS load is not influenced by the down regulation and the decrease in the level of 5% is attributed to IBC and is not affected by their combination. This is due to the fact that this mode presents low aerodynamic damping and is driven by the first tower frequency. The TwrBTOR load is decreased at a level of 8% with the IBC and slightly less with the down regulation. The same behavior is observed in the tower top TTpitch and TTyaw loads. This can be attributed to the reduced aerodynamic damping in higher wind speeds. The low speed shaft torque load is highly reduced due to the reduced rated torque from down regulation to a level of 14% and 11% with P90% and P80% respectively, while when IBC is implemented it is slightly increased. Moreover, the TTroll DEL is highly decreased, proportionally with the down regulation, due to reduced aerodynamic torque.

These two loads (TTroll and LSSTor) are the loads where down regulation contributes the most to, whereas when only IBC is considered they are not influenced. The same applies to a smaller extent for the blade root torsional loads which are also relevant for the damage of the pitch bearings. In general it was shown that the methods complement each other in load reduction and their combination leads to a significant load reduction for all the main system component with the exception of BladeEW loads. Further investigations are required on the switching regions which, although relatively small, was found to influence negatively the combined operation.

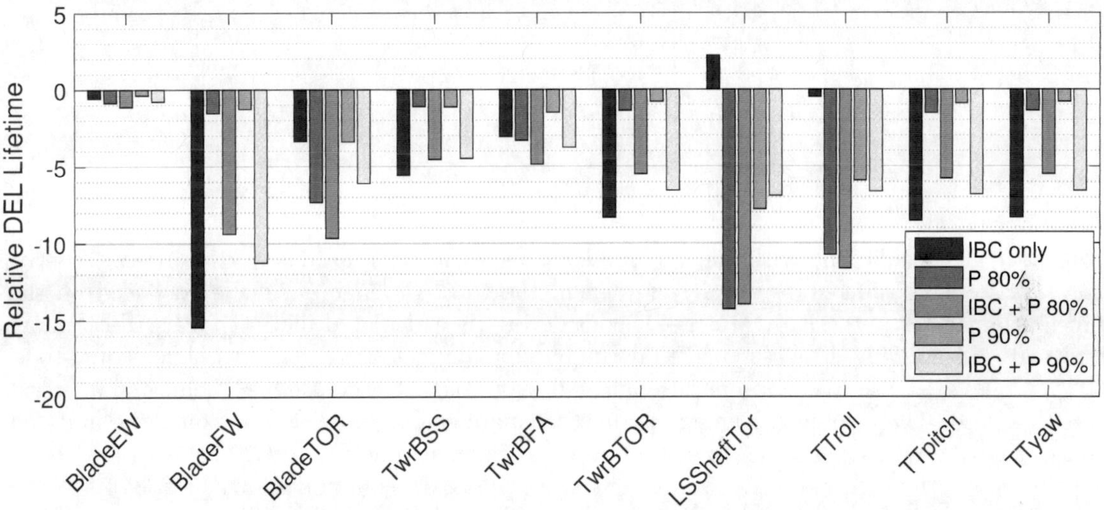

Figure 6. Lifetime (or equally per year) DEL reduction relative to baseline

Figure 7 shows the pitch and performance lifetime metrics. The AEP is not affected by the IBC in any case. Down-regulation of 90% leads to a 6.5 % decrease in AEP and 80% to a 13.5% decrease. The standard deviations of the generator torque and P are decreased proportionally to the down regulation levels showing improvement in the electrical power quality and generator damage. The standard deviation of ω_g is slightly increased to a level lower than 4% with all the methods. The pitch metrics show one of the main benefits in combining the two methods. On one hand, the de-rating approach has no impact on the pitch metrics. On the other hand, when combined with IBC, the pitch actuator demands are decreased substantially, compared to using only IBC. This can be explained by the different operating point compared to the baseline case (see fig. 4). For the updated trajectories the same change in pitch angle affects more the aerodynamic forces.

Although there is no global pitch system damage metric, the results show clearly that the combination of IBC and down regulation reduce the demanded pitch activity from the actuator and the loads acting on the blade bearings. This shows the suitability of the suggested combination for LTE purposes since a wide load spectrum is decreased with reduced requirements from the pitch system. Then, more knowledge from the industry is required to evaluate whether this amount of increase is realistic and at what cost for a commercial pitch system. Of course, the reduction in AEP has to be also evaluated in a wider framework in order to evaluate the feasibility of such a business case.

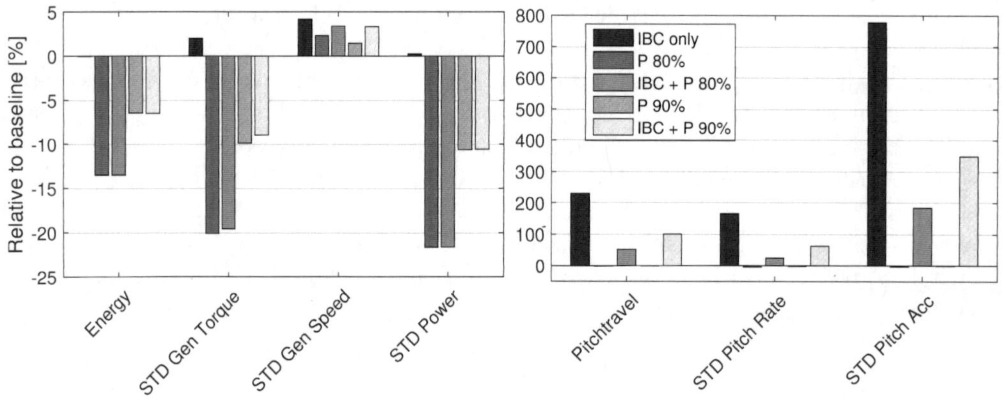

Figure 7. Lifetime (or equally per year) metrics relative to baseline

4.2. Lifetime extension

For the LTE calculations with the retrofit controls some assumptions are required. Firstly, only fatigue is considered as the driver. Then, the total Damage is calculated based on the Linear Damage Hypothesis by Miner [27] considering the calculated lifetime DELs. Since more information is missing, it is assumed that the baseline damage is critical and hence equal to 1. This implies that there are no operational or design fatigue reserves which makes it a 'worst case' scenario. The combined Damage is then calculated taking into account only the updated controller load mitigation performance, in order to calculate the reserves. The key parameters are the application period of the update controller, the DEL reduction and the slope of the S-N curve (i.e. the Wohler exponent m) as seen in equation 1. Based on these reserves and assuming equal distribution of damage over time, the extension period can be calculated. More details on the calculations for life time extension can be found in [12].

$$D_{comb} = \frac{T_{base}}{T_{lifetime}} + \frac{T_{upd}}{T_{lifetime}} \left(\frac{DEL_{life,upd}}{DEL_{life,base}} \right)^m \quad (1)$$

For demonstration purposes, an example case is shown here for a combined 20 year lifetime where the turbine operates for 15 years with the baseline controller and 5 years with the updated. The cumulative damage per load channel is shown in figure 8.

Regarding the cumulative performance metrics for the combined life time, the first metric to be considered is the total energy produced. For the P 90% case the total reduction after 20 years is 1.6% compared to the baseline while for the P 80% case it is 3.4%. The cumulative pitch metrics are presented in figure 9, showing the benefit of applying IBC as retrofit for a limited

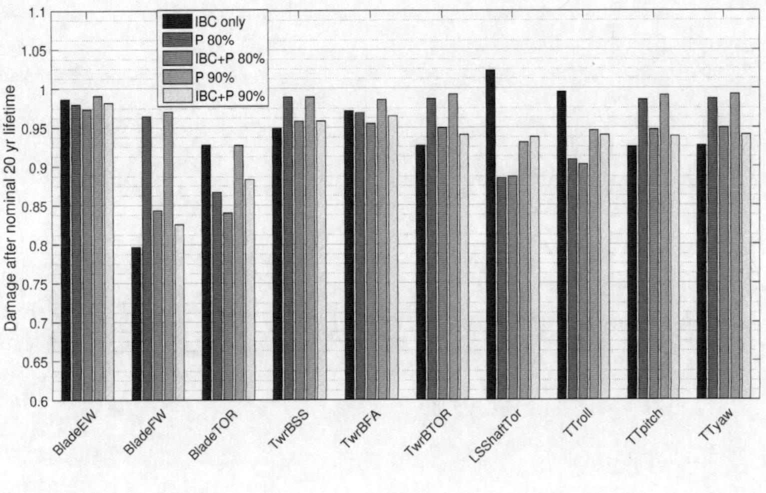

Figure 8. Damage after 20 years of combined operation (15 year baseline CPC)

period over applying it as a baseline feature. Even more, by combining it with down regulation the pitch demands increase to a level lower than 30% for total pitch travel and pitch rate while for the acceleration to a level lower than 100%.

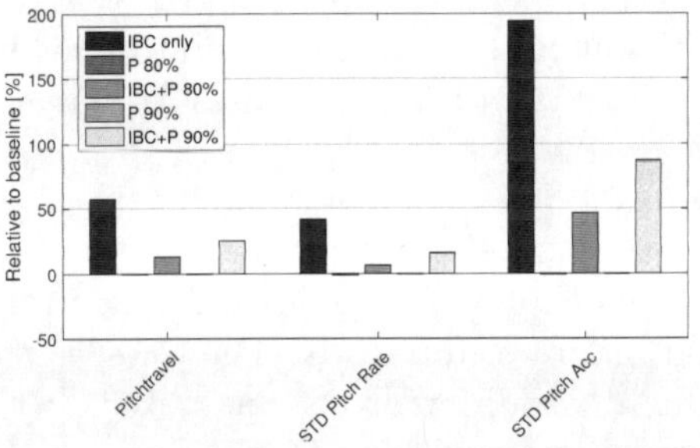

Figure 9. Pitch metrics after 20 years of combined operation (15 year baseline CPC)

The possible extension period in years based on the combined damage and the reduced 'consumption rate' with the new controller method is presented in figure 10. As stated earlier this is a worst case scenario since it is expected that there will be some operational and/or design fatigue reserves. The weight of the Wohler exponent, considered here for the blades, is shown; a damage reserve of 20% can lead to an extension period higher than double the life time. For the tower a period of 1-2 years is calculated. The tower top and shaft channels show a possible extension of 2-4 years. These results raise the question on whether a retrofit control strategy would even be necessary if in reality the operational damage reserves with the baseline controller are high enough. The answer is turbine- and site-specific and can only be answered having real data from operational condition monitoring and exact wind turbine models in order to evaluate each component individually. This discussion shows the limitations of such considerations without further insight.

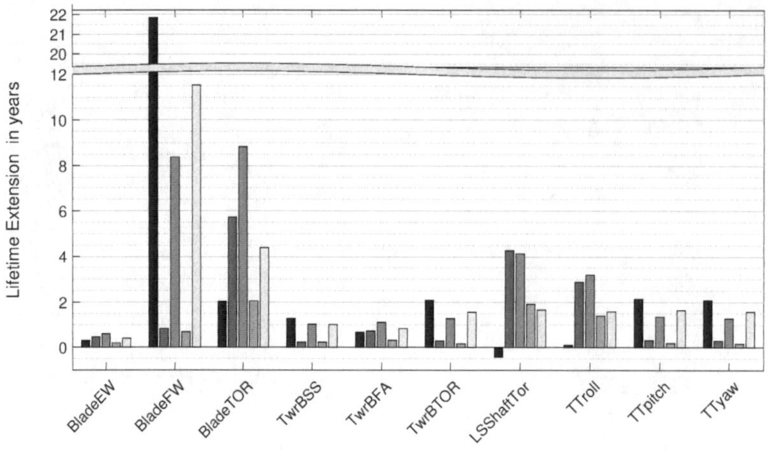

Figure 10. Period of life time extension

5. Conclusions and outlook

In the present work the concept of applying control strategies as retrofit for enhancing the life time extension was investigated. Two suitable control methods, down-regulation and IBC, are suggested, compared and combined. The control algorithms were presented and aeroelastic simulations according to the normal operation case DLC 1.2 were performed in order to evaluate their impact on loads, energy production, pitch and generator metrics. Moreover, a combined 20 year lifetime was considered, where the turbine operates for 15 years with the baseline control and 5 years with the updated scheme.

The results show that there is a benefit in combining down regulation and IBC in full load since they complement each other, in the sense that different loads are targeted with each method. The main focus of the combined scheme are the blade root loads, the low speed shaft torque and to a smaller extend the tower top loads. The tower bottom and blade root edgewise loads are not significantly reduced. Additionally, the actuator duty cycle is reduced when using combined IBC and down regulation compared to applying only IBC, although they are still increased compared to the baseline case. The switching region around rated showed to be a crucial region, dependent highly on the wind turbine and baseline controller design and it was seen that there is room for further optimization. The calculated reductions in AEP were found to be 6.5% and 13.5% for the 90% and 80% down-regulation cases respectively.

Further work on the topic would be to combine the information of Lidar preview in these control algorithms. Adding another, relatively cheap sensor, can give further load reductions combined with the strategies shown here, when a feedforward component can be added on top for both collective and individual pitch demands.

Finally, more information is required in order to have a realistic evaluation of the proposed retrofit control strategies. The need of condition monitoring insight and knowledge of fatigue reserve is prominent, while new methods to evaluate the damage and RUL for subsystems like the pitch system are required. To conclude, insight on the financial aspects of such implementations is critical for realistic feasibility and optimization studies and has to be also evaluated. Despite these, present work demonstrated the possible value of combining IBC and down regulation when applying retrofit control fro lifetime extension,

Acknowledgments

This study was funded by the German Federal Ministry for Economic Affairs and Energy (BMWi) in the framework of German joint research project RAVE-OWP Control (ref. 0324131B) and is part of the research done in the WindForS research cluster. Dr. Andrew Clifton is also gratefully acknowledged for his insightful comments.

References

[1] Ziegler L, Gonzalez E, Rubert T, Smolka U and Melero J J 2018 Lifetime extension of onshore wind turbines: A review covering Germany, Spain, Denmark, and the UK *Renewable and Sustainable Energy Reviews* **82** 1261–1271 ISSN 18790690 URL `http://dx.doi.org/10.1016/j.rser.2017.09.100`

[2] DNV GL 2016 ST-0262: Lifetime extension of wind turbines *Standard*

[3] Megavind 2016 Strategy for Extending the Useful Lifetime of a Wind Turbine Tech. Rep. July

[4] Barlas T K and van Kuik G A M 2010 Review of state of the art in smart rotor control research for wind turbines *Progress in Aerospace Sciences* **46** 1–27 ISSN 03760421

[5] Schlipf D 2016 *Lidar-assisted control concepts for wind turbines* Phd thesis Stuttgart University URL `https://doi.org/10.18419/OPUS-8796`

[6] Bossanyi E A 2003 Wind turbine control for load reduction *Wind Energy* **6** 229–244 ISSN 10954244

[7] Lio W H 2017 *Blade-pitch Control for Wind Turbine Load Reductions* Phd thesis University of Sheffield

[8] Bir G Multi-Blade Coordinate Transformation and its Application to Wind Turbine Analysis 2008 *46th AIAA Aerospace Sciences Meeting and Exhibit* (Reston, Virigina: American Institute of Aeronautics and Astronautics) ISBN 978-1-62410-128-1 URL `http://arc.aiaa.org/doi/10.2514/6.2008-1300`

[9] Bossanyi E A 2005 Further load reductions with individual pitch control *Wind Energy* **8** 481–485

[10] Selvam K, Kanev S, van Wingerden J W, van Engelen T and Verhaegen M 2009 Feedback-feedforward individual pitch control for wind turbine load reduction *International Journal of Robust and Nonlinear Control* **19** 72–91 ISSN 10498923

[11] Laks J, Pao L and Wright A 2009 Combined Feed-forward/Feedback Control of Wind Turbines to Reduce Blade Flap Bending Moments *47th AIAA Aerospace Sciences Meeting including The New Horizons Forum and Aerospace Exposition* 1–16 URL `http://arc.aiaa.org/doi/abs/10.2514/6.2009-687`

[12] Pettas V, Salari M, Schlipf D and Cheng P W 2018 Investigation on the potential of individual blade control for lifetime extension *Journal of Physics: Conference Series* **1037** 032006 ISSN 1742-6588

[13] Ma K, Zhu J, Soltani M, Hajizadeh A and Chen Z Wind turbine down-regulation strategy for minimum wake deficit 2017 *2017 11th Asian Control Conference (ASCC)* vol 2018-Janua (IEEE) pp 2652–2656 ISBN 978-1-5090-1573-3 URL `http://ieeexplore.ieee.org/document/8287595/`

[14] Astrain Juangarcia D, Eguinoa I and Knudsen T 2018 Derating a single wind farm turbine for reducing its wake and fatigue *Journal of Physics: Conference Series* **1037** 032039 ISSN 1742-6588

[15] Mirzaei M, Tuhfe G, Giebel G, Sørensen P E and Poulsen N K 2015 Turbine Control Strategies for Wind Farm Power Optimization *American Control Conference (ACC)* 1709–1714 ISSN 07431619

[16] Aho J, Buckspan A, Pao L Y and Fleming P A 2013 An Active Power Control System for Wind Turbines Capable of Primary and Secondary Frequency Control for Supporting Grid Reliability *AIAA/ASME Wind Symposium* 1–13

[17] Mirzaei M, Soltani M, Poulsen N K and Niemann H H 2014 Model based active power control of a wind turbine *2014 American Control Conference* 5037–5042 URL `http://ieeexplore.ieee.org/document/6859055/`

[18] Engels W P, Marina A, Kanev S and van der Hoek D 2017 Condition based control: Take the load off damaged components ECN-E–017-047 Tech. Rep. September ECN

[19] Richards P, Todd Griffth D and Hodges D 2015 Smart loads management for damaged offshore wind turbine blades *Wind Engineering* **39** 419–436 ISSN 0309524X

[20] Fleming P A, Aho J, Buckspan A, Ela E, Zhang Y, Gevorgian V, Scholbrock A, Pao L and Damiani R 2016 Effects of power reserve control on wind turbine structural loading *Wind Energy* **19** 453–469 ISSN 10954244

[21] DC van der Hoek; and Kanev S 2017 Reducing Wind Turbine Loads with Down-Regulation Tech. Rep. September ECN URL `https://www.ecn.nl/publications/PdfFetch.aspx?nr=ECN-E--17-032`

[22] Aho J, Fleming P and Pao L Y Active power control of wind turbines for ancillary services: A comparison of pitch and torque control methodologies 2016 *2016 American Control Conference (ACC)* (IEEE) pp 1407–1412 ISBN 978-1-4673-8682-1 URL `http://ieeexplore.ieee.org/document/7525114/`

[23] Frost S A, Goebel K and Obrecht L 2013 Integrating Structural Health Management with Contingency Control for Wind Turbines *International Journal of Prognostics and Health Management* **4** 11–20 ISSN 21532648

[24] Bak C, Zahle F, Bitsche R, Yde A, Henriksen L C, Nata A and Hansen M H 2013 De-

scription of the DTU 10 MW Reference Wind Turbine Tech. Rep. July DTU Roskilde URL https://dtu-10mw-rwt.vindenergi.dtu.dk
[25] International Electrotechnical Commision (IEC) 2005 Wind turbines Part 1: Design requirements 61400-1 , 3rd edition Tech. rep. International Electrotechnical Commision (IEC) Switzerland
[26] Jonkman J and Marshall B 2005 FAST User's Guide Tech. Rep. 6 URL http://www.ncbi.nlm.nih.gov/pubmed/21564034
[27] Miner M 1945 Cumulative Damage in Fatigue *Journal of Applied Mechanics* **3** 159–164

Effect of electricity market price uncertainty modelling on the profitability assessment of offshore wind energy through an integrated lifecycle techno-economic model

Anastasia Ioannou[a,*], Andrew Angus[b], Feargal Brennan[c]

[a]Cranfield University, School of Water, Energy and Environment, Cranfield, Bedfordshire, MK43 0AL United Kingdom

[b]Cranfield University, School of Management, Cranfield, Bedfordshire, MK43 0AL United Kingdom

[c]University of Strathclyde, Department of Naval Architecture, Ocean & Marine Engineering, Glasgow G4 0LZ, United Kingdom

*E-mail address of coresponding author: a.ioannou@cranfield.ac.uk

Abstract. According to the Contracts for Difference (CfD) scheme introduced to support the deployment of offshore wind installations, an electricity generation party is paid the difference between a constant "strike price" (determined be means of a competitive auction) and the average UK market electricity price for every MWh of power output produced. The scheme lasts for 15 years, after which the electricity output is sold on the average market price. To this end, estimating the long term profitability of the investment greatly depends on the forecasted market prices. This paper presents the simulation results of future electricity prices based on three different simulation methods, namely: the Geometric Brownian motion (GBM), the Autoregressive Integrated Moving average (ARIMA) and a model combining Mean-Reversion and Jump-Diffusion (MRJD) processes. A number of simulation paths are generated for a time horizon of 10 years and they are introduced to a fully integrated techno-economic model developed by the authors. As a result, joint probability distributions of the NPV derived from the three different methods are presented. This study is relevant to investors and policy makers to check the viability of an investment and to predict its stochastic temporal return profile.

1. Introduction

Offshore wind energy has been rapidly expanding in Europe during the last decade. According to WindEurope annual offshore wind statistics, there are currently 92 wind farms in operation across European countries (4,149 grid-connected wind turbines) [1] with UK accounting for 43% and Germany for 28% of all grid-connected turbines.

In the United Kingdom, there is currently a transition from the Renewables Obligation (RO) scheme to the Contracts for Difference (CfD) scheme, introduced by the recent Electricity Market Reform (EMR). The CfD scheme is a private law contract between a producer and the Low Carbon Contracts Company (LCCC), a government-owned company. An electricity generation party with CfD is paid the difference between a constant "strike price" and the average UK market electricity price ("reference price"). The Generator sells electricity under a Power purchase agreement to a licensed supplier or trader at an agreed reference market price. If the reference price is higher than the strike price, the generation party has to pay back the difference to the LCCC. The bottom line is that company always gets the strike price for the electricity generated. The scheme lasts for 15 years (while the expected lifetime of an OW energy asset is 25 years), after which the electricity output is sold on the average UK electricity market price, hence imposing uncertainty to the revenues yielded by the investment after the 15th year of operation [2]. As such, the forecasting of electricity prices becomes pertinent towards estimating future returns of the offshore wind energy project as well as, the overall profitability of the investment.

This work aims to stochastically assess the impact of volatile market electricity prices on the profitability assessment of offshore wind farms. We develop and formalize a process for more detailed modelling of the future economic cash flow of an offshore wind farm by employing a tested lifecycle techno-economic model which accommodates the uncertainty modelling of electricity prices through employing different forecasting methods in order to test how these can affect the Net Present Value (NPV) distribution analysis of the wind farm. A stochastic cost modelling of the economic profitability is finally derived.

2. Lifecycle techno-economic model

The methodological framework of the techno-economic model is illustrated in Figure 1. The main components of the life cycle cost of a fixed OW farm are the following: (i) the CAPEX module, which includes development and consenting (D&C), production and acquisition (P&A), installation and commissioning (I&C), and decommissioning and disposal (D&D) costs, (ii) the general site characteristics module with details on the weather conditions, site water depth, distance from port, vessels, cost of personnel etc., (iii) the FinEx module with parameters related to the financing expenditures, namely the Weighted Average Cost of Capital (WACC), inflation rate, equity and debt ratio, etc., (iv) the OPEX module, which considers reliability data from literature, cost of personnel, materials, vessels and related maintenance processes, which will provide availability and O&M cost estimates pertinent for the cost analysis and (v) the Revenue model, which considers the net power generation, the strike prices (according to the CfD scheme), and the market electricity price. A detailed description of the model can be found in [3].

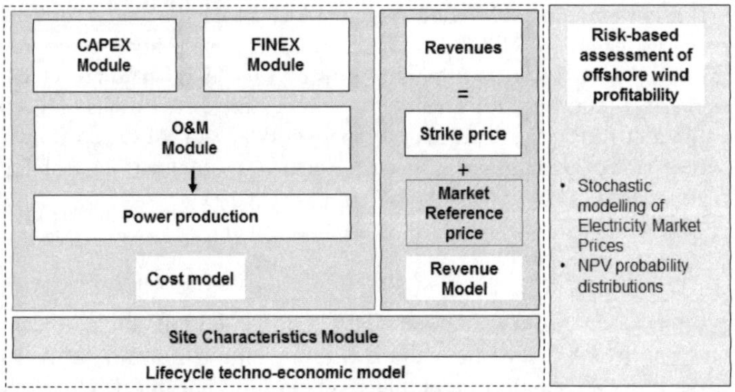

Figure 1 Illustration of integrated techno-economic lifecycle model

3. Modelling & Simulation of future wholesale electricity prices

This section looks at the forecasting methods that were employed to model electricity market prices, towards incorporating the uncertainty and variability in the cash flow model of the analysis. The modelling of market electricity prices has been carried out by numerous authors in literature [4–6], highlighting the advantages and disadvantages of each technique and applying them in various contexts. Time series techniques are usually based on extrapolating a set of historic observations to predict their

IOP Conf. Series: Journal of Physics: Conf. Series **1102** (2018) 012027 doi:10.1088/1742-6596/1102/1/012027

behavior in the future. In [5], electricity price forecast techniques are categorized into: multi-agent, fundamental methods, reduced-form models, statistical approaches and computational intelligence techniques. Statistical methods, including autoregressive models forecast the future value of a time series by applying a mathematical correlation of the previous values with the current values. The ARIMA time series model and the Geometric Brownian motion are among the most cited forecasting techniques [4,7,8]. However, statistical methods cannot capture sufficiently the presence of spikes in the dataset, especially for price-only models, but also for models using fundamental variables. Mean-reverting jump-diffusion (MRJD) processes are more appropriate to reproduce patterns of spikes and reversion to a long term mean level [9].

In this study we have tested above forecasting techniques to identify the effect of uncertainty modelling in electric market prices on the profitability of an offshore wind investment. Daily historical data of electricity market prices (found in [10]) were used to stochastically model the future revenues of the investment.

3.1. Geometric Brownian motion

Financial time series are most commonly based on stochastic differential equations (SDEs) which are the most general descriptions of continuously evolving random variables. Geometric Brownian motion is the simplest and most common financial time series model, according to which the logarithm of the randomly varying quantity follows a Brownian motion with drift.

Brownian motion (also called Wiener process) with drift parameter μ and volatility σ is a kind of Markov stochastic process $\boldsymbol{W} = \{X_t : t \in [0, \infty)\}$ of the form:

$$X_t = \mu t + \sigma W_t \tag{1}$$

The generalised form of the Geometric Brownian motion (GBM) process is specified through a stochastic differential equation (SDE) of the form [11]:

$$\frac{dS_t}{S(t)} = \mu dt + \sigma dW(t) \tag{2}$$

where, S(t) denotes the price of the electricity at time t, μ represents the drift parameter, σ the volatility of the electricity prices and W is the standard Brownian motion.

The Wiener process satisfies the following properties: a) The process starts from 0 $X_0 = 0$ (with probability 1), b) $\boldsymbol{W}$ has Gaussian increments, i.e. for $h \geq 0, X_{t+h} - X_t$ is normaly distributed with $\mu = 0$ and variance σ (same distribution as X_h), c) $\boldsymbol{W}$ has independent increments; that is, for $t_1, t_2, .., t_n \in [0, \infty)$ with $t_1 < t_2 <, .., < t_n$, the random variables $X_{t_1}, X_{t_2} - X_{t_1}, ..., X_{t_n} - X_{t_n - 1}$ are independent, d) X_t has a normal distribution with mean t_n , e) $\boldsymbol{W}$ has continuous paths, namely with probability 1, X_t is continuous on $[0, \infty)$.

Using Itô's lemma and integrating over time, the relationship between an initial value S_t and a later value S_{t+T}:

$$S_{t+T} = S_t \cdot \exp\left[\left(\mu - \frac{\sigma^2}{2}\right)T + \sigma W_t\right] \tag{3}$$

Above equation is the GBM model. This process has the advantage that it always remains positive and it can represent the characteristics of many variables.

3.2. Mean-reverting jump-diffusion (MRJD) process

The jump-diffusion model can be expressed by the following general stochastic differential equation for the increment of the electricity price (after removing seasonality and trend from the dataset):

$$dX_t = \mu(X_t, t) + \sigma(X_t, t)dW_t + dq(X_t, t) \tag{4}$$

Where, dW_t represent the increments of a standard Wiener process (i.e. Brownian motion) and $dq(X_t, t)$ are the increments of a jump process.
When there is a high electricity demand, more expensive power generation technologies need to be brought online to cover the electricity load. During these periods, electricity prices exhibit jumps. In general, spot electricity prices are characterised by high volatility, seasonal cycles and occasional spikes. In mean-reverting jump-diffusion processes, the drift term $\mu(X_t, t)$ can force reversion to long term mean levels. The Ornstein-Uhlenbeck process, which is the most applied mean-reversion process (initially introduced in finance to model interest rate dynamics [12]), is expressed as:

$$dX_t = (\alpha - \beta X_t)dt + \sigma dW_t \tag{5}$$

Where, β the mean-reversion speed and $\frac{\alpha}{\beta}$ is the long term mean reversion level.

3.3. Autoregressive Integrated Moving Average (ARIMA)
ARIMA or Box-Jenkins model [13] is a statistical method standing for autoregressive (AR) integrated (I) moving average (MA) and it is a generalisation of the Autoregressive Moving Average model (ARMA), where "I" (standing for Integrated) is a differencing step that is used to remove trend or seasonality from the time series. ARIMA models use standard notation of ARIMA (p,d,q) and (P,D,Q) for their seasonal counterparts. In power systems applications, ARIMA models have been used for load forecasting [14,15], with good results, as well as to model and forecast day-ahead electricity prices [16,17] and weekly prices [18]. ARIMA method was deemed appropriate for this study considering the ability of the method to take into account the seasonal trend of the dataset of electricity prices.

- The Autoregressive part (p) specifies which previous values from the data series are used to predict the current values or else the number of autoregressive orders.
- The Difference part (d) specifies the order of differencing of the time series before the application of the model. To apply the ARIMA model, the dataset is required to be stationary; if not, a transformation of the series to the stationary form needs to take place. Differencing is one of the simplest ways to achieve this. Box and Jenkins (1976) introduced a model that contains not only the autoregressive and moving average parts, but also the differencing part [19].
- The moving average part (q) specifies the order of moving average orders in the model, namely how the mean values deviation of the previous time series are used to predict the current values.

As such, the mathematical formulation of the ARIMA(p,d,q) model can be described using a lag operator notation (defined as $L^i X_t = X_{t-i}$) as follows:

$$\varphi(L)(1 - L)^d X_t = c + \theta(L)\varepsilon_t \tag{6}$$

where, X_t is the price at time t, c a constant term, d the differencing order, ε_t is the random error at time t; further, $\varphi(L)$ are the parameters of the AR model formulated as:

$$\varphi(L) = 1 - \varphi_1 L - \cdots - \varphi_p L^p \tag{7}$$

where, p refers to the autoregressive terms, while $\theta(L)$ are the parameters of the MA(q) model expressed as:

$$\theta(L) = 1 + \theta_1 L + \ldots + \theta_q L^q \tag{8}$$

where q refers to the moving average terms [13].

4. Simulation of future electricity prices

4.1. Electricity price prediction under the different methods

Monthly and daily data of the wholesale electricity prices were collected from different sources [10,20] to compile a dataset starting from March 2003 to December 2017. The dataset (178 observations) was divided into two parts, the first consisting of 118 observations, which was used to build the model and the second of 60 observations for testing the model.

- The input data required for the Geometric Brownian (GBM) motion model were the mean instantaneous monthly rates of change of the historic prices dataset (equal to 0.0107) and the instantaneous standard deviation of the rates (equal to 0.1626).
- The parameters of the best-fitting ARIMA model for the monthly electricity prices were identified through Expert Modeler of SPSS [21], eliminating the need to identify an appropriate model through a manual trial and error process. The tool indicated an ARIMA(2,1,2)(1,0,1) model consisting of non-seasonal and seasonal parts with a periodicity set to 12. Above notation means that the series was differenced once at lag-1; further, the model includes X_{t-2} and ε_{t-2}, as well as a seasonal lag-12 AR term and a seasonal lag-12 error term.
- The characteristics of the Mean-Reversion and Jump-Diffusion (MRJD) model were modeled by the following relationship:

$$\log(X_t) = f(t) + Y_t \tag{9}$$

where, X_t is the spot electricity market, $f(t)$ is the deterministic and Y_t is the stochastic part of the model. The deterministic part was further modelled through a trigonometric function [22]:

$$f(t) = s_1 \sin(2\pi t) + s_2 \cos(2\pi t) + s_3 \sin(4\pi t) + s_4 \cos(4\pi t) + s_5 \tag{10}$$

where $s_1, \ldots s_5$ are constant parameters, while the stochastic components were modelled as an Ornstein-Uhlenbeck processes, adopted from [23]. Then, the calibration of both the deterministic and the stochastic part was realized through the least square and the maximum likelihood method, respectively.

After determining the input parameters for the best-fitting ARIMA model, the Geometric Brownian motion, and the MRJD processes using historic monthly electricity prices, we, accordingly, simulated 1,000 sample paths, 10 years into the future. Figure 2 illustrates the simulation results of future electricity prices with the MRJD model from 2017 to 2027.

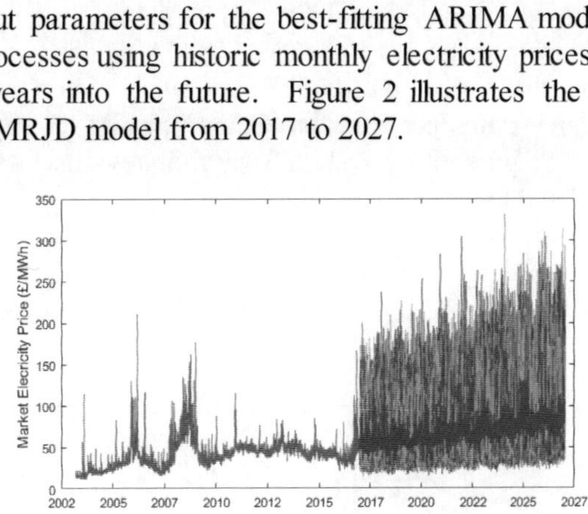

Figure 2 Simulation of future electricity prices with the MRJD model

4.2. Validation of methods

In Figures 3(a)-(c), the 50% upper and lower confidence limits of the resulting model for the first part of the dataset, as well as the observed data for 2015-2017 (2nd part of the dataset) are illustrated for the 3 different forecasting methods. The mean absolute percentage errors and the percentage errors between the observed data and the forecasted values for the four testing years are summarized in Table 1. ARIMA was deemed to have the lowest Mean Absolute Percentage Error in comparison to the other 3 methods namely 14.8%, indicating a relatively better fitness of the ARIMA model to the dataset, followed by the MRJD and the GBM models. MAPE is a measure of prediction accuracy of the forecasting method and is calculated as follows:

$$MAPE = \frac{100\%}{n}\sum_{t=1}^{n}\left|\frac{F_t - A_t}{A_t}\right| \tag{11}$$

Table 1 Validation of methods

	Error (%)				Mean Absolute Percentage Error (MAPE)
	2013	2014	2015	2016	
GBM	12.6%	44.8%	63.9%	70.7%	48%
ARIMA	-3.3%	14.4%	21.2%	20.4%	14.8%
MRJD	10.0%	30.8%	37.1%	35.7%	28.4%

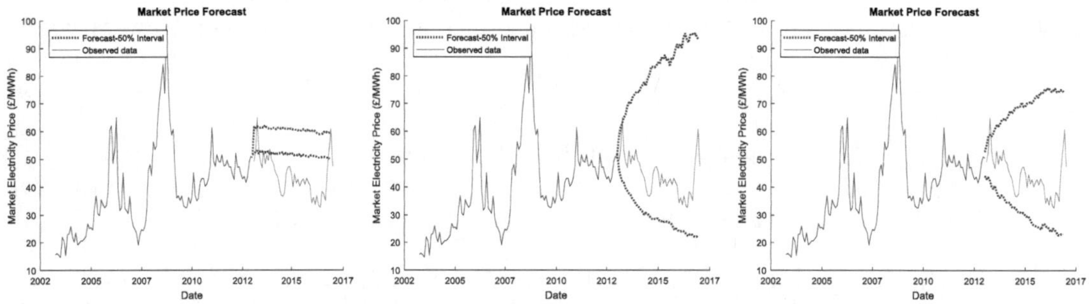

(a) (b) (c)

Figure 3 Test cases for electricity prices forecasting: (a) MRJD, (b) GBM, (c) ARIMA

The variability of the simulated electricity prices can be illustrated in Figures 4(a)-(c). It can be observed that MRJD method demonstrates a lower variability probably due to the drift term which forces reversion to long term mean levels (Figure 4(b)). The ARIMA model has the greatest variability as shown in Figure 4(c), probably due to the complex correlations between the previous values with the current values, leading to diverse paths per each simulation. Finally, as far as GBM is concerned, a positively skewed distribution (Figure 4(c)) with a long tail towards higher electricity prices was compiled from the 1,000 simulated paths.

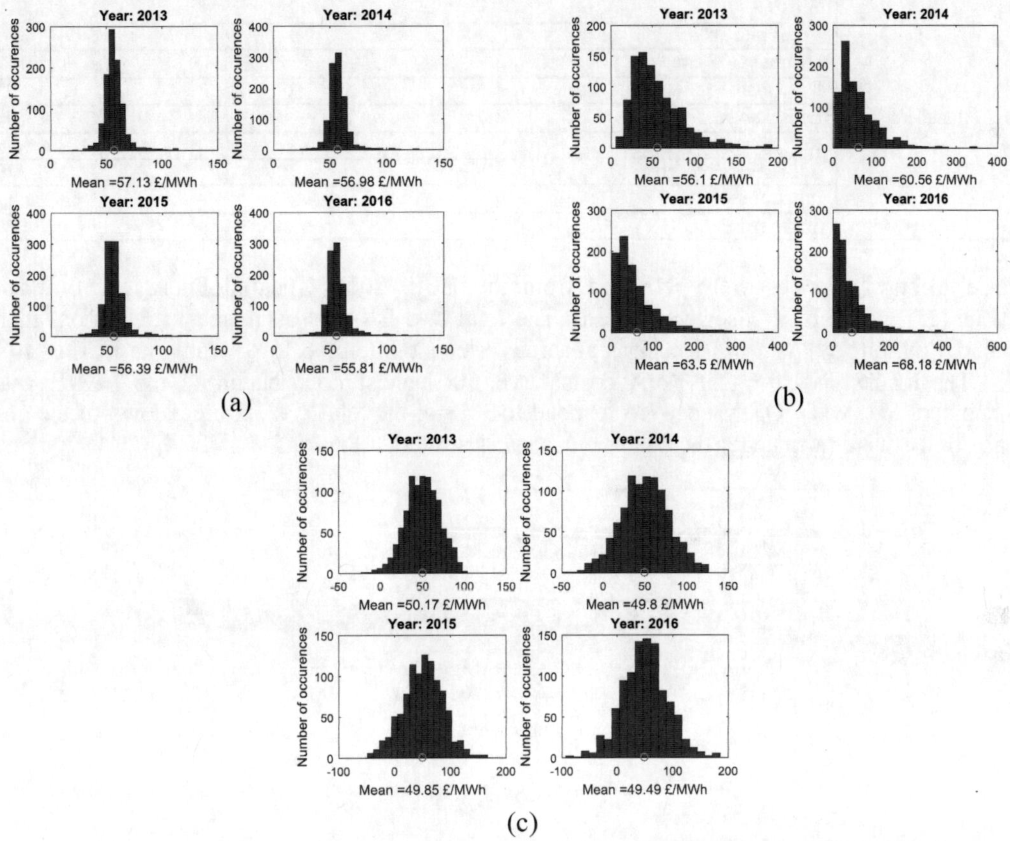

Figure 4 Probability distributions of electricity prices test cases: (a) MRJD, (b) GBM, (c) ARIMA

5. Results

5.1. Case study

Having modelled electricity market prices on a monthly basis, the stochastic modeling, based on distribution of economic outcomes of the investment, is performed. The case study that is used for the application of the method consists of a 504MW capacity wind farm located in the North Sea region, 36km away from shore. Weather data (3-hourly data over a 3-year period) were retrieved from BTM ARGOSS [24] for modelling the operational phase of the asset. The specifications of the wind farm investment are included in Table 2.

Table 2 Base case wind farm investment specifications

Wind farm characteristics		Values
Wind farm	Total wind farm capacity, P_{WT}	504 MW
	Projected operational life of the wind farm, n	25 years
	Construction years, T_{constr}	5 years
	Number of turbines, n_{WT}	140
General Site characteristics	Distance to port, D	36 km
	Water depth, WD	26 m
Wind turbine	Rotor diameter, d	107 m
	Hub height, h	77.5 m
	Pile diameter, D_{pile}	6 m
	Rated power	3.60 MW
	Cut-in speed	4 m/s
	Cut-out speed	25 m/s

IOP Conf. Series: Journal of Physics: Conf. Series **1102** (2018) 012027 doi:10.1088/1742-6596/1102/1/012027

Economic parameters	WACC	8.81%
	Corporate tax	17%
	Depreciation rate	18%
	Debt share	70%
	Equity share	30%
	Return on equity	15.8%
	Interest rate on debt	7%
	Inflation rate	2.5%
	Strike price	140 £/MWh

Wholesale electricity prices were retrieved from the BEIS 2016 Updated Energy & Emissions Projections [25] for the base case investigated. The total CAPEX was estimated £1.67 billion, annual OPEX £56.6 million, NPV=£174.1 million at a real discount rate of 6.15%, while the LCOE=108.9 £/MWh. The results indicate that P&A costs have the highest contribution to the LCOE value, accounting for 46%, while O&M costs correspond to 30% of the total cost. A breakdown of the costs per Phase of the wind farm under the baseline case is illustrated in Figure 5.

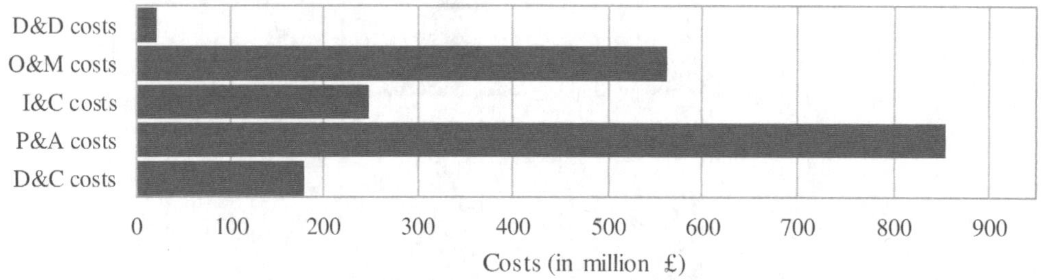

Figure 5 Deterministic lifecycle costs breakdown

5.2. Net Present Value Stochastic Analysis

The joint probability distributions of the NPV derived from the three different methods are plotted in Fig. 6(a)-(c). The assumptions considered for the calculation of the probability distribution curves are the same as the ones compiled in Table 2. The Value at Risk (VaR) is a traditional risk measure approximating the probability that the value of an asset or portfolio will drop below a particular value over a specified confidence level and in the context of a planning horizon. VaR is always specified with a given confidence level α. As such, as shown in Figure 6(a), the NPV of the investment has a VaR 5% of £155million denoting that there is 95% probability that NPV will exceed £155million.

It can, thus, be inferred from the illustrated probability distributions that GBM method has the lowest VaR 5% = £8million, while the MRJD method yields a probability of zero for a negative NPV, as well as the highest mean NPV in comparison to the other three methods. The mean NPV derived from the ARIMA method indicated the closest proximity to the deterministic base case NPV.

As far as the shape of the probability distributions is concerned, following a similar pattern on the electricity prices scatter of data, the resulting NPVs demonstrate a positively skewed normal distribution with a medium variability, when a GBM is employed, implying non-linear relationships between stochastic electricity prices and NPV output. MRJD-derived curve has a low variability, again as a result of the existing drift term, while the opposite applies to ARIMA model which returns a NPV distribution with a high variability.

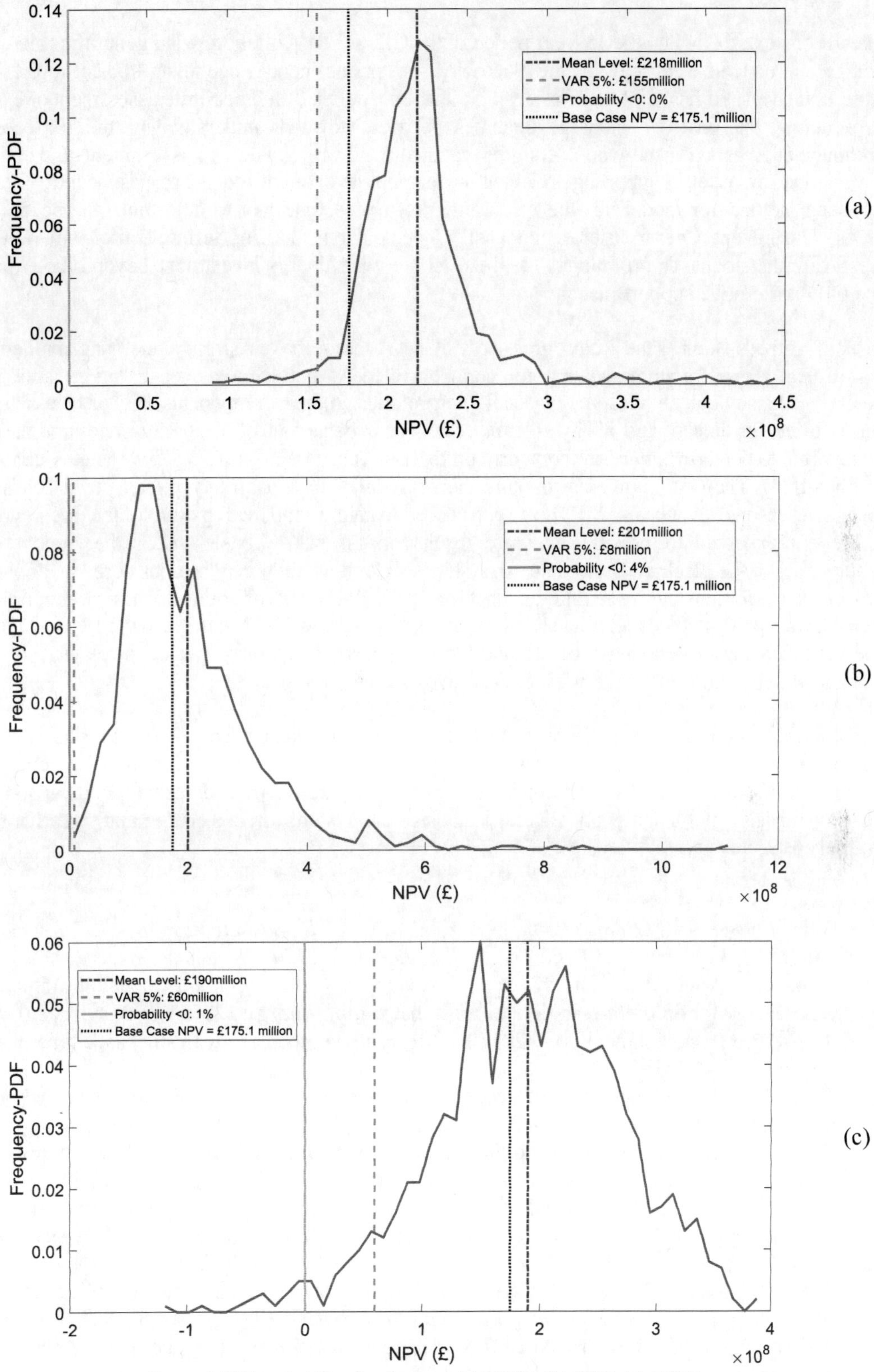

Figure 6 NPV probability distributions: (a) MRJD, (b) GBM, (c) ARIMA

6. Conclusions

In the UK context, following the 15-year period of the CfD scheme, offshore wind generators are obliged to sell their produced electricity on the electricity spot prices; hence, becoming subject to the risk of volatile and uncertain financial cash inflows. To this end, detailed and accurate assessment of expected returns during the post-CfD scheme becomes pertinent towards understanding the real cost and opportunity of investing in new or existing operational wind farms. Such an assessment could facilitate fair valuation of assets, supporting relevant investment/divestment decisions. To identify the best forecasting method for modelling the energy market prices, one has to determine the scope of the analysis. The present paper focuses on stochastically calculating the long-term electricity market prices and estimate the stochastic profitability of the offshore wind energy investment beyond the expiration of the CfD strike price support mechanism.

Statistical methods, such as the Autoregressive Moving Average, have a strong underlying mathematical and statistical theory, accommodating temporal correlations between past observations and current prices; as such, they can attach some physical interpretation to their components. Nevertheless, they are often criticized for their limited ability to capture nonlinear behaviour of electricity prices and they have been reported to perform better for short-term predictions (i.e. forecasts from a few minutes up to a few days ahead) [5]. They can, however, capture the seasonality that electricity prices exhibit on a daily, weekly and seasonal level basis. MRJD are considered to give a simplified picture of the price dynamics and are not expected to provide accurate results on an hourly basis, but rather recover main characteristics of the electricity prices at a daily time scale; thus, they may be considered as appropriate for longer term forecasting, requiring as input only the prior data of a time series to generalize the forecast. Among the methods tested, ARIMA demonstrated the lowest Mean Absolute Percentage Error in the validation cases, denoting a better long-term forecasting capability, which is relevant to service life financial appraisal of offshore wind energy investments.

Other available methods, such as Computational Intelligence techniques can also be considered as more relevant for long term forecasting since they can produce more accurate results, handling complexity and non-linearity. Nevertheless, their application usually requires a larger dataset (in comparison to the price-only models) of fundamental drivers, including the system forecasted demand, weather related data, fuel costs, etc.

References

[1] Wind Europe 2018 *Offshore wind in Europe. Key trends and statistics 2017.*
[2] DECC 2014 *Contract for Difference : allocation process high level summary*
[3] Ioannou A, Angus A and Brennan F 2018 A lifecycle techno-economic model of offshore wind energy for different entry and exit instances *Appl. Energy* **221C** 406–24
[4] Borovkova S and Schmeck M D 2017 Electricity price modeling with stochastic time change *Energy Econ.* **63** 51–65
[5] Weron R 2014 Electricity price forecasting: A review of the state-of-the-art with a look into the future *Int. J. Forecast.* **30** 1030–81
[6] Keles D, Scelle J, Paraschiv F and Fichtner W 2016 Extended forecast methods for day-ahead electricity spot prices applying artificial neural networks *Appl. Energy* **162** 218–30
[7] Conejo A J, Plazas M A, Espinola R and Molina A B 2005 Day-Ahead Electricity Price Forecasting Using the Wavelet Transform and ARIMA Models *IEEE Trans. Power Syst.* **20** 1035–42
[8] Vaienti C, Ioannou A and Brennan F 2017 Cash flow at risk of offshore wind plants *2017 6th International Conference on Clean Electrical Power (ICCEP)* (IEEE) pp 84–93
[9] Seifert J and Uhrig-Homburg M 2007 Modelling jumps in electricity prices: theory and empirical evidence *Rev. Deriv. Res.* **10** 59–85

[10] Institution of Civil Engineers (ICE) 2018 The changing price of wholesale UK electricity over more than a decade
[11] Glasserman P 2004 *Monte Carlo Methods in Financial Engineering* (New York: Springer-Verlag)
[12] Vasicek O 1977 An equilibrium characterization of the term structure *J. financ. econ.* **5** 177–88
[13] Box G E P, Jenkins G M and Reinsel G C 1994 *Time Series Analysis: Forecasting and Control* ed E Cliffs (NJ: Prentice Hall)
[14] Gross G and Galiana F D 1987 Short-term load forecasting *Proc. IEEE* **75** 1558–73
[15] Hagan M T and Behr S M 1987 The Time Series Approach to Short Term Load Forecasting *IEEE Trans. Power Syst.* **2** 785–91
[16] Jakasa T, Androcec I and Sprcic P 2011 Electricity price forecasting - ARIMA model approach *2011 8th International Conference on the European Energy Market (EEM)* (IEEE) pp 222–5
[17] Contreras J, Espinola R, Nogales F J and Conejo A J 2003 ARIMA models to predict next-day electricity prices *IEEE Trans. Power Syst.* **18** 1014–20
[18] Fosso O B, Gjelsvik A, Haugstad A, Mo B and Wangensteen I 1999 Generation scheduling in a deregulated system. The Norwegian case *IEEE Trans. Power Syst.* **14** 75–81
[19] Box G E P and Jenkins G M 1976 *Time Series Analysis: Forecasting and Control* (San Francisco : Holden-Day, c1976)
[20] Ofgem Web Page 2018 Wholesale Market Indicators *Electr. prices Day-ahead baseliad Contract. Mon. Aver.*
[21] IBM Time Series Modeler
[22] Lucia J J and Schwartz E S 2002 Electricity Prices and Power Derivatives: Evidence from the Nordic Power Exchange *Rev. Deriv. Res.* **5** 5–50
[23] MathWorks Documentation Simulating Electricity Prices with Mean-Reversion and Jump-Diffusion
[24] BTM ARGOSS 2017 WaveClimate.com
[25] Department for Business E & I S (BEIS) 2017 Updated Energy & Emissions Projections

IOP Conf. Series: Journal of Physics: Conf. Series **1102** (2018) 012028 doi:10.1088/1742-6596/1102/1/012028

Is the future development of wind energy compromised by the availability of raw materials?

E Gonzalez[1,2], A Ortego[3], E Topham[4,5], A Valero[3]
[1]ORSEIDE, C/ Dublín, 62, 31620 Pamplona, Spain
[2]Universidad de Zaragoza, C/ Mariano Esquillor Gómez, s/n, 50018 Zaragoza, Spain
[3]CIRCE, C/ Mariano Esquillor Gómez, 15, 50018 Zaragoza, Spain
[4]Doctoral Training Centre in Wind and Marine Energy Systems, University of Strathclyde, Glasgow, UK
[5]DNV GL, Edificio Trovador, Plaza de Antonio Beltrán Martínez, 50002 Zaragoza, Spain

E-mail: elena@orseide.com

Abstract. The installation of new electrical power plants from renewable sources is key in the transition towards a low-carbon economy. An important amount of diverse raw materials is required for this development. Due to its current prominence among renewable energy sources, we assess the expected development of wind energy towards the availability of the required raw materials up to 2050. Wind power is found to be in a favourable position, over solar thermal and photovoltaic power. Among the two main wind turbine technologies, the installation of direct drive turbines with permanent magnets faces a more challenging future. Recycling is an important strategy to simultaneously reduce risks and costs.

1. Introduction

The increasing installed capacity of renewable electricity generation is currently considered as one of the most important steps to address climate change. As stated by the European Commission, all sectors must contribute to achieve a minimum reduction of greenhouse gas (GHG) emissions to 80% of 1990's levels [1]. In this context, the electricity sector presents the biggest potential to achieve these European targets and could almost totally eliminate CO_2 emissions by 2050 [2]. According to the International Energy Agency [3], installed capacity from wind energy, solar thermal or photovoltaic power could reach figures of 2400 GW, 980 GW and 4500 GW respectively. Yet this energy transition must be carefully accomplished as a considerable amount of raw materials are required to build these new installations, increasing the pressure on materials availability.

Renewable technologies demand important amounts of elements to build wind turbines, solar cells or solar concentrator surfaces. Besides, current recycling rates of some of these materials are almost negligible due to higher costs of recycling processes [4]. Although recycling presents evident benefits [5], current recycling rates are still very low [6]. As a consequence, the dependency on critical resources could be pronounced as a potential barrier to a wider implementation of emerging renewable energy technologies [7].

The criticality of materials has been extensively studied from different perspectives, such as vulnerability, economic importance, supply or ecological risks [8, 9]. As concerns mid-term renewable energy development, wind is currently playing the most important role as it currently is the more mature and competitive renewable source. As a result, several recent studies have analysed the effect of raw materials (such as rare earth elements) availability towards the expected development of wind power [10–14].

In this paper, we analysed the potential future demand scenarios for 19 metals resulting from the expected new wind power installations up to 2050. Additionally, both solar sources, solar thermal and photovoltaic, are addressed for comparison purposes. The objective is to identify the constraints for future wind development related to raw materials availability.

2. Methodology

In this work, the demand of 19 different materials is assessed up to 2050, given the expected installations of wind power (WP), solar thermal power (STP) and solar photovoltaic (PV) per year. Then, the cumulative demand of each material until 2050 can be obtained. Finally, the comparison of this total demand against the resources and reserves of each material allows any risk to be identified, related to the three studied technology and the different materials considered.

The constraints related to future development of wind power, as well as STP and PV, were identified considering their related demand of the 19 following materials: Silver (Ag), Aluminium (Al), Cadmium (Cd), Chromium (Cr), Copper (Cu), Dysprosium (Dy), Gallium (Ga), Germanium (Ge), Indium (In), Magnesium (Mg), Manganese (Mn), Molybdenum (Mo), Nickel (Ni), Neodymium (Nd), Tin (Sn), Tellurium (Te), Titanium (Ti), Vanadium (V) and Zinc (Zn).

The identification of risks for material scarcity was performed here in different steps, presented in detail subsequently. The core of the methodology relies on the combination of a bottom-up and a top-down approaches. While the bottom-up approach covers the availability in terms of reserves and resources per material, the top-down approach assesses the quantity of each material demanded by each technology.

2.1. Top-down approach

A top-down approach was followed to obtain a cumulative demand of the mentioned materials by the selected renewable energy technologies, in a projected scenario over the period from 2016 to 2050. To do so, the expected yearly installed capacity from wind power [15, 16], solar PV [17–19] and STP [20, 21] was assessed until 2050. The projection of the yearly new installed capacity for the three renewable technologies is illustrated in Figure 1. As can be seen, the repowering of old installations will increase the yearly installed power from 2033 with a special growing from 2039 caused by solar PV mainly.

Then, based on the literature [22], the material requirements per manufactured and installed megawatt (MW) of the three renewable technologies were used to translate the expected power installations into expected cumulative material demand, assuming current state-of-the-art developments and competition with the rest of sectors, in the 2016 - 2050 time period. The manufacturing requirements per material for every renewable technology are presented in Table 1, where the different requirements between wind turbines with geared or direct drive technology with permanent magnets are specified. Indeed, the need of permanent magnets presents in many direct drive wind turbine models increases significantly the requirements of Cu, Dy and Nd.

It is important to mention that not all the direct drive turbine models are built using permanent magnets. One of the largest manufacturer of direct drive wind turbines, Enercon, incorporates an annular generator with separate excitation, that leads to wind turbines being produced completely without permanent magnets. For the sake of simplicity, this has not been considered in the present analysis and only direct drive turbines with permanent magnets have been taken into account. As a result, the mentioned direct drive technology considered hereafter is the one that includes permanent magnets.

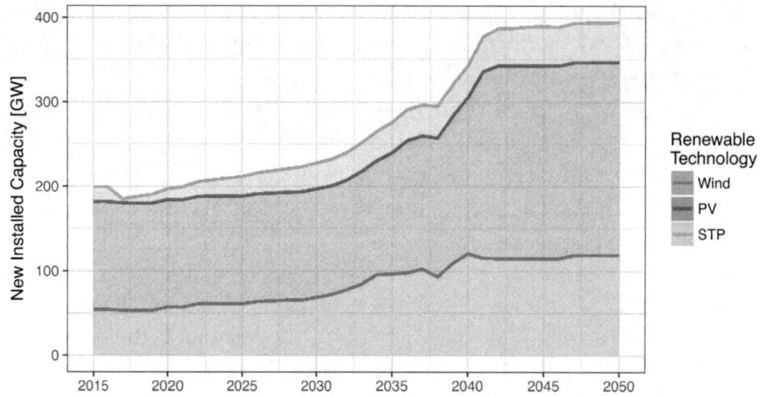

Figure 1: Projection of yearly installed capacity from wind power, solar PV and STP.

Table 1: List of materials used in renewable power installations (in kg/MW).

Material	Wind		PV	STP
	Geared Drive	Direct Drive		
Ag	-	-	113.1	14.2
Al	840.0	560.0	12511.4	9644.0
Cd	-	-	8.5	-
Cr	-	-	2.1	2800.0
Cu	1200.0	5500.0	3551.9	2480.0
Dy	4.9	14.6	-	-
Ga	-	-	0.4	-
Ge	-	-	0.7	-
In	-	-	5.8	-
Mg	-	-	45.5	2840.0
Mn	-	-	-	3480.0
Mo	-	-	9.7	142.4
Nd	60.9	182.8	-	-
Ni	111.0	111.0	0.9	1284.0
Sn	-	-	442.4	-
Te	-	-	7.3	-
Ti	-	-	-	15.0
V	-	-	-	2.0
Zn	-	-	4.3	950.0

As a result, the quantity of a raw material a demanded for all the analysed technologies, t, for a given year y can be obtained (see Equation 1). N_i is the yearly number of manufactured units per technology i, $M_{a,i}$ is the quantity of material a demanded by one functional unit (1 MW) per technology and r_a is the recycled share of material a.

$$d_{a,y} = \sum_{i=1}^{m} N_i M_{a,i}(1 - r_a) \tag{1}$$

In a projected 2050 scenario, renovation and repowering activities must be considered due to lower lifetime of the studied technologies (20-25 years). As a result, the yearly number of

IOP Conf. Series: Journal of Physics: Conf. Series **1102** (2018) 012028 doi:10.1088/1742-6596/1102/1/012028

manufactured units per technology can be defined as the sum of the new units added to the global market N_{ns} and the number of units manufactured to repower old installations N_{rn} (see Equation 2).

$$N = N_{ns} + N_{rn} \tag{2}$$

Additionally, renewable technologies will certainly compete with other sectors in terms of materials' demand. In this work, this material demand for other sectors $d_{a,os}$ has been kept constant until 2050, due to the little information available in the literature. $d_{a,os}$ is estimated based on figures from 2015 as the difference between the total material production $(P_a)_{2015}$ and the material demand from renewable technologies for the same year P_a (see Equation 3).

$$d_{a,os} = (P_a)_{2015} - (d_{a,2015}) \tag{3}$$

Finally, the total demand for a given material a in year y, $(d_{a,T})_y$, and the total cumulative demand $D_{a,T}$ up to 2050 are assessed (see Equation 4 and 5).

$$(d_{a,T})_y = (d_{a,y})_y + (d_{a,os})_{2015} \tag{4}$$

$$D_{a,T} = \sum_{y=2016}^{2050} (d_{a,T})_y \tag{5}$$

2.2. Bottom-up approach

As mentioned, the bottom-up approach covers the availability in terms of reserves and resources per material. While the resources are substrates present in the subsoil, regardless their accessibility or the financial and energy viability of their exploitation, the reserves are deposits that have been proven to be exploitable meeting both technological and financial feasibility. Due to this significant difference, it is important to identify the raw material availability in terms of both reserves and resources. In this case, the assessment on a global basis of the reserves and resources trends from 2016 to 2050 for each material was performed assuming a Hubbert-like production trend. More detailed description of this analysis can be found in [22]. The projected cumulative reserves and resources of the studied materials in 2050 are shown in Figure 2.

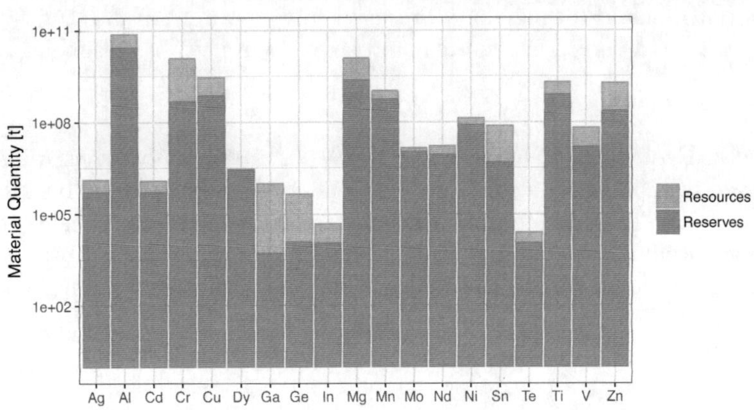

Figure 2: Cumulative reserves and resources of the studied materials in a projected 2050 scenario.

IOP Conf. Series: Journal of Physics: Conf. Series **1102** (2018) 012028 doi:10.1088/1742-6596/1102/1/012028

2.3. Risk identification

Two risk categories are defined considering material demand towards the available resources or reserves. As one can see in Table 2, category A is more restrictive than B, although resources and reserves are dynamic; they can change with extraction technology, prices or discovery of new deposits.

Table 2: Risk definitions related to material availability.

Type	Definition
A	Cumulative demand between 2016 and 2050 $\geq$ current resources ($D_{a,T} \geq RES_{2016}$)
B	Cumulative demand between 2016 and 2050 $\geq$ current reserves ($D_{a,T} \geq RSV_{2016}$)

3. Results

The baseline scenario obtained from the combination of the bottom-up and a top-down approaches is illustrated in Figure 3. This baseline Scenario 1 considers a share of 75% and 25% between geared drive and direct drive wind turbine technology [23], and a constant recycled share per material (r_a), meaning that no increase of the annual recycling rate was deemed. Figure 3B details the material demand share between the renewable technologies studied in the present work.

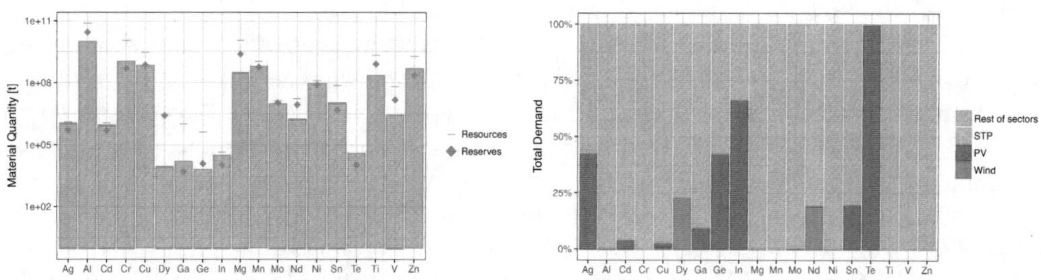

(A) (B)

Figure 3: Cumulative demand (orange bars), reserves and resources of the studied materials in a projected 2050 scenario (A) and material demand share by renewable technology (B) (Scenario 1).

As can be seen in Figure 3, the development of PV will be the most constrained by material availability, followed by STP. On the contrary, wind energy does not seem to be at high risk in the projected 2050 scenario. The unique unfavourable case corresponds to Nickel (Ni) availability. Also, while the availability of rare earth elements (REEs), Dysprosium (Dy) and Neodymium (Nd), does not seem to affect future wind development, a problem may arise related to their supply, almost exclusively from China. The identified risks, together with the most affected technology are summarised in Table 3.

3.1. Impact of annual recycling rate

The baseline Scenario 1 was built with an annual constant recycling rate. Nevertheless, it appears obvious that having increasing recycling rates would allow to overcome bottlenecks in terms of material availability. Recent research has shown the need of recycling rates of up to 5% for some materials to overcome the most severe risks [22]. To illustrate the impact of recycling, a common annual growth of 2% of every material recycling rate has been considered to build the Scenario 2. Results are presented in Figure 4. Additionally, the cumulative demand towards material availability is compared for both Scenario 1 and 2 in Figure 5.

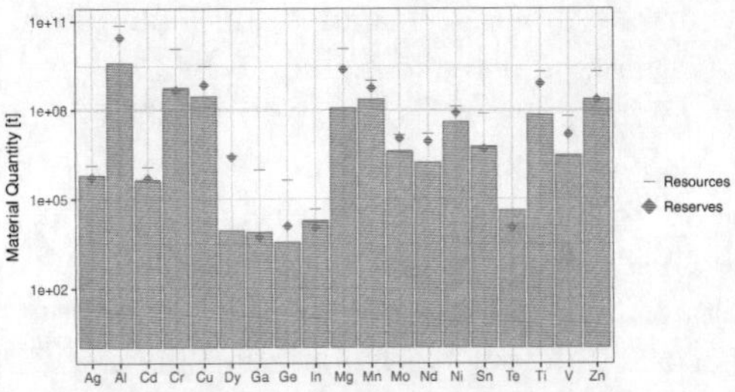

(A)

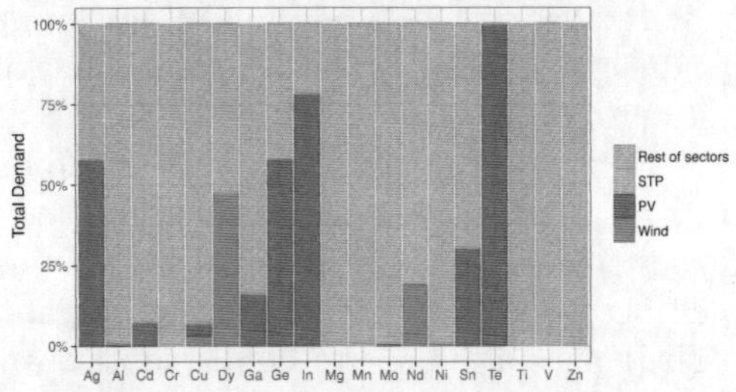

(B)

Figure 4: Cumulative demand (orange bars), reserves and resources of the studied materials in a projected 2050 scenario (A) and material demand share by renewable technology (B) assuming an annual increase of 2% of the material recycling rate (Scenario 2).

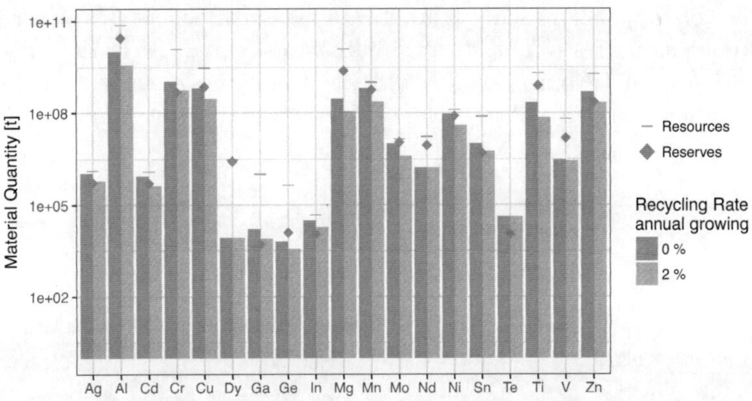

Figure 5: Comparison of the cumulative demand for the scenarios 1 and 2, against the reserves and resources of the studied materials in a projected 2050 scenario.

As illustrated in Figure 5, the cumulative demand is positively affected by increasing the annual recycling rate. Moreover, as presented in Table 3, the occurrence of risks is also significantly reduced.

Although the development of new wind power projects does not seem to be in risk in terms of material availability, recycling appears also as an important topic. Indeed, it is thought that improving the recycling rates during the decommissioning phase of a wind farm would allow to reduce the levelised cost of energy (LCOE) together with a reduction of the environmental impact. The wind industry agrees that recycling metallic components is relatively straightforward; their transportation cost remains a negative factor though [24]. As wind turbines are mainly manufactured using metallic elements, this can be translated into a recyclability rate of about 83%, including mainly big components from the tower, the gearbox, the main shaft, the generator, castings, bearings and parts from the nacelle and the hub. The remaining 17%, non-recyclable, encompasses elements from electronics, oil and coolants, polymers but mainly derives from the blades (nearly of this 13%) [25]. Blades are currently

IOP Conf. Series: Journal of Physics: Conf. Series **1102** (2018) 012028 doi:10.1088/1742-6596/1102/1/012028

the big challenge from a recycling point of view, a part from being a logistical transportation nightmare. Blades are typically made from a composite of glass fibres and epoxy or other thermoset resin, where the cross-linked polymers are difficult to melt down and be recycled [26]. Alongside the decommissioning phase, blades are currently being shredded and then often end up in landfills, while other alternatives such as incineration and recycling could be explored [27]. The first option is becoming less popular due to the willingness of countries to reduce the land fill mass. For instance, Germany banned the disposals of glass fibre reinforced plastics (GRP) in June 2005, due to their high rate (30%) of organics content such as resin and wood. The second most common route is incineration, where the heat is reused to generate electricity in combined heat plants (CHP), but the inorganic loads lead to the emission of hazardous flue gazes. The blades also need to be dismantled and crushed before transportation placing further strain on the environment in terms of energy used and polluting emissions [27]. As a result, recycling should be definitely encouraged, either way it consists of material recycling or product recycling, i.e repowering.

3.2. Impact of wind turbine technology share

The baseline Scenario 1 considered a share of 75% and 25% of geared and direct drive (with permanent magnets) wind turbine technologies respectively. However, direct drive wind turbine technology seems to be privileged from some wind turbine manufacturers, especially if they cover an important share of the offshore wind market. To assess the effect of this technology share, two additional virtual scenarios (3 and 4) have been investigated; Scenario 3 considers a 100% share of geared drive wind turbines, while Scenario 4 assumes no new geared drive turbines installed, so a 100% of direct drive wind turbines. Results from both scenarios are illustrated in Figure 6.

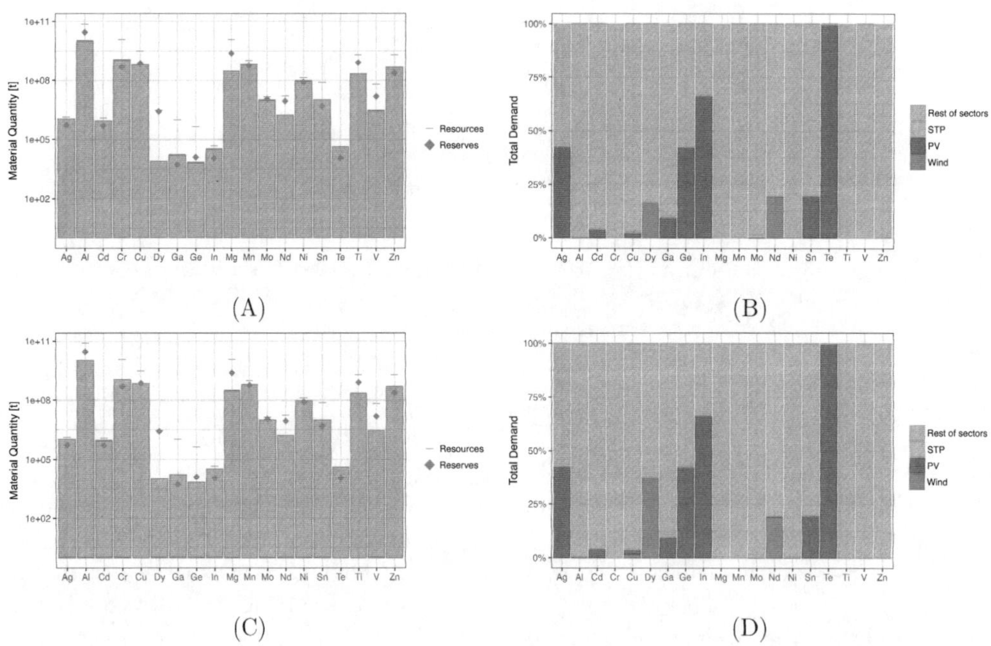

Figure 6: Cumulative material demand given a 100% share of geared drive wind technology (A) - Scenario 3 - and given a 100% share of direct drive wind technology (B) - Scenario 4 -; material demand share by renewable technology for Scenario 3 (B) and Scenario 4 (D).

As can be seen in Figure 6, no significant changes are appreciated in terms of the cumulative demand per material, toward the availability. However, there is a significant increase of the required cumulative quantity of Dy in Scenario 4. As mentioned, this is mainly due to the presence of permanent magnet generators in most direct drive turbines. A more thorough discussion is presented in Section 4.

3.3. Summary of identified risks in each studied scenario

The different risks identified in the four presented scenarios are summarised in Table 3.

Table 3: Type of risk per material in the four different projected scenarios.

	Scenario 1		Scenario 2		Scenario 3		Scenario 4	
	Risk	Most affected technology	Risk	Most affected technology	Risk	Most affected technology	Risk	Most affected technology
Ag	B	RoS	B	PV	B	RoS	B	RoS
Al	-	-	-	-	-	-	-	-
Cd	B	RoS	-	-	B	RoS	B	RoS
Cr	B	RoS	B	RoS	B	RoS	B	RoS
Cu	-	-	-	-	-	-	-	-
Dy	-	-	-	-	-	-	-	-
Ga	B	RoS	B	RoS	B	RoS	B	RoS
Ge	-	-	-	-	-	-	-	-
In	B	PV	B	PV	B	PV	B	PV
Mg	-	-	-	-	-	-	-	-
Mn	B	RoS	-	-	B	RoS	B	RoS
Mo	-	-	-	-	-	-	-	-
Nd	-	-	-	-	-	-	-	-
Ni	B	RoS	-	-	B	RoS	B	RoS
Sn	B	RoS	B	RoS	B	RoS	B	RoS
Te	A	PV	A	PV	A	PV	A	PV
Ti	-	-	-	-	-	-	-	-
V	-	-	-	-	-	-	-	-
Zn	B	RoS	-	-	B	RoS	B	RoS

4. Discussion

While the future development of wind energy should not be constrained by the demanded raw materials, geopolitical problems might tip the balance towards geared drive technologies. Many recent studies have compared the two main wind turbine technologies, i.e. geared and direct drive, in terms of manufacturing costs, reliability and maintainability. Nevertheless, the materials demanded by each configuration have not been addressed. Most direct drive turbines are equipped with permanent magnet generators, relying heavily on REEs (Dy and Nd) [28], provided almost exclusively by China. Additionally, the future demand of permanent magnets will significantly increase from the electrical vehicle industry, creating a direct competition with direct drive turbines manufacturers. Reusing [29] and recycling [30] can be therefore seen as promising strategies. End-of-life recycling could help in the current global supply crisis surrounding REEs so that dependence on China can be decreased.

The risk for future development of wind power projects could be further lowered by promoting alternative strategies at the end-of-life of current installations. Two main options are suggested here. First, lifetime extension should be privileged over repowering, when possible [31]; this would imply evident benefits regarding the materials life-cycle assessment. Secondly, the recycling rates should be increased. In a longer term, recycling could provide a significant secondary supply to reduce the future demand of raw materials [12]. Apart from the environmental benefits, recycling could also contribute to reduce the levelised cost of energy (LCOE). Indeed, this is almost not considered during the decommissioning phase of a wind farm [32]. Higher recycling rates could halve the decommissioning costs, generally underestimated [33], and hence lower the capital expenditures and the LCOE. Also, reusing should be considered as part of the operation and maintenance (O&M) phase. Reusing subcomponents from critical assemblies, such as the gearbox [34], could also reduce the O&M costs.

5. Conclusions

The development of future wind power projects, as other renewable technologies, depends on the availability of certain materials required for the manufacturing and installation of wind turbines.

In this study, the available resources and reserves for 19 different materials have been compared to the expected development of wind energy in a 2050 projected scenario. Two other renewable sources of electricity generation, solar thermal power and photovoltaic power, have also been assessed for comparison purposes. Wind power seems to be the least affected technology given the availability of materials. Among the two main wind turbine technologies, some risks have been identified related to direct drive turbines though. The need of REEs for the production of permanent magnets creates a competition with other industries, as the electrical vehicle, and involves an important risk of shortage due to its geological availability.

Several alternatives, such as lifetime extension and recycling, have been suggested to further lower the risks for future wind development. Nonetheless, these are also valid for the two other explored renewable sources, that are found to be at higher risks.

Finally, if the current materials demands and recycling quotes are maintained, the transition to a low-carbon economy could be threatened by the availability of certain elements. This issue should be deeply analysed to define appropriate strategies that avoid the mentioned constraints.

6. Learning objectives

- A shift towards a low-carbon economy hugely depends on the availability of raw materials.
- The development of wind energy might be compromised by the simultaneous requirement of certain materials from other renewable technologies or other industries.
- While wind energy development seems to be at a lower risk than other renewable sources, direct drive technology using permanent magnets is in a less favourable position.
- Encouraging reusing and recycling of raw materials shows important and diverse benefits.

Acknowledgments

This work has been partially supported by the Spanish Ministry of Economy, Industry and Competitiveness (ENE2017-85224-R).

References

[1] European Climate Foundation 2011 *Power Perspectives 2030: On the road to a Decarbonised Power Sector*
[2] European Commission 2015 2050 low-carbon economy [Online; accessed 30-March-2018] URL https://ec.europa.eu/clima/policies/strategies/2050_en
[3] International Energy Agency 2010 *Energy Technology Perspectives: Scenarios & Strategies To 2050*
[4] Redlinger M, Eggert R and Woodhouse M 2015 *Solar Energy Materials and Solar Cells* **138** 58–71
[5] RDíaz Martín, Trujillo F, JFMorales García, CMayo del Río, EBatuecas Fernández and Adib Guardiola Mouhaffel 2016 *International Journal of Applied Engineering Research* **11** 2990–2995
[6] Wang X, Gaustad G and Babbitt C W 2016 *Waste Management* **51** 204–213

[7] Bradshaw A M, Reuter B and Hamacher T 2013 *Green* **3** 93–111
[8] Achzet B and Helbig C 2013 *Resources Policy* **38** 435–447
[9] Alonso E, Gregory J, Field F and Kirchain R 2007 *Environmental Science & Technology* **41** 6649–6656
[10] Kossakowska K 2017 *17th International Multidisciplinary Scientific GeoConference SGEM 2017* **17** 19–26
[11] Grandell L, Lehtilä A, Kivinen M, Koljonen T, Kihlman S and Lauri L S 2016 *Renewable Energy* **95** 53–62
[12] Habib K and Wenzel H 2014 *Journal of Cleaner Production* **84** 348–359
[13] Harmsen J, Roes A and Patel M 2013 *Energy* **50** 62–73
[14] Alonso E, Sherman A M, Wallington T J, Everson M P, Field F R, Roth R and Kirchain R E 2012 *Environmental Science & Technology* **46** 3406–3414
[15] WindEurope 2016 WindEurope Annual Offshore Statistics 2016 [Online; accessed 02-July-2018] URL https://windeurope.org/wp-content/uploads/files/about-wind/statistics/WindEurope-Annual-Offshore-Statistics-2016.pdf
[16] International Energy Agency 2013 Technology roadmap. Wind energy. [Online; accessed 02-July-2018] URL https://www.iea.org/publications/freepublications/publication/Wind_2013_Roadmap.pdf
[17] Greenpeace 2016 Solar thermal electricity. Global outlook 2016 [Online; accessed 02-July-2018] URL https://www.greenpeace.org/archive-international/Global/international/publications/climate/2016/Solar-Thermal-Electricity-Global-Outlook-2016.pdf
[18] Raccurt O, Disdier A, Bourdon D, Donnola S, Stollo A and Gioconia A 2015 *Energy Procedia* **69** 1551–1557 ISSN 18766102 URL http://linkinghub.elsevier.com/retrieve/pii/S1876610215004130
[19] International Energy Agency 2014 Technology roadmap. Solar thermal electricity. [Online; accessed 02-July-2018] URL https://www.iea.org/publications/freepublications/publication/technologyroadmapsolarthermalelectricity_2014edition.pdf
[20] Parrado C, Girard A, Simon F and Fuentealba E 2016 *Energy* **94** 422–430 ISSN 03605442 URL http://linkinghub.elsevier.com/retrieve/pii/S0360544215015418
[21] Bayod-Rújula A A, Ortego-Bielsa A and Martínez-Gracia A 2011 *Energy* **36** 1996–2010 ISSN 03605442 URL http://linkinghub.elsevier.com/retrieve/pii/S0360544210002161
[22] Valero A, Valero A, Calvo G and Ortego A 2018 *Renewable and Sustainable Energy Reviews* **93** 178–200 ISSN 13640321 URL https://linkinghub.elsevier.com/retrieve/pii/S1364032118303861
[23] Lacal-Arántegui R 2015 *Journal of Cleaner Production* **87** 275–283 ISSN 09596526 URL http://linkinghub.elsevier.com/retrieve/pii/S0959652614009779
[24] ECN 2016 Lifecycle and decommissioning offshore wind [Online; accessed 02-July-2018] URL https://www.ecn.nl/publications/PdfFetch.aspx?nr=ECN-E--16-009
[25] Ceña A and Vazquez I 2018 Webinar: Life extension and repowering of wind farm's main elements and the future trend in the sector (Aemer, InnoEnergy, Rooter)
[26] Bomgardner M M and Scott A 2018 Reclycing Renewables [Online; accessed 02-July-2018] URL https://cen.acs.org/energy/renewables/Recycling-renewables/96/i15
[27] Larsen K 2009 Recycling wind [Online; accessed 02-July-2018] URL https://www.materialstoday.com/composite-applications/features/recycling-wind/
[28] Hogberg S, Pedersen T S, Bendixen F B, Mijatovic N, Jensen B B and Holboll J 2016 *2016 XXII International Conference on Electrical Machines (ICEM)* (IEEE) pp 1625–1629
[29] Hogberg S, Holboll J, Mijatovic N, Jensen B B and Bendixen F B 2017 *IEEE Transactions on Magnetics* **53** 1–9
[30] Rademaker J H, Kleijn R and Yang Y 2013 *Environmental Science & Technology* **47** 10129–10136
[31] Ziegler L, Gonzalez E, Rubert T, Smolka U and Melero J J 2018 *Renewable and Sustainable Energy Reviews* **82** 1261–1271
[32] Topham E and McMillan D 2017 *Renewable Energy* **102** 470–480
[33] Topham E, Mcmillan D, Bradley S and Hart E [Submitted for publication]
[34] Jiang L, Xiang D, Tan Y, Nie Y, Cao H, Wei Y, Zeng D, Shen Y and Shen G 2018 *Journal of Cleaner Production* **180** 846–857

IOP Conf. Series: Journal of Physics: Conf. Series **1102** (2018) 012029 doi:10.1088/1742-6596/1102/1/012029

Lessons from German On-shore wind farm planning

Jinjin Guan
Geography Institute, Ruhr University Bochum
University Street 104, 227, Bochum, 44801, Germany
Email: jinjin.guan@rub.de

Abstract. China continuously maintains the first place of global wind energy with the national installed capacity of 188GW at the end of 2017, accounting for over 34.8% of the total world capacity. With the continuous increase both in number and height of wind turbines, the spatial planning has the task to provide suitable places for wind farms. Thus the rising demand for good locations is increasingly causing conflicts. The "13th Five-Year-Plan" of China aims to optimize the spatial layout of wind energy, resolving wind curtailment problems in North China and encouraging the construction of distributed, low-speed wind farms in South China for consumer near energy production. The ecological and social impacts caused by wind farms become more prominent in the densely populated areas of South China, which brings challenges to spatial planning and land use coordination. As a pioneer of wind industry, Germany is faced with similar problems and very early paid attention on spatial planning issues like wind farm site selection, environmental protection and land use planning. This paper compares the spatial planning systems of two countries and draws experiences from Germany to optimize the Chinese wind farm planning procedures. Planning principles and specific planning procedures for the integration of wind farm and spatial planning both in regional and local levels will be proposed according to the developing status of China. An on-shore wind farm planning framework will be set up implemented at each planning level, and coordinated with other land use and space functions to ensure the synchronous development of renewable energy and surrounding ecological and human environment.

Keywords. On-shore wind farm planning, German spatial planning system, Chinese spatial planning system, wind farm planning procedures

1. Introduction

China maintains the global first place of wind energy continuously with the national installed capacity of 188GW at the end of 2017, accounting over 34.8% of the world [1]. However, new challenges turn up with rapid expansion: serious "wind curtailment" in North China has seriously threatened the wind energy industry by the average "wind curtailment rate" over 17% in 2016 [2]. Chinese government has to slow down the pace and adjust the development strategy. Meanwhile, China has the arduous task of energy transformation by developing renewable energy. The "13th Five-Year Plan" stipulates that at least 210 GW wind energy shall be installed to supply 6% of electricity consumption in 2020 [3]. The newly increased jobs will reach 300,000 from 2016 to 2020, reaching totally 800,000 jobs in wind energy industry of China.

IOP Conf. Series: Journal of Physics: Conf. Series **1102** (2018) 012029 doi:10.1088/1742-6596/1102/1/012029

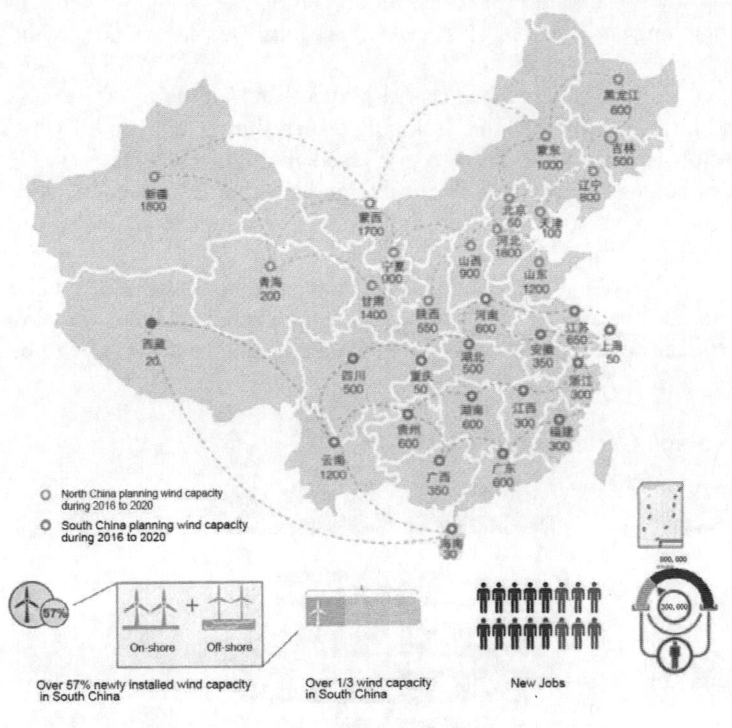

Figure1. Chinese "13th Five-Year-Plan" of wind energy (13th Five-Year-Plan, 2016)

In order to achieve the aim, the future focus shifts from large-scale wind farms in North China to low-speed, distributed wind farms in South China and for near consumption, which brings new issues in spatial planning field. Different from scarcely populated areas of North China, land use conflicts and various environmental impacts caused by wind farms will be more serious in densely populated areas of South China. Researches about specific wind farm planning procedures at the regional and local level as well as their combination with comprehensive planning are of urgent necessity in China.

German wind industry with its advanced wind technologies as well as standard planning regulations and procedures is taken as a reference for China. This paper analyzes the contradiction between existing problems and wind energy targets of China, and difficulties to be encountered in the developing process. Then it elaborates from the aspects of spatial planning to compare the spatial planning systems between China and Germany. The planning procedures and experiences of wind farm in Germany are analyzed to put forward feasible suggestions for optimizing wind farm planning procedures in China.

2. Wind energy development status and difficulties in China

Since the promulgation of "Renewable Energy Law" in 2006, China has accelerated the development speed of renewable energy. As an efficient and clean substitution of fossil fuels, wind energy is stipulated as a priority project and enjoys subsidy policies. Since 2010, China has surpassed the United States and became the biggest wind energy market globally [4]. By the end of 2017, the cumulative wind capacity has reached 188 GW. In terms of quantity, China is undoubtedly a strong country in the wind industry. However, due to the huge domestic electricity demand, wind energy accounts for only 4% of the national electricity consumption. There is still much space for improvement. Figure 2 shows the rapid development period of China from 2010 to 2015 with the peak growth rate over 100%. Since 2016, wind energy development has entered an adjustment phase with a reduced growth rate. Newly emerging issues like wind curtailment in North China

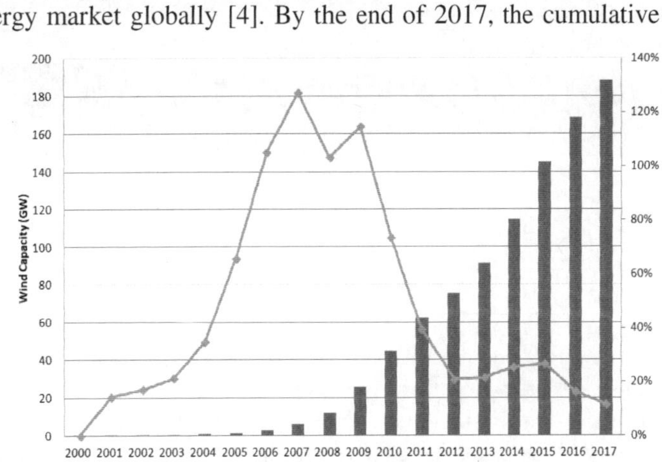

Figure 2. Chinese wind energy installed capacity and growth rate (GWEC, 2017)

need to be solved urgently. Except the technology reasons, improper spatial planning without sufficient analysis and standard planning procedures has caused serious problems like "wind curtailment".

The Chinese "13th Five-Year-Plan" of wind energy has issued policies and measures to develop low-speed, distributed on-shore wind farms in South China. It stipulates that during 2016 to 2020 the newly installed wind capacity in South China should be over 57% and the total capacity over 1/3 (Figure 1). With the new development strategy, wind curtailment rate and grid burden can be reduced by nearby consumption, but more emphasis should be put on the wind farm planning procedures to control the negative influences on local residents' life, visual landscape quality and environment [5]. These negative influences are mainly analyzed and avoided by wind farm planning procedures, while this part is just the weakness of the Chinese wind industry. For establishing detailed planning procedures at regional and local levels, the spatial planning systems of China and Germany have to be compared first.

3. Planning systems comparison between Germany and China

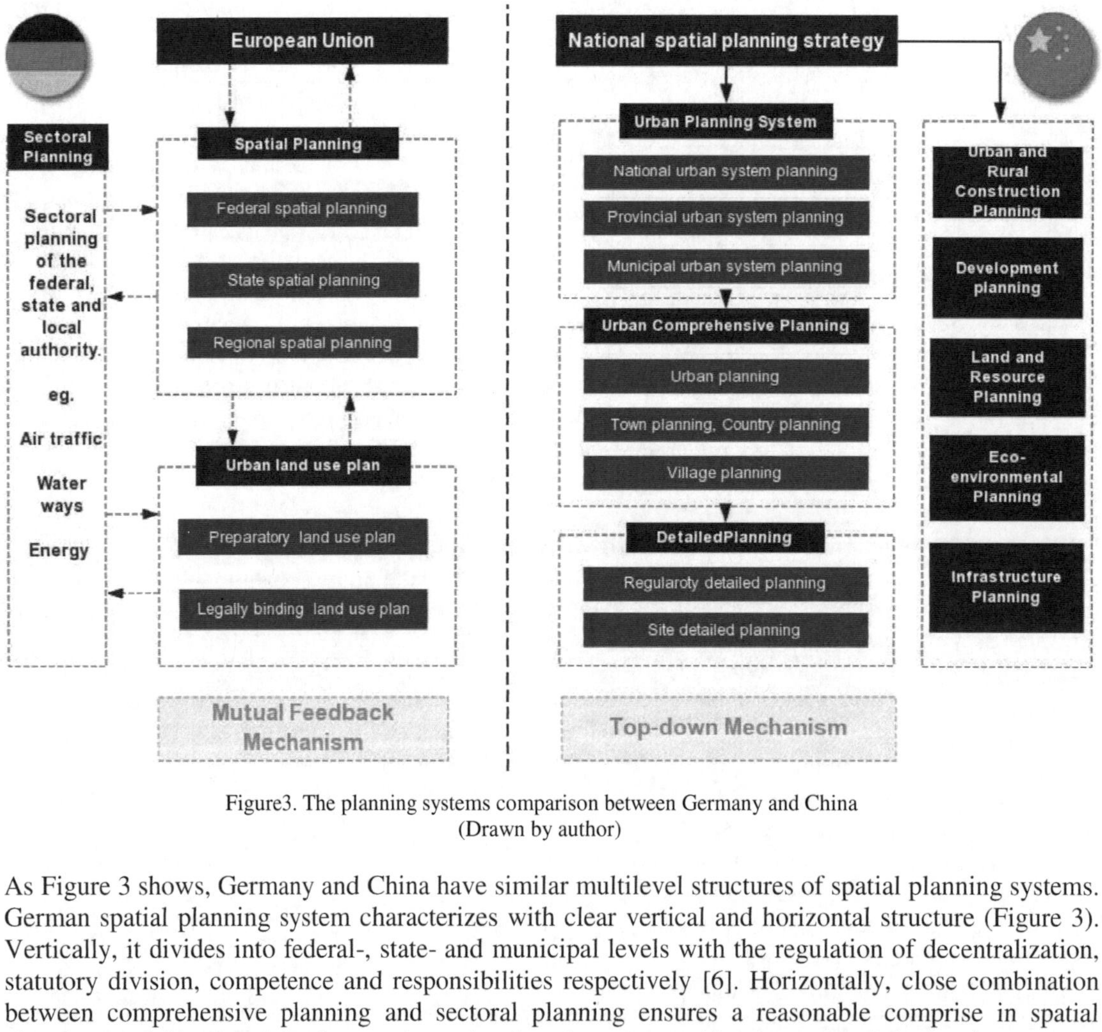

Figure3. The planning systems comparison between Germany and China
(Drawn by author)

As Figure 3 shows, Germany and China have similar multilevel structures of spatial planning systems. German spatial planning system characterizes with clear vertical and horizontal structure (Figure 3). Vertically, it divides into federal-, state- and municipal levels with the regulation of decentralization, statutory division, competence and responsibilities respectively [6]. Horizontally, close combination between comprehensive planning and sectoral planning ensures a reasonable comprise in spatial planning and functional land use. As a developed country with mature spatial planning system, Germany has its own complete planning legislation, authorities and implementation dealing with various kinds of projects. German planning system emphasizes more on statutory division and equal rights in each state, which reflects by a mutual feedback mechanism in spatial planning system. The detailed regulations and implementations dealing with wind farm planning are gradually completed.

More elaborate site selection and planning procedures can be carried out on the basis of a detailed legal system and shared data.

In contrast, China is still under urbanization process with a constantly adjusted and revised spatial planning system for adapting to emerging issues. Because of the huge territory and more government departments, the complex spatial planning system exacerbates the difficulty of integrating wind farm planning into spatial planning system. The Chinese spatial planning system has a clear multi-hierarchy structure with national-, provincial-, regional-, municipal-, town-, village levels. While the horizontal comprehensive planning consists of "Urban and Rural Construction Planning", "Development planning", "Land and Resource Planning", "Eco-environmental Planning" and "Infrastructure Planning", which are self-contained under different departments [7]. Compared with German, Chinese planning system places more emphasis on centralized arrangements and lacks local autonomy, which is reflected in the top-down planning mechanism. The coordination between various planning departments is difficult because of different regulations, management departments and implementation standards. The complexity of the planning system has brought about management conflicts among various departments, information inequality and inconsistent implementation standards, resulting in low planning efficiency and low adaptability to new planning projects.

As a newly emerging planning project, wind farm planning should set up communication and coordination mechanisms with comprehensive planning at each level. Germany can offer experiences and inspiration in dealing with wind farm planning procedures and related spatial planning issues.

4. Lessons from German on-shore wind farm planning

In Germany, wind farm planning belongs to the sectoral planning, which should have accurate correspondence with the comprehensive planning at each level (Figure 4). During the popularization process of wind energy, the wind farm planning will meet more conflicts with other land use and special protection areas. The normalization of wind farm planning procedures and integration into comprehensive planning is of urgent necessary. Several federal states like North Rhine Westphalia and Bavaria have already launched protocol platforms for coordinating wind farm planning and other sectoral planning, especially near the ecologically sensitive areas [8]. Figure 4 shows how the renewable energy planning fits into spatial planning system at each planning level. When meeting conflicts, clear and operational guidance is provided according to related regulations, which is supported by a complete legislation basement.

Figure 5 chronologically lists the five phases of Chinese on-shore wind farm planning, which are effective and operational enough to ensure the totally 188GW wind capacity installation.

	Landscape planning system		Comprehensive planning system		Renewable energy planning system
National-level (Bundesebene)	Landscape programme (Landschaftsprogrammen)	⟷	Spatial planning (Raumordnung)	⟷	Network construction acceleration law (Netzbaubeschleunigungsgesetz)
State-level (Landesebene)	Landscape programme (Landschaftsprogrammen)	⟷	State development programme (Landesentwicklungsprogramme)	⟷	Renewable Energy Sites, Priority Areas, Restricted Areas, Suitability Areas (Standorte Erneuerbarer Energie, Vorrang-, Vorbehaltsgebiete,Eignungsgebiete)
Regional-level (Regionale Ebene)	Landscape framework (Landschaftsrahmenplan)	⟷	Regional plannig (Raumordnungspläne)	⟷	Conflict analysis (Konfliktanalyse)
Local-level (Kommunale Ebene)	Landscape planning (Landschaftspläne)	⟷	Land-use plannig (Flächennutzungsplan)	⟷	Site research (Standorterforschung)
	Greenland planning (Gruendordnungspläne)	⟷	Building design (Bebauungsplan)	⟷	Wind turbines layout (Windkraft Anlagen Layout)

Figure4. The planning system of landscape, renewable energy in Germany [6]

4.1 Macro Site Selection

The first phases is macro site selection by preliminary detecting the wind energy resource, available land, construction condition of the selected sites after getting agreement from local government, and drafting a report for further planning procedures. Before high accuracy wind energy resource conducted, the wind speed and wind direction can be estimated by topography and vegetation status as prejudge. The respective procedures are taken charge by different parties for fairness of project review. In Germany, the implementation of site selection is more refined with a nation-wide wind resource census (Wind Atlas Analysis and Application Program of Germany) and detailed available space classification. Germany has assigned 2% of the country's land area as "concentration zones" for wind farm projects, which are priority areas for wind farm construction with legal effect after approval of the comprehensive planning. Other areas are assigned as suitable areas or restricted areas according to adjustable evaluation criteria for future development. This phase corresponds to federal- and state-level energy strategies, and it has a basic framework with adjustable flexibility.

4.2 Wind Energy Resource Measurement and Assessment

The second phase deals with wind resource measurement and assessment. In China, a wind resource test over 1 year and over 90% effective data is mandatory before wind farm project planning. A series of field work, test tower layout design, installation and data collection are the main task in this phase. Since the growing height of wind hubs, the latest wind resource measurements should be carried out at higher altitudes (e.g. 80 to 120 meters, at least 2/3 of the hub height). The latest equipments for wind measurements are IEC 61400-12 compliant calibrated wind transmitters, wind measuring mast and measuring computer. Wind resource simulation models are developed for high accuracy and less test towers. In Germany, European-wide wind measurement has already been carried out completed and shared by Wind Atlas Analysis and Application Program (WAsP) with high accuracy and long-term corrected statistics [9].

4.3 Wind Farm Project Planning

The third phase of Chinese wind farm planning is the wind farm project planning, which differs from German wind farm planning. Chinese wind farm planning is more engineering-orientation, concerning grid connection, wind capacity distribution, construction safety analysis as well as transportation condition. The environmental impact assessment and staging development are also included in wind farm project planning. But the coordination with other land use and reserve regions is mentioned only in the planning principle without detailed guidance, which is mainly because most already installed wind farms of China are located in wilderness without conflicts with other functions.

While Germany pays more attention to overall spatial planning coordination. Privileged projects like wind farm must be justified by a general plan. The municipal land use plan aims to balance the various aspects to find a reasonable compromise. Wind farm planning will be integrated as part of sectoral planning and should meet the demands of regional comprehensive planning. With the local announcement and Land Use Plan, the legal effect of Federal Building Code occurs, which means that wind farms are only in the designated concentration zones approved.

4.4 Feasibility Research Report

The Feasibility Research Report is investigated and drafted by the third party, usually the consulting company commissioned by the project company. This phase is a preparation for approval by analyzing project feasibility. Three most important indicators are "hours of wind farm operation yearly", "cost of wind farm per KW" and "return on capital". Environmental Impact Assessment is compulsory in this phase for evaluating potential impacts and finding variable solutions. The negative impacts like noise, shadow flicker, impacts on wild life should meet the basic legal requirements. But unified, detailed standards for preventing environmental impacts have not been drafted yet in China. Most current standards are referenced and studied from case study.

4.5 Approval and Construction

Difficulties in approval are usually in the vicinity of nature reserves and inhabited buildings or towns. After approved, a project construction permit will be issued to the project company. In the last phase, the implementation of the plan by construction management determines the project quality and actual impacts on the surrounding environment. It is seen as a part of planning for feedback and adjustment.

These five phases of Chinese wind farm planning are under the responsibility of corresponding planning companies and authorities, with clear tasks and outcomes. The division of wind energy planning procedures also facilitates project evaluation, implementation and financial investment. In terms of comprehensive comparison, the phases of wind farm planning in Germany and China are similar. Germany's strength lies in a more coordinated planning system, shared wind resource measurement data and extensive public participation. Chins is still in the initial stage of wind farm planning development, more attention is paid to technical indicators, production and financial issues. There are insufficient researches on the environmental and planning system issues. With the new strategy of constructing more distributed, low-speed wind farms in South China, the wind farm planning will face more coordination problems in spatial planning systems.

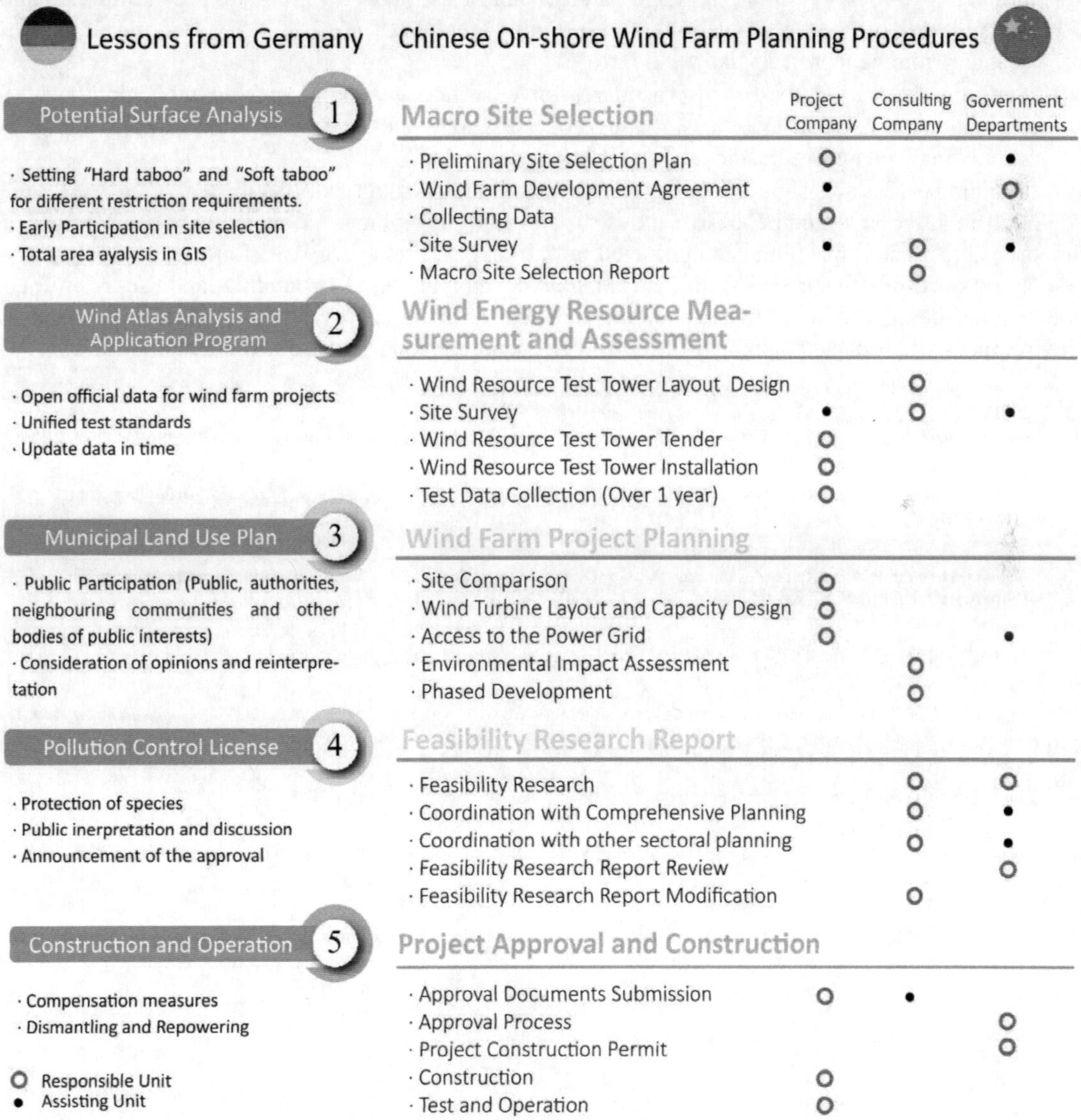

Fig.5 Lessons from German on-shore wind farm planning (Drawn by author)

IOP Conf. Series: Journal of Physics: Conf. Series **1102** (2018) 012029 doi:10.1088/1742-6596/1102/1/012029

5 Conclusion

Due to the new development strategy, China will face more restrictions and challenges in on-shore wind farm planning. In this paper, a problem-oriented comparative study provides some suggestions for solving conflicts in spatial planning, protecting natural resources and environment, improving the efficiency of administrative approval, and optimizing public participation and supervision mechanism.

1) Statistical standards of basic data should be unifies. An open information communication platform in pre-planning phase should be set up to avoid duplication measurements and waste of social resources. A nationwide wind measurement standard and wind atlas can greatly improve the efficiency of macro wind farm site selection.
2) An authoritative, legally effective available land analysis is necessary to divided areas into different application levels like priority, suitable and restricted areas. Refine filter conditions and multi-level outcomes replaced single "wind energy area" is conducive to more rational and orderly development.
3) Coordination mechanism is necessary in land use plan to avoid the conflicts with e.g. natural protection areas, cultural heritage areas, landscape conservation and wildlife habitats. Amendment of related regulations and planning guidelines are necessary for preventing new impacts from new projects like wind farm.
4) Setting universal standards and procedures for wind farm planning, such as publishing laws and regulations at national level, designating concentration zones and buffer zones at regional level, stipulating technical indicators at local level.
5) Planning procedures should be adapted to local conditions and flexibly adjusted.

Wind industry is a comprehensive industry with multidisciplinary cooperation. In this paper, the importance of spatial planning is emphasized in optimizing wind energy distribution, improving wind energy production efficiency as well as sustainable developing with surrounding land use. A profound idea should be kept in mind that we should balance the wind energy profits and the costs paid for environment. Reasonable planning procedures can ensure the long-term development of wind energy.

6. Reference

[1] Global Wind Energy Council .2017. *The Evaluation of Wind Power*. (http://gwec.net/global-figures/interactive-map/) [29.03.2018]

[2] National Bureau of Statistics of China. 2016. *China Statistical Yearbook*. (http://www.stats.gov.cn/tjsj/ndsj/2016/indexeh.htm). [15.03.2018]

[3] National Development and Reform Commission. 2016. *13th Five-Year Wind Energy Plan*. Wind power "13th Five-Year Plan". (http://www.ndrc.gov.cn/fzgggz/fzgh/ghwb/gjjgh/201708/t20170809_857302.html). [18.03.2018]. 7-9.

[4] Global Wind Energy Council. 2010. *Global Wind Report Annual Market Update*. (www.gwec.net).

[5] Manyoky M. et.al. 2016. Evaluating a visual-acoustic simulation for wind park assessment. *Landscape and Urban Planning*. **153**. 180-197

[6] Gerd Turowski. 2002. *Spatial Planning in Germany*. Structure and Concepts. Hannover.11-26

[7] Wang. XD. & Liu. WD. 2012. China's Spatial Planning System: Status Quo, Problems and Reconstruction. Economic Geography. 32 (5): 7-15.

[8] Schöbel. S. 2012. *Windenergie und Landschaftsästhetik*. Zur landschaftsgerechten Anordnung von Windfarmen. München. 128-142.

[9] Heier. S., .2016. *Nutzung der Windenergie*. Bonn. 107-110.

IOP Conf. Series: Journal of Physics: Conf. Series **1102** (2018) 012030 doi:10.1088/1742-6596/1102/1/012030

Human-free offshore lifting solutions

M Leimeister[1,3,4], T Balaam[2], P Causon[1], D Cevasco[1,3], M Richmond[1,3], A Kolios[3] and F Brennan[3]

[1] Offshore Energy Engineering Centre, Cranfield University, Bedfordshire MK43 0AL, United Kingdom

[2] Department of Engineering Science, University of Oxford, Parks Road, Oxford OX1 3PJ, United Kingdom

[3] Naval Architecture, Ocean and Marine Engineering, University of Strathclyde, Glasgow G1 1XQ, United Kingdom

[4] Fraunhofer Institute for Wind Energy Systems IWES, Am Luneort 100, 27572 Bremerhaven, Germany

E-mail: mareike.leimeister@iwes.fraunhofer.de

Abstract. With single elements weighing up to hundreds of tonnes and lifted to heights of 100 meters, offshore wind turbines can pose risks to personnel, assets, and the environment during installation and maintenance interventions. To increase safety during offshore lifts, this study focuses on solutions for human-free lifting operations. Ideas in the categories of logistics, connections, as well as guidance and control, were discussed and ranked by means of a multi-criteria decision analysis. Based upon 38 survey responses weighting 21 predefined decision criteria, the most promising concepts were selected. Logistically, pre-assembled systems would reduce the number of lifts and thus reduce the risk. A MATLAB-based code has been developed to optimise installation time, lifted weight, and number of lifts. Automated bolting and seafastening solutions have high potential to increase safety during the transport of the wind turbine elements and, additionally, speed up the process. Finally, the wind turbine should be lifted on top of the support structure without having personnel being under the load. A multi-directional mechanical guiding element has been designed and tested successfully in combination with visual guidance by cameras in a small-scale experiment.

1. Introduction

Lifting operations in the offshore wind energy industry involve heavy loads in the order of hundreds of tonnes, as well as working heights of around 100 meters, all while the lifting operation is being subjected to the wind and wave conditions at sea. This makes offshore lifting operations and offshore wind turbine installations hazardous. Guidelines and standards for health and safety in lifting operations in general [1–4], as well as specific to the offshore environment [5–7], exist. Despite this, having people directly under the load for guiding and securing is still common practice in offshore wind turbine installations. G+ (formerly G9) [8] provided, in addition to the publicly availably annual incident reports [9–11], the full data for health and safety incidents occurring during offshore lifts from 2014 to the third quarter of 2016. From these reports the total number of incidents and dropped objects per hour of work is determined and shown in figure 1.

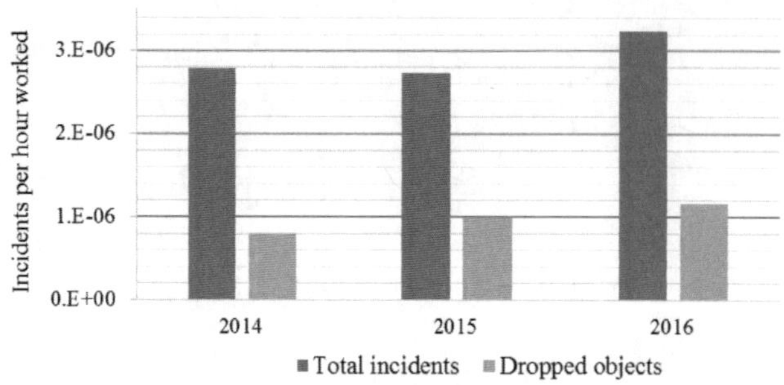

Figure 1. Total incidents and dropped objects per number of hours worked.

Whilst the number of incidents per number of hours worked decreased slightly from 2014 to 2015, it rose again in 2016. Over the same period, normalised incidents due to dropped objects increased year on year. Although small, these increases are important, especially as the installed capacity of offshore wind energy, and thus also the number of lifting operations performed each year are expected to grow significantly, based on the estimations by [12]. A recent study [13], matching safety indicators against the same G+ incident data, draws similar conclusions and the authors advise having safety indicators for dropped objects, particularly during lifting operations, as this is a crucial area for offshore health and safety incidents. The more details in the incident reports [9–11] on the statistics for the reasons, areas, and consequences for incidents substantiate as well the importance of safer and improved handling in offshore lifting operations.

Thus, the motivation for this study was to conduct research into methods and technologies that could reduce the need for personnel near lifting operations, and to assess their feasibility. Different concepts for human-free offshore lifting operations in the categories of assembly and logistics, connections and seafastening, as well as guidance and control, have been collated and investigated, mainly based on broad literature reviews and industry reports on current practices, but also based on patents for new designs and innovative solutions proposed by the authors. The broad range of concepts was assessed with respect to 21 predefined criteria and ranked by means of a multi-criteria decision analysis of 38 survey responses. More details on this analysis can be found in [14]. The concepts most worth developing were investigated in more detail and are presented in this work.

In the following, the most promising solutions are presented for the three categories: assembly and logistics (section 2), connections and seafastening (section 3), and guidance and control (section 4). A short summary of existing concepts precedes each section. At the end of this paper (section 5) the results are concluded.

2. Concepts for logistics and assembly solutions

Installation pre-assembly concepts can be applied as a positive path towards achieving human-free lifting operations offshore, minimizing the number of lifts and reducing the human exposure to lifting hazards. Existing transportation and installation methods are:

- the bunny-ear (BE) method for installation of the rotor in two lifts (figure 2a): first two blades attached to the nacelle are lifted on top of the tower, and then the third blade is connected to the wind turbine in a second lift;
- the rotor star (ROT) method for fully pre-assembled installation of the rotor (figure 2b): the nacelle is mounted in a separate lift on top of the tower and then the entire rotor-hub assembly is lifted as one unit;

- single piece installation (SP) of the rotor-nacelle assembly (RNA) (figure 2c): each blade is lifted individually, as well as the nacelle with pre-assembled hub;
- installation methods for the tower: the tower can be lifted as one piece (1T) or two sections (2T).

Combining the pre-assembly methods for RNA and tower, a wind turbine can be installed with three to six lifts. Installation is also possible in only one or two lifts if the tower section and RNA are fully pre-assembled.

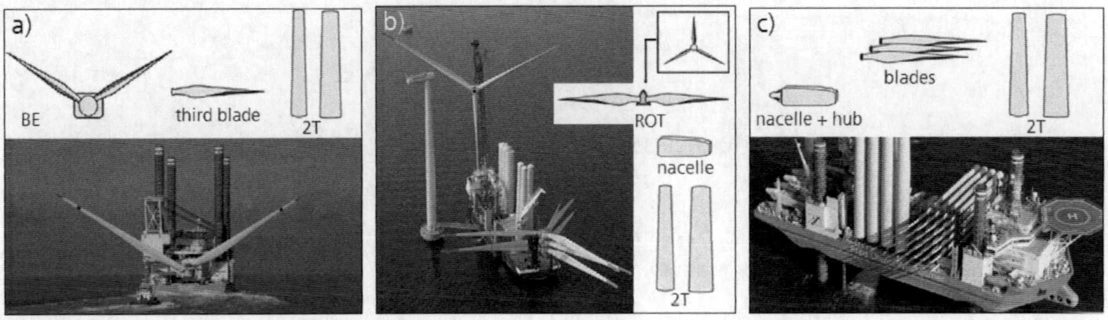

Figure 2. Pre-assembly methods: (*a*) Bunny-ear [15, 16], (*b*) Rotor star [15, 17], (*c*) Single pieces [15, 16].

Even if pre-assemblies of wind turbine components would require fewer lifts and thus reduce the number of hazardous offshore lifting situations, pre-assembled systems are heavier than single pieces and have larger wind exposed areas, which could actually increase the lifting hazards. Furthermore, vessel requirements, lifting capacities, and stability limitations for transportation and installation of pre-assemblies constrain the feasibility and advantages of this pre-assembly method. Thus, a good compromise between number of lifts and lifted weights has to be found.

For this purpose a discrete-event simulation model of lifting operations for the installation of a wind farm is implemented in a MATLAB script file (m-file) and integrated into a graphical user interface, to allow a more user friendly inclusion of the input data. Although the code has been developed for deterministic data input and output, as a future work, the time quantities could be integrated as random variables to model the system stochastically. Due to the coexistence of different criteria the program does not clearly outline the optimal practise, but aims to offer an overview of pros and cons of the different transportation and installation methods.

Figure 3 presents a flow chart of the developed installation optimisation program. Based on data input for the turbine and the vessel, systems-fitting functions derive information on the vessel loading-condition, and the lifting-operation hazard for the previously described BE and ROT configurations, as well as traditional SP transportation. Additionally, special tower and blade transportation methods, such as gathering single blades in a cage and having the tower as one piece or split into two sections, are allowed. Transportation constraints, due to vessel limitations, have been integrated as well. Despite the possibility to include overboard limits, the orientation of the pre-assembly on the vessel (transverse or longitudinal for BE, as well as inside- or outside-pointed transportation for ROT), and other user specified positioning along the deck, simplifying assumptions have been made concerning the vessel's stability. Specifically, asymmetric loading is not allowed and the maximum allowed position of the vessel's centre of gravity cannot be checked. Furthermore, a time counter provides an estimate of the time it

IOP Conf. Series: Journal of Physics: Conf. Series **1102** (2018) 012030 doi:10.1088/1742-6596/1102/1/012030

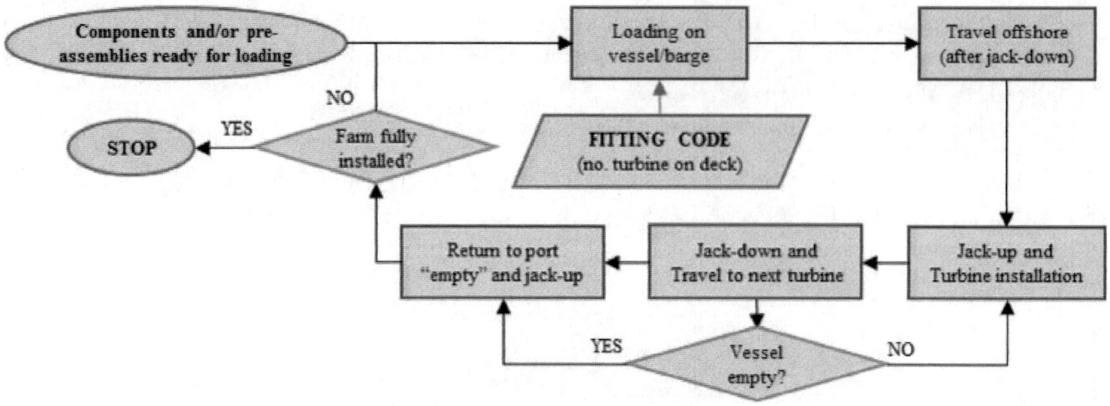

Figure 3. Installation pre-assembly MATLAB code flow chart.

would take to install the turbines based on the specified configurations. This estimation, however, is affected by time variables which have to be assumed, such as the time to load components and pre-assemblies on the vessel. As it was also necessary to account for the effect of weather windows and vessel's availability, a time variable for the average working hours per day is defined.

At this preliminary stage, the program was validated against two real case studies: Thanet and OWEZ (Egmond aan Zee) wind farms. For the latter, three possible configurations, along with characteristic properties, are presented in figure 4. The 36 wind turbines of OWEZ wind farm were transported in a BE-2T configuration, starting from Ijmuiden port, and installed by the A2SEA Sea Power jack-up vessel within around 67 days [15, 18]. The assumed loading and offshore installation times for each component were taken from [19], while a sensitivity analysis has been performed on the working hour, as shown in figure 5. As expected, in general the SP transportation would have led to a considerable increase in installation time. Although, as verified by the simulations, the adopted BE-2T configuration has the highest time saving, a ROT solution could reduce the amount of the maximum weight lifted by the crane, with the same number of lifts and slightly increased time required to install (about 10 days).

		Occupied area [m²]	Net load on deck [tons]	No. turbines per voyage	Total no. of lifts	Max. lifted load [tons]	Configuration visualization
SP-2T	4 blades in cage	1007	1339	5	216	88.0 (nacelle with hub)	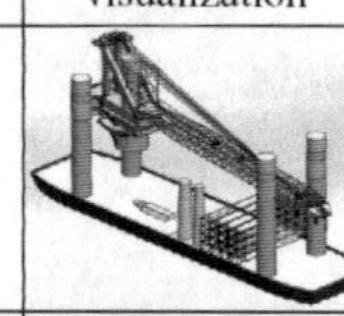
BE-2T	Transverse	969.1	535.6	2	144	101 (BE pre-assembly)	
ROT-2T	Inside-pointed	325.8	267.8	1		83.6 (upper tower element)	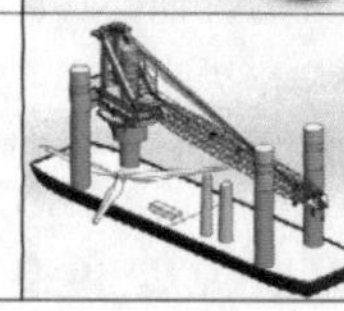

Figure 4. Exemplary transportation possibilities for OWEZ wind farm case study.

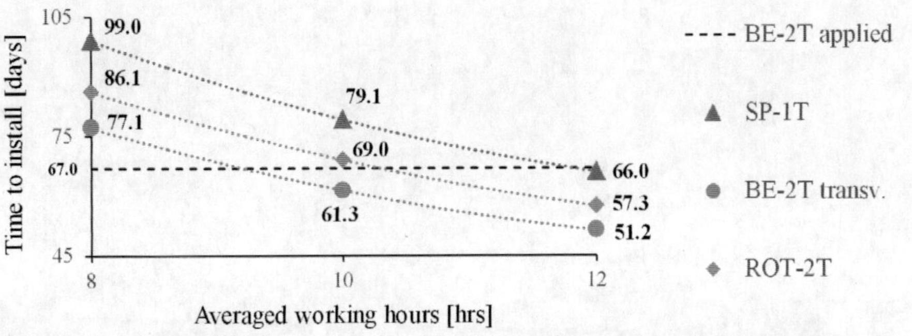

Figure 5. Sensitivity study on the average working hours per day, exemplarily for the OWEZ wind farm case study.

In regard to Thanet wind farm, installed as a SP-1T configuration by the MPI Resolution jack-up vessel [15, 18], it emerged that, although pre-assembly configurations could have reduced the installation time by approximately a month, the highest criticality for lifted weight remains with the tower. As the transportation of the tower in a single piece is generally preferred, the tower is indeed usually the heaviest component, and thus presents the greatest lifting hazard. However, due to vessel limitations, transporting the tower in two sections is increasingly required for larger turbines. This may be more cost effective than designing and building specialised "ad hoc" vessels. Subsequently, the following technologies and concepts for transportation and installation are focussed on the tower assembly.

3. Concepts for connections during installation and seafastening during transport

Connections play an important role in offshore wind turbine installation: for the final connection of single lifted pieces, but also for seafastening the turbine components on the transportation vessel. Ring-flange connections are most common for the major components of an offshore wind turbine. Typically more than 120 high-strength, pre-stressed bolts of large sizes (around M64 or M72 [20]) are used at each flange connection. Mostly, the bolts are already positioned on the wind turbine elements, which saves the labour of manually placing the bolts in the holes. Still, every single bolt needs to be tightened accurately, requiring special equipment such as hydraulic bolt tensioners.

An extensive review has been done on new connection and seafastening methods and designs. Innovative connection solutions include concepts based on friction [21] or the recently developed technology BLUE Wedge [22]. Besides the latter technology, which can also be used for seafastening, hydraulic systems and an internal jack system are proposed as seafastening solutions [23]. Staying with the traditional bolted flange connection, auxiliary systems for accurately tightening the bolts are required to ensure that no failure or hazard occurs due to loose bolts. Automated bolting systems exist [24, 25]; however, in the case of offshore wind turbine installation, the robot has to get on site and up the turbine as well.

The greatest possibility for improvement was found to be in enhanced seafastening methods, particularly regarding tower sections. Hydraulic seafastening would offer both safer and quicker seafastening, and thus increase the weather window compared to the conventional method of manual bolting. In addition, a robot arm employment for automated seafastening bolting is suggested (figure 6), which would retain the reliability of manual bolting while still reducing the risk and increasing the speed. However, the technology and logistical complexity set this study at a preliminary analysis phase.

IOP Conf. Series: Journal of Physics: Conf. Series **1102** (2018) 012030 doi:10.1088/1742-6596/1102/1/012030

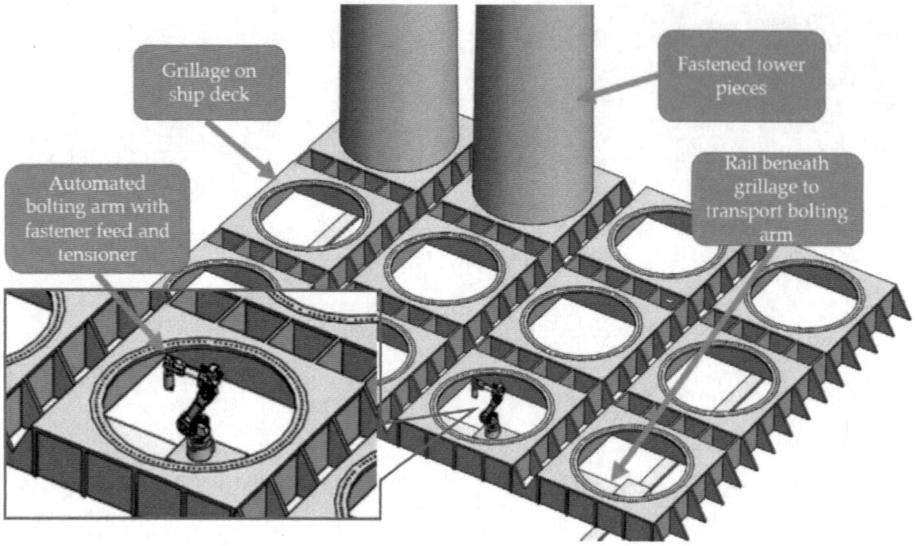

Figure 6. Automated bolted seafastening, concept illustration [14].

4. Concepts for guidance and control auxiliaries for lifting and installation

Guidance and control equipment for global hoisting, the mostly vertical lift from the ground to the aimed level, are already widely employed in lifting operations for wind turbine installations. Large movements during lifting are controlled via taglines. Developed systems, as presented in figure 7, are the Boom Lock [26], the Tagline Master [27], which can be controlled remotely, and the Blade Dragon [28], a remote controlled blade yoke.

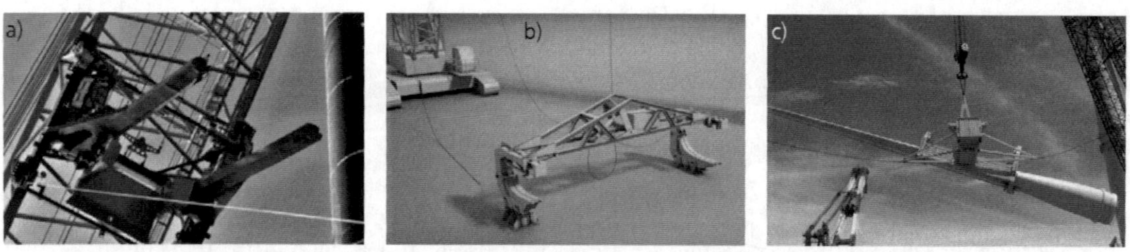

Figure 7. Global hoisting systems: (*a*) Boom Lock [26], (*b*) Tagline Master [27], (*c*) Blade Dragon [28].

When bringing the load close to the counterpart, correct positioning for the final installation has to be ensured. Thus, auxiliaries for centring and rotational alignment are required. Guide pins (figure 8a) or guide rods with corresponding socket sections help to align bolts and bolt holes, whilst cones or funnels (figure 8b) center the lifted section with respect to the counterpart. The latter method has already been used for a human-free met mast installation [29], but it is quite limited in size. Guide pins on the other hand are commonly used for installation of wind turbine elements; however, as it can clearly be seen in figure 8a, people are still directly involved in the lifting procedure.

To remove personnel from beneath the payload, innovative solutions, such as camera systems together with novel mechanical guidance designs for rotational alignment and centralisation, are

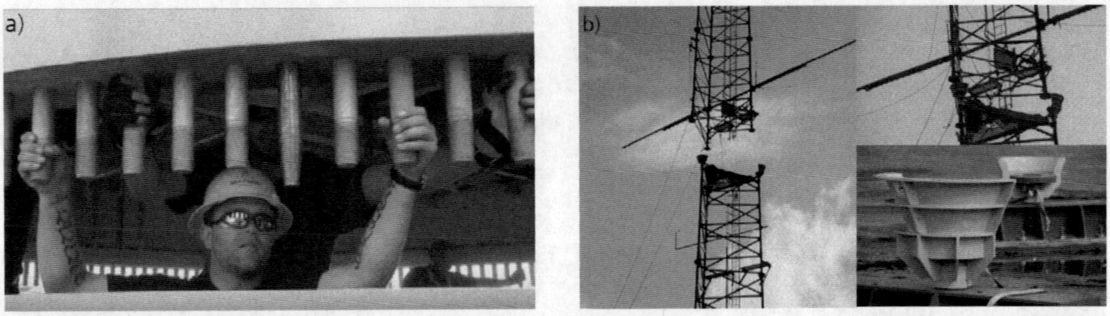

Figure 8. Current use of fine hoisting systems: (*a*) Guide pins [30], (*b*) Guiding funnel [29].

required. In this study, simplified small-scale tests were performed to quantify changes in accuracy, operability, and time when banksmen were moved from beneath the payload and visual and/or mechanical guiding elements were employed instead. Thus, three results were of interest: time taken for the lift, centrality, and rotational alignment.

The experiments were performed with a cylindrical steel section of 1.314 m outer diameter, 0.022 m wall thickness, and 0.55 m height. This was lifted by means of a standard gantry crane with three axes of movement and a freely rotating boom. Each lift started at the same position, 3.05 m from the target destination - a cylinder cross section drawn on the ground, with markings for bolt locations -, with the same crane operator standing behind a screen and listening for instructions from a banksman. Each test was performed five times and two banksmen alternated instructing the lifts for a more meaningful average result.

Visual guidance was tested for two types of camera - standard cameras (Swann 650TVL CCTV with night vision and 55° field of vision) and a 360-degree camera (Samsung gear-360) - in four different configurations:

(i) three standard cameras equally distributed on the circumference of the lifted cylinder section;

(ii) three standard cameras as in (i) and a fourth attached to the top of the lifting chains;

(iii) one 360-degree camera located in the middle of the target position, as shown in the two videos [31, 32];

(iv) three standard cameras as in (i) combined with the 360-degree camera as in (iii) (figure 9a).

For mechanical guidance, a holistic design was developed, bringing together the technical ideas of existing guiding systems, such as cones for centralisation or guide pins and socket sections for rotational alignment. This mechanical guiding element is removable and reusable as it consists of two parts: a guide rod attached to the lifted part and paired with a socket section within a conical extension attached to the fixed section, as visualized in figure 9b. Three guiding element pairs, equally distributed on the circular flanges, were considered to be sufficient. Adding threads to three sets of two or three bolt holes, the guiding elements can directly be screwed to the top and bottom turbine sections, respectively. After each lift, the elements can be removed again and reused for another lift. The experiments were performed with a simplified model of the mechanical guiding system design, as shown in figure 9c and video [33].

The first experiments focused only on centralisation of the lifted section on top of the (drawn) bottom cylinder with the help of cameras, while the second test campaign focused on centrali-

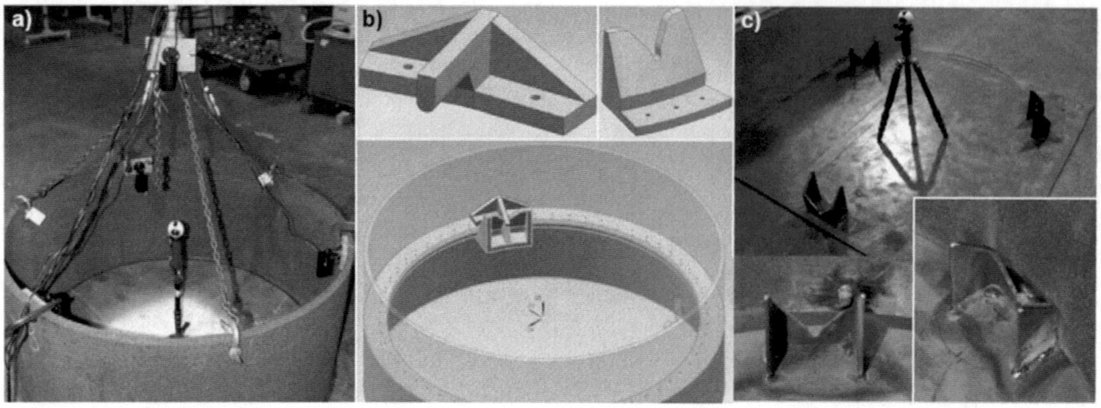

Figure 9. (*a*) Cameras for visual guidance, (*b*) Holistic mechanical guiding system, (*c*) Visual and mechanical guidance experiment setup.

sation as well as correct rotational positioning, firstly using the best camera setup found from the first set of experiments, then utilizing in addition the mechanical guiding system as shown in figure 9c.

The results are shown in table 1, representing the accuracy with respect to centralisation and rotational alignment, as well as the time taken, including a comparison to manual handling (with banksmen under the load). The first test campaign showed that with the combined setup (case iv) of three circumferential standard cameras and one central 360-degree camera the least eccentricity could be achieved. However, considering the additional criterion of rotational positioning happens at the expense of time and accuracy. Bringing together visual and mechanical guiding systems speeds up the lifting process enormously, while at the same time yielding a perfect fit. The best camera setup (case iv) allows rough and fast positioning of the lifted element above the target position, while the mechanical guiding system automatically positions the two pieces perfectly above each other as soon as the guide rod and bottom guiding element come in to contact (figure 9c).

Table 1. Comparison of the average results for all the configurations tested.

Configuration tested	Time taken [s]	Centralisation error [mm]	Rotation error [deg]
Centralisation only			
Banksmen under load	175.6	4.4	
Case (i): 3 cameras	551.2	15.6	
Case (ii): 4 cameras	496.0	5.2	
Case (iii): 360° camera	349.4	9.0	
Case (iv): 3+360° camera	391.0	4.2	
Centralisation and rotational alignment			
Banksmen under load	225.6	7.8	0.48
Case (iv) (3+360° camera)	498.4	4.9	1.36
Case (iv) + mech. guidance	202.0	0.0	0.00

Although, one has to keep in mind the time for connecting and removing the guiding element, the manually guided lift would take much longer than stated in the experiment, as it would have to be performed as slow enough to eliminate any eccentricity and rotational misalignment with the help of guiding hands. Therefore, the use of cameras in combination with the developed mechanical guiding elements could speed up an offshore lifting process by a factor of two, based on not too optimistic estimations, without posing risk to any person.

The wind turbine installation time can be related to the corresponding costs, which make up around 2% of the total offshore wind farm installation cost, based on information from Offshore Design Engineering Limited [34] for an offshore wind farm with around £1.6M/MW installed. Assuming that performing the lifts makes up just 20% of the total time for the wind turbine installation due to the long travelling distances, it can be estimated that lifting operations account for 0.4% of the capital expenditure for an offshore wind farm. Depending on conditions and size of a specific wind farm, it can be expected that the fabrication costs of the reusable mechanical guiding elements, as well as the acquisition costs for the cameras will at least pay for itself and may even enhance economic gains during the installation of one offshore wind farm.

5. Conclusion

The present work provides feasible solutions for human-free offshore lifting operations. Guidance and control systems, automated bolting or hydraulic seafastening, as well as a tool for optimised planning of offshore wind turbine logistics and installation, can help to remove people from beneath loads and thus contribute to increased safety in offshore lifting operations. In particular:

- The developed program for installation logistics outlines advantages and possible sensitivities of different installation procedures.
- A novel, although preliminary, concept for seafastening has been suggested.
- Supplementing existing tools, such as Boom Lock and taglines, by a holistic solution for visual and mechanical guidance, consisting of three circumferential cameras and one 360-degree camera, as well as innovative reusable guiding elements, can lead to faster and highly accurate offshore lifting procedures without putting personnel in dangerous areas. This will not only speed up the lifting opertion itself, but is also expected to entail financial benefit in offshore wind farm installation.

Regardless of the additional economic benefit, cameras and mechanical guiding elements are easy to implement and would crucially reduce the number of incidents during installation work by removing people from the high-risk areas. Increased safety and protection of life should be of highest priority.

Acknowledgments

This work was supported by grant EP/L016303/1 for Cranfield University and the University of Oxford, Centre for Doctoral Training in Renewable Energy Marine Structures - REMS (http://www.rems-cdt.ac.uk/) from the UK Engineering and Physical Sciences Research Council (EPSRC) and the G+ consortium. Special thanks go to Kate Harvey, Bir Virk, and Andrew Sykes from G+ for providing data on incidents and for their support within this health and safety study, as well as to David Delamore, Peter Geddes, and Joe O'Toole from Ørsted for their insights into the industries's current practices and needs.

References

[1] Statutory Instruments 1998 The Lifting Operations and Lifting Equipment Regulations 1998
[2] IOGP 2006 Lifting & hoisting safety recommended practice
[3] British Standards 2006 Code of practice for safe use of cranes - Part 1: General

[4] British Standards 1998 Code of practice for safe use of cranes - Part 11: Offshore cranes
[5] G9 Offshore Wind Health & Safety Association 2014 Good practice guideline: Working at height in the offshore wind industry: Working at height in the offshore wind industry URL https://www.gplusoffshorewind.com/work-programme/guidelines
[6] renewableUK 2013 Guidelines for the Selection and Operation of Jack-ups in the Marine Renewable Energy Industry
[7] DNV GL 2016 Marine operations and marine warranty
[8] G+ Offshore Wind Health & Safety Association 2018 G+ Global Offshore Wind Health and Safety Organisation URL https://www.gplusoffshorewind.com/
[9] G9 Offshore Wind Health & Safety Association 2014 *2014 Incident Data Report* URL https://www.gplusoffshorewind.com/work-programme/hse-statistics/g9-2014
[10] G9 Offshore Wind Health & Safety Association 2015 *2015 Incident data report* URL https://www.gplusoffshorewind.com/work-programme/hse-statistics/g9-2015
[11] G+ Offshore Wind Health & Safety Association 2016 *2016 Incident data report* URL https://www.gplusoffshorewind.com/work-programme/hse-statistics/gplus-2016
[12] WindEurope 2018 *Offshore Wind in Europe: Key trends and statistics 2017*
[13] Seyr H and Muskulus M 2016 Safety Indicators for the Marine Operations in the Installation and Operating Phase of an Offshore Wind Farm *Energy Procedia* **94** 72–81 ISSN 18766102
[14] Richmond M, Balaam T, Causon P, Cevasco D, Leimeister M, Kolios A and Brennan F 2018 Multi-Criteria Decision Analysis for Benchmarking Human-Free Lifting Solutions in the Offshore Wind Energy Environment *Energies* **11** 1175 ISSN 1996-1073
[15] Uraz E 2011 *Offshore Wind Turbine Transportation & Installation Analyses: Planning Optimal Marine Operations for Offshore Wind Projects* Master Thesis Gotland University Sweden URL http://www.diva-portal.se/smash/get/diva2:691575/FULLTEXT01.pdf
[16] Harrington K 2015 US Wind Energy Industry Develops around Block Island Turbines URL https://www.aiche.org/chenected/2015/08/us-wind-energy-industry-develops-around-block-island-turbines
[17] OffshoreWindbiz 2015 Royal HaskoningDHV Team Studies OW Financing Side URL https://www.offshorewind.biz/2015/07/06/royal-haskoningdhv-team-studies-ow-financing-side/
[18] 4C Offshore 2018 Offshore Wind Farms URL https://www.4coffshore.com/windfarms/
[19] Ahn D, Shin S c, Kim S y, Kharoufi H and Kim H c 2017 Comparative evaluation of different offshore wind turbine installation vessels for Korean west–south wind farm *International Journal of Naval Architecture and Ocean Engineering* **9** 45–54 ISSN 20926782
[20] SKI 2018 Ring flange connections URL http://www.ski-consult.de/1/verbindungstechnik/ringflanschverbindungen.html
[21] Heistermann C, Husson W and Veljkovic M 2009 Flange connection vs. friction connection in towers for wind turbines *NSCC* 296–303
[22] Krabbendam R 2018 Fistuca BLUE Wedge Connection URL http://www.heavyliftnews.com/wind-energy/video--fistuca-blue-wedge-connection
[23] Hoeksema W 2014 *Innovative Solution for Seafastening Offshore Wind Turbine Transition Pieces During Transport* Master Thesis Delft University of Technology Delft
[24] Sydenham M W and Brown T 2015 Robotic Installation of OSI-Bolts *SAE Technical Paper Series*
[25] Müller R, Hörauf L, Vette M, San Martin J L, Alzaga A, Hohmann J, Althoefer K and Würdemann H 2014 Robot guided bolt tensioning tool with adaptive process control for the automated assembly of wind turbine rotor blade bearings *Prod. Eng. Res. Devel.* **8** 755–764
[26] High Wind NV 2014 The Boom Lock URL http://www.high-wind.eu/boomlock/
[27] AH Industries 2018 Tagline Master URL http://www.ah-industries.dk/index.php/lifting-equipment/
[28] Liftra 2017 LT975 Blade Dragon URL https://liftra.com/product/blade-dragon/
[29] Forewind 2013 *HSE case study 2: The "human free" met mast installation*
[30] SCHEUERLE Fahrzeugfabrik 2012 Transport solutions for Offshore Wind Power Plants URL https://www.youtube.com/watch?v=oHXN9qq-S6Y
[31] Balaam T and Leimeister M 2017 REMS_Group Project_A5.2_Global View URL https://www.youtube.com/watch?v=mme1c4Kr4tU{\&}t=2s
[32] Balaam T and Leimeister M 2017 REMS_Group Project_A5.2_360 View URL https://www.youtube.com/watch?v=nASOm9QKLcU{\&}t=4s
[33] Balaam T and Leimeister M 2017 REMS_Group Project_C1.5_360 View URL https://www.youtube.com/watch?v=oOL1FNmdcZ4
[34] Offshore Design Engineering Limited 2007 *Study of the Costs of Offshore Wind Generation: A Report to the Renewables Advisory Board & DTI*

Reactive power and voltage control interaction and optimization in the Danish largest wind power plant at Kriegers Flak

Vladislav Akhmatov[1], Qiuwei Wu[2], Tibor Takarics[2]

[1] Energinet, Transmission System Operator of Denmark, Tonne Kjærsvej 65, 7000 Fredericia, Denmark
[2] Technical University of Denmark, Dept. of Electrical Engineering, Ørsted Plads, Building 348, 2800 Kgs. Lyngby, Denmark

Abstract. With the total 600MW power production capacity, Kriegers Flak will be the largest offshore wind power plant (OWPP) in Denmark. Kriegers Flak will utilize the two unequal sections, 400MW and 200MW, on the two offshore platforms interconnected via a 9 km 200kV cable and connected to the onshore substation via two 80 km 220kV cables. From the onshore 220kV substation, the 220kV connections continue to the two different 400/220kV substations of the Danish transmission grid. From the 400MW offshore platform, the connection continues to the 150kV offshore infrastructure of Germany. The Kriegers Flak grid connection resamples a meshed offshore grid (MOG) which increases complexity of and requirements to its overall voltage and reactive power control. This presentation describes how the overall voltage and reactive power control system is proposed and designed for optimized operation and reduction of the stress and efforts on the control equipment of the Kriegers Flak 220kV system.

1. Introduction

The Kriegers Flak 600MW offshore wind power plant (OWPP) will be among the largest electrical power generation units in Denmark East, which is the Danish transmission system being synchronous with the Nordic system and asynchronous with the Continental European system, see Fig. 1. Accounting for the wind turbines at Kriegers Flak, the nominal wind power capacity will correspond to 65% and 210% of the maximum and minimum consumption in Denmark East, respectively.

The Kriegers Flak grid-connection will be different from the already existing wind power plant grid-connection systems, see Fig. 2. The Kriegers Flak grid-connection will be a meshed 220kV cable network combining the two unequally sized offshore collection platforms, KFA with 200MW and KFB with 400MW, and feeding into the two different 400/220kV transformation substations in the Danish onshore transmission system [1], while all exiting OWPP connections are radial. The 220kV network becomes part of the meshed onshore transmission system of Denmark.

Through the 220/150kV transformation offshore platform, the connection will continue to the German offshore infrastructure and, then, to the German onshore transmission grid, while all existing connections do not have such interconnections [2]. The 220kV grid-connection system shall already be in operation by December 2018 due to utilization of the full Danish-German interconnector, and the 600MW wind power will be established by 2021.

IOP Conf. Series: Journal of Physics: Conf. Series **1102** (2018) 012031 doi:10.1088/1742-6596/1102/1/012031

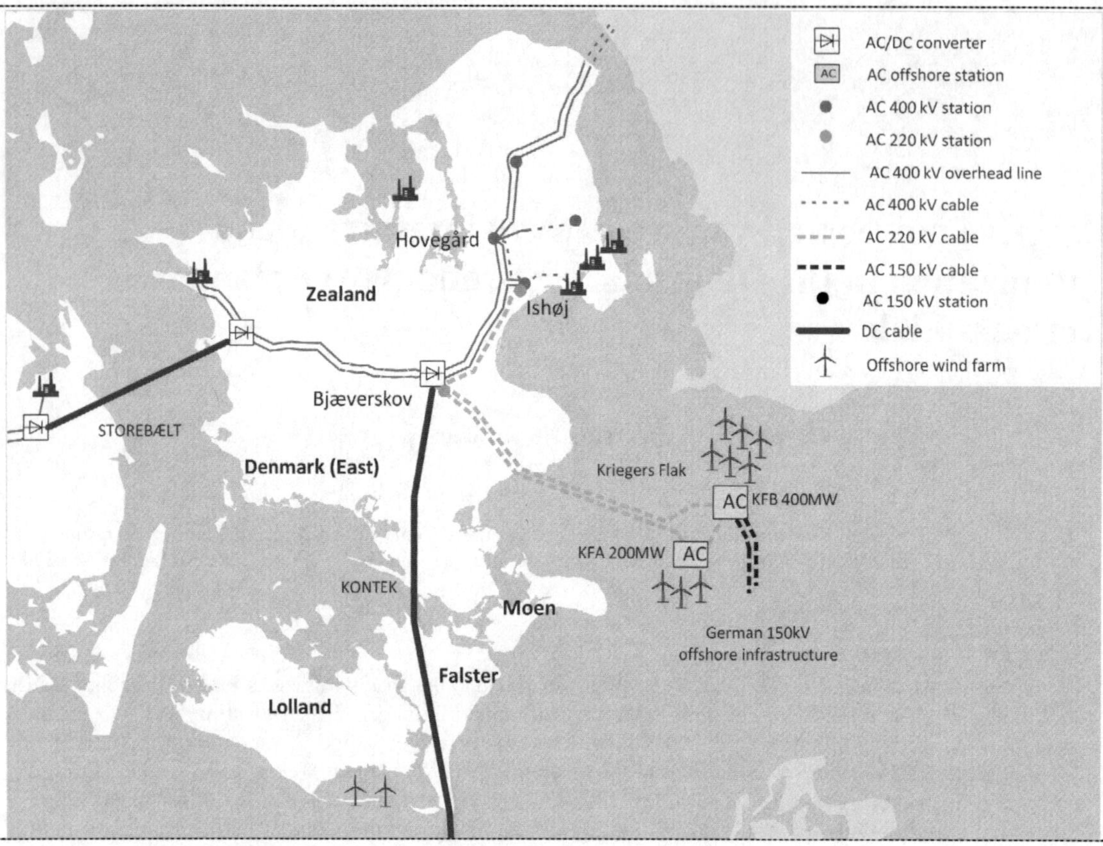

Figure 1. Grid-connection of Kriegers Flak 600 MW OWPP to Denmark.

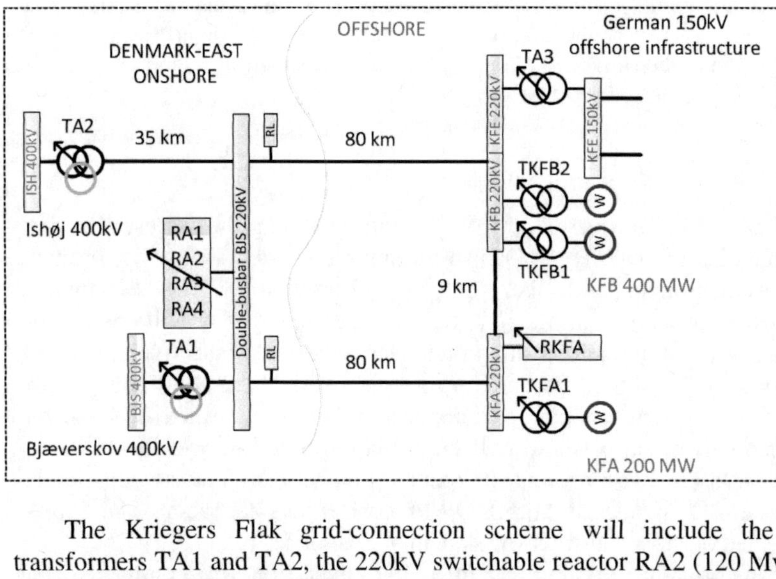

Figure 2. Meshed 220kV grid-connection and voltage and reactive power control equipment.

The Kriegers Flak grid-connection scheme will include the two 400/220/10kV onshore transformers TA1 and TA2, the 220kV switchable reactor RA2 (120 Mvar) and variable reactors RA1, RA3, and RA4, (each 60-120 Mvar) in the compensation substation BJS 220kV, the 220kV switchable reactor RKFA (120 Mvar) on the KFA platform, and the 220/150kV offshore transformer on the KFB/KFE platform. The two onshore transformers are with 500 MVA each and the offshore transformer is with 450 MVA power capacity. The 220/33kV offshore transformers TKFA1, TKFB1 and TKFB2 will be on the KFA and KFB platforms collecting the wind power production to the

IOP Conf. Series: Journal of Physics: Conf. Series **1102** (2018) 012031 doi:10.1088/1742-6596/1102/1/012031

220kV transmission system. The 220kV cable lengths (rounded up) will be ISH-BJS 35 km, KFB-BJS 80 km, KFA-BJS 80 km, KFA-KFB 9 km.

The technical complexity of the Kriegers Flak grid-connection system and variety of its operation regimes are ever seen [2], which includes the normal n-0 and numerous n-1 operation regimes and utilization of the dynamic line ratings of the 220kV cables in congested regimes. Furthermore, the active power may also direct from the onshore grid to the platforms due to the transport over the 220/150kV interconnection in low wind. Among numerous technical topics which are addressed to the Kriegers Flak grid-connection, this presentation deals with the voltage and reactive power control of the meshed 220kV grid and specifically to:

- Optimization of the reactive power and voltage control of the OWPP, such as possibility of the energy losses reduction in the OWPP 33kV collection networks.
- Utilization of the onshore voltage and reactive power control, such as possibility of the stepping and switching minimization of the 220kV compensation equipment.
- Yet unknown interaction dynamics with the 150kV offshore infrastructure over the 220/150kV transformation offshore substation.
- Proposals for overall coordination and optimization of the reactive power control using the wind power plant controllers (WPPC).

The proposal for coordination of the reactive power control will be presented by simulations.

2. Operation and control principles

The Danish Transmission System Operator (TSO) Energinet is establishing the Kriegers Flak grid-connection from the 33kV terminals of the 220/33kV transformation substations on the offshore platforms to the 400kV onshore substations. The OWPP operator will establish and connect the offshore wind turbines to the 33kV collection network, i.e. via the 33kV submarine cables, all the way to the 33kV terminals of the 220/33kV transformation substations.

In Denmark, the common practice has been that the OWPP point-of-settlements (POS) are the 33kV terminals [3]. This means that the delivery of the active power to the transmission grid and provision of the ancillary services, such as the voltage and reactive power control, are completed at the POS. The common practice has been that the offshore transformers are set in operation at fixed tap-positions securing the 33kV voltage delivery within a defined operation band from the transmission system to the OWPP collection network.

For the Kriegers Flak OWPP, the new common practice will be introduced. The tap-position control of the 220/33kV offshore transformers will be given to the OWPP operator. This means that the tap-position control shall lead to better operation and energy supply optimization from the wind turbines. In the new common practice, the voltage and reactive power control reference points may be moved from the POS at the 33kV terminals to the 220kV terminals on the offshore platforms. This control measure shall reduce variations in the voltage and reactive power interaction with the tap-position control of the transformers and reactors at the BJS 220kV onshore substation.

Since the 220/33kV offshore transformers remain property of Energinet and offshore maintenance efforts are expensive, the optimization goal shall be achieved at minimizing utilization of the tap-position shifts.

3. AVR/RPC system

Energinet establishes the Automatic Voltage Regulation / Reactive Power Control (AVR/RPC) system for coordinated voltage and reactive control within the Kriegers Flak 220kV network using the tap-changers of the TA1, TA2 and TA3 transformers and switches and tap-changers of the switchable and variable reactors RA1, RA2, RA3, RA4 and RKFA [1]. The AVR/RPC system will maintain the reactive power exchanges in the 400/220kV substations within defined bands, e.g. around 0 Mvar, and maintain the voltages on the offshore platforms within defined bands, e.g. around 235kV. The AVR/RPC system has been designed and tested by simulations for all expected operation regimes and power flow patterns of the Kriegers Flak [1].

IOP Conf. Series: Journal of Physics: Conf. Series **1102** (2018) 012031 doi:10.1088/1742-6596/1102/1/012031

The AVR/RPC system controls neither the tap-changers of the offshore transformers TKFA1, TKFB1, TKFB2, nor the WPPC of the KFA and KFB OWPP, see Fig. 3. This decision is made because the wind turbines will be commissioned few years after the grid-connection system with the Danish-German interconnector have been taken in operation and also for better redundancy of the vital voltage and reactive power control.

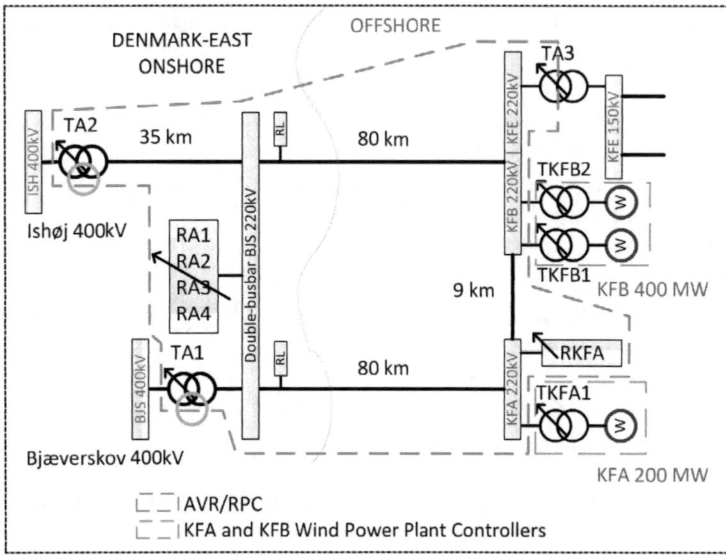

Figure 3. Areas of control: AVR/RPC of TSO and WPPC of the OWPP developer.

The AVR/RPC system and the tap-changers of the offshore transformers TKFA1, TKFB1 and TKFB2, will be influenced by several factors which are not present in the conventional wind power plant connection systems. There will be some active and reactive power transport over the 220/150kV transformation substation.

The static range of the reactive power via the 220/150kV transformation substation shall be maintained at ±40Mvar [2]. Thus, the reactive power exchange with the German 150kV offshore infrastructure may vary dynamically within 80Mvar. The initial studies of Energinet indicate a strong sensitivity of the reactive power transport from the operation conditions within the 150kV infrastructure, which practically speaking means a high risk of dynamically varying reactive power transport with a not yet known pattern.

Since the Kriegers Flak is also the Danish-German interconnector, the active power will be transported in both directions between the onshore grid and the offshore platforms. This influences the voltage profiles over the 220kV system. The voltage drop between the platforms and the onshore substation BJS 220kV will increase when the transport directed to the land and reduce or even reverse when directed to the platforms.

Combinations of the n-1 operation regimes and utilization of the dynamic line ratings may load the 220kV cables over their static line ratings and cause specific conditions where the cables deliver much less reactive power to the grid than in normal operation. In such loaded conditions the current is high and the cable reactance is consuming more of the reactive power generated by the cable capacitance. Thus, the AVR/RPC system will be forced to step down on the higher, small steps use the first, large steps of the tap-changers causing more abrupt control of voltage and reactive power [1].

With no other control available between the offshore platforms and the onshore grid, the dynamic reactive power imbalance and voltage fluctuations will be "absorbed" by the AVR/RPC system. This may increase the stepping and switching shifts within the controlled equipment. Energinet wishes to foresee and mitigate the causes of unnecessary activation of the AVR/RPC system.

4. Power and energy losses reduction

In cooperation with the Technical University of Denmark, Energinet has conducted a study on potential evaluation of the power and energy losses reduction in the 33kV cable network using the tap-changers of the 220/33kV offshore transformers [4]. The study has been based on the operation scenarios with different wind power production, superimposed by different active power and different reactive power transports via the 220/150kV offshore transformation substation, and superimposed by the OWPP optimization schemes.

This presentation deals with the part of the study where the connection regime is n-0 with all 220kV components are in operation. The two presented optimization schemes deal with the control of the 220/33kV transformer tap-changers:

- Not-optimized tap positions and 0Mvar at the POS 33kV terminals (the common practice).
- Optimized tap positions (the new practice for Kriegers Flak) and 0Mvar at the 220kV terminals.

The four wind power production scenarios which are 100% (high), 50% (medium), 5% (low) and Disconnected (none), multiplied by three active power transport scenarios which are Maximum Denmark→Germany, Maximum Germany→Denmark and No Transport, multiplied by the five reactive power flow scenarios which are -40Mvar, -20Mvar, 0Mvar, 20Mvar and 40Mvar positive from 150kV to 220kV system, and multiplied by the two optimization schemes become 120 main operation scenarios, which are included in this presentation. Attention shall be paid to the transport volatility of the wind power production. For example, the combination of the maximum Germany→Denmark transport and 100% wind means 0MW transport over the 220/150kV offshore transformation substation because of the 220kV system congestion on the Danish side. The same 100% wind and the opposite directed transport means 80MW Denmark→Germany transport over the 220/150kV offshore transformation substation because there is still available transport capacity through the 150kV infrastructure.

For the energy losses assessment, a model of the 33kV cable collection networks using the generic cable data has been included [4]. The 33kV generic cable data are: R=0.07 ohm/km, X=0.1 ohm/km, C=0.4 μF/km, i.e. a single, averaging cable type is applied to represent all cable types in the entire 33kV cable network. This consideration works sufficient and fine since the study does not include any multi-parameter 33kV cable network layout/routing optimization. The energy losses, the tap positions of the 220/33kV offshore transformers and those of the AVR/RPC components have been calculated for each main operation scenario.

The study has shown that utilization of the 220/33kV offshore transformers for the OWPP operation optimization (the new practice of Kriegers Flak) may reduce annual energy losses within the 33kV collection network by 10…15%. The potential losses reduction range is given in comparison to the scheme with no optimization (the common practice before Kriegers Flak).

The study has shown that the annual energy losses reduction is not sensitive to the reactive power transport over the 220/150kV offshore transformation substation. The result of the losses reduction is also less sensitive from the reactive power and voltage operation conditions in the 220kV system as far as the AVR/RPC maintains the 220kV system within the defined reactive power and voltage ranges.

The study has shown that the tap-position shifting of the 220/33kV offshore transformers slightly increases, while the switching and tap-position shifting of the equipment controlled by the AVR/RPC are less sensitive to the utilization of the 220/33kV offshore transformers for OWPP optimization. On the other hand, the switching and tap-position shifting of all equipment are very sensitive to the reactive power transport through the 220/150kV offshore transformation substation.

5. Overall reactive power control optimization

The overall reactive power control optimization includes the AVR/RPC system, the 220/33kV offshore transformers with their tap-changers, and the WPPCs. The optimization targets outbalancing of the reactive power exchanges via the 220/150kV offshore transformation substation and minimizing of the AVR/RPC activation by implementation of a superior control and utilization of all available control within the 220kV network of Kriegers Flak.

IOP Conf. Series: Journal of Physics: Conf. Series **1102** (2018) 012031 doi:10.1088/1742-6596/1102/1/012031

5.1. Wind Power Plant Controllers

The two KFA and KFB windfarm sections shall be established with the two individual WPPCs in order to deliver the required and agreed ancillary services to the grid. The model structure of the WPPC of the KFB section is illustrated in Fig. 4. The WPPC receives the measurements from the 33kV and 220kV terminals of the offshore substation and the available wind power measurement from the wind model, calculates and distributes order signals to the individual wind turbine generator models. The order signals to the wind turbine generators include the active power as well as the individual reactive power or voltage signals.

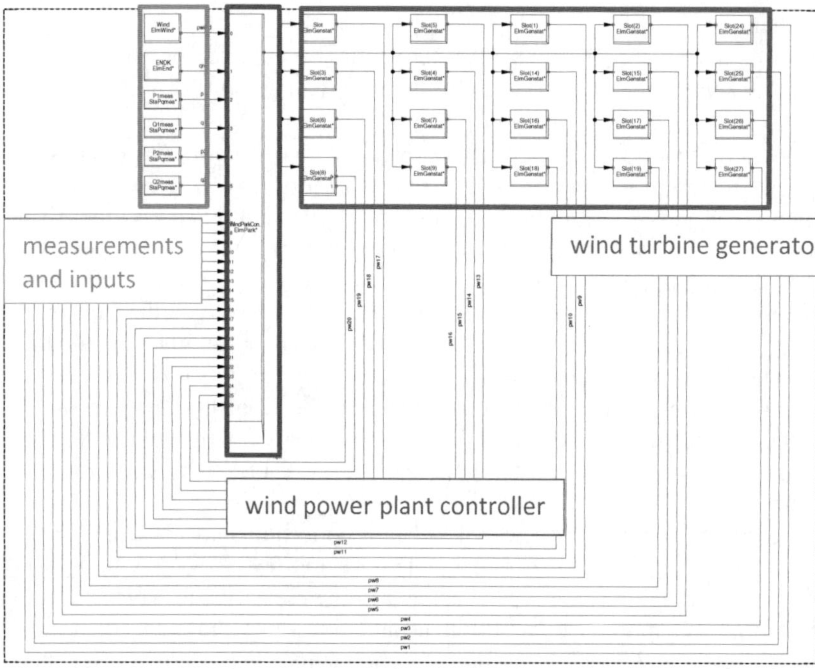

Figure 4. Model structure of WPPC of the KFB wind farm section.

In this modelling structure, all wind turbine generators are always available and in operation. All wind turbine generators receive the same active power and voltage order signals from the WPPC model and deliver the same response in the means of active power and voltage to their 33kV terminals. This approach is seen in the model implementation in Fig. 4. In this approach, all wind turbine generators send their measurements to the WPPC, but only a single wind turbine generator sends the control feedback to the WPPC, see Fig. 4. This modelling reduction is sufficient because all wind turbines shall, in this approach, have the same control feedback. If another approach were used in which some wind turbines were out of operation or should have different operation conditions, then all wind turbine generators should also have individual control feedbacks sent to the WPPC.

5.2. 220/33kV transformer control

The tap-changer control of the 220/33kV offshore transformers is implemented for securing the operation voltage around the optimized reference voltage at the 33kV terminals. The tap-changer control is independent from the AVR/RPC control.

5.3. Proposed superior control

The target of the proposed superior control of Kriegers Flak is illustrated in Fig. 6. With no superior control, the reactive power exchange via the 220/150kV offshore transformation substation will directly interact with the AVR/RPC and its equipment behind the BJS 220kV substation. Further, this reactive power exchange will influence the voltages on the KFA and KFB platforms activating the tap-changer controllers of the 220/33kV offshore transformers.

Establishment and utilization of the superior control shall activate the WPPCs of the KFA and KFB wind farms absorbing some of the reactive power imbalance from the 150kV offshore infrastructure, see Fig. 5, which reduces the switching and tap-position shifting of the AVR/RPC control equipment, and adjusting the 33kV voltages, which minimizes the tap-changing of the 220/33kV offshore transformers.

The proposed superior control is shown in Fig. 6. The characteristic time of the AVR/RPC is in a range of tens of seconds. Therefore, the characteristic time of the superior control shall be in a range of several seconds in order to prevent unnecessary activation of the AVR/RPC equipment. Considering the characteristic time, the proposed superior control shall be an automatic control, and not manually-interfaced.

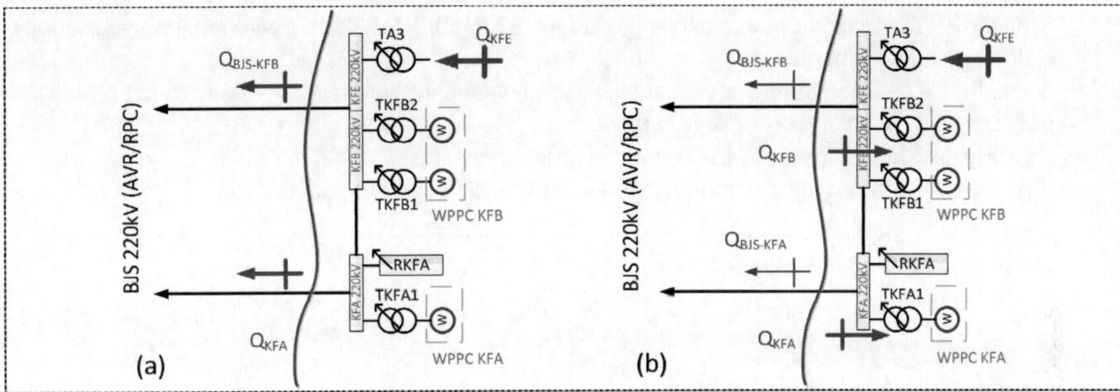

Figure 5. Reactive power flow between the 220/150kV offshore substation and the BJS 220kV substation (AVR/RPC equipment): (a) – no superior control, (b) – proposed superior control.

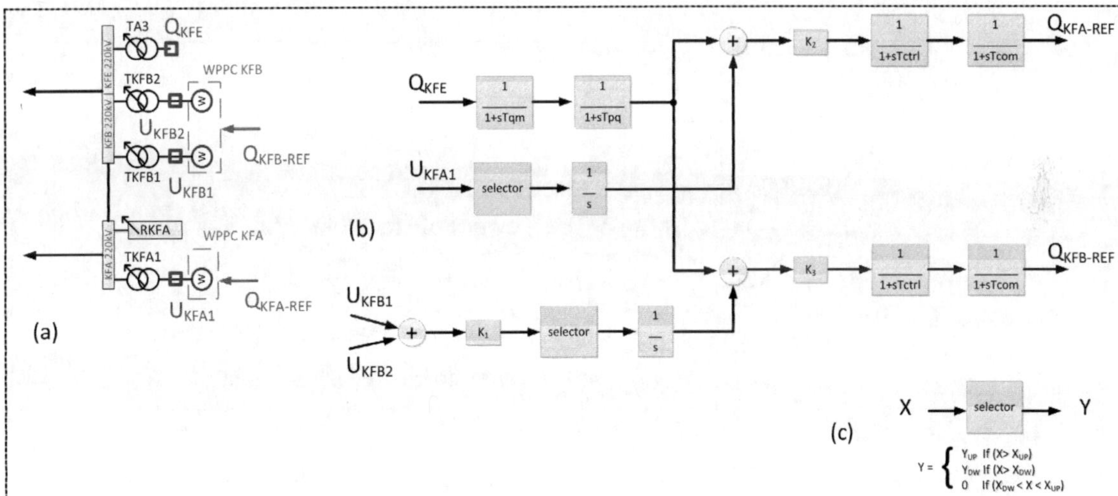

Figure 6. Proposed superior control (a) –measurements, (b) –main block-diagram, (c) –selector.

In this presentation, the input signals are the measurements of the voltages U_{KFA1}, U_{KFB1}, and U_{KFB2}, and the measurement of the reactive power Q_{KFE}. The output signals are the reactive power references to the WPPCs of the KFA windfarm, $Q_{KFA\text{-}REF}$, and the KFB windfarm, $Q_{KFB\text{-}REF}$. The gain K_1 =½ because of the two available voltage measurements, U_{KFB1} and U_{KFB2} at the two 33kV terminals of the 220/33kV offshore transformers. The gains K_2 =0.333 and K_3 =0.667 representing the 200 MW and 400 MW power capacities of the KFA and KFB windfarms in relation to the total 600 MW power capacity of both WPPC control systems. The time constants are T_{QM} =0.1 s, T_{PQ} =0.1 s and T_{COM} =0.1 s, and the control time is T_{CTRL} =2.5 s. The selectors and integrators are securing that the proposed superior control stabilizes the 33kV voltages and, then, does not trigger excessive activation of the tap-changers of the 220/33kV offshore transformers.

Attention is paid to that the proposed superior control, as shown in Fig. 6, does not include the necessary measures of redundancy, such as replacement of an invalid measurement, lost communication and equipment outage. Such measures shall always be part of the implementation.

5.4. Simulation example

The presented simulation example includes the three operation scenarios. The three scenarios are with the same wind power production and the same active power transport via the 220/150kV offshore transformation substation, which are shown in Fig. 7. The difference between the three operation scenarios is:

- Scenario 1: The reactive power exchange via the 220/150kV offshore transformation substation is 0Mvar, and no superior control is applied.
- Scenario 2: The reactive power exchange varies dynamically within +/-40Mvar, which is shown in Fig. 8, and no superior control is applied.
- Scenario 3: The reactive power exchange varies dynamically within +/-40Mvar, as shown in Fig. 8, and the proposed superior control, as shown in Fig. 6, is included.

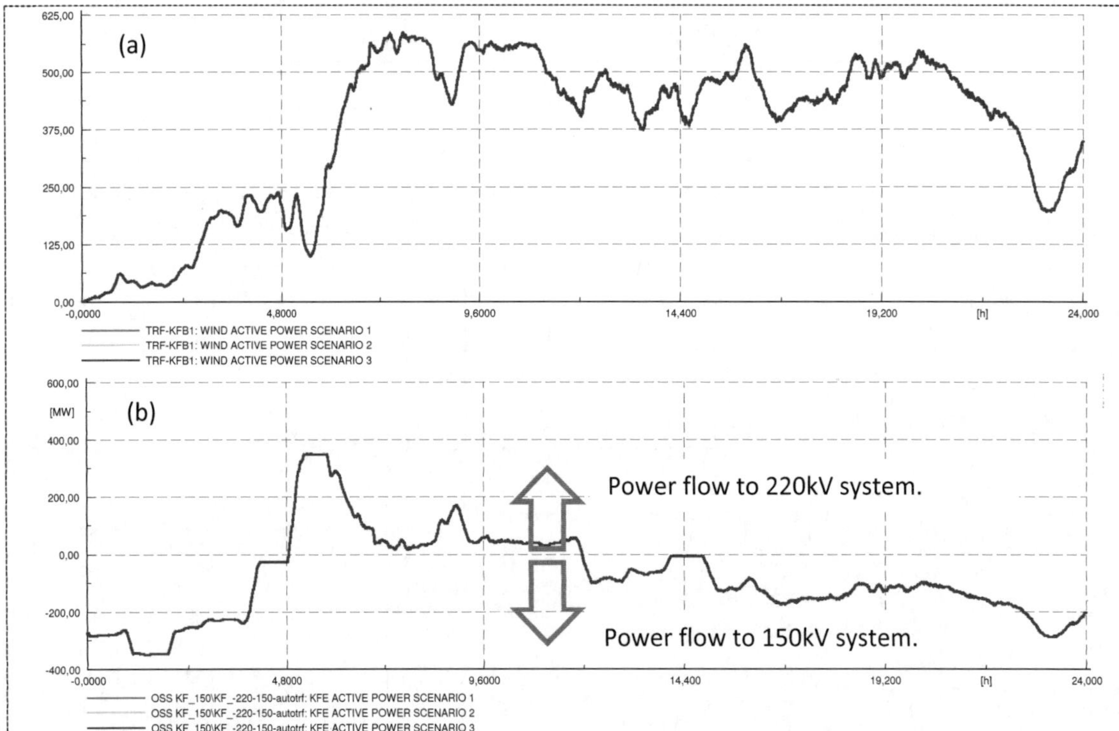

Figure 7. Active power in MW over 24 hours (the same for the three scenarios): (a) – wind power production at Kriegers Flak, (b) – transport over the 220/150kV offshore transformation substation.

Attention is paid to that the magnitude and characteristic time of the reactive power variations are not yet known and to be determined when the Kriegers Flak interconnector will be brought in operation. Therefore, the two scenarios with the reactive power variations over the 220/150kV offshore transformation substation are generated as periodical in this simulation example. The response of the AVR/RPC system is presented in Fig. 9 for the scenario 1, in Fig. 10 for the scenario 2, and in Fig. 11 for the scenario 3.

The AVR/RPC system response in Fig. 9 is due to the varying active power flow with corresponding variations of the reactive power and voltage within the 220kV system. Introduction of the varying reactive power flows, and with no control countermeasures, leads to excessive activation of the AVR/RPC equipment and also to the tap-changers of the 220/33kV offshore transformers, see Fig. 10. This response may call for more frequent maintenance periods of the equipment and, for the offshore and onshore transformers, for the power delivery congestion. Especially, increased maintenance needs of the offshore transformers may cause additional loss to all involved parties.

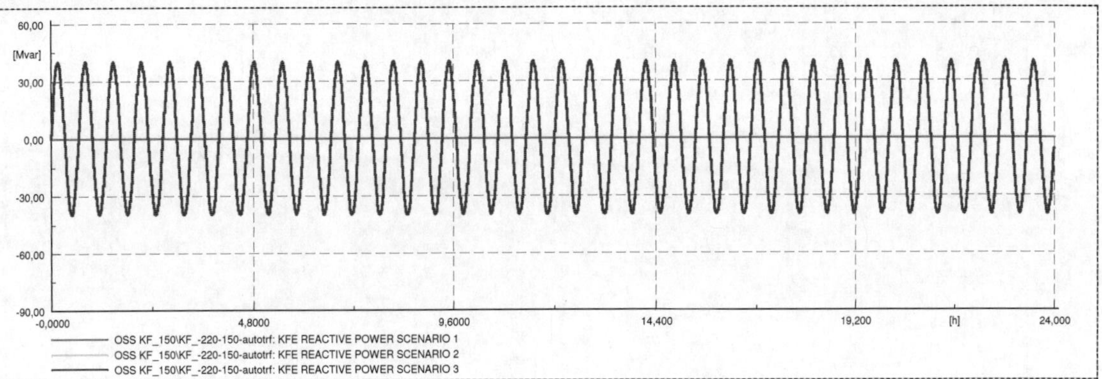

Figure 8. Reactive power in Mvar over 24 hours over the 220/150kV offshore transformation substation: 0Mvar for scenario 1 and dynamically varying within +/-40Mvar for scenarios 2 and 3.

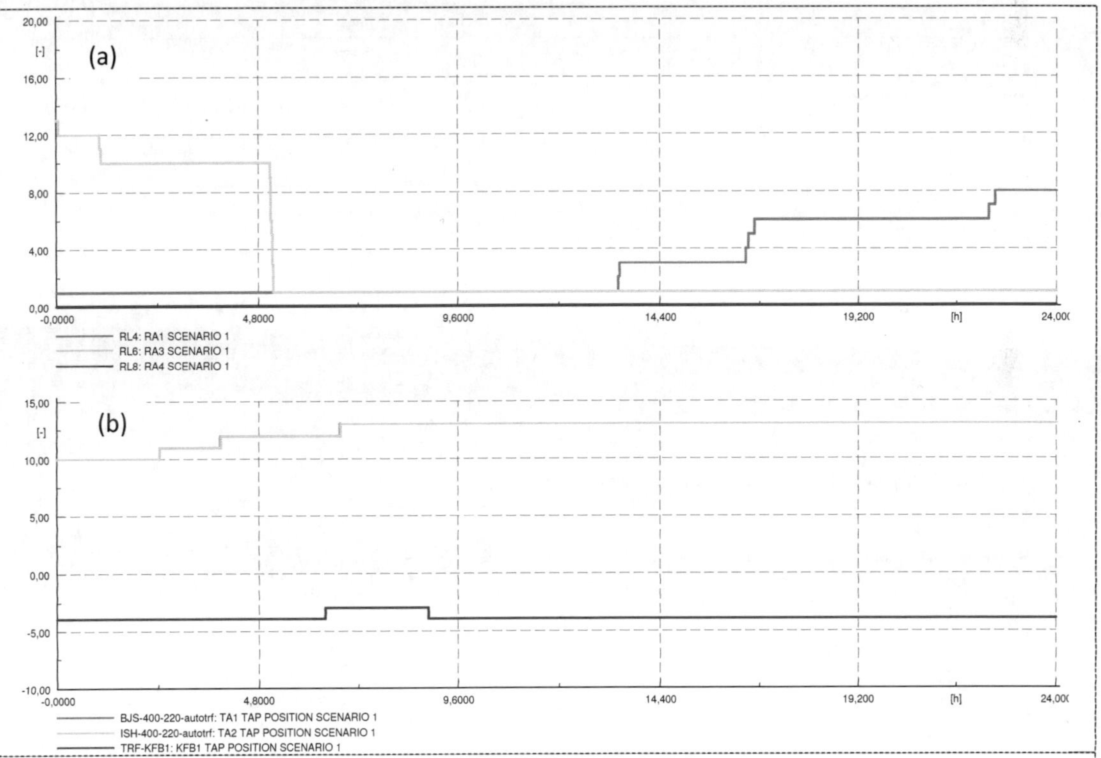

Figure 9. Scenario 1 response of the control equipment tap-positions: (a) –reactors, (b) –transformers.

Utilization of the proposed superior control, which results are shown in Fig. 11, may significantly reduce activation of the control equipment. The proposed superior control seems very promising for relaxing of the AVR/RPC control efforts and for activation of the tap-changers of the 220/33kV offshore transformers.

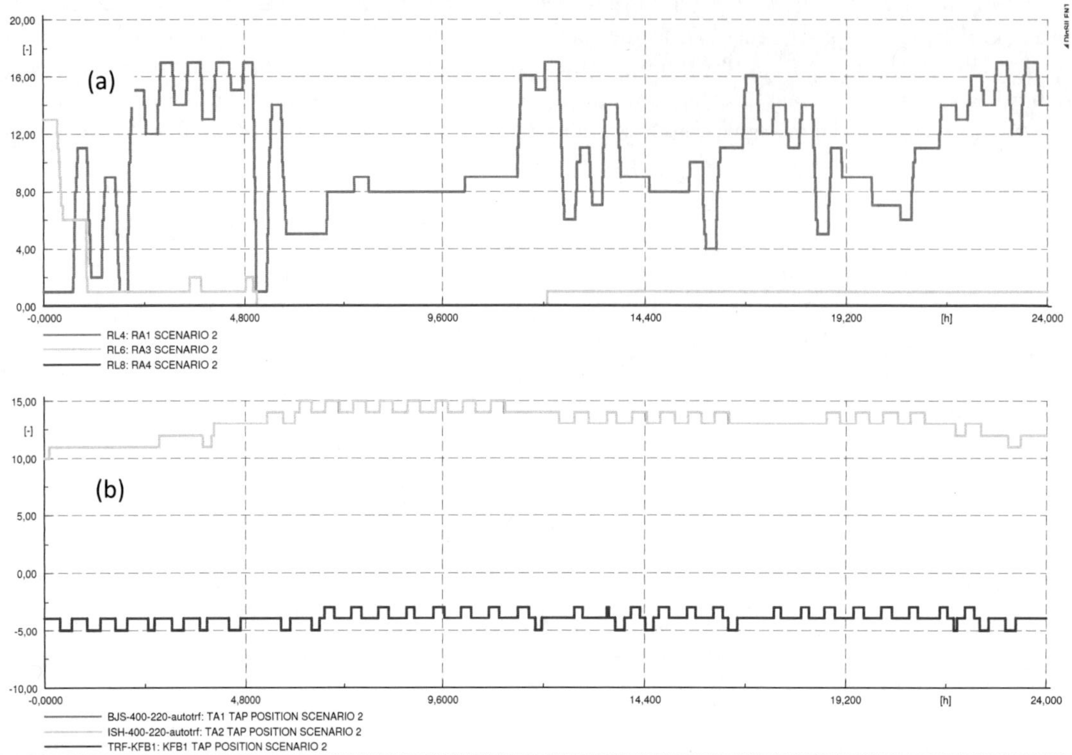

Figure 10. Scenario 2 response of the control equipment tap-positions: (a) –reactors, (b) –transformers.

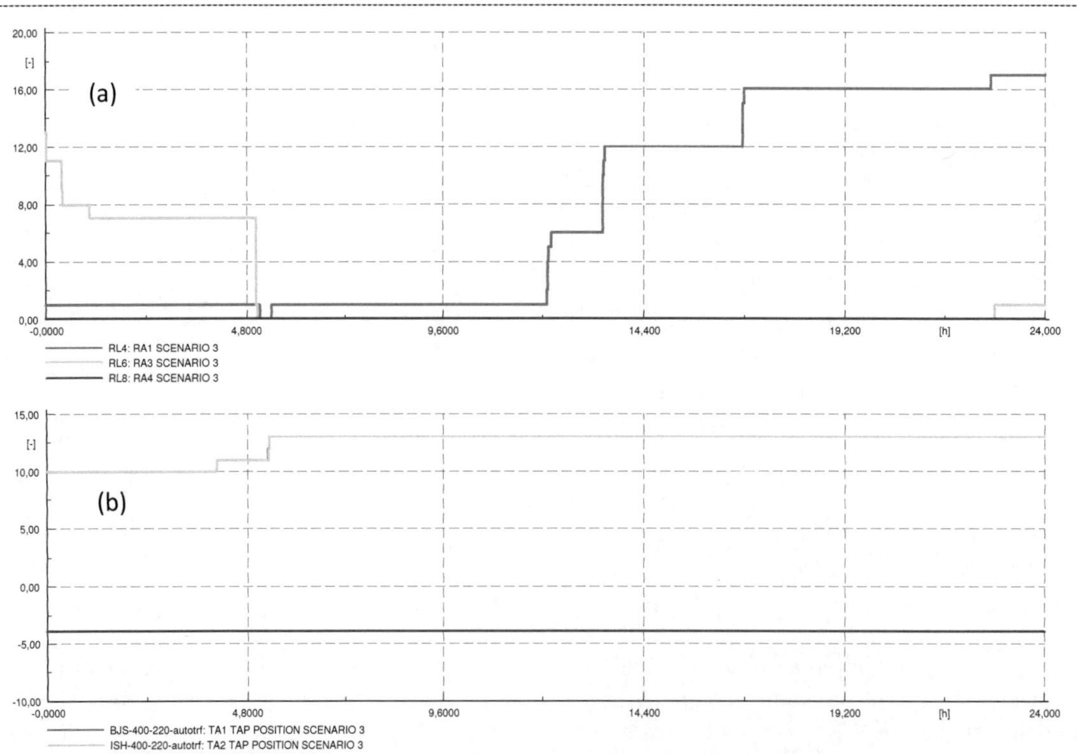

Figure 11. Scenario 3 response of the control equipment tap-positions: (a) –reactors, (b) –transformers.

6. Conclusion

The Kriegers Flak OWPP grid-connection will utilize the meshed 220kV cable system and include an interconnector to the German 150kV offshore infrastructure via the 220/150kV offshore transformation substation. The voltage and reactive power control within the 220kV system will primarily be conducted by the AVR/RPC system with coordinated control of the switchable and variable 220kV reactors and the tap-changers of the 400/220/10kV onshore transformers and the 200/150kV offshore transformer.

Further, the new practice of Kriegers Flak will be utilization of the tap-charger control of the 220/33kV offshore transformers for optimized operation of the 33kV collection cable networks. The study has shown that the tap-changer control may significantly reduce the power and energy losses in the 33kV networks practically without increase of the AVR/RPC activation.

On the other hand, excessive usage of the AVR/RPC equipment and the tap-changers of the 220/33kV transformers can be excited due to dynamic variations of the reactive power exchange via the 220/150kV offshore transformation substation. The study has shown that inclusion of the superior control may reduce such negative influences and remove stress on the AVR/RPC equipment and the tap-changers of the 220/33kV offshore transformers. The proposed superior control includes the WPPCs in the overall voltage and reactive power control scheme, which compensates for unwished reactive power exchanges and stabilizes the operation voltage in the 33kV collection cable networks. The study has shown a significant reduction in the control equipment usage and stress when the proposed superior control has been activated.

7. References

[1] V. Akhmatov, "Automatic Voltage and Reactive Power Control of Kriegers Flak Offshore Wind Power Plant Grid-Connection System", Offshore Wind conference 2017, PO-0255, London, UK, June 6-8 2017. 10 p.

[2] A. K. Marten et al.," Kriegers Flak – Combined Grid Solution: Coordinated Cross-Border Control of a Meshed HVAC/HVDC Offshore Wind Power Grid", IET Renewable Power Generation DOI: 10.1049/iet-rpg.2017.0792, Online ISSN 1752-1424 Available online: 19 February 2018.

[3] "Technical regulation 3.2.5 for wind power plants above 11 kW". Energinet Transmission System Operator of Denmark. Published UK edition. 22.07.2016.

[4] T. Takarics, "Voltage Control Optimization by using the Wind Power Plant Controllers at Kriegers Flak AC", Technical University of Denmark, Dept. of Electrical Engineering, Dec. 2017, 16.p.

IOP Conf. Series: Journal of Physics: Conf. Series **1102** (2018) 012032 doi:10.1088/1742-6596/1102/1/012032

Developing, implementing and testing up and down regulation to provide AGC from a 10 MW wind farm during varying wind conditions

Eldrich Rebello, David Watson and Marianne Rodgers

Wind Energy Institute of Canada 21741 PE-12, Tignish, Prince Edward Island

E-mail: eldrich.rebello@weican.ca

Abstract. With increasing levels of electrical energy generated by intermittent sources such as wind turbines, their participation in grid ancillary services is becoming a necessity. Typically, all generated energy from variable generators is absorbed by the electric grid and balancing is left to traditional generators. Wind turbine technology has matured to the level where a large wind generator is capable of providing ancillary services such as up- and down- active power regulation (secondary frequency regulation). The up-regulation capacity of a variable generator is constrained primarily by external factors such as the prevailing wind speed in the case of a wind turbine. This work uses the Wind Energy Institute of Canada's (WEICan) 10 MW Wind R&D Park (Type 5 generators) in Prince Edward Island, Canada, to test and evaluate a simple algorithm to provide up- and down-regulation services from a wind park. The developed algorithm uses a 10-minute averaged wind speed to estimate the available park generation potential. A fixed power curtailment is applied to provide room for up-regulation. An historical, external AGC signal is then applied to the wind park's active power set-point and the resulting park performance is evaluated. Results of the 4.5 hour test prove the technical capability of the wind farm in participating in the regulation market. A performance score of 64 % was calculated according to the PJM method, averaged across the test duration.

1. Introduction

Wind energy and other intermittent renewable sources (such as solar PV) are increasingly being used throughout the world to generate electricity. Electrical energy production from wind in some jurisdictions has increased to supply a significant percentage of their annual consumption. For example, in 2017 Spain, Germany, and Denmark produced 18%, 21%, and 43% of their annual electrical demand from wind power [1], [2], respectively. In Spain wind power supplied 60.7% of the demand at one specific moment [1].

The operation of any power system involves balancing generation and load, which has traditionally been done with conventional, synchronous generators. Traditional generators have been used for energy production as well as providing additional grid functions such as regulation of grid frequency or voltage [3]. Collectively, these services are referred to as ancillary services [4]. Providing ancillary services such as regulation from intermittent renewable energy sources is uncommon. As wind and solar displace traditional generators they, or other sources such as storage, will be required to provide ancillary services. The grid rules in high wind penetration regions require that turbines accept active and reactive power set-points as well as active power ramp rate limits. Wind turbines are capable of changing their active power output in response to several factors. One set of factors is meteorological and includes changes in wind speed. This is not strictly a controlled response as the maximum power production is limited by the prevailing wind speed. A second set of factors is grid conditions such as grid frequency changes. Turbines respond through means such as power-frequency response curves [5]. This is typically done to satisfy grid requirements. A third set of conditions are in response to external set-points such as network operator limits or automatic generation control (AGC) signals. Often, network operators enforce maximum power export limitations on wind farms. As an example of wind generation being limited, in 2016 in Ireland, 2.9% of the available wind energy production was curtailed [6] in response to local transmission constraints and bulk system limitations. Although all new wind farms in Ireland (and several other jurisdictions) are required to respond to external power set points, few, if any, use this capability to provide secondary frequency regulation (AGC).

Frequency changes on the grid are a good representation of the balance between supply and demand. The response of the grid frequency to events such as load or generation changes can be divided into regions such as primary, secondary and tertiary response as shown in Figure 1 which is based on [7]. AGC is a means of secondary frequency regulation and is achieved by temporary increases (regulation-up) or decreases (regulation-down) in generator active power output. Regulation-up and regulation-down services are sometimes collectively referred to as regulation services (Reg-services for short) [8]. Regulation is done on a time scale faster than the fastest market clearing schedule and is used by system operators to meet performance standards (e.g. NERC). These standards are typically based on time averages of a balancing areas' Area Control Error (ACE). Area Control Error is the difference between the scheduled and actual electrical generation within a control area on the grid while accounting for frequency bias and power exchanges [9]. AGC is one method of providing Reg-services using generators that can respond to fluctuations in the MW/min range [7]. The dispatch centre monitors and controls generators across the power system to balance generation and load thereby maintaining a constant system frequency [9]. The AGC signal is derived from the ACE signal and is sent as a regulation signal to Market Participants' or Qualified Schedule Entities' AGC controllers. In essence, if system frequency falls below the target value (60 Hz), the AGC system sends a Reg-Up signal, calling for an increase in active power generation. Conversely, if system frequency increases above the target value, the AGC system sends a Reg-down signal to reduce generation.

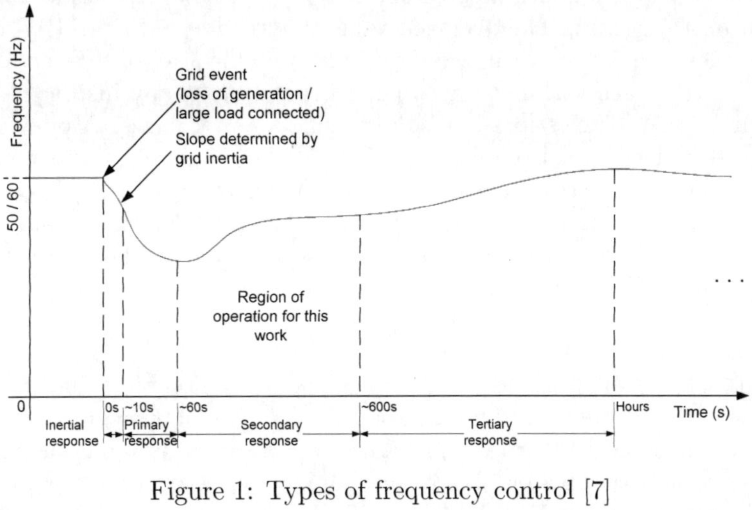

Figure 1: Types of frequency control [7]

Providing AGC through a wind turbine has been investigated on an NREL CART3 600 kW

IOP Conf. Series: Journal of Physics: Conf. Series **1102** (2018) 012032 doi:10.1088/1742-6596/1102/1/012032

turbine in [10] where reasonable accuracy is shown between the produced power and the target. As discussed in [11] either a delta strategy (% of available power) or a fixed reserve can be used as to provide the base power upon which to add the system operators AGC signal. In this investigation a fixed reserve of 1 MW was used while the wind park was generating between 1 MW and 10 MW (rated farm power). The fixed reserve of 1 MW represents 10% of rated farm power and is chosen to be consistent with previous tests performed with the same turbines [12].

1.1. Technical Objective
The objective of this investigation is to test the ability of a wind farm to provide secondary frequency regulation in order to understand the technical challenges and limitations of providing this service from a wind farm. This work demonstrates the potential of a wind farm to provide AGC in a manner similar to a conventional generator. We use a power set-point update frequency of 4 seconds. This contrasts with typical wind turbine curtailments which do not change for minutes or hours. This work builds upon [12] and examines providing AGC in the region between cut-in and rated wind speed (see Figure 5).

1.2. Test Site Information
This test is performed at the Wind Energy Institute of Canada's 10 MW Wind R&D Park located in Prince Edward Island, Canada. It has 5 DeWind D9.2 turbines, each of which have a 2 MW rating, 93 m rotor diameter and a hub height of 80 m. They are unique turbines in that they have a synchronous generator that is directly grid connected. The generator is connected to the variable speed rotor through a hydraulic torque converter (Voith WinDrive) and a two-stage gearbox [13]. It is considered a Type 5 wind generator [14]. The generators generate at 13.8 kV, which is stepped-up to 69 kV to tie into the local network. The wind park has an active power set-point target and a voltage set-point target, both of which are controlled by the local utility. During the testing period, the voltage set-point target was constant at 71.0 kV and the active power set-point target was varied. The active power set-point was disconnected from utility control and was updated every 4 seconds.

2. Algorithm Methodology and Formulation
2.1. Algorithm
An overview of the implemented algorithm is presented in Figure 2. We begin by using a 10 min rolling average of the hub height wind speed measured at each turbine. The averaging is done by the turbine's control system. The averaged values are read on a SEL RTAC 3530 controller via DNP3. Here, these are used to estimate the power available at each turbine. The theoretical power is the power that each turbine is capable of generating at any instant of time based on the prevailing wind speed. The estimation is done using a power curve. A power curve for each wind turbine is determined based on the historical nacelle wind speed and power production data. This is done to avoid errors when using the manufacturer's design power curve (free-field wind speed versus measurements down wind of the rotor). Between cut-in and rated wind speeds, the power curve is modeled by a second order polynomial fit to data from December 2017 and January 2018. The averaged wind speed value is used with each turbine's power curve to estimate the power production. The sum of these powers across the wind park produces the "theoretical power" value (Figure 3 (b)).

To accommodate the regulation bid (1 MW), half the regulation bid value is subtracted from the calculated theoretical power. This creates room for 0.5 MW of up-regulation. This produces the baseline power value (See Figure 2 and Figure 3). We provide regulation in the range of 1 MW to 10 MW of net park power. Below net park power levels of 1 MW, the AGC signal is ignored. In the regulation range of 1 MW to 10 MW, we provide 1 MW of AGC room, i.e. 500 kW of up-regulation and 500 kW of down-regulation.

The historical AGC signal data is read from a CSV file by a simple LabVIEW script. This script runs on an independent computer that acts as a MODBUS server. The AGC signal is sent to the RTAC via MODBUS. This AGC value is summed with the baseline power value to produce the park MW set-point value (Equation 1). The AGC signal acts as a bias value for the baseline power value. This value is sent from the RTAC to the DeWind park control system via DNP3. A closed loop power controller is implemented in the DeWind park controller, which distributes the park power set-point across the turbines. We do not control the active power setpoints for individual wind turbines.

$$Park\,power\,target = Theoretical\,power - 500\,kW + AGC\,Local \tag{1}$$

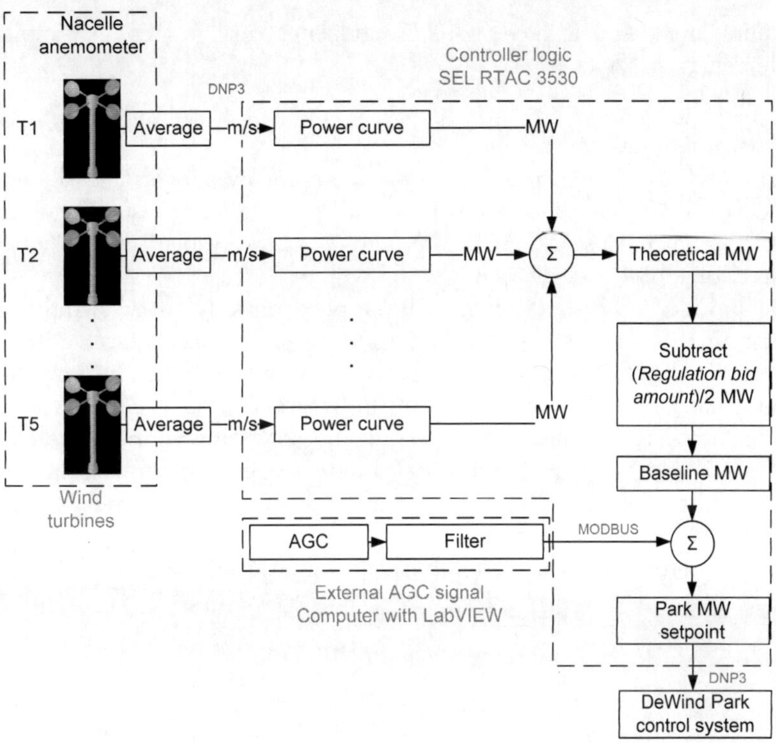

Figure 2: Developed curtailment algorithm

2.2. Historical AGC Signal and regulation room

This work uses a 4.5 hour (270 minute) AGC signal (in MW) extracted from historical data provided by the Alberta Electric System Operator (AESO). A new AGC set-point is calculated and generated every 4 seconds. It is important to emphasise that this AGC signal did not have any correlation with prevailing meteorological or grid conditions during the test. AESO send a time-varying AGC signal to generators over and above an energy dispatch value (MW). The energy dispatch value is relatively constant. As illustrated in Figure 3, the key difference in the interpretation of the regulation room in the case of variable generators is in the changing energy dispatch level—the dashed red line in Figure 3 (a). This figure illustrates how the energy dispatch level translates to this work. This variation is a consequence of the power output of a wind farm being dependent on a variable input (wind speed).

This work uses a 1 MW regulation market offer. This translates to a maximum power change of 0.5 MW above or below the calculated "baseline power" at any given time (See Figure 3).

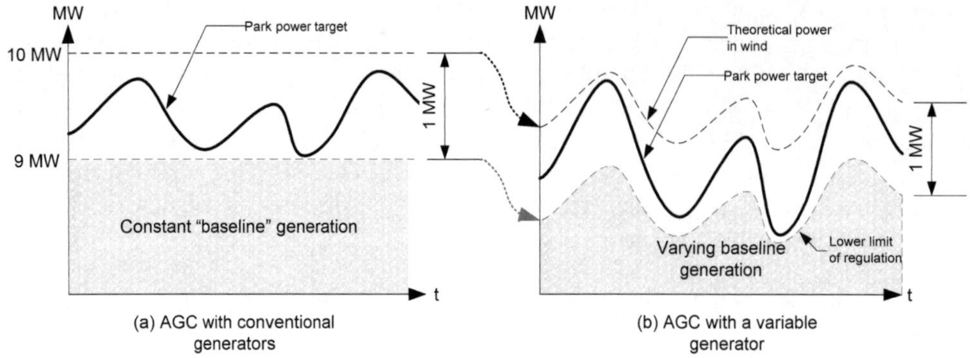

Figure 3: Difference in the appearance of AGC regulation room in a conventional generator (a) versus a variable generator (b)

2.3. AGC signal scaling and filtering

The AGC signal is extracted from the AGC MW set-point by subtracting the energy dispatch value. This AGC signal is then scaled to fit in a 1 MW range using a scaling factor. The signal maximum and minimum were set to +0.5 MW and -0.5 MW respectively. The result of this is shown in Figure 4 (blue plot).

The scaled & limited AGC signal cannot be applied directly to the wind park for several reasons. Primary among these is the stochastic nature of the park power output. This could result in drowning out the effect of the AGC signal change if the change in prevailing wind speed produces a greater change in the park power output. Other factors include the finite response time of the park controller, turbines response time and communication delays.

To avoid repeated, small changes to the park power target, a differential magnitude filter is implemented. The filter output changes only when the difference between the new AGC set-point value and the existing value is greater than 100 kW as shown in Equation 2. The 100 kW value for Δ_{filter} was determined through empirical evaluation.

$$\begin{aligned} if\ |x[n] - x[hold]| > \Delta_{filter} \\ then\, y = x[n] \\ else,\ y = x[hold] \end{aligned} \tag{2}$$

The AGC signal before and after filtering is shown in Figure 4. It is important to note that the filtered AGC signal shown in Figure 4 is not used for performance score calculations. Signal filtering is performed internally and is not seen by the network operator. Performance score calculations must be done using the network operator's AGC signal. In this case, this is the scaled and limited signal as shown in Figure 4.

2.4. Operating region of the wind turbines

This work seeks to demonstrate that a wind park can be used to provide regulation services when operating below rated power. This is in continuation to the work in [12], which performed a similar test but with the wind turbines operating at rated power. These two situations are contrasted in Figure 5. The result of operating below rated power is that the output of the wind farm is not steady at 10 MW but varies depending on the prevailing wind speed. The "theoretical power" thus calculated is constantly varying.

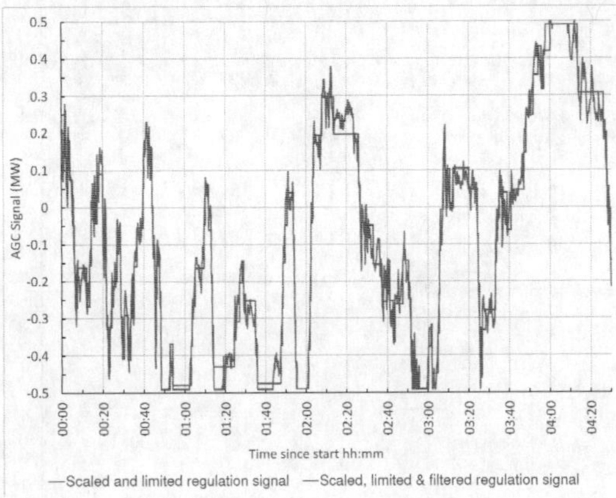

Figure 4: Historical AGC signals and filtered AGC signal as used for park power setpoint

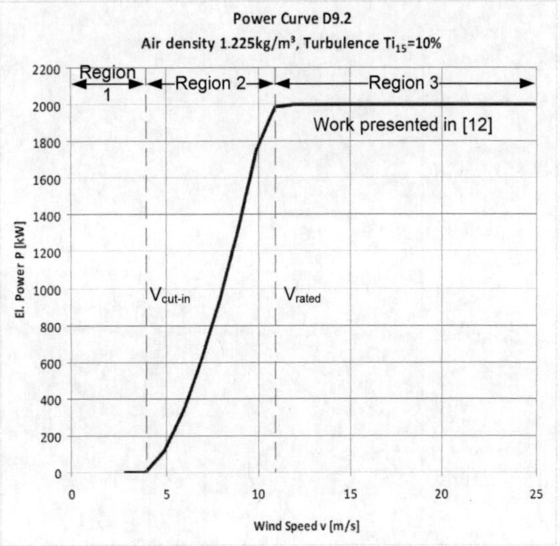

Figure 5: Turbine operating region of this work (Region 2) versus that of [12] (Region 3)

Table 1: Experiment statistical summary

	AGC local filt. (kW)	Power Park Target (kW)	Net park power (kW)	Power Error (kW)	Abs. Power Error (kW)	Pitch T1 ()	Pitch T2 ()	Pitch T3 ()	Pitch T4 ()	Pitch T5 ()	T1 Pow. (kW)	T2 Pow. (kW)	T3 Pow. (kW)	T4 Pow. (kW)	T5 Pow. (kW)
Max.	490	6366	6877	1381	1899	13.1	12.0	12.9	12.8	11.8	1805	2040	2023	1728	1942
Avg.	-79	4058	4059	1	235	2.7	4.4	4.2	3.6	3.7	802	830	845	818	825
Min.	-490	1553	1593	-1899	0	0.7	0.8	0.7	0.7	0.7	90	219	258	41	173

3. Results

The test was performed on March 7th, 2018 from 12:00 until 16:30 during a period when the 10-minute average wind speed ranged from 6.7 to 9.5 m/s. This is between the cut-in wind speed of 4.5 m/s and the rated wind speed of 10.5 m/s. A statistical summary of the experiment period is presented in Table 1 where it is important to note that this 4.5-hour test had a range of net park power from 1600 kW to 6800 kW, which is half of the entire 0 to 10 MW range of the wind park. Figure 6 shows the test results as a time series.

3.1. Park Performance Scores

A performance score quantifies how well the generator's output follows the AGC signal. There are multiple methods to calculate the performance score. We use two different methods to calculate the performance score. Results from both methods are summarised in Table 2 and the two methods are detailed below.

3.1.1. NRC Method The first performance scoring method used is the National Research Council Canada (NRC) method developed for AESO conventional generators [15]. This method

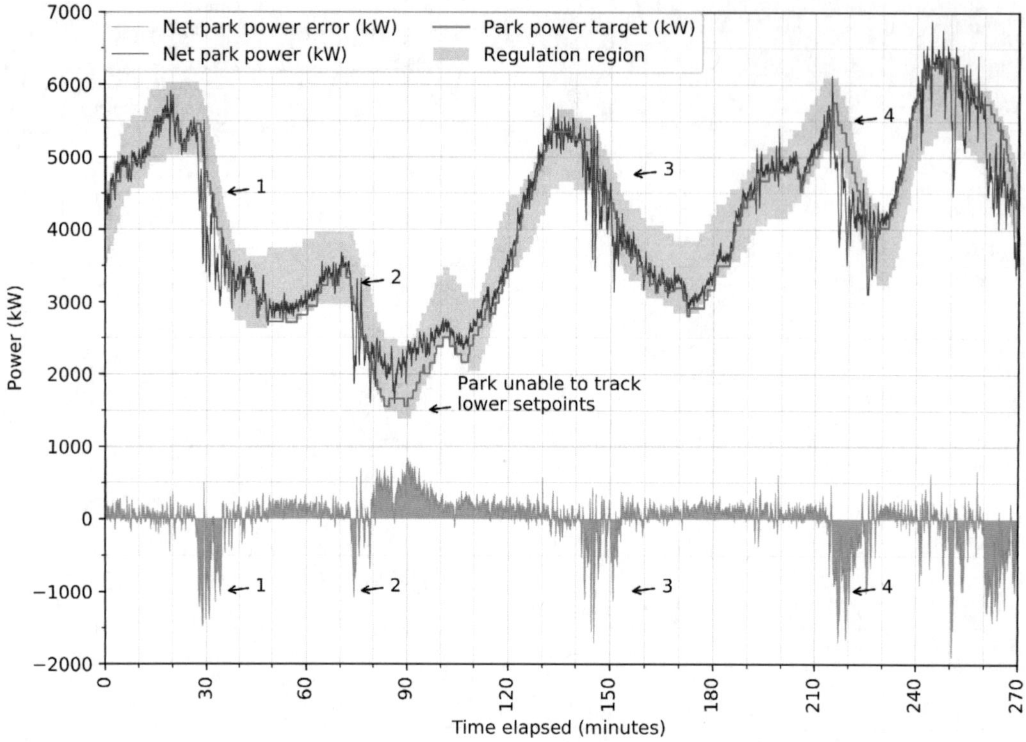

Figure 6: Results from experiment. Numbered points indicate instances of decreasing average wind speeds.

Table 2: Performance score comparison with other technologies [15] [16]

	NRC η	PJM η
Coal	67%	76% (Reg A)
Hydro	86%	78% (Reg D)
Gas	75%	91% (Reg D)
Energy Storage	-	92% (Reg D)
This work[a]	28%	64%
Rated Power[a] [12]	94%	74%

[a]PJM reg signal not used

is based on that developed in [17]. The performance score (η) is between 0 and 1 with 1 being perfect performance. To begin, we first calculate a value of δ which is a ratio of two sums. The numerator sum is the deviation of the net park power from the AGC setpoint for each 2 second interval. The denominator sum is the deviation between the economic dispatch value and the AGC setpoint. This is shown in Equation 3 [15].

$$\delta = \frac{\sum_t |AGC\,setpoint - Net\,output|}{\sum_t |AGC\,setpoint - Economic\,dispatch|} \tag{3}$$

Converting Equation 3 to our terminology results in Equation 4 where ($Theoretical\,power - Regulation\,bid$) represents the lower limit of regulation. In our case, the lower limit of regulation is equal to the economic dispatch value.

$$\delta = \frac{\sum_t |Park\,Power\,target - Net\,park\,power|}{\sum_t |Park\,Power\,target - (Theoretical\,power - Regulation\,bid)|} \tag{4}$$

The Performance Score is then calculated with a sigmoid function of δ. The best performance is when the net park power is equal to the park power target, which would result in $\delta = 0$ and $\eta = 1$.

$$\eta = \frac{erfc(a \cdot \delta - b)}{erfc(-b)} \tag{5}$$

Where $erfc$ is the complementary error function and a and b are 3.0 and 1.25 as in [15].

We calculate an NRC performance score of 28% for our test. The performance scores at different net park powers is shown in Figure 7 for this test period along with the results from [12] (full power operation). The dashed line indicates power ranges where data is not available. Note how quickly the performance improves for power levels above 4000 kW.

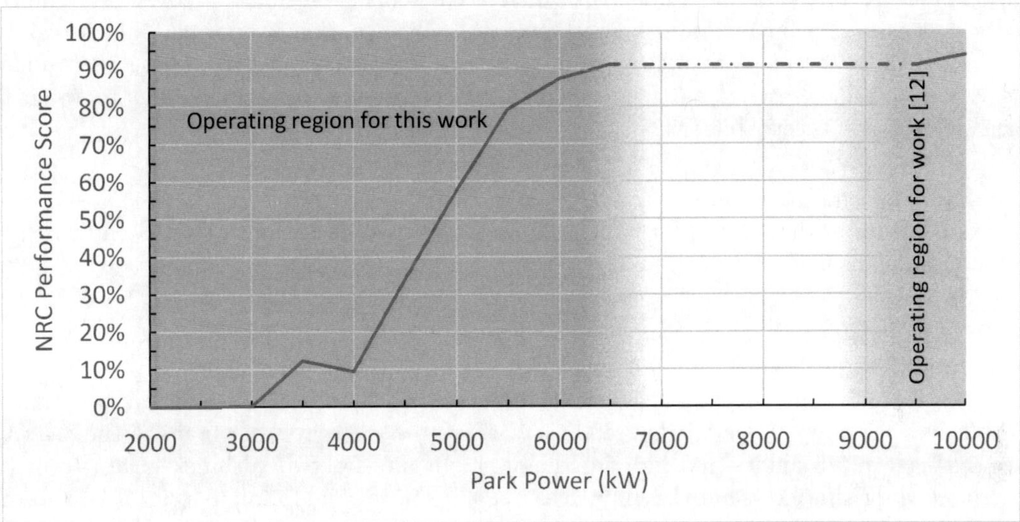

Figure 7: NRC performance score collated across this work and [12]

3.1.2. PJM Method The second performance scoring method is that of PJM (Pennsylvania-New Jersey-Maryland). PJM's performance score calculation method [18] [19], is the average of three performance metrics: Accuracy, Delay, and Precision. A performance score for each of the five hours of the test was calculated. The last hour had only 30 minutes. The weighted average of each hour's performance scores was calculated to find the PJM performance score for the test period resulting in a net performance score of 64%. The values of Unit AREG and Fleet TREG were set to 0.5 MW and a park ramp rate of 32 MW/min was used. The results are in Table 3.

3.1.3. Performance Summary As shown in Table 2, the net performance score calculated according to the NRC method is 28%. This is in contrast to the 94% reported in [12]. The reason for this difference is the amount of time spent in the low power range (< 4000 kW) and the large error when the wind speed was decreasing (See Figure 6). This is a consequence of the NRC score being based entirely on precision. The PJM method has a lower weighting for

Table 3: Hourly PJM Performance scores over test period

Hours	1	2	3	4	5	Weighted avg
Accuracy Score	0.63	0.75	0.53	0.62	0.57	62.8%
Delay Score	0.92	0.92	0.90	0.92	0.75	89.9%
Precision Score	0.43	0.37	0.46	0.32	0.25	38.3%
Composite Score	0.66	0.68	0.63	0.62	0.53	63.7%

precision [18] and therefore, the difference between the rated power test (calculated from [12] data) and this test's results is not as stark.

Based on the results presented in Table 2, the net NRC performance score implies that a wind park is not effective at providing secondary regulation. This would appear to be a consequence of the wind park operating below 4000 kW for 46% of the test period. The turbine technology results in poor accuracy in this region, as seen in Figure 7. It is evident that performance improves as the wind park's power output increases. Although the PJM performance score is below that of traditional generators, some of the difference can be explained as above. One limitation of our results is limited data with which to calculate performance scores. The solution would be to perform a longer duration test over the whole park power range. This test will produce more data to calculate performance scores at each power level. These performance results can be used to restrict participation in the secondary regulation market to when the performance score is acceptable.

3.2. Examining performance scores

This section examines the reasons for poor performance scores at low power levels and when the wind speed is decreasing. Both explanations rely on Figures 6, 7 and 8.

3.2.1. Poor performance at low power levels As visible in Figure 6, the net park power follows the park power target closely but deviates during a few intervals when the power error is significant. From Figure 7, the effect of the power error on the performance score is most noticeable at park power levels below 4000 kW. Figure 8 presents one hour of the test and focuses on blade pitch angles and individual power outputs for two turbines. Data from two turbines is used for simplicity but the inferences made here are valid for all five turbines. Observe the large difference in the park power target and the park net power in Figure 8 (a) - Region II where the generated power is greater than the target power. This would indicate that a lower limit of curtailment has been reached. As visible in Figure 8 (c), the turbines do not curtail their individual power outputs to below 500 kW, resulting in a large, positive power error. Observe also that the turbine power reduces to below 500 kW only when the wind speed is insufficient to produce 500 kW. When the wind speed permits, each turbine's power output is actively controlled to a minimum of 500 kW by the pitch system. This is visible in Figure 8 (b), Region II. The rotor speed stays almost constant over the entire period presented in Figure 8. It is unclear why the turbines do not curtail their active power outputs to below 500 kW. This lower curtailment limit applies to each turbine individually and not the park as a whole. Each turbine sees a different instantaneous wind speed and reaches this 500 kW limit independent of the other turbines. This causes the net park power error to increase below 4000 kW and not simply 2500 kW (5×500 kW). A possible solution is to limit the active power region in which the wind park participates in the regulation market.

3.2.2. Poor performance during periods of decreasing wind speed The theoretical power is based on the 10 min average wind speed. The averaging process acts as a filter and adds a delay between the instantaneous and average trends. When the instantaneous wind speed drops, the power

IOP Conf. Series: Journal of Physics: Conf. Series **1102** (2018) 012032 doi:10.1088/1742-6596/1102/1/012032

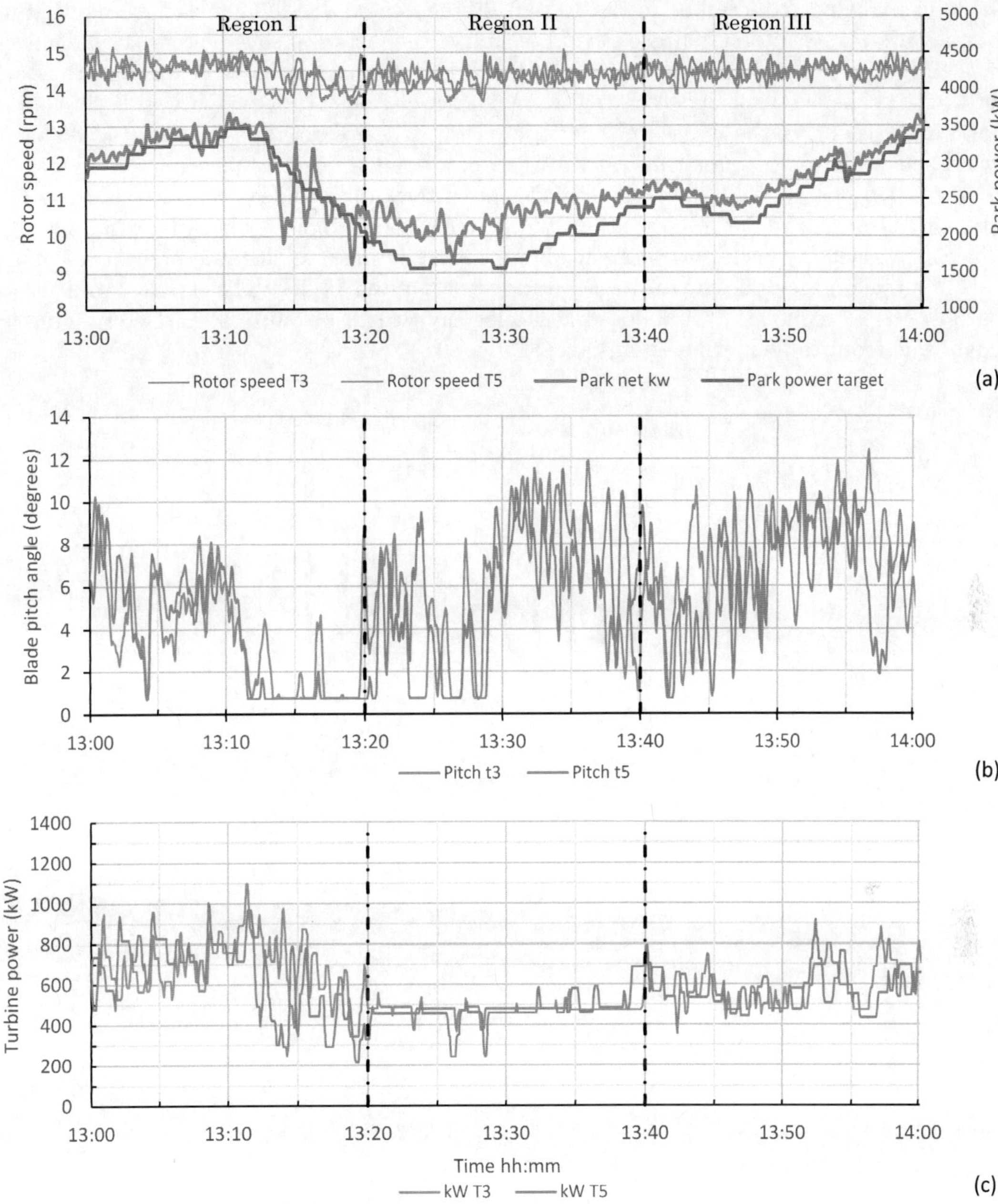

Figure 8: Turbine performance during different operating situations. Region I illustrates performance during decreasing wind speeds, Region II illustrates the power error in the low power region (See Section 3.2.1) and Region III illustrates performance during increasing wind speeds.

IOP Conf. Series: Journal of Physics: Conf. Series **1102** (2018) 012032 doi:10.1088/1742-6596/1102/1/012032

available in the wind reduces before the 10 min average value does. This leads to a situation where the Park Power Target is higher than the sum of the instantaneous powers available from the five turbines. This is the cause of the mismatch between the net park power and the park power target as the wind speed decreases (See numbered regions of Power Error in Figure 6 and Region I in Figure 8). In these cases, the park power target is greater than the power available in the wind and the turbines cannot meet the target, producing a negative power error. In Region I in Figure 8 (b) observe that the turbine blades pitch towards zero degrees to maximise power production. Over the whole test duration, the power error is highest when the wind speed is insufficient to meet the park power target. From Figure 9, observe that the percentage power error is negative and has the greatest magnitude at zero pitch angles. The simplest solution to this problem is to reduce the wind speed averaging interval to below 10 min. Determining an optimal value requires further investigation.

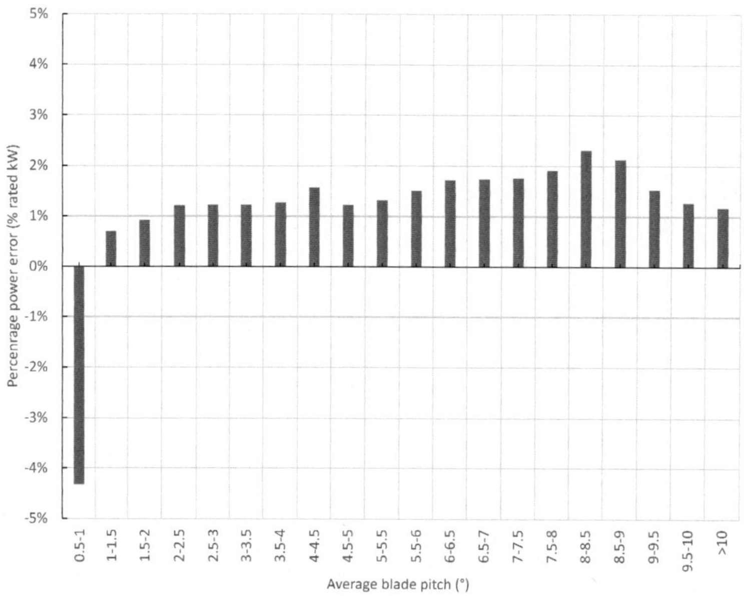

Figure 9: Blade pitch versus power error

4. Conclusion

Obtaining secondary frequency regulation services (i.e. up- & down- AGC) from wind plants is technically viable. This work presented a simple algorithm to allow for wind park participation in the secondary regulation market. The algorithm presented allows for 1 MW of regulation room on a 10 MW wind farm through dynamic curtailment. An external, historical AGC signal is used as a bias for the park power set-point. Park performance was evaluated using two calculation methods, NRC's method and the PJM method. This demonstration was carried out on a wind park with Type 5 wind turbines, a design that is uncommon. The algorithm developed results in poor performance during periods of decreasing wind speeds. We also observe poor performance at lower park power levels. This is likely a consequence of the control system of the specific turbines tested.

Acknowledgements

The authors would like to thank Natural Resources Canada (NRCan) for their financial support and the Alberta Electric System Operator (AESO) for data inputs. This work made extensive

IOP Conf. Series: Journal of Physics: Conf. Series **1102** (2018) 012032 doi:10.1088/1742-6596/1102/1/012032

use of OSIsoft's PI data management system for data logging and analysis.

References

[1] *RED Electrica De Espana: 2017 The Spanish Electricity System Preliminary Report 2017* p 14 URL http://ree.es/sites/default/files/downloadable/avance_informe_sistema_electrico_2017_eng.pdf
[2] Poul L 2018 2017: A record year for wind in germany and denmark Online report URL https://news.vattenfall.com/en/article/2017-record-year-wind-germany-and-denmark
[3] Lai L L 2001 *Power system restructuring and deregulation: trading, performance and information technology* (John Wiley & Sons)
[4] *California Independent System Operator: Market processes and products* URL https://www.caiso.com/market/Pages/MarketProcesses.aspx
[5] Altin M, Göksu Ö, Teodorescu R, Rodriguez P, Jensen B B and Helle L 2010 Overview of recent grid codes for wind power integration *Optimization of Electrical and Electronic Equipment (OPTIM), 2010 12th International Conference on* (IEEE) pp 1152–1160
[6] Eirgrid 2017 *Annual Renewable Energy Constraint and Curtailment Report 2016* URL http://www.eirgrid.ie/site-files/library/EirGrid/Annual-Renewable-Constraint-and-Curtailment-Report-2016-v1.0.pdf
[7] NERC 2011 balancing and frequency control: A technical document prepared by the nerc resources subcommittee Tech. rep. North American Electric Reliability Corporation
[8] Kirby B, Milligan M and Ela E 2010 Providing minute-to-minute regulation from wind plants: Preprint Tech. rep. National Renewable Energy Laboratory (NREL), Golden, CO.
[9] IESO 2018 Independent electricity system operator: Markets and related programs URL http://www.ieso.ca/en/sector-participants/market-operations/markets-and-related-programs/ancillary-services-market
[10] Aho J, Buckspan A, Laks J, Fleming P, Jeong Y, Dunne F, Churchfield M, Pao L and Johnson K 2012 A tutorial of wind turbine control for supporting grid frequency through active power control *American Control Conference (ACC), 2012* (IEEE) pp 3120–3131
[11] Ziping W, Wenzhong G, Tianqi G, Weihang Y, Zhang H, Shijie Y and Xiao W 2018 *Journal of Modern Power Systems and Clean Energy* **6** 1–16
[12] Nasrolahpour E, Houston C, Harper S, Rodgers M, Zareipour H and Rosehart W 2017 Bidding strategy in energy and regulation markets for a wind power plant *PROCEEDINGS OF IREP'2017 SYMPOSIUM* pp 1–5
[13] Dewind d9.2 wind turbine specifications URL http://www.dewindco.com/eng/product/prod01_01.asp
[14] Habash R Wind turbine generators for wind power plants URL http://www.site.uottawa.ca/~rhabash/ELG4126WindGenerators.pdf
[15] Kabiri K and Song D 2018 *Regulating reserve performance assessment for the Alberta electric system operator*
[16] Croop D 2017 Regulation update URL https://www.pjm.com/~/media/committees-groups/committees/oc/20170711/20170711-item-19-regulation-update.ashx
[17] Papalexopoulos A D and Andrianesis P E 2014 *IEEE Transactions on Power Systems* **29** 441–449
[18] Croop D 2016 Performance scoring URL http://www.pjm.com/~/media/committees-groups/task-forces/rmistf/20160413/20160413-item-02-performance-scoring.ashx
[19] Benner S 2018 Performance score template URL http://pjm.com/~/media/markets-ops/ancillary/mkt-based-regulation/performance-score-template.ashx?la=en

IOP Conf. Series: Journal of Physics: Conf. Series **1102** (2018) 012033 doi:10.1088/1742-6596/1102/1/012033

Hybrid Auxiliary Power Supply System for Offshore Wind Farm

Xing Huang [1], Yao Chen [2]

[1] BEZIT Park Block 103 Level1, Jiuxianqiaobei Road, NO.10, Chaoyang District, Beijing, China. Post Code: 100015
[2] Forskargränd 7, SE-721 78, Västerås, Sweden

[1] helen-xing.huang@cn.abb.com
[2] cathy-yao.chen@se.abb.com

Abstract. Diesel Generators (DG) are commonly adopted to supply the critical auxiliary loads for Offshore Wind Farm (OWF) during islanding operation. To reduce fuel consumption, avoid over-sizing of DG, and further enable a smooth resynchronization to main grid, a hybrid Auxiliary Power Supply (APS) system is designed with a combination of DGs, Wind Turbine Generators (WTGs), and battery Energy Storage System (ESS). Different from the typical industrial solution, this hybrid APS system uses not only DGs, but also WTGs to supply part of the auxiliary load when there is wind blowing for reducing fuel consumption and the corresponding refuelling works; moreover, the wind turbine converters can provide reactive power for balancing the charging power from the cable array inside the OWF, so as to avoid the oversizing of DG itself. ESS are also introduced to further reduce the dependence on the DGs. With enough capacity of ESS, it is even possible to remove DGs while still be able to fulfil the requirement on auxiliary load supply during islanding mode.

In this paper: firstly, the cost effectiveness of such hybrid APS is discussed via techno-economic analysis; secondly, two types of hybrid APSs with coordinated hierarchical control architectures are introduced, as well as the control strategy design for both centralized and distributed controllers; thirdly, simulation models are developed in MATLAB/Simulink, where two types of models are included. They are phasor-type model for centralized control strategy verification, and average-type model for distributed control strategy verification. The performance of hybrid APS models is validated under typical working scenarios, and the simulation results show that the control strategy has no high requirement on the communication speed between central controller and distributed controllers.

1. Introduction

As defined in the GL Rules for Classification and Construction IV (Industrial Services for Offshore Technology) [1], for all offshore installations, an independent emergency source should be provided, which is used for electrical power supply to the wind turbine auxiliary load, when the main power supply fails, and the supply time period should meet the requirement from half an hour to 4 days for different loads. Similarly, the German Transmission System operator has set a requirement of APS system to provide power for 12 hours or even longer time under extreme emergencies [2]. It can be noticed that, the original UPS system installed in wind turbine is not enough to meet these requirements.

To fit this requirement, usually DGs are installed as the major power supply under emergencies. However, using DGs only for APS has disadvantages:

- Firstly, fuel consumption is unavoidable. Take one export cable failure as an example, for a 500MW OWF, the fuel consumption can be as high as 1000tons per year. Given the diesel price at 150USD/bbl [3], the fuel cost is above 1MUSD per year.
- Secondly, according to Reference ([2]), the DGs have to be oversized to cover both the active power and reactive power demands. Take a 500MW OWF with 82 WTs for example. In the worst case, the DGs have to be oversized by above 200%.
- Thirdly, the DG requires additional space and weight (e.g. 2x50 tons [4]) in the platform for fuel storage, on top of that, periodical maintenance and refuelling are needed.

In order to solve these problems, this paper proposed a hybrid APS system with two alternative solutions, as shown in Figure 1. Different from typical industrial solution, this hybrid APS uses not only DGs, but also WTGs to supply part of the auxiliary load when there is wind blowing for reducing fuel consumption and the corresponding refuelling works; moreover, the wind turbine converters can provide reactive power for balancing the charging power from the cable array inside the OWF, so as to avoid the oversizing of DG itself. ESS can also be introduced to further reduce the dependence on the DGs. With enough capacity of ESS, it is even possible to remove DGs while still be able to full fill the requirement on auxiliary load supply during islanding mode.

In this paper: firstly, the cost effectiveness of such hybrid APS system is discussed via techno-economic analysis; secondly, the hierarchical control architecture of the hybrid APS system is introduced, as well as the control strategy design for both centralized and distributed controllers; thirdly, simulation models are developed in MATLAB/Simulink and the performance of the hybrid APS system is validated under typical working scenarios.

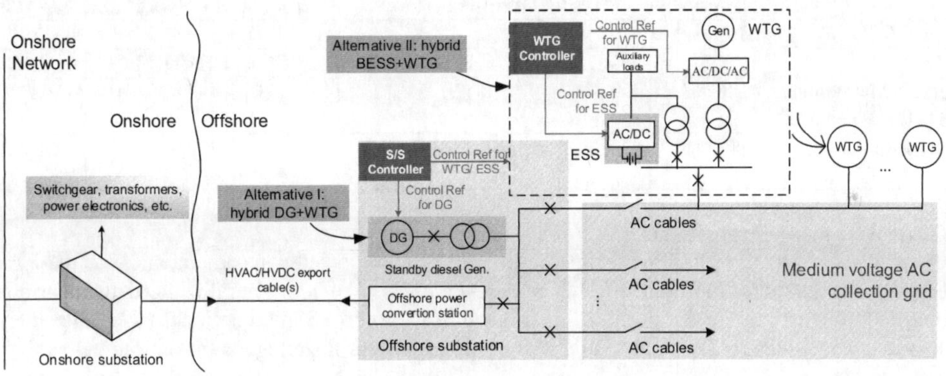

Figure 1 Diagram of Hybrid APS

2. Techno-economic analysis

The techno-economic analysis in this report mainly focuses on the cost saving generated by hybrid APS solution. Cost saving analysis results show that, this hybrid APS solution is able to achieve rational economic benefits for OWF, compared with original APS (only DG for APS). Detailed calculations are introduced briefly in Appendix A. Some basic conclusions for this hybrid APS solution are summarized here:

- Alternative I - Hybrid DG + WTG:

  This solution demonstrates rational economic benefits in terms of both capital cost saving and life time maintenance cost saving. Take a 500MVA OWF as an example. The final capital cost saving is around 14.6MUSD, which is around 4% of electrical system cost. Assuming WTGs can share 50% of auxiliary power, the annual maintenance cost saving is in the range of 76.5~645.4kUSD per year considering different transmission technology and shipment distance, which is around 1% of annual operation and maintenance cost for a HVDC VSC OWF. The total life time cost saving of this hybrid APS in 30 years is in the range of 17~34MUSD, as shown in Table 1.

Furthermore, sensitivity study shows that, the increase of fuel price will bring significant economic benefit for hybrid APS on maintenance cost saving in the following years. Assume ~38% increase of fuel price to 2040, the maintenance cost saving of the new hybrid APS solution can increase ~25%. Solutions like multi-substations and platforms can help to decrease the unavailability of OWF, thus save the maintenance cost for fuel and refuelling. However it will introduce a much higher capital investment as well, which is less cost effective than the hybrid APS solution in this perspective.

- Alternative II - Hybrid ESS + WTG:

This solution only demonstrates higher economic benefits on life time maintenance cost saving, with little contribution to capital cost saving. Also take the 500MVA OWF as an example, this solution can save annual maintenance cost in the range of 152~1290kUSD per year considering different transmission technology, as shown in Table 1, while the capital investment is more or less the same. This is due to that the application of ESS can avoid the fuel consumption and refuelling requirement, however the capital cost of battery is also high.

In this case, the sensitivity study on battery price is done for the economic analysis of ESS+WTG solution. The prediction from DOE shows fast decrease of battery price in the following years. Compare with the price in 2010, the battery price may decrease to 10% in 2030. [3] Along with this decrease, the total life time cost of a 500MVA offshore wind farm may decrease to 11 million USD. The decrease of battery price will make the hybrid ESS + WTG solution more attractive.

Table 1 – Cost Saving (CS) results of a 500MW OWF with distance to shore from 50km to 300km

	Alternative I - Hybrid DG + WTGs			Alternative II - Hybrid ESS + WTGs		
Capital CS	15 MUSD			~ 0 kUSD		
	HVAC	HVDC LCC	HVDC VSC	HVAC	HVDC LCC	HVDC VSC
Annual Maintenance CS (kUSD/Year)	[76, 135]	[78, 162]	[420, 645]	[152, 270]	[154, 324]	[842, 1290]
Total Life CS MUSD/20Years	[17, 19]	[17, 20]	[27, 34]	[3, 5]	[3, 6]	[17, 26]

3. Control strategy development

Two types of systems have been developed to design the coordinated control strategies for the hybrid APS system, as illustrated in Figure 2(a) where the APS consists of both centralized DG and distributed WTGs, and (b) where the APS consists of distributed WTGs with ESS and optionally centralized ESS. Both system adopt hierarchical control architecture. Details control strategy is introduced below.

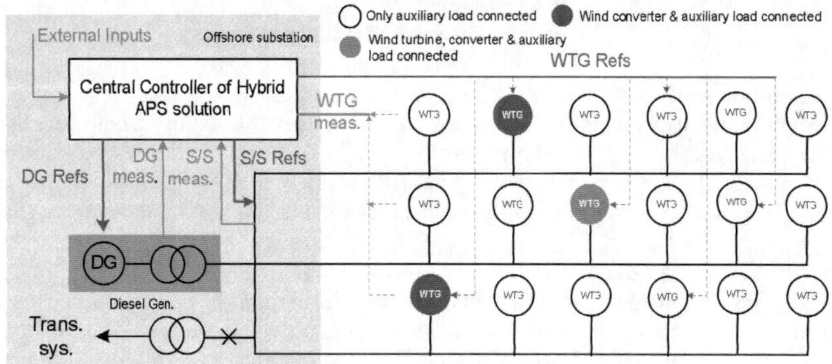

(a) Centralized DG and distributed WTGs

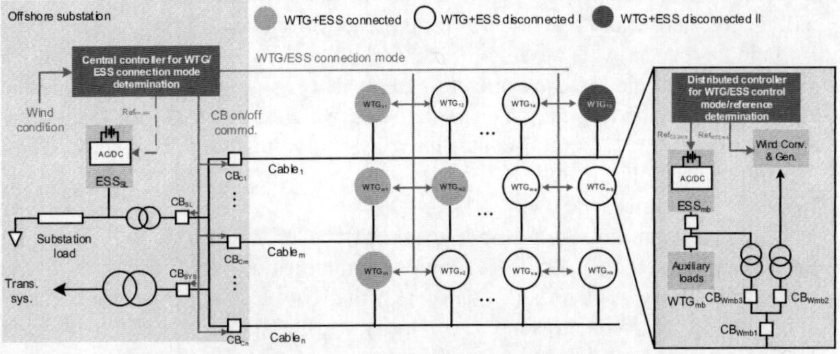

(b) Distributed WTGs with ESS and optional centralized ESS

Figure 2 Control system architecture

In Figure 2(a), three different connection modes of WTGs are defined:

- Mode 1 (White): only the auxiliary loads of WTG are connected to the collection grid, which can be seen as a pure load in the islanding system;
- Mode 2 (Blue): both wind converter and the auxiliary loads are connected, which can be seen as a reactive power source in the islanding system;
- Mode 3 (Green): the whole WTG is connected to the collection grid, which can be seen as an active/reactive power source in the islanding system.

Two layers of control coordination are defined, considering the active and/or reactive power contribution from the WTGs based on the wind conditions.

- First tier coordination between DG and WTG: aiming to balance the power while still maintaining the minimum fuel consumption. The input factors include acceptable frequency/voltage variation range, wind speed, etc., the possible outputs include power/voltage references for DG, active/reactive power commands for group of WTGs, etc.
- Secondly tier coordination among multiple WTGs: Since the total capacity of WTGs of an OWF is far larger than the auxiliary loads, the WTGs should be selectively switched on, to take care of the active/reactive power commands allocated by the first tier coordination and meanwhile achieve optimal efficiency within WTGs and the collection grid. The input factors include generator type, location, WTG condition, etc., the possible outputs include on/off and active/reactive power commands for multiple WTGs, etc.

In Figure 2(b), also three connection modes of WTGs are defined:

- Mode 1 (Green): WTG+ESS connected, where circuit breakers CB_{Wmb1} to CB_{Wmb3} are closed;
- Mode 2 (White): WTG+ESS disconnected I, where only CB_{Wmb2} and CB_{Wmb3} are closed;
- Mode 3 (Blue): WTG+ESS disconnected II, where circuit breakers CB_{Wmb1} to CB_{Wmb3} are open.

Correspondingly, two layers of controllers are designed and coordinated: central level controller which is located in the offshore platform, and distributed level controller which is located in the individual WTG.

- Central controller to determine the connection mode of the multiple WTGs/ESSs according to the wind speed, wind turbine conditions, loading level and the layout of the OWF.
- Distributed controller to determine the control mode and corresponding control references for individual WTG/ESS according to the connection mode and system stabilization requirements. There are three control modes:
  - ■ Active/reactive power (P/Q) mode

    WTG in P/Q mode can operate as current source, and provide active and reactive power for the auxiliary load, but cannot establish the system voltage and frequency in islanded mode. Two sub-control modes are included: with or without droop characteristics in P/Q mode. With droop characteristics shown in Figure (3), multiple power sources can operate together in an autonomous way. Without droop characteristics, the distributed controller will receive active and reactive power references from central controller.

- Voltage/frequency (V/f) mode
  WTG in V/f mode can operate as synchronous voltage source with Primary Frequency/Voltage Regulation characteristics, which is similar to the droop characteristics shown in Figure (3). Central controller will send voltage and frequency references to the distributed controller in WTGs.
- Resynchronization mode
  WTG in resynchronization mode can operate as synchronous voltage source with Secondary Frequency/Voltage Regulation characteristics, which can be used to regulate the system frequency/voltage, and finally help the wind turbine to be re-synchronized & re-connected to other voltage source (i.e. utility grid) with little or even zero power injection, which is called seamless resynchronization. In this mode, central controller will send additional frequency/voltage offset to distributed controllers in WTGs, so that their droop characteristics are modified, and the system frequency/voltage are modified in sequence.

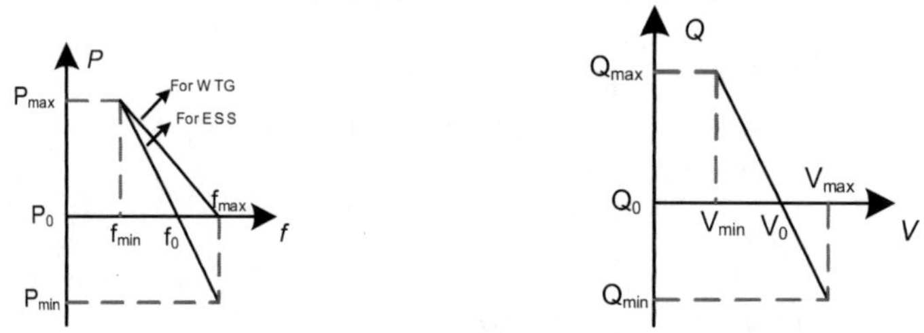

Figure 3 P/f and Q/V characteristics under P/Q control mode

4. Simulation verification

Take hybrid APS solution with ESS and WTGs for example. To verify the performance, two types of models are developed in MATLAB/Simulink.

- Phasor-type model
  This model is used for centralized control strategy verification, as shown in Figure 4(a). Four WTGs with ESS are modeled as controllable power sources, connected to the cable array modeled with reactance and capacitance, which are then connected to the platform. In order to simplify the internal modeling and primary control for WTG and ESS, which will be address by the average-type below. Both WTG and ESS in phasor-type model are controllable power sources. And the three-phase series RLC branch is inserted just as a high resistance to avoid two current sources connected directly. And corresponding measurement and PI control methods of WTG and ESS are also the same with the average-type model.
- Average-type model
  This model is used for distributed control strategy verification, as shown in Figure 4(b). Two WTGs with ESS as well as their control systems are developed, which are connected to the power source. As we mainly focus on the control strategy design of grid-side WTG/ESS, the influences caused by motor-side wind turbine converter or energy storage equipment are ignored here. This model is designed according to the mathematical equations in dq synchronous rotating reference frame, and is connect into the OWF collection grid by using controllable voltage source blocks.

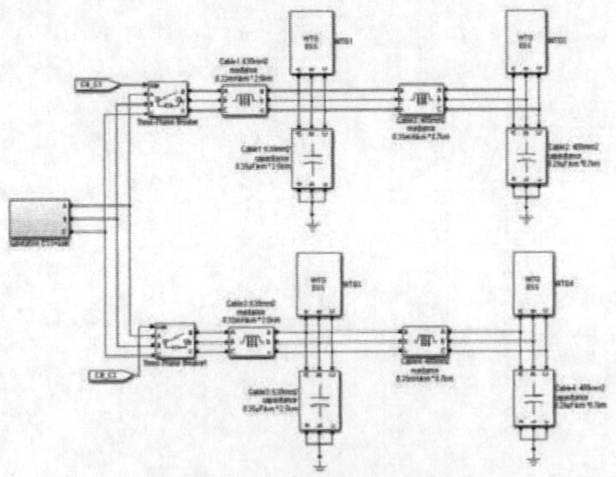

(a) Phasor-type model of wind farm

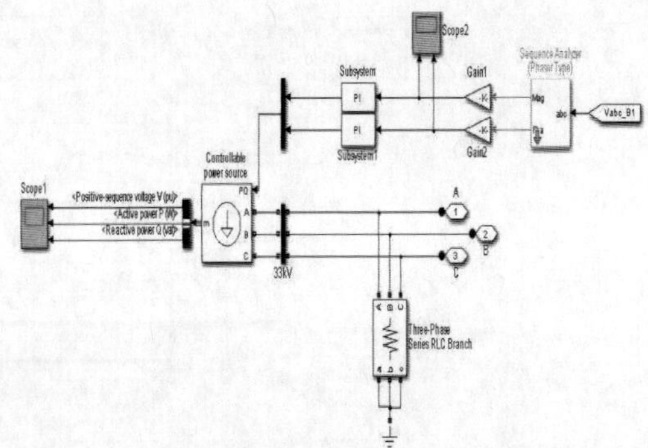

(b) Phasor-type model of WTG

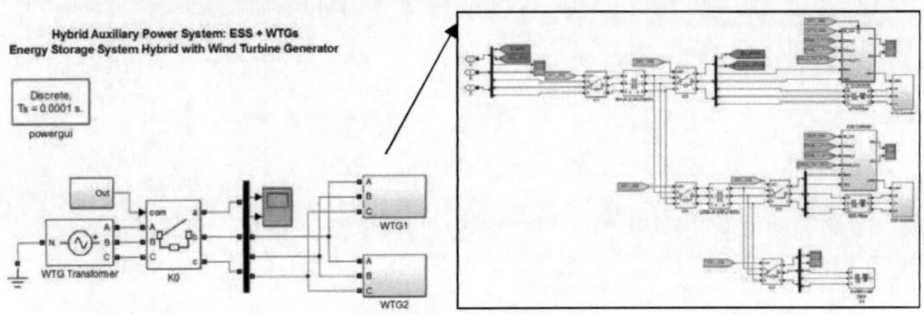

(c) Average-type model

Figure 4 Simulation model for hybrid ESS+WTG APS system control strategy verification

Figure 5 exemplifies one simulation case from the centralized control strategy under the scenario of wind speed changes. The central controller decides the connection mode (CM) of individual WTG according to the wind speed and the WTG availability, as shown in Figure 5(a). The active/reactive power references from the WTGs and ESSs are computed and adjusted in order to balance the load from the auxiliary systems and the cable array, as shown in Figure 5(b) and Figure 5(c). The total simulation time period is set to 100s, which can be divided into five time period as listed below:

- $t\in$[0s, 10s], all WTGs in connection mode 3, because the WTGs are not ready for operation yet. Thus the output power of WTG1 is zero, while ESS1 supplies 100% of local load.
- $t\in$[10s, 20s], all WTGs in connection mode 2, because wind power is not high enough to balance local auxiliary load and ESS charging demand. Since the wind speed is only 3m/s, P_{WTG1} is only about 70kW, and consequently P_{ESS1} is reduced to 10kW. WTG1 can fully compensate reactive power need from auxiliary load, so that the Q_{WTG1} increases to 60kVar, and Q_{ESS1} reduces to zero;
- $t\in$[20s, 60s], all WTGs in connection mode 1, because the wind power is high enough to balance local power demand, but the required number of wind turbine to balance the substation level power demand is not less than the available wind turbine for grid connection. WTG1 starts to connect to cable array as the first wind turbine, and at steady-state, P_{WTG1} is 166kW when the wind speed is 4m/s, and increase to 207kW when the wind speed increases to 4.3m/s. Besides, WTG1 takes care of balancing the reactive power demand of the small grid, which means Q_{WTG1} is 675kVar, including about 735kVar capacitive charging power from the two cables, and also 60kVar inductive power from the local auxiliary load.
- $t\in$[60s, 80s], only WTG1 and WTG2 are needed to operate as connection mode 1, because the wind speed is further increased, so that only two WTGs can fully compensate the substation level auxiliary load and ESS charging demand.
- $t\in$[80s, 100s], only WTG1 is needed to operate as connection mode 1, because the wind speed has further increased so that only one wind turbine is needed to connect to the cable array.

IOP Conf. Series: Journal of Physics: Conf. Series **1102** (2018) 012033 doi:10.1088/1742-6596/1102/1/012033

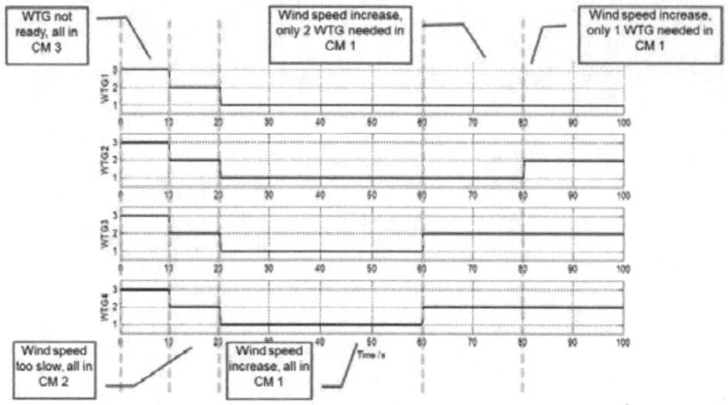

(a) WTG connection mode determination

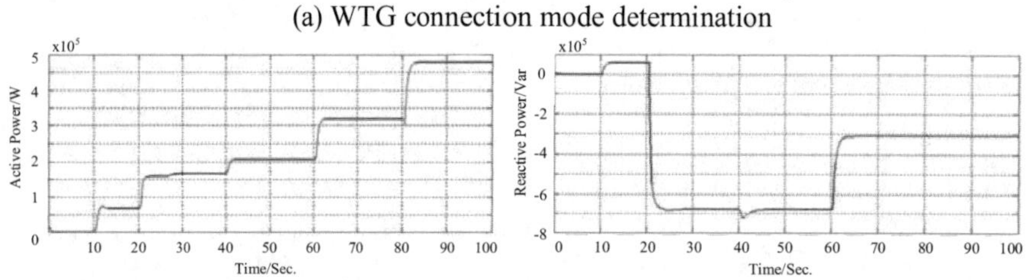

(b) WTG1 active/reactive power output

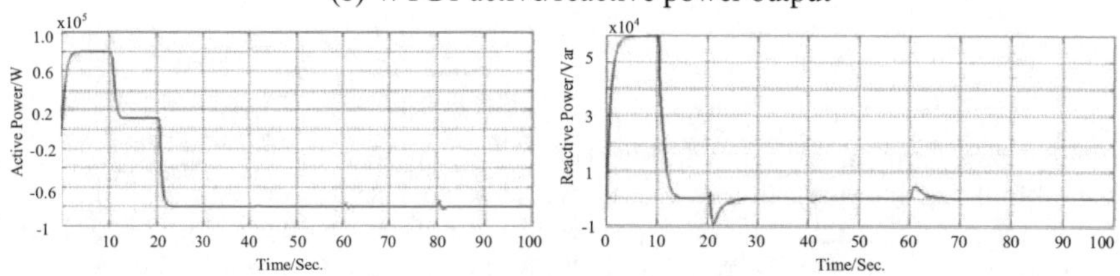

(c) ESS1 active/reactive power output

Figure 5 Simulation results from the phasor-type model

Figure 6 exemplifies the other simulation case from the distributed control strategy when WTG with ESS is in islanding mode operation from the beginning, through black-start and local system restoration get re-connected back to the grid. The V/f and P/Q control modes can be switched between WTG and ESS smoothly, and the re-connection to the grid is seamless with little power injection to grid. The total simulation time period is set to 60s, which can be divided into four time period as listed below:

- t∈[0s, 16s], Black Start Operation. Firstly, to build the system voltage and frequency, ESS1 and ESS2 are started with no load, and then supply the auxiliary load by closing the corresponding Circuit Breakers (CB). In this case two wind turbines are in Mode3. Secondly, to save the energy consumption of ESS, WTG1 and WTG2 are started in Mode2.
- t∈[16s, 32s], Local System Restoration. To help the auxiliary power supply for OWF, central controller may select several WTs (≥1) to form a local islanding system, which means to pick several WTs to work in Mode1. Assume WT1 and WT2 are both selected to be in Mode1, then WT1 will build the voltage of AC collection grid firstly, and WT2 will re-synchronize and re-connected to the AC collection grid by using Re-synchronous control method.
- t∈[32s, 48s], Seamless Grid Connection Operation. When the grid power is recovered, WT1 and WT2 should switch from islanding mode to grid-connected mode. To realize the seamless switch with less influence to the auxiliary power supply, and provide the OWF restoration with fast re-connection speed, the central controller will coordinate the WTs in a bottom-up way: firstly, WT1 and WT2 are in Mode2 to support their own AS load; secondly, central controller

IOP Conf. Series: Journal of Physics: Conf. Series **1102** (2018) 012033 doi:10.1088/1742-6596/1102/1/012033

closes the CB to recover the voltage of AC collection grid; and thirdly, each WTs control their own WTG/ESS to re-synchronize and re-connected to the collection grid.

- t∈[48s, 60s], the distributed controllers control WT1 and WT2 to transfer from the emergency control mode for auxiliary power supply to normal grid-connected mode for wind power generation. At t=52s, ESS1 will stop working, while ESS2 will switch into P/Q control without Droop characteristics for power charging. At t=54s WTG1 and WTG2 will switch into P/Q control without Droop characteristics.

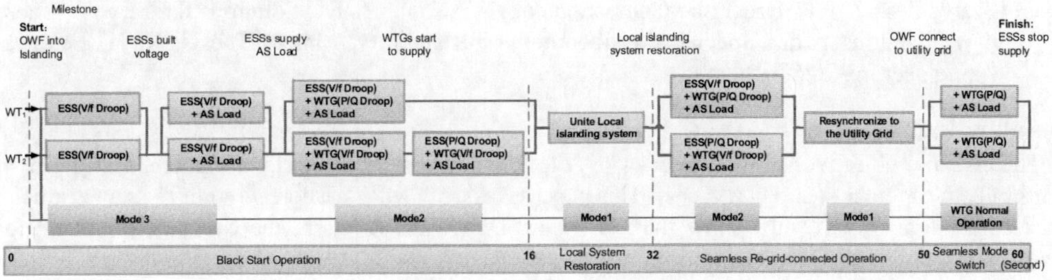

(a) Control sequence definition

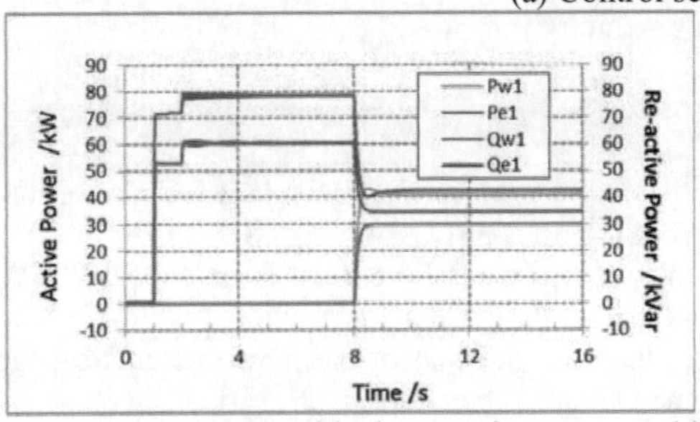

(b) P/Q of WT1 during black-start phase

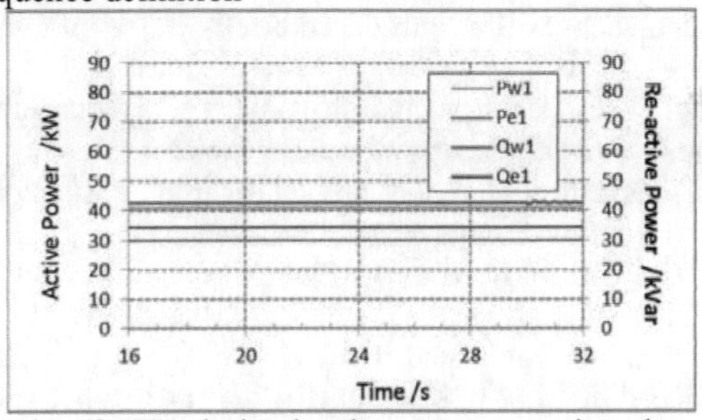

(c) P/Q of WT1 during local system restoration phase

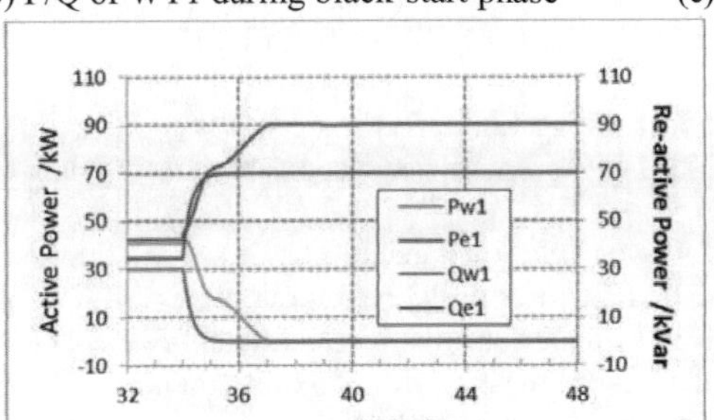

(d) P/Q of WT1 during grid connection phase

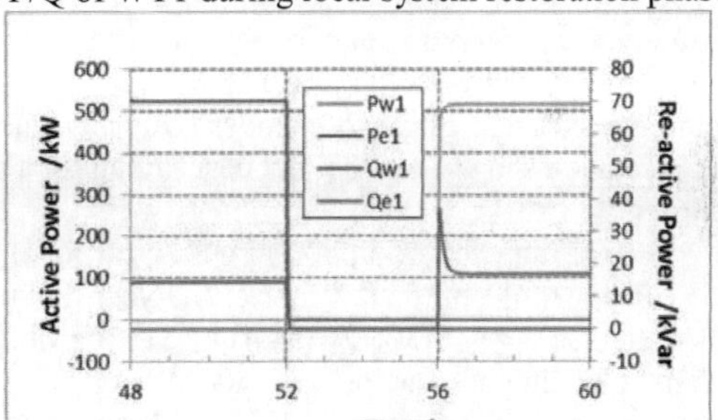

(e) P/Q of WT1 during seamless mode transition

Figure 6 Simulation results from average-type model

5. Conclusions

A hybrid APS system for OWF with DG+WTGs or ESS+WTGs is provided, which uses WTGs and/or ESS to supply critical auxiliary loads, so that the power capacity of DG and its fuel consumption can be reduced or even replaced. A hierarchical control system with two levels are proposed thereafter, to realize the coordination of multiple devices in the OWF. Simulation results show that the designed control strategy can deal with different cases while still maintaining the stable operation of OWF; the wind power can be fully utilized to share the power demand for both auxiliary loads and ESS charging. Plus, the control strategy has no high requirement on the communication speed between central controller and distributed controller.

References

[1] GL Rules for Classification and Construction, IV Industrial Services, Offshore Technology, Electrical Installation, Aug. 15th, 2007.

[2] Joakim Berggren, "Study of auxiliary power systems for offshore wind turbines - an extended analysis of a diesel gen-set solution," Master thesis, Univ. Uppsala. June 2013.

[3] Annual Energy Outlook 2013, reported by DOE/EIA, 0383, 2013. http://www.eia.gov/forecasts/aeo/pdf/0383(2013).pdf

[4] Lazaros P. Lazaridis. Economic Comparison of HVAC and HVDC Solutions for Large Offshore Wind Farms under Special Consideration of Reliability, Master Thesis, KTH Electrical Engineering, 2005.

Appendix A – Cost Saving Analysis

The essential objectives of the hybrid APS solution is to reduce the cost of offshore wind farm, in terms of both capital cost and O&M cost. This paper takes a 500MW offshore wind farm as an example. The economic analysis results show that: By using DG+WTG solution, the capital cost saving for electrical systems can reach 4%, and the O&M cost saving can reach 1%. By using the ESS+WTG solution, the capital cost remains the same, however the O&M cost saving can be doubled to ~2%. Calculations will be introduced briefly in the following sections.

➢ *Cost saving analysis*

Capital cost saving includes two parts: diesel generator cost saving, and offshore platform cost saving due to avoid oversizing of diesel generator set.

This paper found that, the total capital cost saving of one 500MW offshore substation platform = Platform Cost Saving + DG Set Cost Saving = 14,225kUSD +360kUSD =14,585kUSD.

Given the cost of 0.58MEUR/MW for the electrical system for offshore wind farm, for a 500MW offshore wind farm, the total cost for the electrical systems is 290MEUR.

It can be calculated that by using the hybrid APS solution, the cost of electrical systems can be reduced by 14.6MUSD/290MEUR x 100% = 4% (exchange rate to be 1EUR=1.3USD).

For ESS+WTG solution, the capital cost saving from diesel generator set, fuel tank and offshore substation is roughly the same to the capital investment for the battery ESS system (22MUSD).

➢ *Maintenance cost saving*

Maintenance cost saving also includes two parts: fuel cost saving, and refuelling cost saving.

Calculation shows that, the fuel cost saving for a 500MW OWF with distance to shore from 50km to 300km is:

- In the range of [55.0, 80.3] kUSD/year for HVAC transmission technology;
- In the range of [55.9, 96.5] kUSD/year for HVDC LCC transmission technology;
- In the range of [303.2, 384.4] kUSD/year for HVDC VSC transmission technology.

The refuelling cost saving for a 500MW OWF with distance to shore from 50km to 300km is:

- In the range of [21.4, 54.5] kUSD/year for HVAC transmission technology;
- In the range of [21.7, 65.6] kUSD/year for HVDC LCC transmission technology;
- In the range of [117.9, 261.1] kUSD/year for HVDC VSC transmission technology.

It should be noticed that all figures listed above are obtained with an assumption that only 50% of the energy can be supplied by WTGs.

Given the cost of 1.5MEUR/MW for 20 years of O&M cost for offshore wind farm, for a 500MW offshore wind farm, the total cost for O&M in 20 years is 750MEUR.

Take 500MW offshore wind farm with HVDC VSC transmission technology, and 100km distance to shore for example, it can be calculated that by using the hybrid APS solution, the cost of operation and maintenance can be saved about 20years x (319.4kUSD + 126.8kUSD) / 750MEUR x 100% = 1% (exchange rate to be 1EUR=1.3USD). Considering the increase of the fuel price and labor cost, the costing saving percentage will go higher.

For ESS+WTG solution, since fuel consumption will be totally removed, the total maintenance cost saving will be doubled compared with DG+WTG solution.

➢ *Sensitivity analysis*

To cope with the potential inaccuracy of input parameters, the economic evaluations about five input parameters are concluded here:

- Fuel price

The situation of fossil fuels running out quickly will lead to the increasing of fuel price in the following years. To 2040, this fuel price increasing may lead to around 38% increase, and makes the OWF annual maintenance cost saving of this new hybrid APS solution increase around 25%.

- Fuel tank size

A small tank can lead to a smaller annual maintenance cost for OWF, but will increase the capital cost for platform. Base on the analysis results, 50~150Ton is relatively optimal size of fuel tank for a 500MW OWF with distance from service station to offshore substation platform 100km.

- WTGs power contribution

The more power contribution from WTs means more cost saving for the OWF. In this case, the OWF operator should choose the WTGs contribution with higher priority than DG sets for APS.

- Unavailability of OWF with multi-substations

Multi substations and platforms can help to decrease the unavailability of OWF, save the diesel consumption, and the maintenance cost for fuel consumption and refuelling. But it does not increase the total lifetime cost saving due to high capital investment. Take double substations and platforms (2*350MW OWF) as an example, about 30% decrease of total lifetime cost saving is shown, compared with single substation and platform topology (1*500MW OWF).

- Shipment distance per refuelling operation

As the fast operation vessel is rented at a time unit of 24hrs, the rent time of vessel will change in step-wise. In this way, if the increase of rent time has not been step changed, exp. from 1day to 2 days for one refuelling operation, the total life cost saving of OWF will not change rapidly with difference only within 1%; if the rent time of vessel has been step increased, exp. from 1day to 2 days for one refuelling operation, then the total life cost saving of OWF will change rapidly with difference around 10%.

- Battery Price

Besides the evaluation results above, which are suitable for both DG+WTG solution and ESS+WTG solution, battery price is one key parameter for the economic analysis of ESS+WTG solution. The prediction of DOE shows the fast decrease of battery price in the following years. Compare with the price in 2010, the battery price may decrease to 10% in 2030. Along with this decrease, the total life cost of a 500MVA offshore wind farm may decrease to 11 million USD. This decrease of battery price will make the hybrid ESS + WTG solution more attractive.

Large-scale wind generation simulations: From the analysis of current installations to modelling the future

Matti Koivisto[1], Petr Maule[1], Poul Sørensen[1], Lukas Galdikas[2], Nicolaos Cutululis[1], Simone Biondi[2]

[1]Technical University of Denmark, Department of Wind Energy, Roskilde, Denmark

[2]European Network of Transmission System Operators for Electricity (ENTSO-E), Brussels, Belgium

E-mail: mkoi@dtu.dk

Abstract. Modelling and understanding variability in wind generation will be increasingly important in the future with growing shares of wind power in energy systems. Crucially, the modelling needs to be extended to future scenarios, also considering the expected technological development of installations. Reanalysis data is often used in large-scale simulations to model the variability in wind. Wind power plant (WPP) data is also required, but may be only partially available. In this paper, a methodology for estimating missing hub height data is presented, using multiple regression models and large WPP and turbine datasets. The resulting estimated hub heights are presented on a pan-European level, and a scenario with capacity factor development until 2050 for two example countries is shown in detail.

1. Introduction

As the amount of installed wind power increases, there is a growing need to model variability in wind generation in detail. As the degree of interconnection in Europe has grown significantly, the modelling of wind generation variability needs to be seen as a pan-European, rather than a country-specific issue. In addition to considering current installations, the modelling needs to be extended to future scenarios with changing wind power technologies. Such modelling is needed for databases such as the ENTSO-E's Pan-European Climate Database (PECD) which is used for electricity market and network studies.

There exists a large amount of literature on the use of meteorological reanalysis data in wind generation simulations, e.g., [1]-[4]. To take into account the differences between mesoscale modelling output and the local wind speeds wind power plants (WPPs) experience, scaling of the reanalysis wind speeds is often required [1], [2].

To apply reanalysis wind speeds in wind generation simulations, information about WPP installations is needed. In the dataset used in this paper, many of the technical parameters of the WPPs are missing. More than half of the installed onshore wind generation capacity has missing hub height information. As hub height is a crucial parameter when applying reanalysis data, a methodology for estimating the missing hub heights was created.

This paper shows the resulting estimated hub heights on a pan-European level, with focus on the differences between regions. Results for two countries with different existing WPP installations, namely Denmark and Finland, are presented in detail. The importance of estimating the missing hub height data when applying the reanalysis wind speed scaling is demonstrated. In addition, the different expected future capacity factor (CF) developments for these two countries are presented.

2. Reanalysis data in wind generation simulation

This section presents the meteorological reanalysis data used in this study. Scaling, which is used to calibrate the simulated wind speed time series to meet historical CF data, is also described. The modelling is carried out using the CorRES tool developed at the Technical University of Denmark, Department of Wind Energy [5].

2.1. The meteorological reanalysis data

The simulated wind generation time series used in this paper are based on reanalysis data obtained from the Weather Research and Forecasting (WRF) model, which is a mesoscale modelling system [6]. To generate the meteorological time series of interest, most importantly wind speeds, the downscaling presented in [7], [8] has been used.

The WRF model gives hourly meteorological time series on a 10 km x 10 km grid, as shown for an example geographical area in Figure 1. Covering the area under modelling, the grid of data can be used to simulate what wind generation output would be with a given set of installations with given technical WPP parameters. The capabilities of the approach for modelling important spatiotemporal dependencies in wind generation have been demonstrated, e.g., in [3], [4]. 35 years of WRF data were available for simulating the wind generation time series used in this study.

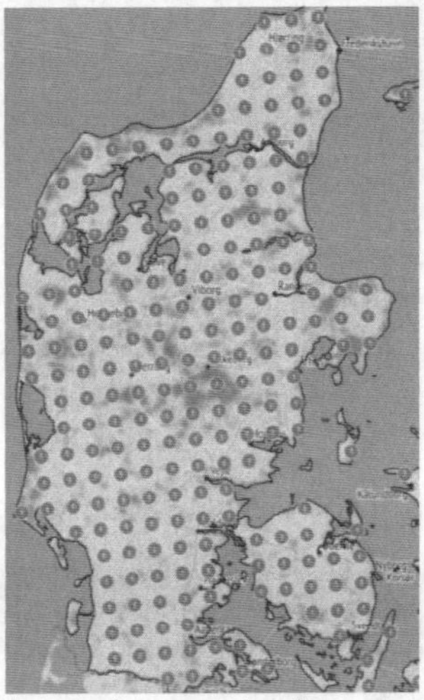

Figure 1. The meteorological simulation points used when modelling onshore wind generation in western Denmark.

2.2. Using wind speed at hub height

When wind speeds are obtained from the WRF model as explained in the previous subsection, they are given for a specific height. When modelling wind power output, they are retrieved at the representative hub height of the generation location being modelled (if modelled turbines have different hub heights, the mean hub height weighted by installed capacity is used). Especially on lower heights, the mean wind speeds are very different on different hub heights, as can be seen in Figure 2. This emphasizes the importance of having correct hub height data in reanalysis modelling.

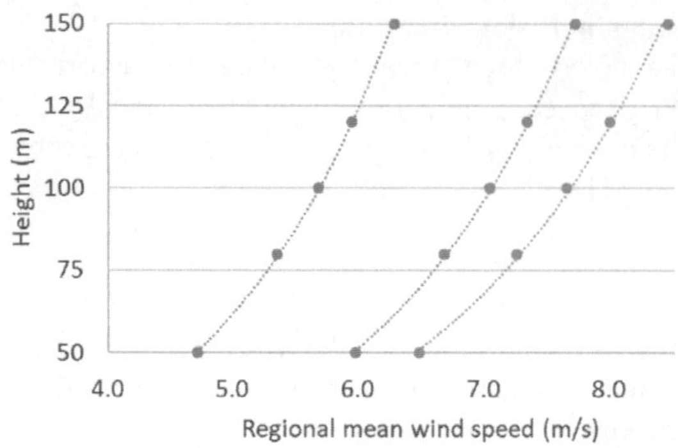

Figure 2. Mean wind speeds (regional averages) at different hub heights for three example regions with different wind resources.

2.3. Scaling of reanalysis wind speeds

Although reanalysis data is very suitable for modelling large-scale wind generation, it has been shown that using reanalysis wind speeds directly can cause errors in estimated CFs [1], [2], [4]. To remedy this, scaling of the reanalysis wind speeds have been proposed in [1], [2]. A scaling similar to [2] is used in this paper. Wind speeds acquired from WRF modelling are scaled until a historical target CF is reached. The historical data are either CFs directly (if available), or CFs calculated from installed capacities and annual energy generation data.

3. Estimating today's hub heights

This section starts by describing datasets used in this study. Then, a methodology for estimating the significant amount of missing hub height data is explained by first describing the general procedure, and then explaining the required regression modelling. Finally, the resulting hub heights on European level are presented.

3.1. Datasets used

The WPP dataset acquired from [9], with e.g., WPP locations and installed capacities, was used as the basis for the reanalysis simulations. The dataset, and a turbine dataset from the same source, are also used in the estimation of the missing hub heights. Although a majority of the onshore WPPs do not have hub height in the available data, 6090 WPPs with hub height data were still available for the regression modelling needed in the missing data estimation.

3.2. Overview of hub height estimation

The hub height estimation procedure consists of several stages, as shown in Figure 3. Stage 1) means simply that if the dataset has a hub height given for a WPP, it is used. If hub height data does not exist, but turbine model is known, the turbine model's rotor diameter is used in regression modelling to estimate the hub height in stage 2). A hub height specified for a turbine type in the turbine dataset is not used, as the ranges of possible hub heights for many turbine types are very wide.

If turbine type is not known, in stage 3) knowledge of turbine rated power is used in the estimation of the WPP hub height. Turbine power can be calculated from the WPP's installed capacity and number of turbines. If even turbine power is not known, commissioning year is used in estimating the hub height in stage 4).

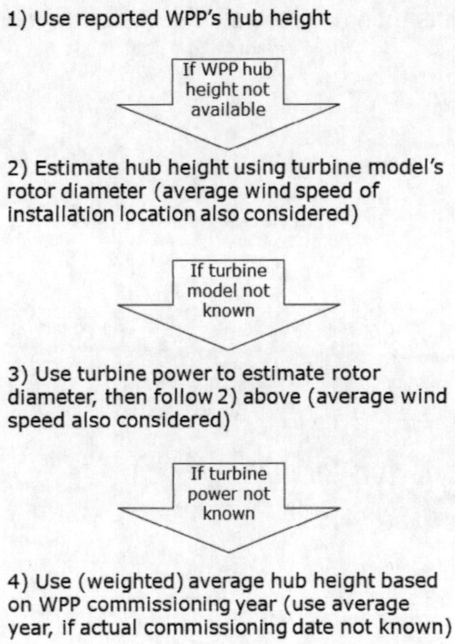

Figure 3. The different stages in the hub height estimation procedure. A later stage is reached only if the hub height cannot be estimated in the previous stage. Stage 4) can be carried out for any WPP, so ultimately all WPPs will have a hub height estimate.

3.3. Regression modelling

In the hub height estimation procedure shown in Figure 3, regression modelling is needed in both stages 2) and 3). Figure 4 shows the main explanatory variables when estimating hub height in stage 2). Larger rotor diameter clearly indicates a higher hub height, which makes sense as the turbine gets physically larger. However, there is significant variance in hub height with a given rotor diameter, which indicates that additional explanatory variables may be needed.

As can be seen Figure 4, with a given rotor diameter, a location with a lower mean wind speed seems to be associated with a higher hub height. Mean wind speeds were taken from Global Wind Atlas (GWA) [10] using the latitude and longitude of each WPP. In addition, it was noticed that all differences in hub heights between the different countries cannot be explained only by rotor diameter and the location's mean wind speed, so a dummy variable describing the country the WPP is located in was added to the regression model. The final model to be estimated for stage 2) is

$$y_{\mathrm{hh}} = b_0 + b_{\mathrm{dRot}} x_{\mathrm{dRot}} + b_{\mathrm{ws}} x_{\mathrm{ws}} + b_{\mathrm{dRot\_ws}} x_{\mathrm{dRot}} x_{\mathrm{ws}} + b_1 x_1 \ldots + b_{\mathrm{k}} x_{\mathrm{k}} + e, \quad (1)$$

where y_{hh} is hub height, x_{dRot} is rotor diameter, x_{ws} is mean wind speed at the location, e is the error term of the model and the b parameters are the coefficient to be estimated. The part $b_1 x_1 \ldots + b_{\mathrm{k}} x_{\mathrm{k}}$ of (1) is the dummy variable system signalling the different countries analysed.

The most important estimated coefficients of (1) are shown in Table 1. It can be seen that an in increase in rotor diameter predicts higher hub height, which makes sense. The coefficient for x_{ws} is zero (it was not found statistically significant in the regression modelling); however, the coefficient for $x_{\mathrm{dRot}} x_{\mathrm{ws}}$ is negative. This means that an installation location with a higher mean wind speed predicts a lower hub height for a given rotor diameter, but the magnitude of this effect depends on the diameter. These results are in line with the visual analysis of Figure 4.

For stage 3), the regression model is built based on the dependencies shown in Figure 5. It can be seen that rotor disk area and turbine rated power are highly correlated, which makes sense as a large rotor is usually associated with high rated power. The modelling for stage 3) is built so that turbine rated power can be used to estimate disk area, while also considering the mean wind speed of the installation location. The estimated disk area is then used to calculate rotor diameter, and the diameter is used in (1) to estimate the hub height.

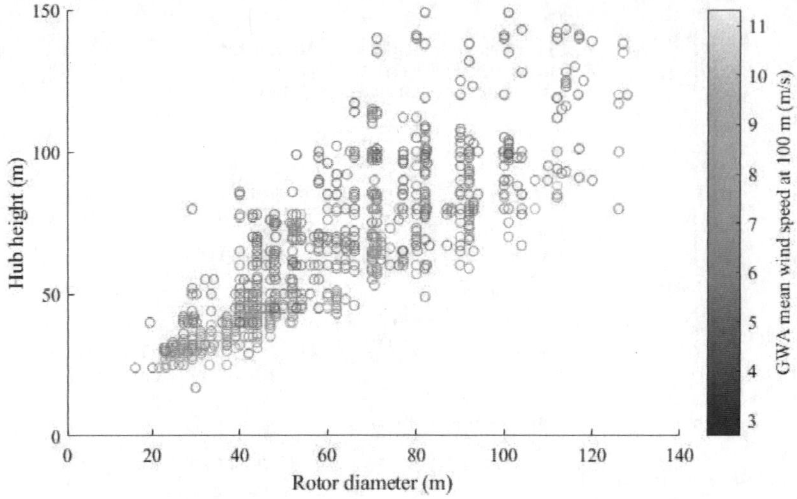

Figure 4. Rotor diameter and hub height of WPPs with colouring using Global Wind Atlas (GWA) average wind speed [10] at installation location.

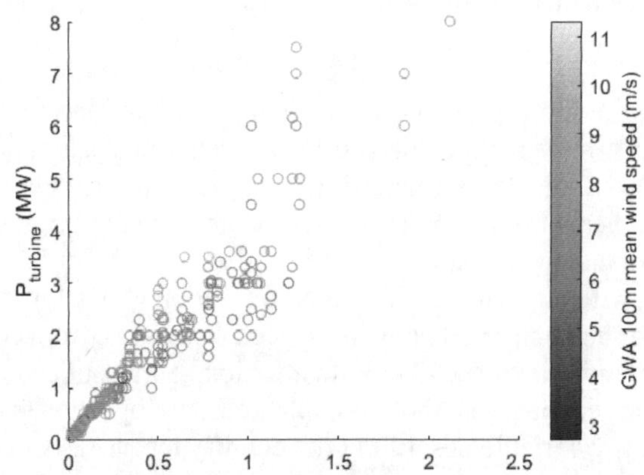

Figure 5. Rotor disk area and turbine rated power with colouring using Global Wind Atlas (GWA) average wind speed [10] at installation location.

Table 1. Most important estimated coefficients of (1)

	Estimated coefficient
b_{dRot}	1.132
b_{ws}	0
b_{dRot_ws}	-0.0509

3.4. Results on pan-European level

Utilising the missing hub height estimation methodology described in the previous subsections, hub heights were estimated for each WPP in the dataset. Figure 6 shows how the different hub height estimation stages were used in giving a hub height for each WPP. It can be seen that stage 2) was used the most: this means that many WPPs did not have hub height data, but had the turbine type with a rotor diameter specified, which was used in estimating the hub height in (1).

A pan-European view of the resulting average onshore wind hub heights is shown in Figure 7, with distinct differences between the regions. E.g., in Germany the highest hub heights are found in the south where wind speeds are on average lower. For Denmark, UK, Spain and Italy, average hub heights are generally lower; for Nordic countries, with relatively new installations, hub heights are generally quite high.

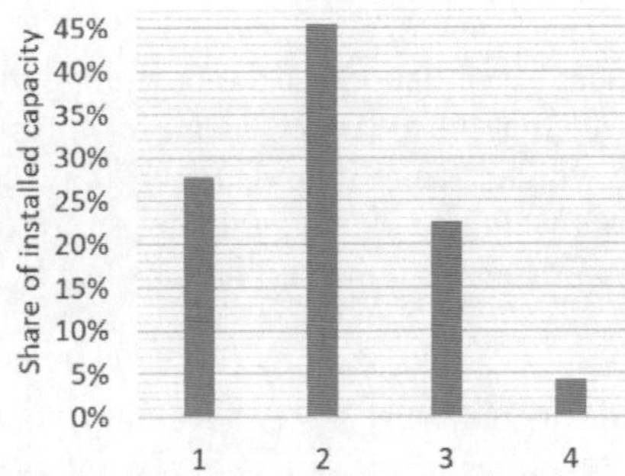

Figure 6. Shares of WPP hub heights that were estimated using the four different stages (as shown in Figure 3), as percentages of installed onshore wind capacity.

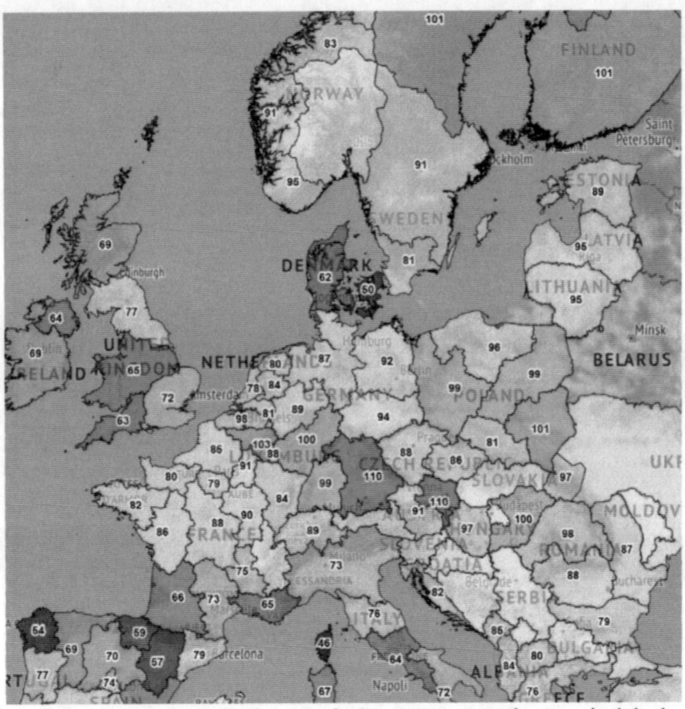

Figure 7. Estimated weighted (by installed capacity) average onshore wind hub heights in meters representing installations at the beginning of 2015. Regions with less than 10 MW of installed wind power are left blank.

4. Effects on future capacity factors in simulations

This section shows how the modelling of today's hub heights is utilized in estimating CF development in a future scenario for two example countries, namely Denmark and Finland. The resulting onshore wind CFs are compared, and the importance of estimating the missing hub height data is discussed.

IOP Conf. Series: Journal of Physics: Conf. Series **1102** (2018) 012034 doi:10.1088/1742-6596/1102/1/012034

4.1. Scenario assumptions

To analyse the expected development of CFs in the future, a scenario until 2050 was created. It is assumed that onshore hub heights increase towards an average of 120 m by 2050, and specific power decreases so that by 2050 it is on average 30 % lower than today. Results for Denmark (DK) and Finland (FI) for onshore wind are shown in Table 2. Average onshore wind hub height in Finland raises to 120 m already by 2030, but is not assumed to increase further. In Denmark, the increase of the average onshore wind hub height is assumed to take longer, consisting of both new installations and repowering.

4.2. Comparing capacity factor development

The onshore wind CFs for today (installations at the beginning of 2015) for Denmark and Finland are shown in Table 2. It can be seen that the today's CF for Finland is slightly higher than for Denmark, although Denmark is considered windier than Finland. This difference is explained by the newer WPP fleet in Finland, resulting in a high weighted average hub height for today (104 m), as seen in Table 2. In scenario year 2030 the two countries have quite similar CFs, but by 2050 the CF of Denmark is much higher than that of Finland.

Table 2. Onshore wind hub height and capacity factor developments in the modelled scenario

	DK hub height (m)	DK CF	FI hub height (m)	FI CF
Today (2015)	60	0.24	104	0.25
2030	87	0.34	120	0.33
2040	103	0.39	120	0.35
2050	120	0.45	120	0.37

Hub heights are averages weighted by installed capacities. CFs are averages of 35 simulated meteorological years. Calibration of today is based on data from [11] for Denmark, and [12], [13] for Finland.

4.3. Issues if missing hub heights are not estimated

Finland shows a weighted average hub height of 104 m for today (Table 2), which explains the CF being higher in Finland than in Denmark; however, without estimating the missing hub heights, the weighted average would have been only 62 m for Finland (as many newer installations had missing data). Thus, missing hub height estimation was crucial in understanding the today's CFs in the two countries.

It is important to also consider the effect of scaling, as explained in Section 2.3. If the incorrect average hub height of 62 m would have been used for Finland for the today's scenario, the CF of 0.25 shown in Table 2 would still have been met: by scaling wind speeds from the assumed height of 62 m up significantly (because scaling can always meet any CF target). As scaling is assumed to stay the same in the future scenario, the incorrect scaling would have been used in the scenario up to 2050; this would have created an incorrect CF development for Finland towards 2050. This amplifies the importance of modelling the current installations in detail when applying the reanalysis wind speed scaling.

4.4. Modelling forest regions

Finland and Denmark are quite different countries in terms of forest areas, and this needs to be taken into account in the modelling to draw reliable conclusions on the differences between the countries. In the modelling presented in this paper, this has been carried out in two ways. Firstly, forest regions are considered in the WRF modelling. Even though in the WRF modeling at 10 km x 10 km spatial grid spacing it is not possible to explicitly resolve the effect of most forest regions in the Nordic countries, the surface roughness length of grid boxes with forest as dominant land type has been increased to 0.9 m, following the recommendation of [14]. Secondly, the calibration of wind speeds to meet historical wind power generation, as explained in section 2.3, is used to take into account local environment that may not be fully modelled in WRF.

5. Conclusion

In this paper, a methodology for estimating missing hub height data was developed. Regression modelling using large WPP and turbine datasets were utilised to estimate a hub height for each analysed WPP. The resulting hub heights show interesting differences on European level. The different expected future CF developments for two countries with different existing WPP fleets, namely Denmark and Finland, were presented. The importance of estimating the missing hub heights when applying the scaling of reanalysis wind speeds was demonstrated. The modelling can be applied on a pan-European level, as required, e.g., in the simulations for the PECD database.

6. Acknowledgements

The authors would like to acknowledge Andrea Hahmann from Technical University of Denmark, Department of Wind Energy for the WRF modelling and giving information and advice related to it. The authors from Technical University of Denmark, Department of Wind Energy acknowledge support from the NSON-DK project (Danish Energy Agency, EUDP grant 64018-0032).

References

[1] I. González-Aparicio, F. Monforti, P. Volker, A. Zucker, F. Carcri, T. Huld and J. Badger, "Simulating European wind power generation applying statistical downscaling to reanalysis data", *Applied Energy*, vol. 199, pp. 155-168, August 2017.

[2] I. Staffell and S. Pfenninger, "Using bias-corrected reanalysis to simulate current and future wind power output", *Energy*, vol. 114, pp. 1224-1239, November 2016.

[3] M. Marinelli, P. Maule, A. N. Hahmann, O. Gehrke, P. B. Nørgård and N. A. Cutululis, "Wind and Photovoltaic Large-Scale Regional Models for Hourly Production Evaluation", *IEEE Transactions on Sustainable Energy*, vol. 6, no. 3, pp. 916-923, July 2015.

[4] E. Nuño, P. Maule, A. Hahmann, N. Cutululis, P. Sørensen and I. Karagali, "Simulation of transcontinental wind and solar PV generation time series," *Renewable Energy*, vol. 118, pp. 425-436, April 2018.

[5] M. Koivisto, K. Das, F. Guo, P. Sørensen, E. Nuño, N. Cutululis and P. Maule, "Using time series simulation tool for assessing the effects of variable renewable energy generation on power and energy systems", *WIREs Energy and Environment* (accepted for publication).

[6] W. Skamarock, J. Klemp, J. Dudhia, D. Gill, D. Barker, M. Duda, X. Huang, W. Wang and J. Powers, "description of the advanced research WRF version 3", Boulder, Colorado, USA, 2008.

[7] A. N. Hahmann, D. Rostkier-Edelstein, T. T. Warner, F. Vandenberghe, Y. Liu, R. Babarsky and S. P. Swerdlin, "A Reanalysis System for the Generation of Mesoscale Climatographies," *Journal of Applied Meteorology and Climatology*, pp. 954-972, May 2010.

[8] A. N. Hahmann, C. L. Vincent, A. Peña, J. Lange and C. B. Hasager, "Wind climate estimation using WRF model output: method and model sensitivities over the sea," *International Journal of Climatology*, vol. 35, no. 12, p. 3422–3439, October 2015.

[9] The Wind Power: http://www.thewindpower.net/

[10] Global Wind Atlas, DTU Wind Energy in partnership with the World Bank Group, utilizing data provided by Vortex: http://globalwindatlas.com

[11] Danish Energy Agency, Data about Danish wind generation and installations: https://ens.dk/service/statistik-data-noegletal-og-kort/data-oversigt-over-energisektoren

[12] ENTSO-E, Statistical Factsheet 2014: https://www.entsoe.eu/Documents/Publications/Statistics/Factsheet/entsoe_sfs2014_web.pdf

[13] The Europen Wind Energy Association (now WindEurope), Wind in power: 2014 European statistics: http://www.ewea.org/fileadmin/files/library/publications/statistics/EWEA-Annual-Statistics-2014.pdf

[14] E. Dellwik, J. Arnqvist, H. Bergström, M. Mohr, S. Söderberg and A. Hahmann, "Meso-scale modeling of a forested landscape", *Journal of Physics: Conference Series (Online)*, vol. 524, no. 1, 012121, 2014.

IOP Conf. Series: Journal of Physics: Conf. Series **1102** (2018) 012035 doi:10.1088/1742-6596/1102/1/012035

Analysing the effect of the sliding feed-in premium scheme to the revenues of a wind farm with the use of mesoscale data - case study of Greece

K Loukidis, A Vlamakis and P Papastamatiou

ENTEKA S.A. 2 Tychis str, 152 33 Chalandri Greece
kloukidis@enteka.gr

Abstract. One of the main challenges of the wind industry is the accurate prediction of the revenues that a developing project is expected to produce during its operation lifetime. The European Commission's Guidelines on State Aid for Environment Protection and Energy 2014-2020 (2014/C 200/01) (European Commission, 2014), enforced the transition from the Feed-in Tarif (FiT) scheme to a market-oriented Feed-in Premium remuneration mechanism. This evolution creates additional uncertainty when evaluating the revenues of a wind energy project, since the remuneration of the produced energy is now market depended instead of stable, as it was under the FiT scheme. The present study attempts to estimate the level of uncertainty that the change of support scheme creates. The expected revenues under both schemes are calculated for three (3) continuous years in the Greek market with the assumption that the strike price of the new FiP scheme coincides with the fixed tariff of the FiT scheme and the results are compared in an annual and total period. The study takes into account the revenues from the day-ahead market (i.e. the only wholesale market currently exists in Greece) and from the support mechanism. The results of the comparison reveal small fluctuation in the annual revenues of the wind stations examined between the two schemes, which does not seem to increase when examining the total 3-year period

1. Introduction

Greek wind energy market was operating under a feed-in tariff scheme (hereinafter "FiT") up to the end of 2015. The FiT scheme formulated a protected environment for investors, since wind energy plants were compensated according to a constant, pre-defined tariff for the 20-year validity period of the Power Purchase Agreement (PPA) of each project.

The European Commission's Guidelines on State Aid for Environment Protection and Energy 2014-2020 (2014/C 200/01) (European Comission, 2014), hereinafter "EEAG 2014-2020", enforced the transition from the FiT scheme to a market-oriented feed-in premium mechanism (hereinafter "FiP"). Law 4414/2016 (OG 149A, 9.8.2016) - which enforced the new supporting scheme in Greece - determined that new wind projects will receive a premium on top of their income from the market, in the form of a variable (sliding) premium. In practice the sliding premium is calculated, on a monthly basis, as the difference between a Reference Value ("RV") and the hourly market price (the System Marginal Price or "SMP"). Moreover, from 1.1.2017 onwards, the RV is determined through auctions announced and executed by Regulatory Authority for Energy (RAE).

The present study investigates the abovementioned schemes for the case of the Greek wind energy market, with respects to estimating wind farm revenues under both schemes.

The scope of the present study is to evaluate and quantify the effect that the change of the supporting scheme will have on the revenues of wind projects in various locations in Greece.

2. Basic principles of the two remuneration schemes

The following paragraphs briefly describe the remuneration schemes, previous and existing, that have been applied in the Greek wind energy market.

2.1. The previous remuneration scheme: Feed-in tariff (FiT)

Until 2015 the development of wind energy in Greece was governed by a feed-in tariff (FiT) scheme, according to which the investors were compensated a by-law pre-defined tariff for a 20-year period, according to a Power Purchase Agreement (PPA) between project company and the market operator.

The FiT had a value of 73 €/MWh on 2006, where at the end of 2015 was 105 €/MWh. Both values refer to projects located in the interconnected system. Further, the value of 2015 refer to projects that don't receive state subsidy for their construction.

2.2. The new remuneration scheme: Feed-in premium (FiP)

In compliance with the European Commission's Guidelines on State Aid for Environment Protection and Energy 2014-2020 (Official Journal of EU, 2014/C 200/01, par. 124-131), wind projects in Greece which had not signed a Power Purchase Agreement (PPA) until the end of 2015, will be remunerated through a Feed-in premium (FiP) mechanism. The basic provisions for the operation of the FiP scheme in Greece, have as follows:

i) The wind projects receive a premium in the form of a variable (sliding) premium, on top of their income from the market. The premium is calculated on a monthly basis and its sum with the hourly market price, practically targets a strike price, the so-called Reference Value (RV). For year 2016, the RV for wind projects was defined by law at 98€/MWh. From 1.1.2017 onwards the RV is defined per project through auctions organized by the Regulatory Authority of Energy (RAE). The first auction, held on 2.7.2018, determined RVs between 68,18 €/MWh and 71,93 €/MWh. As long as the RV is being defined for a project, it remains valid for 20 years (contract with the Electricity Market Operator).

ii) The wind projects will sell their electricity directly to the market, while they will be subject to balancing responsibilities, unless no liquid intra-day market exists. In order to meet these market obligations, the wind projects receive, on top of their remuneration, a management fee equal to 2€/MWh. Finally, the wind projects have the option to outsource their market obligations to other entities, the Aggregators, while the Electricity Market Operator has been defined as a Last-resort Aggregator.

3. Methodology applied

In order to calculate the revenue of a wind farm under the FiP regime, energy production information for each and all of the wind projects of the country would be needed. Since production data from the existing wind farms are not available to the public, the approach was to investigate whether the use of mesoscale wind data in representative locations (where wind farms are located) would provide accurate enough results so that a comparison between the two schemes could be made.

Given that the main purpose of the study is the comparative assessment of the two different remuneration schemes, the absolute energy production values are not indispensable, as long as the wind resource assessment provides valid and consistent results in both cases to be compared. Moreover, in the case of the FiP calculation, it is essential to know the variation of the energy production in an hourly basis, since this is needed to be combined with the hourly System Marginal Price (SMP) (Hellenic Transmissions System Operator, n.d.) values in order to estimate the premium for every station. Hence, it was decided to use mesoscale wind data (EMD International A/S;, n.d.) because of the high correlation factors they have shown in the past, when compared with actual wind measurements (Foussekis & Gkarakis, 2014).

The regions of the country where wind farms are located have been identified from the official data from the Greek Regulatory Authority for Energy. A total of 45 representative locations were eventually selected for which the mesoscale data sets were acquired. In each of these locations a representative nominal capacity was assigned, derived from the annual wind statistics report of the Hellenic Wind Energy Association (HWEA) (Hellenic Wind Energy Association, n.d.), so that eventually each of these locations represent a virtual wind station (see Appendix 1) which is representative of the actual spatial distribution of the total operating wind capacity in Greece.

A calculation model was created, in order to match, for each wind station, the hourly wind data with the power curve of a typical 3 MW wind turbine. The calculated total energy production was then compared to the official data from the Hellenic Electricity Market Operator (Hellenic Electricity Market Operator, n.d.) on a monthly and yearly basis, in order to verify the accuracy of the results. Based on that comparison, a correction factor was applied to the calculations, so that the calculated production was adjusted to match the official total wind energy production on an annual basis.

The results of the energy calculations confirm the initial assumptions, since the monthly calculated results present indeed high correlation with the actual production figures on an hourly basis (R^2~ 0,9, see following graphs).

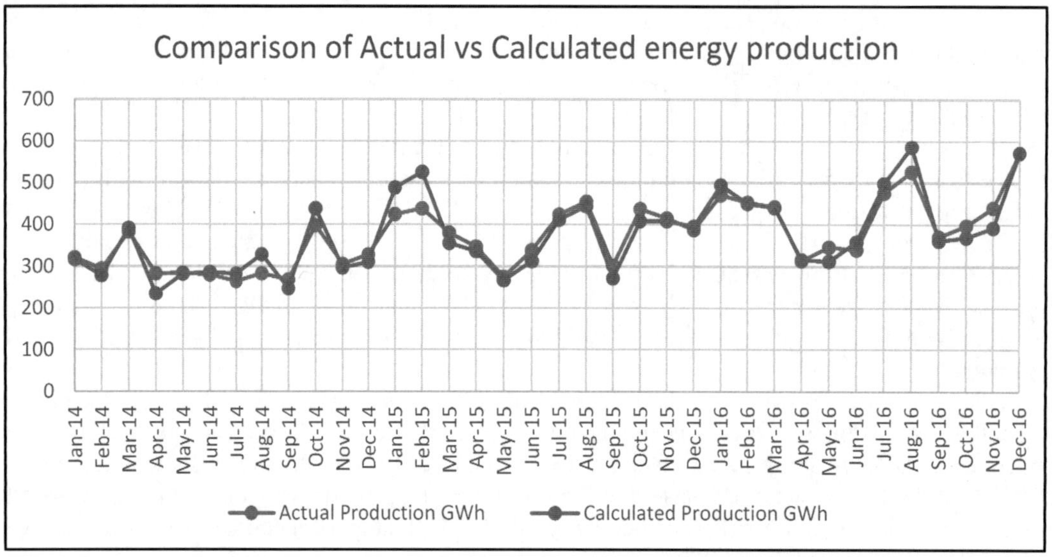

Figure 1. Comparison of Actual vs Calculated energy production.

Finally, the calculated energy production values were used to estimate the revenue of each wind station for both FiT and FiP schemes, with the fair assumption that the tariff of the FiT coincides with the strike price of the FiP scheme.

3.1. Selection of representative locations for wind stations

The first step was to determine representative wind stations for which to estimate the revenues under both FiT and FiP schemes. Due to the lack of analytical public data for the energy production of the operating wind farms in Greece, it was inevitable to define virtual wind stations, hereinafter the "wind stations", for each one of them the remuneration was then calculated for the above-mentioned schemes.

In order to obtain accurate results, a large number of candidate locations were selected, spread in all over the country in representative location according to where the operating wind farms are located. Particularly, 51 locations were initially selected for the virtual wind stations.

For each one of the locations, hourly mesoscale wind speed data at 100 m a.g.l. from EMD & ConWx were used, covering the period 01.01.2014-31.12.2016. Due to the fact that 6 locations presented extremely low wind potential (<4m/s @ 100 m a.g.l.), they were finally rejected and no further analysis was conducted from them. Thus, 45 locations constitute the final list of wind stations' locations used for the present study. The basic information (coordinates, wind speed and map) for the abovementioned locations are presented in Appendix 1.

3.2. Power capacity allocation to wind stations

The next step was to allocate the power capacity for each one of the 45 selected wind stations at the locations of Table 1. In order to match a representative capacity for each wind station with regards to the actual distribution of the total installed wind capacity in Greece, the following methodology was performed (it is noted that the methodology described represents year 2014, while similar methodology was used for the years 2015 and 2016 accordingly).

The first step was to calculate the annual average wind capacity for all the country.

In order to estimate an annual average wind capacity in Greece, the monthly reports of EMO (LAGIE) for the RES Special Account were used. The following Table illustrates the total installed wind capacity in Greece (interconnected system & non-interconnected islands) at the end of each month of the year 2014. The corresponding data for whole examined period are presented in Appendix 2.

Table 1. Total installed wind capacity in Greece for 2014.
Source: EMO's report for RES Special Account for January 2015.

Jan-14	1827
Feb-14	1847
Mar-14	1847
Apr-14	1847
May-14	1847
Jun-14	1884
Jul-14	1902
Aug-14	1902
Sep-14	1902
Oct-14	1910
Nov-14	1933
Dec-14	1978
Annual Average	**1885,50**

The second step was to allocate the wind capacity to each wind station accordingly.

In order to allocate the aforementioned average capacity to the wind stations, the annual "wind statistics report" of Hellenic Wind Energy Association (HWEA) for the period 2014-2016, were used. Due to the fact that HWEA reports present the installed capacity per administrative region, as an intermediate step, the annual average capacity of step 1 was primarily allocated per region. The actual capacity per region derives from the data of HWEA report for 2014 (see Table 2).

Subsequently, the values of actual capacity per region were normalized according to the mean annual capacity of 1885,50 MW calculated. Hence, new normalized values for wind capacity per region were calculated for year 2014. Finally, it was assumed that the normalized wind capacity of each region is equally distributed to the virtual wind stations selected in that region. Table 2 consolidates the aforementioned calculations.

Table 2. Estimation of the wind capacity for each wind station

Region	Actual capacity at the end of the year	Normalised capacity [MW]	Number of wind stations	Capacity of each wind station of Region [MW]
Attica	83,32	79,35	3	26,45
North Aegean	36,04	34,32	3	11,44
Western Macedonia	52,9	50,38	3	16,79
Western Greece	130,35	124,14	5	24,83
Ionion Islands	83,7	79,71	2	39,86
Central Macedonia	41	39,05	2	19,52
Crete	193,4	184,19	7	26,31

IOP Conf. Series: Journal of Physics: Conf. Series **1102** (2018) 012035 doi:10.1088/1742-6596/1102/1/012035

South Aegean	88,81	84,58	3	28,19
Peloponnese	367,95	350,42	6	58,40
Central Greece	602,8	574,08	6	95,68
Thessalia	17	16,19	2	8,10
East. Macedonia & Thrace	282,55	269,09	3	89,70
Total	**1979,82**	**1885,50**	**45**	

The complete HWEA reports and relevant calculations for the capacity of each wind station for the whole period (2014-2016) are presented in Appendix 3.

3.3. Estimation of wind energy production

The energy production for each virtual wind station was calculated on hourly basis for the period 01.01.2014-31.12.2016. For this purpose, the following data were used:

i) hourly mesoscale wind speed data for each wind station at 100 m a.g.l covering the period 2014-2016
ii) the power curve of a 3 MW wind turbine of a dominant manufacturer in the Greek wind industry
iii) the capacity in MW for each wind station derived from the previous paragraph for each year of the examined period

Particularly, the hourly wind speed values were matched to the power curve of the selected WTG type and the value derived was then adjusted according to the capacity assigned for each wind station in order to calculate the hourly energy production for each one of the 45 wind stations.

The total wind energy production for 2014 for the amount of the wind stations was estimated at **4.254 GWh**. This number is quite close to the value of **3.689 GWh** (13,28% deviation) which was the actual wind energy production of that year, considering that mesoscale data were used instead of actual wind measurements and without taking into consideration availability and grid losses.

In order to adjust the calculated wind energy production (4.254 GWh) to the actual annual energy production (3.689 GWh), a correction factor of 86,72% was applied resulting in the **normalized** annual energy production for each wind station. Finally, the cumulative energy production per region was calculated, which is presented in the following table.

Table 3. Estimated annual energy production per Region for 2014

Region	Normalised capacity [MW]	Estimated annual energy production [MW]	Normalized annual energy production [MW]
Attica	79,35	204,11	177,00
North Aegean	34,32	97,56	84,60
Western Macedonia	50,38	52,41	45,45
Western Greece	124,14	181,07	157,02
Ionion Islands	79,71	132,84	115,20
Central Macedonia	39,05	42,07	36,48
Crete	184,19	554,03	480,44
South Aegean	84,58	256,75	222,65
Peloponnese	350,42	584,38	506,76
Central Greece	574,08	1393,37	1208,28
Thessalia	16,19	30,77	26,68
East. Macedonia & Thrace	269,09	724,73	628,46

Total	**1885,50**	**4.254,09**	**3689,00**

3.4. Calculation of remuneration under both schemes

3.4.1. Remuneration under the FiT scheme. In order to estimate the compensation of each wind station for the scenario of a FiT scheme, it was assumed a stable tariff of 98 €/MWh which is actually the **Reference Value (RV)** that Law 4414/2016 (OG 149A) introduced for wind projects installed either in interconnected system or non-interconnected islands. It is mentioned that according to the Ministry of Energy, the Reference Value determined for each RES technology is actually reflecting the Levelized Cost of Energy (LCOE) of the corresponding technology. Subsequently, the compensation of each wind station under a FiT scheme was estimated on an hourly basis, by multiplying the assumed Feed-in Tariff, i.e. the Reference Value with the hourly energy production **(Q_h)**, as follows:

$$\boldsymbol{Total\ Income_{FiT,h} = RV * \sum_{h=1}^{n} Q_h} \qquad (1)$$

3.4.2. The concept of the FiP Scheme. According to the new RES/CHP support scheme, effective from 01.01.2016, the Total Income of the amount of the operating stations of a given RES technology is calculated on a monthly basis (calculation period). Most of the aforementioned Total Income is secured from the energy market, in the form of a Market Income. In addition to thc Market Income, a new type of operating aid has been introduced for the electricity generation from RES in the form of a **sliding premium (sFiP)**. The total monthly premium for all the stations of a given RES technology is estimated as follows:

$$\boldsymbol{Total\ Premium_{m,RES\ tech.} = Total\ Income_{m,RES\ tech.} - Market\ Income_{m,RES\ tech.}} \qquad (2)$$

The Total Income and the Market Income are calculated with the following equations, subject to the assumption that all power plants enjoy the same Reference Value (RV):

$$\boldsymbol{Total\ Income_{m,RES\ tech.} = RV * \sum_{ws=1}^{k} \sum_{h=1}^{n} Q_{ws,h}} \qquad (3)$$

$$\boldsymbol{Market\ Income_{m,RES\ tech.} = \sum_{ws=1}^{k} \sum_{h=1}^{n} AVE_h * Q_{ws,h}} \qquad (4)$$

where:
***m:** month (calculation period)*
***k:** the total number of operating stations of a given RES technology*
n**: the total hours of the month in which there was participation of a RES technology in the **Day-Ahead Scheduling (DAS)
***h**: the hour that a RES technology participated in the DAS*
***$Q_{ws,h}$**: the produced energy of a given wind station (ws) injected in the electrical grid at hour h*
***AVE_h**: The Average Value of Energy at hour h*

It is obvious from the above equations that the total monthly income for all stations of a given RES technology with the FiP scheme coincides with the expected total monthly income of those stations with the FiT scheme, as long as it is actually the product of the Reference Value with the total energy produced in the examined period (month).

However, it is underlined that while the FiP scheme distributes the same total amount, to the total RES producers, as if they were operating under a FiT scheme, the income of each one RES producer is differentiated in the FiP scheme, as analyzed in the following paragraph.

3.4.3. Calculating the remuneration under the FiP Scheme. The Total Income of a wind station (ws) is calculated on a monthly basis (calculation period) and consists of two separate amounts: The Market Income and a state aid (premium) granted on top of the Market Income.

$$\boldsymbol{Total\ Income_{m,ws} = Market\ Income_{m,ws} + premium_{m,ws}} \qquad (5)$$

all amounts in Euros (€), m: month, ws: wind station
The Market Income of each wind station is estimated as follows:

$$\boldsymbol{Market\ Income_{m,ws} = \sum_{h=1}^{n} AVE_h * Q_{ws,h}}\ (6)$$

where:
AVE_h: the Average Value of Energy at hour h
h: the hour that a RES technology participated in the DAS
The calculation of the hourly Average Value of Energy (AVEh) is based on the hourly System Marginal Price (SMP_h) and the hourly value of the Wholesale Energy Market mechanisms (WEM_h):

$$\boldsymbol{AVE_h = SMP_h + WEM_h}\ (7)$$

all values in €/MWh

Given that:

i) The AVE_h is basically determined by the **SMP_h**

ii) the exact methodology for the calculation of the **WEM_h** is still pending

it is assumed in the present study that the AVE_h coincides with the **SMP_h,** i.e.:

$$\boldsymbol{AVE_h \cong SMP_h}\ (8)$$

Thus, for the present study the Market Income is actually calculated with the following equation:

$$\boldsymbol{Market\ Income_{m,ws} = \sum_{h=1}^{n} SMP_h * Q_{ws,h}}\ (9)$$

On top of the Market Income, each wind station also collects a premium from the Special RES account. The amount of the premium (in €) is based on the calculation of a sliding feed-in premium in €/MWh (sFiP), which is calculated on a monthly basis. For this reason, a new variable is introduced, the **MArket Special Price** ("MASP"), which is calculated by EMO (LAGIE) for each RES technology on a monthly basis (in €/MWh). The calculation of the monthly sFiP has as follows:

$$\boldsymbol{sFiP_m = RV_{ws} - MASP_{wind-energy,m}}\ (10)$$

where:
RVws: *Reference Value of each wind station (defined by auctions from 01.01.2017 onwards)*
MASP: *Market Special Price calculated per month for each RES technology (common for all wind stations)*

The MASP actually reflects the weighted Average Value of Energy (AVE). According to the provisions of article 3 of Ministerial Decision RF187480/7.12.2016 (OG 3955 B), the MASP for non-controllable RES, like wind energy, is calculated as follows:

$$\boldsymbol{MASP_{RES\ tech,m}} = \frac{\sum_{h=1}^{n} AVE_h * Q_{RES\ tech,h}}{\sum_{h=1}^{n} Q_{RES\ tec,h}} \approx \frac{\sum_{h=1}^{n} SMP_h * Q_{RES\ tech,h}}{\sum_{h=1}^{n} Q_{RES\ tec,h}}\ (11)$$

where:
n: *the total hours of the month in which there was participation of a RES technology in the DAS*
h: *the hour that a RES technology participated in the DAS*
$Q_{RES\ tech,h}$: *the total produced energy for a given RES technology injected in the electrical grid at hour h*
AVE_h: *the Average Value of Energy at hour h*
SMP_h: *the System Marginal Price at hour h*
*Note: the **AVE_h** was assumed to be equal with the **SMP_h,** for the reasons mentioned before.*

It is stressed that for the calculation of the total injected energy of a given RES technology per hour ($Q_{RES\ tech,h}$), the amount of the power plants of that RES technology are taken into account, regardless their supporting scheme (FiT or FiP).

The above equation for the calculation of the MASP illustrates the necessity for market responsiveness for the wind stations participating in a FiP scheme. Particularly, the wind stations should maximize their energy production during the periods of high value of energy price, i.e. they should have a positive correlation with the SMP.

With respects to the input data for the above equation, it is mentioned that the hourly SMP values (SMP_h) for the examined period were taken by the relevant reports of Hellenic Transmission System Operator (HTSO, "ADMIE" in Greek), while the hourly produced energy for the case of wind energy ($Q_{wind\text{-}energy,h}$) has already been estimated according to the methodology presented.

Taking the estimated values of the MASP, the monthly values of the sFiP are calculated with the use of equation 10. The following Table presents the estimated values of the Market Special Price and $sFiP_m$ for each month of the year, for the case of wind energy and for the year 2014.

Table 4. Estimated Market Special Price (MASP) for wind energy according to the input data (production of wind stations, SMP_h) for year 2014.

Month	$MASP_m$ [€/MWh]	$sFiP_m$ [€/MWh]	SMP_m [€/MWh]
Jan	65,15	32,85	64,63
Feb	65,03	32,97	66,64
Mar	46,99	51,01	47,61
Apr	49,25	48,75	51,29
May	44,56	53,44	45,97
Jun	48,67	49,33	49,50
Jul	76,79	21,21	70,03
Aug	56,73	41,27	57,80
Sep	52,07	45,93	55,98
Oct	52,00	46,00	53,41
Nov	59,40	38,60	62,06
Dec	62,70	35,30	65,15
Average	**56,61**	**41,39**	**57,51**

[a] The monthly SMP is also presented for comparison reasons

From the above analysis derives that the Market Special Price (€/MWh) is not actually a price that each RES producer is compensated. It is a price introduced with the sole purpose to estimate the monthly value of the $sFiP_m$ (also in €/MWh).

It is mentioned that in the case of the present study, where all wind stations are assumed to have the same Reference Value of 98 €/MWh, the value of $sFiP_m$ for each month is common for all wind stations.

Finally, the premium (in €) each wind station receives is determined according to the provisions of Law 4416/2016 and MD 187480/7.12.2016 (OG3955B). Particularly, the premium, which depends on the produced energy of each wind station over the calculation period (month), is calculated with the following equation:

$$\boldsymbol{premium} = (\boldsymbol{RV_{ws}} - \boldsymbol{MASP_{wind-energy,m}})\boldsymbol{X}(\boldsymbol{Q_{ws,m}} - \boldsymbol{Q_{ws,m,SMP=0|t>2h}}) \quad (12)$$

where:
RV_{ws}: *the RV that a wind station has secured*
$MASP_{wind\text{-}energy,m}$: *the MASP value for month m for wind projects*
$Q_{ws,m}$: *the total produced energy of a wind station for month m*
$Q_{ws,m}$, $SMP=0|t>2h$: *the energy produced by the wind station over the month m, where the SMP=0 for over 2 hours (cf. Law 4416/2016, article 5, par.10 & MD 187480/7.12.2016, article 5, par.3)*

The monthly income under a FiP scheme was calculated per wind station, as well as per region for the examined period (2014-2016). The results are presented in Appendix 6. As expected, the total

IOP Conf. Series: Journal of Physics: Conf. Series **1102** (2018) 012035 doi:10.1088/1742-6596/1102/1/012035

monthly income for all wind stations is the same for FiT and FiP schemes (see last column of Tables in Appendices 4 and 6).

4. Results from the comparison

The results of the comparison reveal small fluctuation in the annual revenues of each wind station between the two schemes, which does not seem to increase when examining the total 3-year period (analytical results per station per year are presented in Appendix 7)

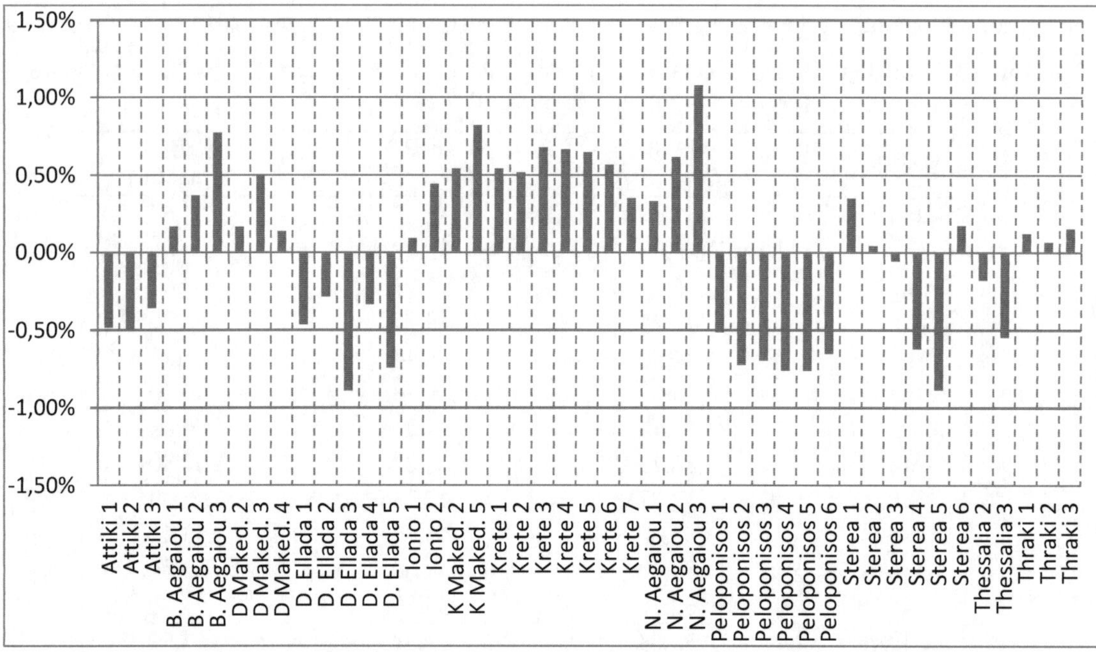

Figure 2. Comparison of revenues under FiT & FiP schemes for the period 2014-2016 for 45 representative locations of wind farms in Greece

5. Conclusions

Combining hourly mesoscale wind data for selected locations (representative to the actual installed capacity in Greece) with hourly SMP data can provide accurate results regarding the energy production and hence the total revenues of wind stations for the case of Greece, as these are derived by the day-ahead market and the support mechanism. The present study shows an affordable impact of the support scheme on the annual revenues of a wind plant for the period examined (years 2014-2015-2016), under the assumption that the strike value of the FiP scheme coincides with the fixed tariff of the FiT scheme. The annual income fluctuation for wind plants operating under different remuneration schemes were calculated to have a maximum value of ±1,5% per year, while for the total 3-year period the income variation does not seem to increase under the basic scenario.

6. Learning objectives

The present study proposes a methodology based on the use of mesoscale wind data and answers the question whether the change of the support scheme from FiT to sliding FiP in Greece creates significant uncertainty on the evaluation of the revenues of a windfarm. More calculations should be performed with longer periods of data and in other countries in order to better assess the accuracy of the energy production calculation and level of uncertainty. Moreover, the typical market obligations (e.g. balancing responsibilities) should also be taken into account per specific project in order to conclude a final investment decision.

IOP Conf. Series: Journal of Physics: Conf. Series **1102** (2018) 012035 doi:10.1088/1742-6596/1102/1/012035

References

[1] EMD International A/S;, n.d. *https://www.emd.dk.* [Online]
Available at: https://www.emd.dk/windpro/mesoscale-data/subscribe/emdconwx-mesoscale-data/

[2] European Comission, 2014. Guidelines on State aid for environmental protection and energy 2014-2020. *Official Journal of the European Union.*

[3] Foussekis, D. & Gkarakis, K., 2014. *Comparison of reanalysis data with long-term measurements in complex and coastal terrains.* Barcelona, EWEA.

[4] Hellenic Electricity Market Operator, n.d. *http://www.lagie.gr.* [Online]
Available at: http://www.lagie.gr/systima-eggyimenon-timon/ape-sithya/miniaia-deltia-eidikoy-logariasmoy-ape-sithya/

[5] Hellenic Transmissions System Operator, n.d. *www.admie.gr.* [Online]
Available at: http://www.admie.gr/en/operations-data/electricity-power-market-participation/market-data/

[6] Hellenic Wind Energy Association, n.d. *www.eletaen.gr.* [Online]
Available at: http://eletaen.gr/wp-content/uploads/2018/07/2017-hwea-wind-statistics-greece.pdf

Appendices available at: Link for Appendices

Innovative methods for wind turbine power curve upgrade assessment

Ludovico Terzi[1], Andrea Lombardi[1], Francesco Castellani[2], Davide Astolfi[2]

[1] Renvico s.r.l., Via San Gregorio 34, 20124 Milano, Italy. www.renvicoenergy.com
[2] University of Perugia - Department of Engineering, Via G. Duranti 93 - 06125 Perugia, Italy

E-mail: ludovico.terzi@renvico.it

Abstract. Wind turbine power curve upgrades have recently been attracting considerable investment in the operation of wind farms and noticeable attention in the wind energy literature. Due to the non-stationary conditions to which wind turbines are subjected, the most consistent strategy for quantifying the production improvement from the installation of an upgrade is comparing, after at least some months of operation, the post-upgrade production against a model of the pre-upgrade production under the same conditions. Formulating adequate models for the power of the upgraded wind turbines is in general non-trivial and it can be difficult to achieve the precision that wind turbine practitioners typically require for the production assessment. In the present work, a multivariate linear method for selecting the most appropriate input for modeling a given output is presented and applied to a test case. The test case is a multi-megawatt wind turbine owned by Renvico, on whose blades vortex generators and passive flow control devices have been installed. Applying the proposed method, it is possible to compute with precision the production improvement in the first five months of post-upgrade operation (purely aerodynamic upgrade) and in the subsequent three months (aerodynamic and control upgrade). It is therefore possible to appreciate the different contributions to the production enhancement from the aerodynamic and control improvement. A non-upgraded wind turbine from the same wind farm is also studied and the precision of the results inspires the use of the proposed method for performance control and monitoring in general.

1. Introduction

Multi-megawatt wind turbine is nowadays a mature technology: this implies, on one side, that public subsides are gradually declining and, on the other side, that there is a continuous improvement in the efficiency of wind kinetic energy conversion and in the control and monitoring of operating wind farms.

Wind turbine power curve upgrades have therefore recently attracted a considerable attention from wind turbine practitioners. One subtle point about wind turbine power curve upgrades is that quantifying realistically their impact on the power production is tricky, because their efficiency might depend considerably on the features of the wind flow at the microscale level [1]. It is evident that it doesn't make much sense to compare the production of a wind turbine pre and post upgrade, because wind turbines operate under non-stationary conditions. The most appropriate approach is comparing the post-upgrade production, after at least some months of upgraded operation, against a simulation of how much the wind turbine would have produced if the upgrade hadn't been installed.

Recently, this topic has been addressed in a certain number of studies in the wind energy literature. There are basically four studies. The first one is [2]: this manuscript contains an interesting discussion motivating why the binning method for the study of the power curve is insufficient for facing satisfactorily wind turbine power curve upgrades. Basically, as supported also in [3], precision modeling of the power of a wind turbine calls for multivariate dependencies. Therefore, on these grounds, in [2] a novel method for modeling wind turbine power output and quantifying power upgrades is proposed. It is a generalization of the conventional kernel regression method (which is, in essence, a weighted average of all the data points with the weighting coefficient associated to each input measurement decaying to zero when the euclidean distance between the evaluation point and the input measurement increases). Two issues are faced: dimensionality and bias. The first problem is approached by allowing hybrid kernels (additive-multiplicative). The second problem is tackled by selecting a calibration data set for the model that has similar features with respect to the target data set: this is done defining a distance conceptually similar to the Mahalanobis distance. In [2], this method is applied to the study of two test case: a real one (vortex generators installation) and an artificial one (pitch angle optimization: the data are created fictitiously basing on the logic of pitch angle control).

In [1], the problem of quantifying the production improvement from vortex generator installation is studied using SCADA data and time-resolved data having respectively ten minutes and seconds of sampling time. The point is that, using the kernel plus method and the SCADA data, the computation of the energy improvement results to be affected by considerable uncertainty. To enlarge the data set and therefore improve the statistical significance, the time-resolved data are employed: the authors of [1] show that the power-power approach is sufficient to provide reliable results, when this kind of data is used.

The power-power approach is an intuitive method and it is based on quantifying how the power difference between the target upgraded wind turbine and a reference non-upgraded wind turbine changes after the upgrade of the wind turbine of interest. Possibly, as in [1], data can be pre-treated by removing measurements characterized by wind turbine operation under the wake of nearby ones. Doing this, operation conditions are selected that resemble as much as possible the ideal ones in real environment. The main result of [1] is that vortex generators can have an efficiency that considerably depends on the flow conditions at the microscale level.

This method has been generalized in [4]: the test case (the same as in the present work) is a multi-megawatt wind turbine sited in a very complex terrain. Vortex generators and passive flow control devices have been installed on this test wind turbine. The idea of [4] is modeling the power of the upgraded wind turbine as a function of the powers of a certain number of nearby wind turbines (possibly and realistically more than one, while the power-power approach employs only one reference wind turbine). Since the relation between the input and the output is unknown, a data-driven approach has been employed: a feed-forward Artificial Neural Network is the selected model. In [4], it is shown that a bootstrap technique on the training and validation data sets is useful for improving the precision of the results. Other wind turbine power curve upgrades (including a real test case of pitch angle optimization) are studied, according to the same philosophy, in [5].

This study moves from [4] and [5]. It is a collaboration between the University of Perugia and the Renvico company, owning and managing 335 MW of wind turbines in Italy and France (www.renvicoenergy.com). At least three considerable steps forward with respect to [4] are addressed.

The first step forward regards the fact that the model of [4], despite more general than the power-power approach, still contained a certain degree of arbitrariness in the selection of the input for modeling the power of the upgraded wind turbine. One of the objectives of this work is overcoming this limitation: an appropriate multivariate linear model has therefore been identified by means of stepwise regression. Since the stability and the quality of the results were required features, particular care has been devoted to the validation of the model. Therefore, the outcome of this work is not only the computation of the production improvement, but also the procedure for selecting the appropriate inputs and validating the model. The method is generalizable to the study of whatever kind of wind turbine upgrades and the precision obtained in this very complex test case is impressive. As regards the test case of interest, it is important to notice that three input variables (from three nearby wind turbines) are selected for modeling the power of the upgraded wind turbine: this implies that using only one, as in the power-power approach, might not be robust enough.

The second improvement with respect to [4] regards the fact that the upgrade has been installed in two steps: first, the purely aerodynamic part has been installed and the wind turbine has been operating for five months. Subsequently, the control of the revolutions per minute has been optimized and the wind turbine has been operating with the entire upgrade for three months up to when this study has been performed. Therefore, with this work it is possible to separate the effects of the two sub-upgrades and quantify them.

The third important point of this work is that the adopted method provides sufficient precision in the computation of the energy production in order to employ it for performance monitoring in general. In this work, this is supported by modeling the power output of a non-upgraded wind turbine and assessing the precision of the model.

Summarizing, the structure of the manuscript is therefore the following: first, the test case and the data sets are briefly described in Section 2; subsequently the model is formulated in Section 3; the results are summarized in Section 4; conclusions and further directions are sketched in Section 5.

2. The wind farm and the data set

The layout of the wind farm is reported in Figure 1. From the contour lines, it is possible to appreciate the complexity of the terrain. T7 (in red in Figure 1) is the wind turbine that has been upgraded with vortex generators and passive flow control devices. The wind turbines are multi-megawatt and the hub height is 80 meters above ground level.

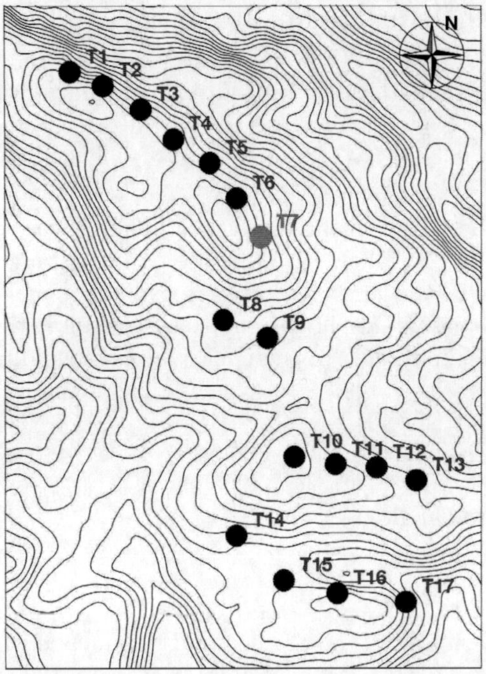

Figure 1. The layout of the wind farm.

The vortex generators installed on the blades are injection moulded thermoplastic components made out of a compound polymeric material and they are mounted both inboard at the root section of the blade as well as outboard at the tip of the blade. The adopted passive flow control devices are flaps that are glued onto the trailing edge of blade root and of the blade tip. All these devices produce an increase of the lift on the basis of different principles: the vortex generators delay the airflow separating from the blade surface, the blade tip flow control devices extend the blade chord. For detailed discussions on the physics of the different types of flow control devices and on their use as smart rotor control for loads reduction and therefore wind turbine life-cycle increase, see for example [6, 7, 8, 9].

In Figure 2, a photo is reported of a blade of T7 equipped with the flow control devices. It has been reported for a qualitative comprehension of the kind of devices. For a detailed discussion of the technology, see [10].

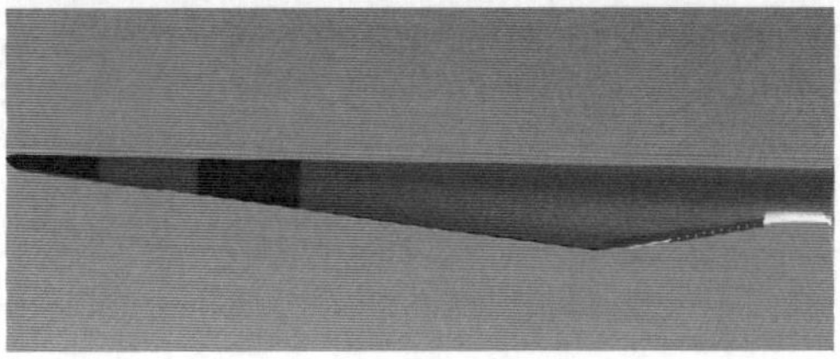

Figure 2. A picture of a wind turbine blade of T7 equipped with the flow control devices.

The data sets at disposal are the following:

- D_{bef} goes from 01/01/2016 to 01/07/2017. It is a period during which the standard blade configuration was adopted.

IOP Conf. Series: Journal of Physics: Conf. Series **1102** (2018) 012036 doi:10.1088/1742-6596/1102/1/012036

- D^1_{aft} goes from 01/09/2017 to 01/03/2018. It is a period during which T7 has been operating with the improved blade configuration.
- D^2_{aft} goes from 01/03/2018 to 01/06/2018. It is a period during which T7 has been operating with the improved blade configuration and the optimized revolutions per minute control.

3. The method

The proposed approach is based on linear regression. The objective is to estimate the power output of T7, denoted as y in the following. This has to be estimated based on the data from the other 16 wind turbines in the farm because, as explained in [4], the nacelle wind speed measurements from T7 are biased (in [4], it is hypothesized that the nacelle transfer function hasn't been updated after the upgrade). In particular, for each wind turbine the data at disposal are:

- the nacelle wind speed,
- the power output,
- the individual blade pitch angles,
- the rotor revolutions per minute,
- the high speed rotor temperature.

The conditions for data filtering are that all the wind turbines are producing output and that the power of T7 is below rated: this second condition is requested because the power doesn't upgrade when it is already rated. The filter on the request of power production has been based on the appropriate counter of grid production, available in the SCADA data set.

The procedure for selecting the input to the model is based on the stepwise regression algorithm [11], which is a data-driven iterative algorithm for automatically selecting the most appropriate inputs for a multilinear regression task.

Starting from an initial model, at each step a potential regression term is added to or removed from the model. This is done by calculating the $p-value$ of an $F-statistics$ and testing the performance of the model with and without that potential term. Only the regressors whose probability of not having coefficients in the model is less then a specified threshold on the $p-value$ are included and maintained. The threshold on the $p-value$ is denoted as p_{remove}. Due to its crucial role in the input selection process, the value for the p_{remove} parameter must be chosen carefully. To this end, an extensive study has been carried one on the best value for p_{remove}. The tested values are $10^{-\gamma}$ with $\gamma = 1, \ldots, 15$. For each value of p_{remove}, the input selection and estimation pipeline is subjected J times to K-fold cross-validation [12]. In particular, the data set $\mathrm{D_{bef}}$ is divided J times randomly in two subsets: $(K-1)/K$ of the data are used for training and the remaining $1/K$ are used for validation. K has been selected to be 10 for this study, because the objective is having a robust model and therefore short validation folds are selected. J has been set to 300 to increase the statistical significance of the study.

Given a value γ for p_{remove}, for each of the $j = 1, \ldots, J$ runs of the cross-validation, the most significant inputs are selected and the estimated power of T7 $\hat{\boldsymbol{y}}_{valid_{j,\gamma}}$ is modeled via linear regression. The performance of the linear model for the given value γ for p_{remove} is estimated in terms of the mean absolute error, averaged over the $j = 1, \ldots, J$ folds:

$$\bar{\delta}_\gamma = \sum_{j=1}^{J} \frac{\overline{|\hat{\boldsymbol{y}}_{valid_{j,\gamma}} - \boldsymbol{y}_{valid_{j,\gamma}}|}}{J} \tag{1}$$

where $\boldsymbol{y}_{valid_{j,\gamma}}$ is the real power of T7 for the validation subset of that run.

From the feature selection process, it arises that $\bar{\delta}_\gamma$ is of the same order of magnitude ($\simeq$ 100 kW) for each value of p_{remove}: the improvement in choosing $p_{remove} = 10^{-1}$ instead of $p_{remove} = 10^{-15}$ is averagely only 3 kW in $\bar{\delta}_{\gamma}$, at the cost of employing many more regressors (13 against 3) and having a less robust model since, furthermore, the number of possible regressor configurations is much higher.

For whatever choice of $p_{remove} \leq 10^{-10}$, instead, one obtains that the selected inputs are the same (except for a small number of outliers) and they are:

- the power output of T6,
- the power output of T9,
- the rotor revolutions per minute of T8.

Therefore, the decision is modeling the power output of T7 as a function of the three above inputs. Besides the statistical significance of the method adopted for identifying the appropriate inputs for modeling the power of T7, support to the decision comes from the fact that it is definitely plausible. In fact, the wind turbines selected for the inputs are the nearest to T7 and one input per nearby wind turbine is selected, therefore there is no redundant information. Furthermore, it is reasonable that in such a complex terrain the power output or the rotor revolutions are more stable than the wind speed (because the wind turbine acts like a filter) and are preferable to model the power of T7.

The power of T7 is modeled via linear regression as function of the inputs selected as described above. The input measurements are normalized and organized in a matrix $\underline{x}_{train}$. The pseudo-inverse of $\underline{x}_{train}$ is used to compute the weight matrix of the linear regression model as:

$$\underline{W} = \underline{x}_{train}^{-1} \cdot \boldsymbol{y}_{train} \tag{2}$$

Finally, the estimated power output $\hat{\boldsymbol{y}}_{\boldsymbol{valid}}$ on the test data set is computed as a linear combination of the input at test time, weighted using the model weight matrix $\underline{W}$:

$$\hat{\boldsymbol{y}}_{\boldsymbol{valid}} = \boldsymbol{x}_{valid} \cdot \underline{W} \tag{3}$$

3.1. Modeling the power of a non-upgraded wind turbine

Consider applying the same method for a wind turbine that has not been retrofitted. Wind turbine T4 has been selected for this test, but any other non-upgraded turbine in the farm could have been selected. The inputs for modeling the power production of T4 have been selected with the same procedure described above and are:

- the power of T2,
- the power of T3,
- the power of T5,
- the rpm of T5.

4. Results

In the following, the procedure is reported for estimating the energy improvement. The data sets at disposal are employed as follows:

- D_{bef} is randomly divided in two subsets: D0 (1 year of data) and D1 (6 months of data). D0 is used for training the model and constructing the weight matrix $\underline{W}$, D1 is used for validating the model.
- D_{aft}^{i} with $i = 1, 2$ (also named $D2^i$ for simplifying the notation in the following) is used in its entirety to quantify the performance improvement.

The union of $\mathrm{D}^1_{\mathrm{aft}}$ and $\mathrm{D}^2_{\mathrm{aft}}$ is named as $D2^{tot}$.

The residuals between the measurement y and the simulation $\hat{y}$, for the data sets D1 and D2, are studied. In particular, the interest is in how the residuals vary after the upgrade with respect to before. Therefore, consider Equation 4 with $i = 1, 2$.

$$R(\boldsymbol{x}_i) = y(\boldsymbol{x}_i) - \hat{y}(\boldsymbol{x}_i). \tag{4}$$

For $i = 1, 2$, one computes

$$\Delta_i = 100 * \frac{\sum_{\boldsymbol{x} \in Data_i} (y(\boldsymbol{x}) - \hat{y}(\boldsymbol{x}))}{\sum_{\boldsymbol{x} \in Data_i} y(\boldsymbol{x})} \tag{5}$$

Since Δ_i is constructed with the relative discrepancies of power data each having the same sampling time (10 minutes), the quantity $\Delta = \Delta_2 - \Delta_1$ provides a percentage estimate also of the energy improvement.

The above procedure has been repeated several times with several random choices of D0 and D1 for training and validating the model. Doing this, it is possible to obtain an average estimate of the energy improvement, as well as the standard deviation (and therefore reasonable lower and upper limits for the energy improvement).

4.1. T7: aerodynamic upgrade

In Figure 3, for a sample selection of D0 and D1, the sets $R(\boldsymbol{x}_1)$ and $R(\boldsymbol{x}_2^1)$ are plotted after being averaged within intervals having amplitude of the 10% of the rated power.

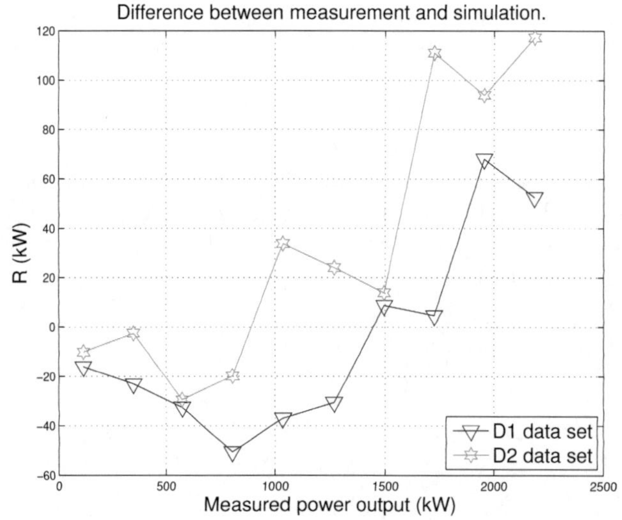

Figure 3. The average difference between measurements and simulation (Equation 1), for data sets D1 and $D2^1$ and for a sample run of the model.

The average energy improvement is computed as $\Delta = 3.6\%$. In other words, the estimate is that T7 has produced, during data set $D2^1$ and below rated power, the 3.6% more than it would without aerodynamic improvement. The standard deviation is computed as $\sigma_\Delta = 0.4\%$: therefore, reasonable upper and lower limits of the energy improvement are $\Delta_+ = 4.0\%$ and $\Delta_- = 3.2\%$. This corresponds to an improvement in the whole production during $\mathrm{D}^1_{\mathrm{aft}}$ given by $\Delta_E = 1.8\% \pm 0.2\%$.

4.2. T7: aerodynamic and control upgrade

In Figure 4, for a sample selection of D0 and D1, the sets $R(\boldsymbol{x}_1)$ and $R(\boldsymbol{x}_2^2)$ are plotted after being averaged within intervals having amplitude of the 10% of the rated power.

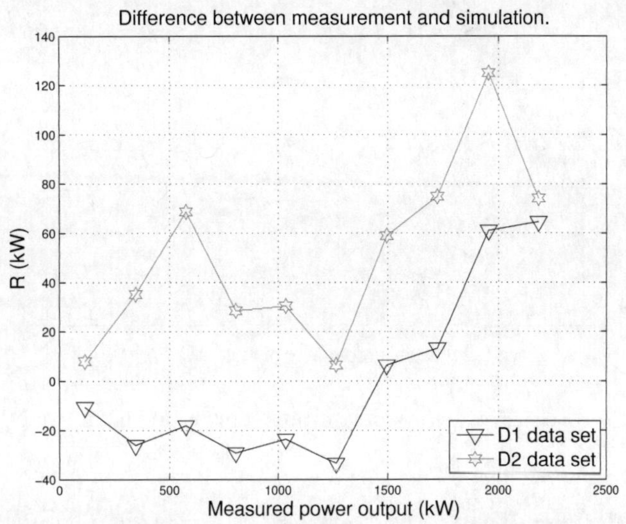

Figure 4. The average difference between measurements and simulation (Equation 1), for data sets D1 and $D2^2$ and for a sample run of the model.

The average energy improvement is computed as $\Delta = 5.2\%$. The standard deviation is computed as $\sigma_\Delta = 0.2\%$: this corresponds to an improvement in the whole production during D^2_{aft} given by $\Delta_E = 2.6\% \pm 0.1\%$.

This result allows to appreciate, comparing against the estimate in subsection 4.1, that the control optimization provides a production improvement of the order of $\Delta_E^{control} = 0.8\%$. This is compatible with the estimates in [5] for other test cases: the typical order of magnitude of energy improvement from control optimization is the percent.

Furthermore, it is interesting to study the characteristic curves before and after the control system upgrade. T7 is analyzed in comparison with a sample wind turbine from the rest of the wind farm (T1). The curves after the aerodynamic upgrade and before the control upgrade (data set D^1_{aft}) are reported in Figure 5 and it arises that for T1 and T7 they are basically indistinguishable.

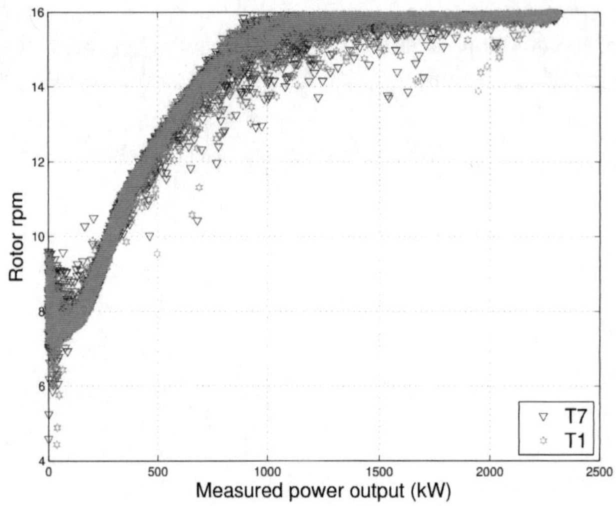

Figure 5. The power - rotor revolutions per minute curve, for data set D^1_{aft}. Turbines T7 and T1.

The curves after the aerodynamic upgrade and the control upgrade (data set D^2_{aft}) are reported in Figure 6 and it arises that the curves for T1 and T7 are distinguishable. For a given value of the number of rotor revolutions per minute, T7 produces more power than T1.

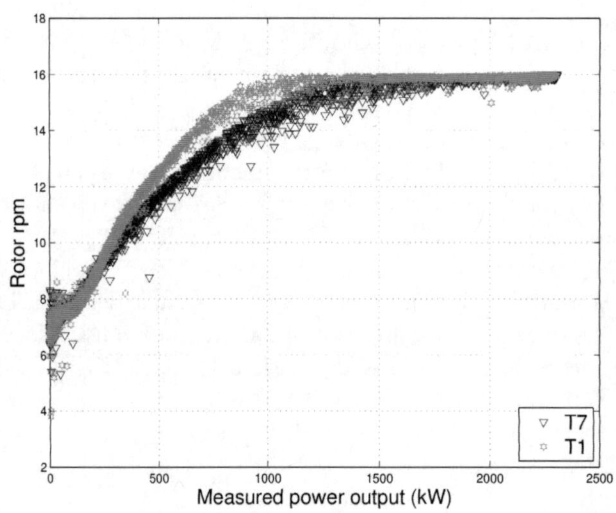

Figure 6. The power - rotor revolutions per minute curve, for data set D^2_{aft}. Turbines T7 and T1.

4.3. T4

Figure 7 is, similarly to Figures 3 and 4, a plot of $R(\boldsymbol{x}_1)$ and $R(\boldsymbol{x}_2)$ on a sample model run, once the data are averaged in power intervals of 10% of the rated. The difference with respect to Figures 3 and 4 is evident: in Figure 7, the average residuals $R(\boldsymbol{x}_1)$ and $R(\boldsymbol{x}_2)$ as a function of the power are almost identical the ones with respect to the others.

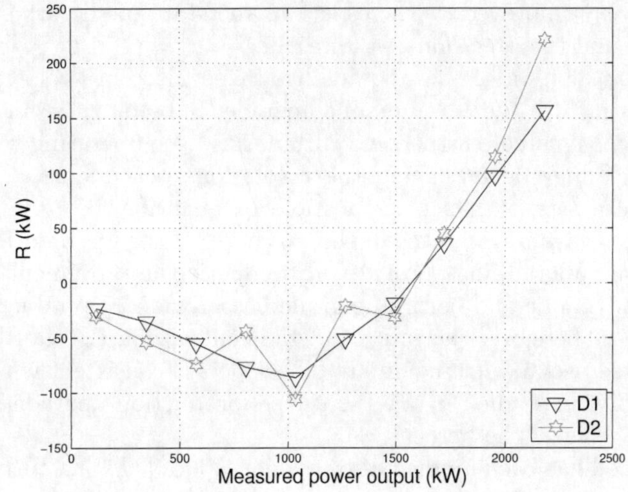

Figure 7. The average difference between measurements and simulation (Equation 1), for turbine T4. Data sets D1 and $D2^{tot}$ and sample run of the model.

The average production variation (basing on Equation 5) is estimated to be $\Delta_E = 0.05\%$: practically zero. This result supports that, even for a wind farm sited in a very complex terrain as the selected test case, it has been possible to model the production of a wind turbine with remarkable precision. This opens the perspective of employing the proposed method not only for upgrade assessment, but also for ordinary control and monitoring of the performance of operating wind turbines.

5. Conclusions and further directions

The Renvico company and the University of Perugia have been cooperating for years and recently they have devoted particularly to the study of wind turbine power curve upgrades. Actually, this topic has become relevant in the operation and management of wind farms and requires devoted techniques that are commonly in the toolbox of the academia rather than industry. For this reason, the literature about the data-based study of wind turbine power upgrades after at least some months of operation is somehow at its early stages, but it is boosting rapidly [2, 1, 4, 5].

In this work, a test case has been selected because it has been considered the most instructive between those analyzed jointly by Renvico and UniPG and the study has been performed in depth. This has allowed a methodological approach, opening the perspective of generalizing the use of the proposed techniques, and has provided interesting insight about how a wind turbine power upgrade works.

The test case has been a multi mega-watt wind turbine from a wind farm sited in a very complex terrain. On the blades, vortex generators and passive flow control devices have been installed. As discussed in [4] and [5], a qualitative assessment of the upgrade through the study of the power curve is impossible because the nacelle wind speed measurements after the upgrade are presumably biased. This limitation of the data set at disposal has been turn into a power, because it has motivated the formulation of a general method for modeling a given output (in this case, the power of the upgraded wind turbine) as a function of the most appropriate input.

An automatic feature selection algorithm based on stepwise multivariate regression has been adopted and it has allowed to select the most meaningful input variables for a linear model whose output is the power of the upgraded wind turbine. Furthermore, a bootstrap technique on the pre-upgrade data set at disposal has allowed to artificially produce the repeatability of the tests and this has considerably improved, with respect to the state of the art in the literature, the statistical significance and the precision of the results.

Furthermore, the installation of the upgrade has taken place in two steps: for five months, the wind turbine has run with the aerodynamic upgrade. Subsequently, for three months up to when this study has been conducted, the wind turbine has been operating with the aerodynamic upgrade and a control upgrade managing the revolutions per minute. Therefore, studying separately the two data sets, it has been possible to quantify the production improvement from the complete upgrade and from each of the two parts of the upgrade. The estimate is that the purely aerodynamic upgrade has provided a production improvement that is slightly less than the 2%; the aerodynamic and control upgrade has provided a production improvement of the order of the 2.6%. Therefore, the impact of the control upgrade can be estimated as the difference between these two and it amounts to 0.8%. This is consistent with other test cases of control system optimization studied in [5]: the typical production upgrade from control system optimization is of the order of the percent.

The proposed method has been applied also to a non-upgraded wind turbine as a consistency check. The result are remarkably precise: once one has trained the model on a data set, the residuals between model and measurements in two different data sets behave impressively similar. The obtained precision opens the perspective of employing this kind of method not only for the study of wind turbine power upgrades, but also for the control and monitoring of the performances in normal operation.

Two are the main further directions of this work: the first one is employing the proposed method for the study of other test cases. The second one is testing other kinds of method (for example, non-linear) on this test case and analyzing how the precision of the results changes.

Bibliography

[1] Hoon Hwangbo, Yu Ding, Oliver Eisele, Guido Weinzierl, Ulrich Lang, and Georgios Pechlivanoglou. Quantifying the effect of vortex generator installation on wind power production: An academia-industry case study. *Renewable Energy*, 113:1589–1597, 2017.

[2] Giwhyun Lee, Yu Ding, Le Xie, and Marc G Genton. A kernel plus method for quantifying wind turbine performance upgrades. *Wind Energy*, 18(7):1207–1219, 2015.

[3] Bruce Stephen, Stuart J Galloway, David McMillan, David C Hill, and David G Infield. A copula model of wind turbine performance. *IEEE Transactions on Power Systems*, 26(2):965–966, 2011.

[4] Davide Astolfi, Francesco Castellani, and Ludovico Terzi. A scada data mining method for precision assessment of performance enhancement from aerodynamic optimization of wind turbine blades. In *Journal of Physics: Conference Series*, volume 1037, page 032001. IOP Publishing, 2018.

[5] Davide Astolfi, Francesco Castellani, and Ludovico Terzi. Wind turbine power curve upgrades. *Energies*, 11(5):1300, 2018.

[6] Thanasis K Barlas and GAM Van Kuik. Review of state of the art in smart rotor control research for wind turbines. *Progress in Aerospace Sciences*, 46(1):1–27, 2010.

[7] Kuo-Chang Tsai, Cheng-Tang Pan, Aubryn M Cooperman, Scott J Johnson, and CP Van Dam. An innovative design of a microtab deployment mechanism for active aerodynamic load control. *Energies*, 8(6):5885–5897, 2015.

[8] Unai Fernández-Gámiz, Clara Marika Velte, Pierre-Elouan Réthoré, Niels N Sørensen, and Eduard Egusquiza. Testing of self-similarity and helical symmetry in vortex generator flow simulations. *Wind Energy*, 19(6):1043–1052, 2016.

[9] Iñigo Aramendia, Unai Fernandez-Gamiz, Jose Antonio Ramos-Hernanz, Javier Sancho, Jose Manuel Lopez-Guede, and Ekaitz Zulueta. Flow control devices for wind turbines. In *Energy Harvesting and Energy Efficiency*, pages 629–655. Springer, 2017.

[10] J. Laursen, P. Fuglsang, and P.B. Enevoldsen. Dinosaurs, racecars and blades: Discover the connection. Technical report, Wind Power World, 2012.

[11] PT Pope and JT Webster. The use of an f-statistic in stepwise regression procedures. *Technometrics*, 14(2):327–340, 1972.

[12] Payam Refaeilzadeh, Lei Tang, and Huan Liu. Cross-validation. In *Encyclopedia of database systems*, pages 532–538. Springer, 2009.

Comparative analysis of binning and Gaussian Process based blade pitch angle curve of a wind turbine for the purpose of condition monitoring

Ravi Kumar Pandit and David Infield

University of Strathclyde,16 Richmond St, Glasgow G1 1XQ, UK.
Corresponding author e-mail: ravi.pandit@strath.ac.uk

Abstract. Several studies have used the power curve as a critical indicator to assess the performance of wind turbines. However, the wind turbine internal operation is affected by various parameters, particularly by blade pitch angle. Continuous monitoring of blade pitch angle can be useful for power performance assessment of wind turbines. The blade pitch curve describes the nonlinear relationship between pitch angle and hub height wind speed which to date has been little explored for wind turbine condition monitoring. Gaussian Process models are nonlinear and nonparametric technique, based on Bayesian probability theory. Such models have the potential give results quickly and efficiently. In this paper, we propose a Gaussian Process model to predict blade pitch curve of a wind turbine for condition monitoring purposes. The obtained Gaussian Process based blade pitch curve is then compared with a conventional approach based on a binned blade pitch curve for identifying operational anomalies purposes. Finally, the weaknesses and strengths of these methods are summarised. SCADA data from healthy wind turbines are used to train and evaluate the performance of these techniques.

1. Introduction

The wind energy had witnessed exponential growth in past decades and became the fastest among the available alternative energy sources. Global Wind Energy (GWEC) on its Global Wind Report [1], found that more than 54 GW of a wind turbine was installed across the global market in 2016, which now includes more than 90 countries and cumulative capacity grew by 12.6% to reach a total of 486.8 GW [1]. However, expensive operation and maintenance (O&M) cost makes wind energy costly affairs and therefore, reducing O&M costs and improving the reliability has become an emerging research interest. Unexpected failures of the wind turbines components cause high repair costs and result in machine downtime and loss of revenue. K. Fischer and A.et al. [2], suggested that O&M cost is responsible for 20-30% of the life cycle cost of onshore and up to 30% of offshore wind farms. To protect turbine components from premature failures, improve reliability and reduces O&M costs, SCADA based condition monitoring is considered as an effective technique and are expected to grow in coming years. Several literatures are focused on the power curve to assess the performance of a wind turbine for the purpose of condition monitoring, see [3,4]. However, the internal operation of turbines is affected by various turbine parameters that make difficult to visualize the turbine operation with the help of power curve alone. The turbine performance parameter (e.g., blade pitch angle) can play an important role in analyzing the internal operations of wind turbines and described in the upcoming section.

The wind turbine blade pitch curve describes the nonlinear relationship between pitch angle and wind speed and can be useful for robust wind turbine condition monitoring. For example, Singh [5] used a misaligned wind vane as a case study for wind turbine performance assessments where both power curve and blade pitch angle were constructed to identify the performance change due to misaligned vane. The comparative studies of these curves suggest that the blade pitch curve detects performance change while it is unidentified by the power curve. Furthermore, in ref [6], the wind turbine blade curve used in which multivariate outlier detection approach based on k-means clustering and Mahalanobis distance are applied. Using this curves, kurtosis, and skewness calculated for identifying underperformance associated with wind turbines. Thus, pitch curve monitoring is significant for identifying the abnormal behavior due to failures (e.g., pitch failures).

There are several statistical methods mostly confined to power curve of a wind turbine fitting and are broadly divided into the parametric and nonparametric methods. The nonparametric approach does not impose any pre-specified model unlike parametric method and so nonparametric models performance is better as compared to parametric models [7]. Examples of the nonparametric techniques include artificial neural networks [8], Gaussian Process [9,10], and k-nearest neighbor [11]. A comprehensive review of parametric and nonparametric approach related to wind turbines can be found in [3,7].

A Gaussian Process (GP) is a nonparametric, data-driven approach and widely used in solving problems related to classification and regression. GP is defined as a collection of random variables, any finite number of which have a joint Gaussian distribution [12]. GP model is a robust approach for approximating the complex nonlinear model, and one of its advantages is that estimated values come with confidence intervals. The confidence intervals are calculated separately by estimating the magnitude of the associated uncertainty and play a vital role in uncertainty analysis [9].

The recent expansion of wind power system leads to the development of accurate methods for their parameter identification such as the wind turbine blade pitch curve. Recent wind turbine condition monitoring studies have mostly focused on the power curve for evaluating performance. However, this cannot reflect the complete turbine operation since the operational behavior of the wind turbines profoundly influence by a parameter such as rotor power, torque, and pitch angle. Valid assessments of these parameter improve the power performance of a wind turbine. In this study, blade pitch angle impact on wind turbine performance is analyzed using the blade pitch curve that reveals the nonlinear relationship between pitch angle and the hub height wind speed that can be useful for analyzing wind turbine performance and the detection of faults.

In this paper, we propose a Gaussian Process approach to estimate the blade pitch curve of a wind turbine for application in identifying potential faults. The obtained GP based blade pitch curve is then compared with a conventional approach based on a binned blade pitch curve together with individual bin probability distributions to identify operational anomalies. Finally, the weaknesses and strengths of these methods are summarised. SCADA data from healthy wind turbines are used to train and evaluate the performance of these techniques.

This paper is structured as follows: Section 1 is the introduction. Section 2 describes the wind turbine blade pitch curve. Section 3 describes the SCADA dataset and its pre-processing. Section 4 outlines the methodologies and this section further divided into subsections explaining Gaussian Process (GP) and the binning approach to estimate the wind turbine blade pitch curve. Section 5 discusses the comparative analysis of proposed techniques, and Section 6 concludes the paper with intended future work.

2. Operational Curves of a Wind Turbine

The power curve widely used by wind operators to assess power performance, warranty formulations, energy assessment, and fault diagnosis applications. Power curve shows the power output strong dependence on wind speed (Figure 1) and mathematically expressed by following cubic relation [13],

$$P = 0.5\,\rho A C_p(\lambda, \beta\,)\,v^3 \tag{1}$$

where ρ is air density (kg/m^3), A is swept area (m^2) , C_p is the power coefficient of the wind turbine and v is the hub height wind speed (m/sec). The shape of the power curve governed by the cubic relation of these parameters.

A typical blade pitch curve described the nonlinear relationship between turbine pitch angle and wind speed and shown in Figure 2. Blade pitch angle customarily set to maximize the power production at below rated wind speed and at above-rated wind speed, pitch angle continuously adjusts to restrict power production to rated wind speed.

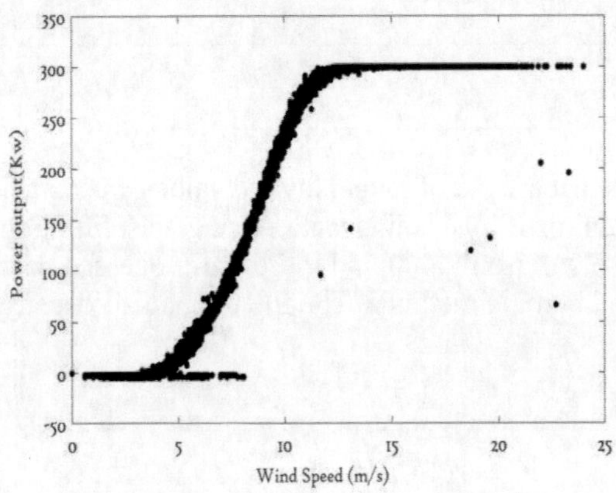

Figure 1: Measured power curve

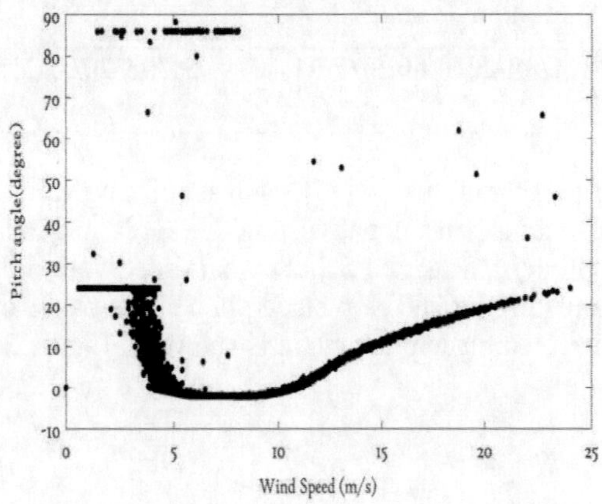

Figure 2: Measured pitch curve

The measured power curve and pitch curve (Figure 1 and 2) uses 10 minute average SCADA data obtained from a pitch controlled wind turbine. The IEC standard (61400-12-1) [14], advocate air density correction before any further analysis in which a corrected wind speed V_C is calculated using equations (2) and (3) as shown below,

$$\rho = 1.225\left[\frac{288.15}{T}\right]\left[\frac{B}{1013.3}\right] \quad (2)$$

and,

$$V_C = V_M\left[\frac{\rho}{1.225}\right]^{\frac{1}{3}} \quad (3)$$

where, V_C and V_M are the corrected and measured wind speed in m/sec and the corrected air density is calculated by equation (2) where B is atmospheric pressure in mbar and T the temperature in Kelvin for which 10 minute average values obtained from SCADA data are used. The corrected wind speed (V_C) from equation (3) will be used for constructing blade pitch curve based on Gaussian Process and binning methods in upcoming sections.

3. SCADA data description and pre-processing

The wind farms linked with SCADA data that contains enormous data sets which is useful for continuous condition monitoring of a wind turbine. SCADA data may be used efficiently to 'tune' a wind farm operation which is useful in identifying early warnings of possible failures and improving power production across many turbines in all conditions. SCADA can be used with any resolution, but 10-minute intervals generally recommended to reduce transmitted data bandwidth. SCADA data records historical and operational data of individual turbine where each data set contains average values of more than 100 parameters such as torque, wind speed, wind direction, drive train acceleration, and tower acceleration and so on. These parameters materialized with maximum, minimum, standard deviation. The datasets used in this study comes from a wind farm located in southern Europe. Monthly data used for model training and testing purposes contains 4464 data points beginning with time stamp ''1/08/2010 00:00 AM'' and ending at time stamp ''31/08/2010 23:50''. Using these data points, measured power and blade pitch curves are shown in Figure 1 and 2 respectively. It is worth to note that the monthly SCADA data used in this study is having an average monthly temperature of 29.779°C. However, the malfunction of sensors, mechanical systems, and the data collection system makes SCADA data subject to error and this affects the overall SCADA based condition monitoring of a wind turbine. Hence, pre-processing of SCADA is a must for minimizing the impact of these errors in order to extract useful information from the data.

IOP Conf. Series: Journal of Physics: Conf. Series **1102** (2018) 012037 doi:10.1088/1742-6596/1102/1/012037

Table 1. SCADA datasets description.

Start timestamp	End timestamp	Measured data	Filtered data
1/08/2010 00:00 AM	31/08/2010 23:50 PM	4400	2068

Criterion outlined in [15], such as missing values, values that are out of range, invalid values, mismatch timestamp, and negative power values are taken to minimize misleading data. Using this filtration approach, measured SCADA data sets were reduced by 2068 from 4400 and these pre-processed data points used to develop blade pitch curve based on GP and binning methods. The filtered and air density corrected blade pitch curve is shown in Figure 3.

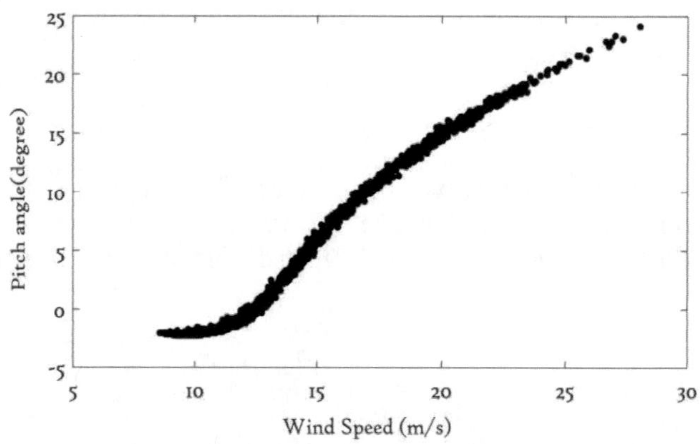

Figure 3: Pre-processed blade pitch curve

4. Methodologies

The two methods namely; binning and Gaussian Process used to build an effective blade pitch angle curves for the purpose of condition monitoring of a wind turbine are described as follows.

4.1. Gaussian Process based blade pitch curve

A Gaussian Process (GP) is a nonparametric stochastic approach widely used in classification and regression related problems. A GP describes a distribution over functions and is a collection of random variables, any finite number of which have a joint Gaussian distribution, [12]. A GP model is specified by a mean function and covariance function using the following equation,

$$f(x) \sim GP\big(m(x), k(x,x')\big) \tag{4}$$

where $m(x)$ is the mean function and $k(x,x')$ is the covariance function.

The covariance function describes the similarity between any set of data points relates to a multivariate Gaussian distribution as per the GP assumption. Covariance function selection determines the GP model effectiveness. Various covariance functions are available depending upon the need and application and are well described in [12]. The squared exponential covariance function (k_{SE}) with a fitted hyper-parameter is used in this study to predict the blade the pitch curve and is defined as:

$$k_{SE}(x,x') = \sigma_f^2 \exp\left(-\frac{(x-x')^2}{2l^2}\right) \tag{5}$$

In a Gaussian process, the incoming SCADA data can be considered noisy which affects the covariance function, k_{SE}, hence it is necessary to add a noise term to the covariance function in order to minimize the effect of noisy data, and hence equation (5) is modified to:

$$k_{SE}(x,x') = \sigma_f^2 exp\left(-\frac{(x-x')^2}{2l^2}\right) + \sigma_n^2\delta(x,x') \tag{6}$$

where σ_f^2 and l are defined as the hyper-parameters. σ_f^2 describe the signal variance and l is a characteristic length scale which describes how quickly the covariance decreases with the distance between points. The squared exponential covariance function is technically a smooth sample function being infinitely differential. The modified squared exponential covariance function equation is used to estimate the blade pitch curve of a wind turbine and then this is compared with a traditional pitch curve in terms of uncertainty and accuracy, and is used to identify how quickly faults can be identified. The GP pitch curve with an estimated 95% confidence intervals (CIs) have been used in this paper, as shown in Figure 4. Furthermore, Figure 5 suggest that the estimated GP pitch curve follows the expected pattern.

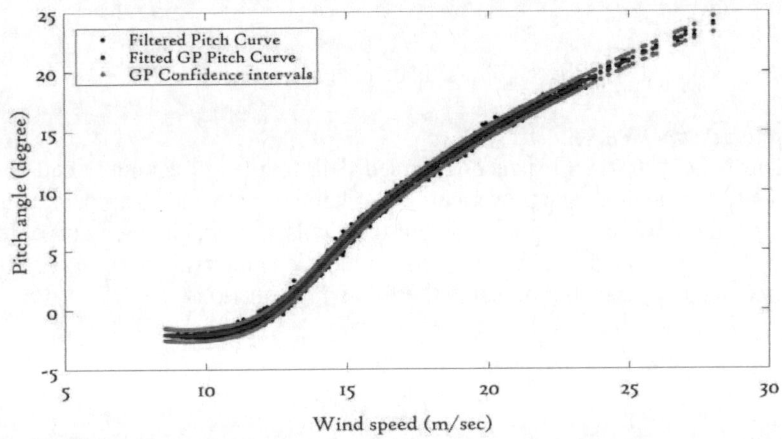

Figure 4: Estimated GP pitch curve with confidence intervals

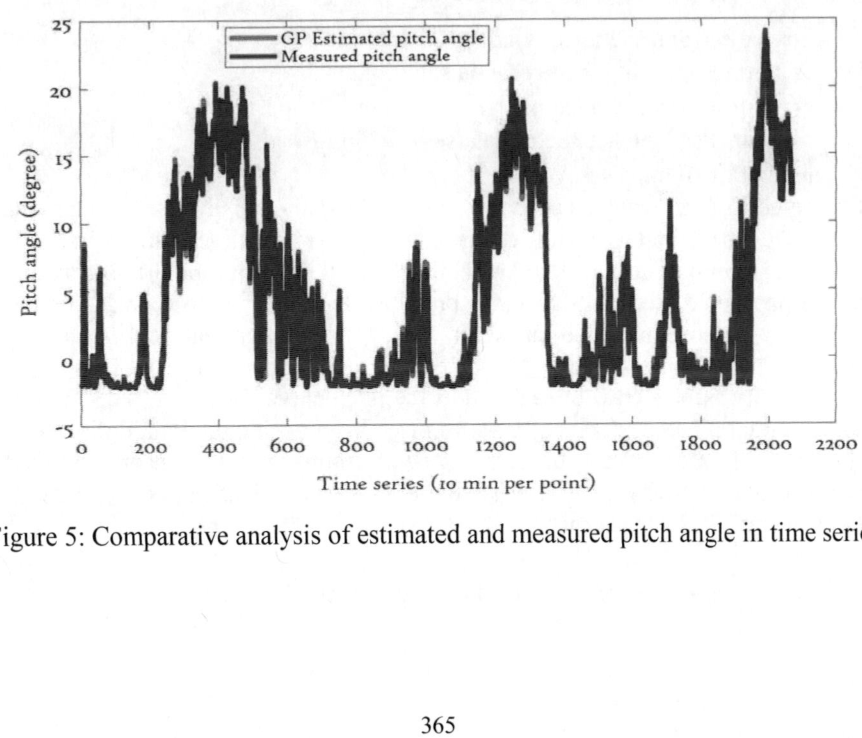

Figure 5: Comparative analysis of estimated and measured pitch angle in time series

Due to the nonparametric nature of the GP model, the residual analysis is essential to find out whether GP distribution is Gaussian or not. Residual is the difference between measured and estimated value. The frequency distribution of the residuals is shown in figure 6 together with a fitted Gaussian distribution. As expected, the distribution of residuals is closely following Gaussian.

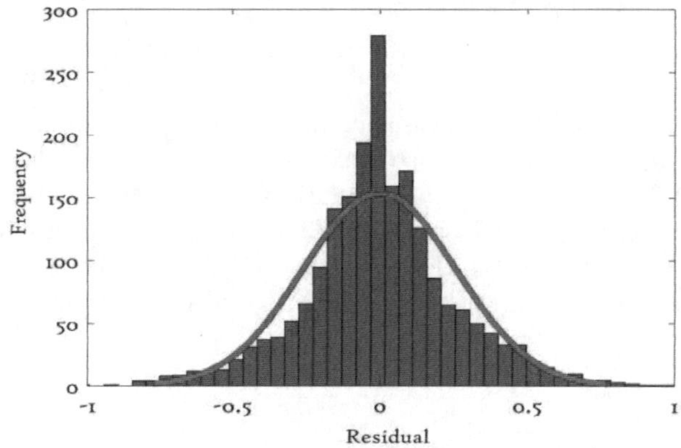

Figure 6: GP residuals histogram fitting

4.2. Binned based blade pitch curve

A Gaussian The IEC 61400-12 [14] recommended data reduction technique called the 'binning' which is generally used to calculate the power curve of a wind turbine and its associated uncertainty. In this paper, the 'method of bins' technique is applied to calculate the blade pitch curve. In the 'method of bins' 10-min average SCADA data is grouped into wind speed intervals of 0.5 m/s in order to get an average output pitch angle for each bin by using the following equations,

$$V_i = \frac{1}{N_i}\sum_{j=1}^{Ni} V_{n,i,j}$$

$$B_i = \frac{1}{N_i}\sum_{j=1}^{Ni} B_{n,i,j}$$

where, V_i = normalised and averaged wind speed in bin i.
$V_{n,i,j}$ = normalised wind speed of data sets j in bin i.
B_i = normalised and averaged pitch angle in bin i.
$B_{n,i,j}$ = normalised pitch angle of data set j in bin i.
N_i = number of 10 min average data sets in bin i.

It should be noted that the wind speed is the most significant source of uncertainty and including more data points give more certainty to the average value in the pitch curve. Type B uncertainties would be difficult to treat in a consistent manner without greater knowledge of the instrumentation used. Therefore, in this paper, we used the statistical spread evident in the binned data. The pitch curve is calculated together with error bars and shown in Figure 7. The two standard deviations (i.e., 95% confidence intervals) of measured pitch angle values are used to obtain the error bars which is used to measure the uncertainty associated with each bin of the pitch curve. However, the 'binning' is not the most effective technique since its accuracy compromised by choosing bin width of 0.5 m/sec because within each bin the measured power or pitch will depend strongly and non-linearly on wind speed and a wide bin would result in systematic bias. Also, there is a need in practice to get sufficient data points in each bin to be of statistical significance.

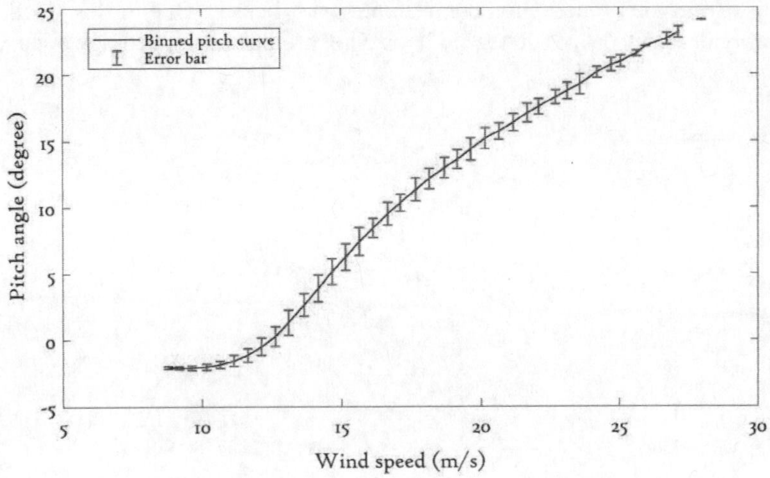

Figure 7: Binned blade pitch curve with error bars

5. Comparative studies

Conventional IEC Standard binned pitch curves described above are compared with a GP pitch curve in Figure 8; this shows that GP pitch curve model closely follows the binned pitch curve. This observation confirms that a GP is competent in estimating the wind turbine pitch curve. However, the accuracy of a GP model depends upon the quantity and quality of the data, as well as the appropriate method used. A low number of pitch angle-wind speed pairs may not give a smooth pitch curve while a high number is also not desirable because of the mathematical challenge (due to cubic inversion issue) of calculating a Gaussian Process for a large number of data points, [12].

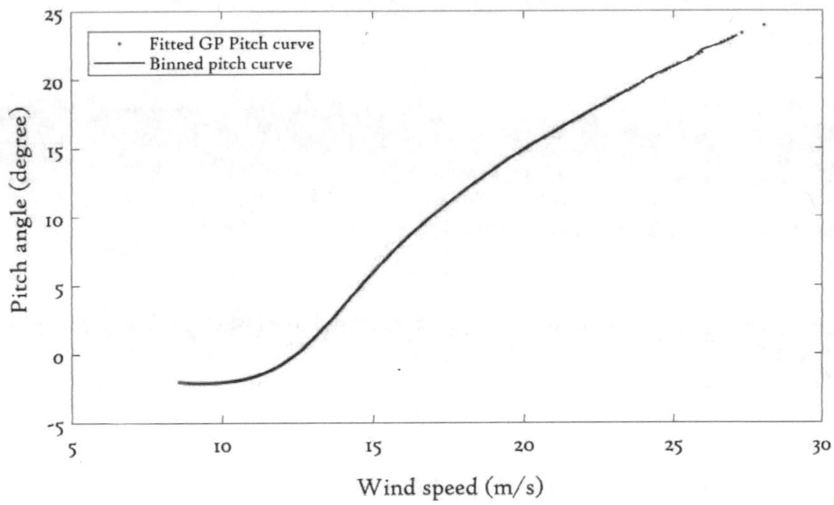

Figure 8: Comparison between binned and GP based pitch curve

A GP model comes with intrinsic confidence intervals (CIs), and this is key to wind turbine anomaly detection. These GP confidence intervals give significant information on the uncertainty surrounding an estimation, but they are also model-based estimates, [16]. Data points that lie outside of the confidence intervals can be considered anomalous, signifying a potential malfunction of the wind turbine. The GP estimated pitch curve compared with the binned pitch curve (with 95% confidence interval of error bars) in terms of uncertainty and is presented in Figure 9. This comparative analysis suggests that GP CIs are smaller than those from the binned pitch curve in almost all cases. Due to this smaller CIs, ability to reject the unhealthy or faulty data is better in GP as compared to the binned power curve, see Figure 9.

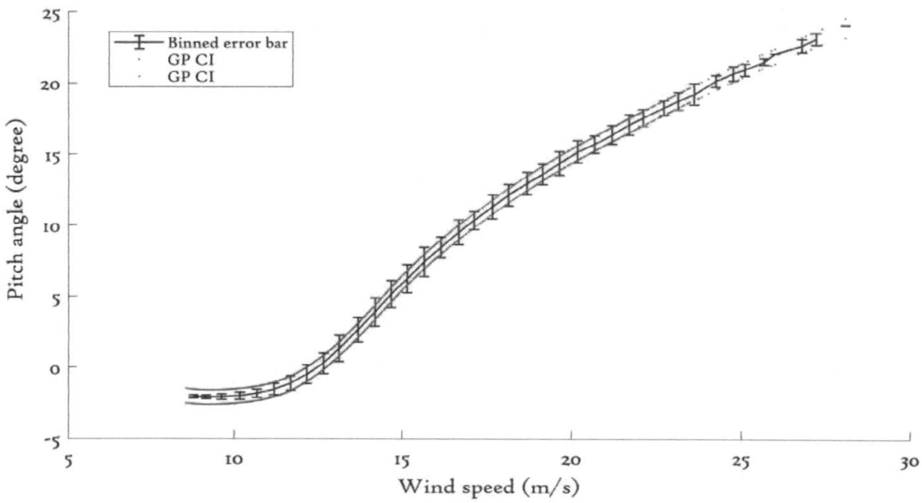

Figure 9: Comparative studies of pitch curve uncertainty based on binning and GP

6. Conclusion and Discussion

In this paper, a novel approach using a Gaussian Process to estimate blade pitch curve has been presented and assessed in comparison with a binning approach. The comparison suggests that the GP model is an effective approach, and can produce accurate results based on limited data. For a binned pitch curve, the data variation is highest between a cut in and cut out wind speed range of the pitch curve. The GP confidence limits are useful for uncertainty analysis and are crucial for robust anomaly detection. By comparing a binned pitch curve with a GP pitch curve, it has been found that the uncertainty is smaller as compared to the binned pitch curve in the cut in to cut out wind speed range.

Future work constructs a robust anomaly detection algorithm based on a Gaussian Process.

Acknowledgments

This project has received funding from the European Union's Horizon 2020 research and innovation programme under the Marie Sklodowska-Curie grant agreement No 642108.

References

[1] Global Wind Energy Council. Global Wind Report Annual Market Update. 2016. (Acessed on 1st June 2018)

[2] K. Fischer, F. Besnard, L. Bertling. Reliability-centered maintenance for wind turbines based on statistical analysis and practical experience. IEEE Trans Energy Convers, **27** (1) (2012).

[3] M. Lydia, S.S. Kumar, A.I. Selvakumar, G.E. Prem Kumar. A comprehensive review on wind turbine power curve modeling techniques. Renew. Sustain. Energy Rev., **30** (2014), pp. 452-460.

[4] J. Gottschall, J. Peinke. How to improve the estimation of power curves for wind turbines. Environ. Res. Lett., **3** (2008), Article 015005.

[5] Singh, P. (2013). Analytical Techniques of SCADA Data to Assess Operational Wind Turbine Performance. Glasgow: University of Strathclyde.

[6] A. Kusiak and A. Verma. Monitoring Wind Farms with Performance Curves. IEEE Transactions on Sustainable Energy, vol. **4**, no. 1, pp. 192-199, Jan. 2013. doi: 10.1109/TSTE.2012.2212470.

[7] S. Shokrzadeh, M. Jafari Jozani and E. Bibeau. Wind Turbine Power Curve Modeling Using Advanced Parametric and Nonparametric Methods. IEEE Transactions on Sustainable Energy, vol. 5, no. **4**, pp. 1262-1269, Oct. 2014. doi: 10.1109/TSTE.2014.2345059.

[8] A. Marvuglia and A. Messineo. Monitoring of wind farms power curves using machine learning techniques. Appl. Energy, vol. **98**, no. 0, pp. 574–583, 2012.

[9] R. K. Pandit and D. Infield. SCADA-based wind turbine anomaly detection using Gaussian process models for wind turbine condition monitoring purposes. IET Renewable Power Generation, vol. **12**, no. 11, pp. 1249-1255, 20 8 2018. doi: 10.1049/iet-rpg.2018.0156.

[10] R. K. Pandit and D. Infield. Gaussian Process Operational Curves for Wind Turbine Condition Monitoring. Energies, vol 11,no 7, pp. 1631, 2018. doi: 10.3390/en11071631.

[11] A. Kusiak, H. Zheng, and Z. Song. Models for monitoring wind farm power. Renew. Energy, vol. **34**, no. 3, pp. 583–590, 2009.

[12] C.E. Rasmussen, C.K.I. Williams. Gaussian processes for machine learning. the MIT Press (2006). ISBN 026218253X.

[13] T. Burton, D. Sharpe, N. Jenkind, E. Bossanyi. Wind Energy Handbook. Wiley (2001)

[14] Wind Turbines—Part 12-1: Power Performance Measurements of Electricity Producing Wind Turbines, British Standard, IEC 61400-12-1, 2006.

[15] M. Schlechtingen and I. F. Santos. Comparative analysis of neural network and regression based condition monitoring approaches for wind turbine fault detection. Mech. Syst. Signal Process., vol. **25**, no. 5, pp. 1849–1875, 2011.

[16] R. K. Pandit and D. Infield. Performance Assessment of a Wind Turbine Using SCADA based Gaussian Process Model. The International Journal of Prognostics and Health Management (IJPHM), vol 9, no 2, pp. 6, 2018. doi: https://www.phmsociety.org/node/2492.

IOP Conf. Series: Journal of Physics: Conf. Series **1102** (2018) 012038 doi:10.1088/1742-6596/1102/1/012038

Condition monitoring of 88km long Offshore HVDC Power Cable: comparison of DTS and as built data

Etienne Rochat, Sanghoon Chin, Fabien Ravet

OMNISENS SA, Riond Bosson 3, 1110 Morges, Switzerland

etienne.rochat@omnisens.com

Abstract. Distributed temperature sensing (DTS) has been deployed for a field trial on a high voltage direct current (HVDC) cable in the North Sea. Temperature measurements using a fibre optic sensing cable bundled onto the HVDC cables are compared with both the cable characteristics and the cable route demonstrating an excellent agreement. Horizontal Directional Drilling (HDD), land and offshore cable joints, transitions as well as hot spots and excess burial are clearly visible. Thermal data thus helps in assessing the cable condition.

1. Introduction

Over recent years, distributed temperature sensing (DTS) systems have been installed on many wind farm export and inter-array cables, as well as on some on-shore buried cables [1]. DTS was originally used as a hot spot detection tool to identify known and unknown thermal bottlenecks; it is now increasingly used as a data input to a dynamic cable rating tool in order to optimize the cable load [2].

In this paper, we report on a DTS measurement campaign of an operational 88km long HVDC submarine export cable in the North Sea. The purpose of the test was to investigate performances over the full cable length. For this reason, two Brillouin optical time-domain analyzers (BOTDA) known as DITEST [3], (one on each side of the cable), were deployed. The required loop configuration was installed on the far end with respect to each DITEST. Both instruments could measure the full cable length, but only half of the sensing distance per instrument was considered, in order to have better performance over the full length. (Note that temperature measurement performance decreases with distance when using any fibre optic based DTS instrument.)

Temperature data are compared with the cable characteristics and the cable laying conditions. Land cable sections with their junction boxes, horizontal directional drilling (HDD) sections and cable transitions are clearly identified through their respective thermal response, as it was also the case for the variation in burial depth and transition in covering material.

Separately, and in order to push the technology to its current limits, the sensing scheme is repeated and extended using optical amplifiers in separate laboratory conditions to reach a 180km total sensing distance.

2. Monitoring concept

The HVDC power cable is the concatenation of a 17.5km section (of which a 1.5km long land section), with a cross-section of 1200mm^2 and a 70.5km section, with a cross-section of 875mm^2.

IOP Conf. Series: Journal of Physics: Conf. Series **1102** (2018) 012038 doi:10.1088/1742-6596/1102/1/012038

A fibre optic cable (FOC) for telecommunication purposes was laid bundled with the two DC power cables. The fibre loss was measured at approximately 18dB with a total loop optical budget of 36dB, which corresponds to the upper limit of the commercially available BOTDA.

Two commercially available BOTDA-DTS, known as DITEST, based on the so-called double-sidebands configuration [3] were used for full cable length temperature measurement (**Figure 1**). The optical power of each system was tuned such as to remain below non-linearity thresholds, as discussed in [4]. Using 5m spatial resolution and a 20min measurement time, each interrogator produced a full temperature profile over the full distance. Repeatability at the far end from the interrogator was around 8°C. Note that the repeatability is computed according to IEC 61757-2-2 standard [5]. Repeatability is a statistical measurement of the "precision of temperature data based on repeated temperature traces at a given location, expressed by twice the standard deviation of corresponding temperature sample points in each temperature trace, with the fibre optic sensor held at constant temperature [5]." By definition, the repeatability is given at twice the standard deviation (2σ) although commercial documentation often provides results at one standard deviation (1σ), which makes the result look twice better…

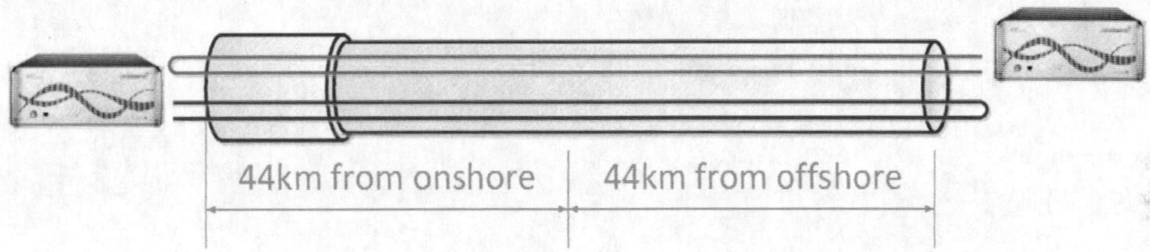

Figure 1. Monitoring scheme featuring two DITEST at each side with a fibre loop at the opposite end of the cable.

To meet the project requirement, only half of the FOC was measured by each interrogator, in 15min, with a spatial resolution of 4m. Using this configuration with post processing, the temperature repeatability in the middle of the cable was below 0.6°C (at 2σ) as shown in **Figure 2**.

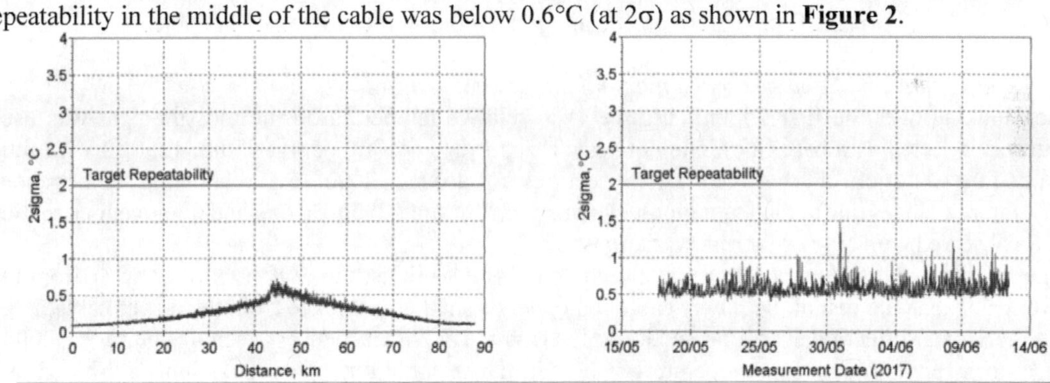

Figure 2. Temperature repeatability (2sigma, 4m spatial resolution, 15min measurement time) along the distance (left) and repeatability variation over 1 month in the middle position (right).

Note that long term repeatability graph (**Figure 2**-right) is made on a live cable with significant thermal fluctuations related to load variations, hence some spikes.

Taking into account the difference between the required repeatability (2°C at 2σ) and the achieved repeatability (0.6°C at 2σ), there is some margin for improvement on the spatial resolution. Reducing it from 4m to 2m would result in a repeatability of 1.5°C (at 2σ), still well within the required performance, but with much finer granularity when it comes to hot spot localization, burial depth variation or scouring localization.

IOP Conf. Series: Journal of Physics: Conf. Series **1102** (2018) 012038 doi:10.1088/1742-6596/1102/1/012038

No cable absolute temperature calibration was performed prior to installation. Field calibration consisted in a reference temperature point in the substation, together with a standard Brillouin to temperature coefficient of 1MHz/°C.

3. Measurement results

3.1. Full length temperature measurement
Full temperature measurement is shown in **Figure 3**. The profile may look fairly inhomogeneous, but represents accurately the cable conditions; some features are related to the cable itself and some to its installation and sea bed conditions They are discussed in the following paragraphs.

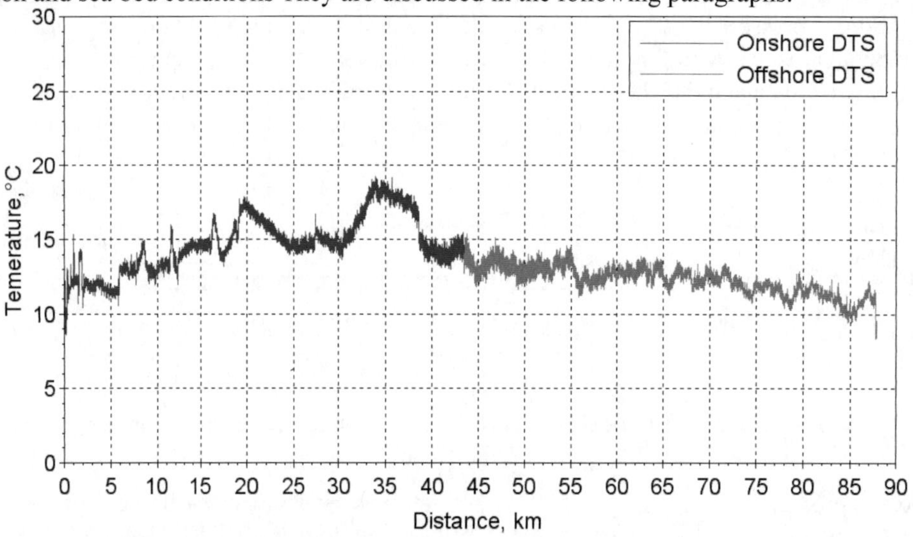

Figure 3. Temperature profile along the 88km-long subsea power cable

3.2. Land sections, junction box and horizontal directional drilling
When looking over the first 1500m (**Figure 4**) in relative temperature (one measurement was used arbitrarily selected as a base line), one notices clearly 4 dips (450m, 650m, 950m, 1250m) with little temperature variation. In between are 5 sections of around 200m each, showing homogenous temperature changes due to load variations. Between 1500m and 1900m, a long and homogenous section can be observed with similar thermal variations.

Dips correspond to junction boxes between short land sections; the FOC is further away from the power cable than the rest of the power cable route, hence smaller influence from the power cable.

A horizontal directional drilling section (HDD) is clearly identified between 1500m and 1900m, thanks to a very homogenous temperature zone with large variations (**Figure 4**, right) attributed to a deeper burial depth than the remaining part of the long section together with a poor heat exchange with the environment.

A lead-in section (up to 200m) shows a slow but constant varying temperature (no load pattern, see **Figure 4**, right). This section is not in contact with the power cable, and is therefore not influenced by the load. The small temperature increase is likely due to seasonal soil temperature variation (daily oscillation pattern is filtered out at this FOC depth).

These findings agree well with the as built information.

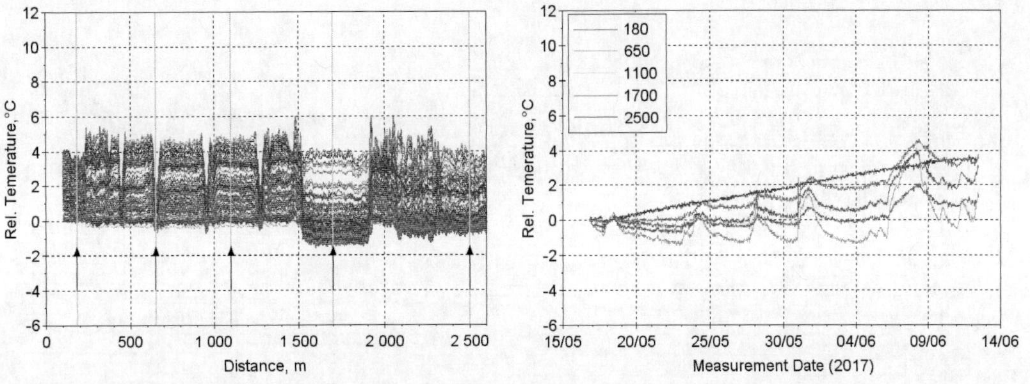

Figure 4. Temperature variation of land section and HDD over distance (left) and time (right) at given distance (corresponding to arrows on the left) as shown.

3.3. Cable transition

Further away, the cable joint between the two cross-sections is clearly visible (**Figure 5**). Larger variations are seen on the smaller cross section cable, which make sense; the smaller the cable the larger the loss, at constant current. In addition, the smaller cross section cable's burial depth is shallower than for the larger cross section cable, thus resulting in larger thermal variation as well.

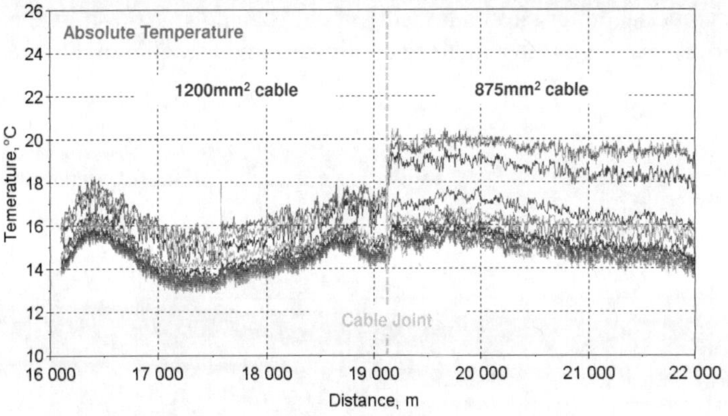

Figure 5. Cable joint with corresponding thermal behaviour.

3.4. Depth of burial

DTS measurement can also provide very interesting information on burial depth, as shown in **Figure 6**. Based on concatenated survey data (bottom lines), one sees a large change in burial depth at 4.5km; the additional depth directly results in a temperature increase along the cable.

Similarly, dips in the seabed (post-installation) are identified at KP7, KP10 and KP15; they correspond exactly to temperature peaks and the logical conclusion is that those dips have now been filled up by adjacent sea bed material, resulting in a deeper burial depth.

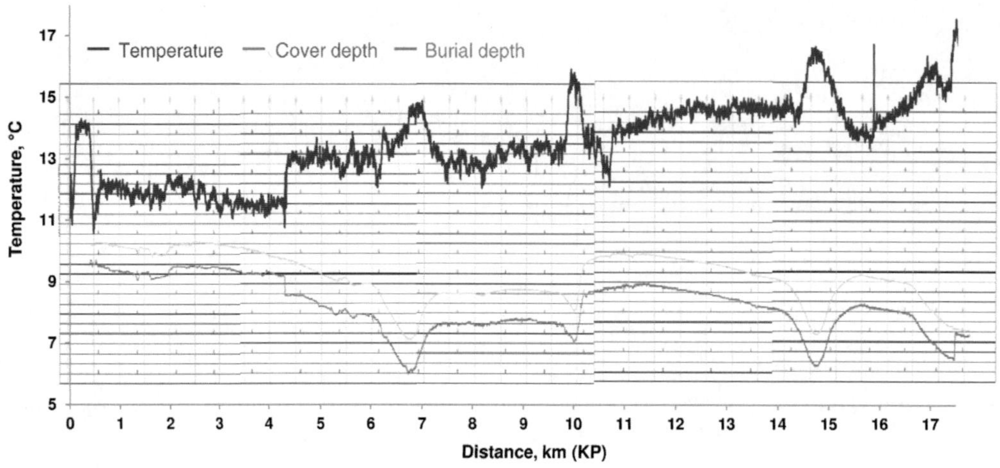

Figure 6. Comparison between temperature profile and the cable depth at the beginning of the cable

3.5. Change of seabed material

Another fascinating example is illustrated in **Figure 7**; around 38.5km, a noticeable ~100m long 2.5°C cooling effect is measured. Such a smooth transition can only be related to a variation of the cable environment. It matches sand / organic silt / silty spots to sand transition in the seabed survey. As expected, with finer material, the water induced cooling effect is smaller and the temperature is higher [6].

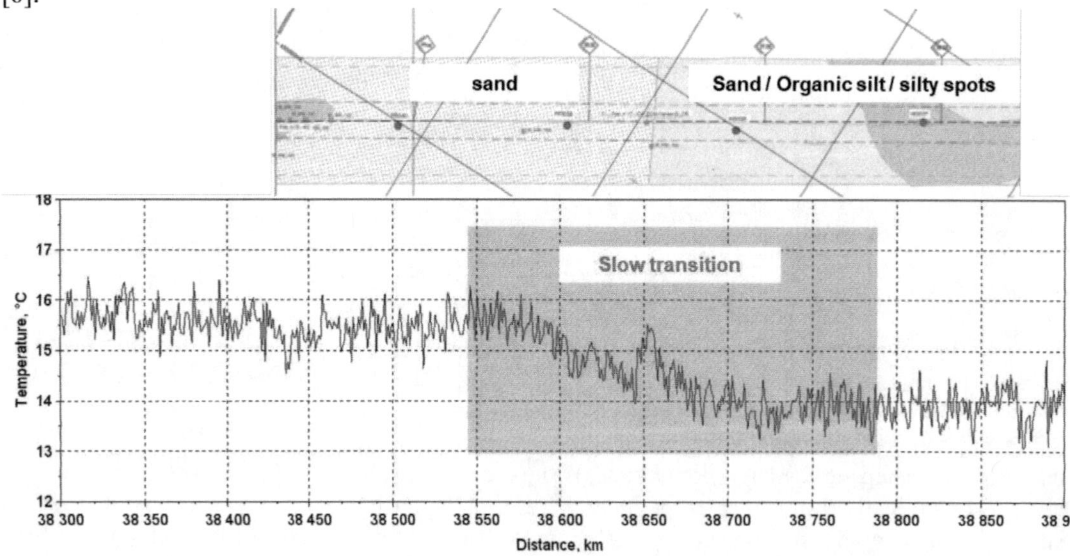

Figure 7. Thermal change due to a change in the cover material.

3.6. Distance between FOC and power cable

When looking at the minimum and maximum temperature over distance, a well localized dip was observed around 70.2km (**Figure 8**-left). Its temperature variation does not follow the load pattern (**Figure 8**-right). This is only possible when the FOC is not influenced by the cable anymore, that is, when is it located further away from the cable than at other positions. With the FOC being bundled to the HVDC, it is likely that the FOC came loose over a few meters

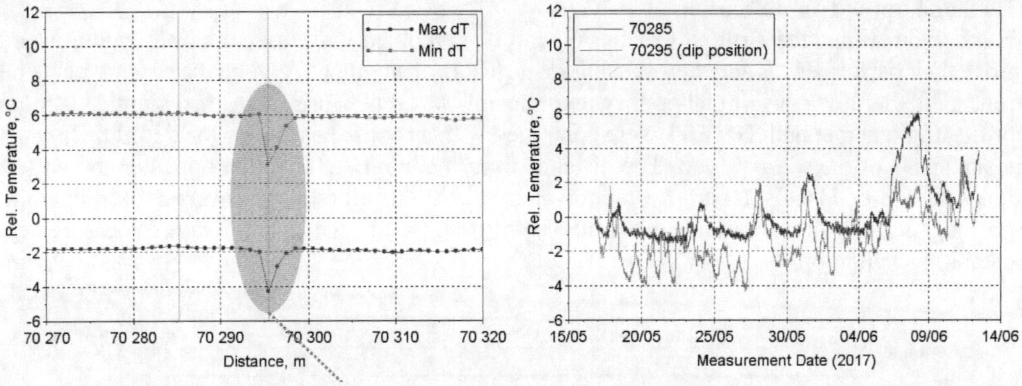

Figure 8. Loose FOC in the bundle.

4. Strain compensation

BOTDA are intrinsically sensitive to both temperature and strain. This is often criticised as being a show stopper for using Brillouin DTS in the field, albeit the only option for longer cables such as this one

Having deployed almost 2500 kilometres of Brillouin based DTS in the power market, much experience has been gained on the presence or the absence of strain and the means of correcting it. In particular, comparison between the cable's "as built" and measured Brilloiun profile, as well as large and sharp temperature fluctuations are potential signatures of strain coupling. Figure 9 shows a temperature profile (red) with obvious variation of a couple of degrees that are related to sharp transition of the burial depth (bottom black curve). For such strained sections, temperature before and after can be assumed to be nearly identical so that strain is the difference between the measured Brillouin profile and a constant temperature. The strain component is then simply subtracted from the temperature profile to provide a clean signal (blue). Note that the strain-compensated temperature follows the load pattern, as expected.

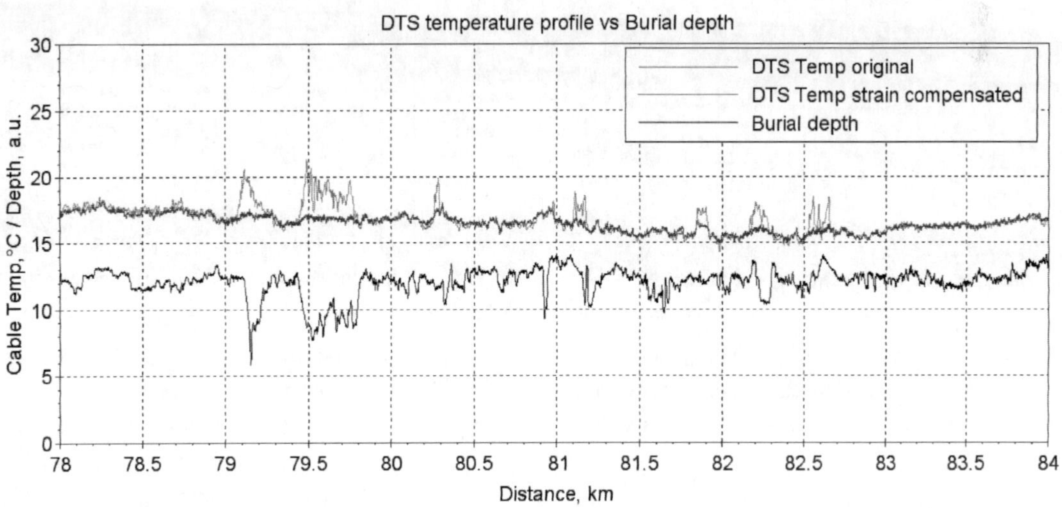

Figure 9. Example of strain cross-sensitivity related to large burial depth variation with strain-compensated temperature.

IOP Conf. Series: Journal of Physics: Conf. Series **1102** (2018) 012038 doi:10.1088/1742-6596/1102/1/012038

5. DTS and optical amplification

DTS based on stimulated Brillouin scattering can be combined with optical amplification in order to improve their performances and their sensing range [7]. In an attempt to extend the sensing range beyond current field deployments, a laboratory experiment was built using an "as simple as possible" amplification scheme: both DITEST were combined with an optical amplifier located at their respective opposite side's of the sensing loop. The distance from the interrogator to the opposite end was 180km with a total loss of 37.1dB, featuring an equivalent 0.21dB/km fibre loss (**Figure 10**). In the middle, a 100m long fibre section was laid in a water bath to allow for stable temperature measurement and repeatability evaluation.

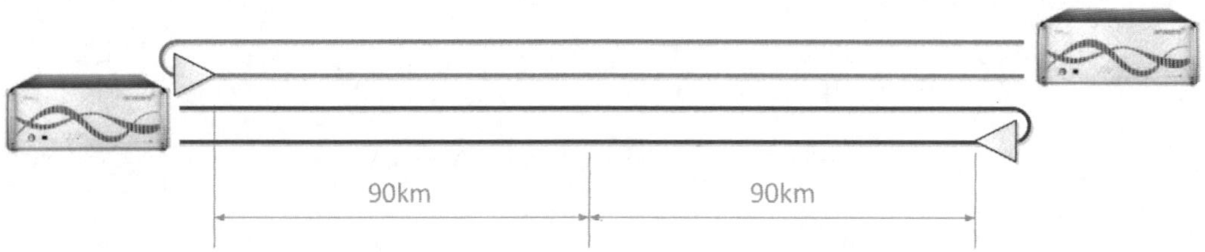

Figure 10. 180km sensing featuring two DITEST and two optical amplifiers at the respective opposite end.

Using a spatial resolution of 5m and a measurement time of 15 minutes, a repeatability of less than 3°C (2σ) was achieved in the middle of the cable, as can be seen in **Figure 11**.

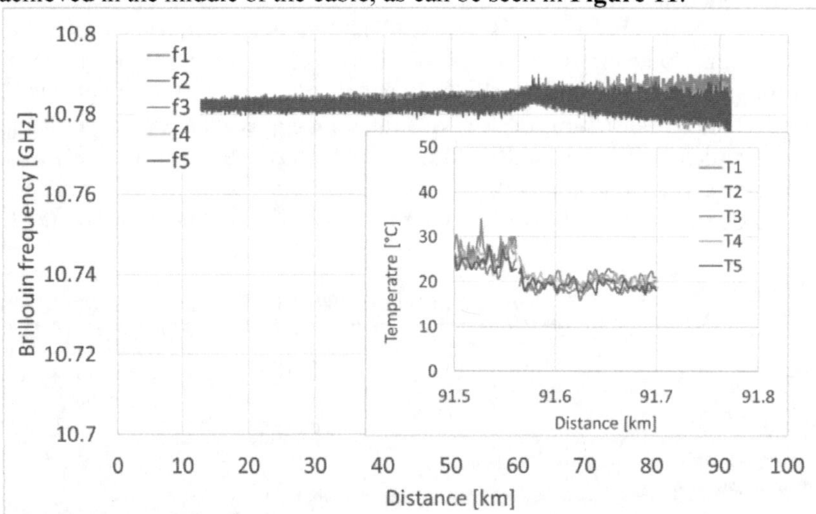

Figure 11. 180km sensing featuring 3°C repeatability (5m spatial resolution, 15min measurement time, 2σ) at the middle of the cable.

Without amplification, a BOTDA scheme would require a loop in the middle of the cable; for instance in a subsea joint asalready demonstrated in the field [8]. Alternative schemes using single end interrogation either cannot match the distance [9] or cannot achieve the short spatial resolution required by the application [10].

6. Conclusion

DTS measurement was performed on an 88km-long wind farm HVDC submarine power cable using two commercially available BOTDA known as DITEST, one at each end of the cable. A loop was made at the respective other side.

Single side measurement was possible, thanks to the interrogator optical budget matching the 18dB (one way) loss, however by combining data from each interrogator up to the middle of the cable, a repeatability of 0.6°C (at 2sigma) with 4m spatial resolution was achieved in 15 minutes; this is well below the 2°C target. In retrospect it was seen that spatial resolution could have been set to 2m, providing a better granularity and hence higher probability of detecting small events without exceeding the temperature repeatability target.

Beyond simple temperature monitoring, the DTS measurement allowed clear identification of cable features such as junction boxes within the land sections, HDD's, transition between land and offshore section as well as cable joints

In addition to cable characteristics, the DTS measurement was used to identify burial depth variation and sea bed material accumulation, as well as material transition. This demonstrates a clear link between the as built and survey data and the temperature data, that is key to assess the cable condition.

Finally, and separate to this actual cable measurement, the use of an optical amplifier with the DITEST BOTDA, demonstrated effective monitoring over 180km long sensing distance was in the laboratory. This configuration is currently commercially deployed over shorter distances.

7. Acknowledgment

The authors thank Dr. Christian Freitag, TenneT TSO, for the extremely useful discussions on comparing DTS data with the "as built" information.

References

[1] E. Rochat, B. Matvichuk, M. Nikles and J. Rowsell, "Long power cables: exposing incipient faults and optimizing performance using extra-long fibre optic distributed temperature monitoring.," in *9th International Conference on Insulated Power Cables*, 2015.

[2] P. Leemans and B. Mampaey, "Belgian experience with real time thermal rating system in combination with distributed temperature sensing techniques," in *CIGRE 2016*, 2016.

[3] S. Diaz, S. F. Mafang, M. Lopez-Amo and L. Thevenaz, "A High-Performance Optical Time-Domain Brillouin Distributed Fiber Sensor," *Sensors,* vol. 8, pp. 1268-1272, 2008.

[4] M. Alem, M. A. Soto and L. Thévenaz, "Analytical model and experimental verification of the critical power for modulation instability in optical fibers," *Opt. Express,* vol. 23, pp. 29514-29532, 11 2015.

[5] I. 61757-2-2, *Fibre optic sensors – Part 2-2: Temperature measurement – Distributed sensing,* IEC, 2016.

[6] T. Hughes, T. Henstock, J. Pilgrim, J. Dix, T. Gernon and C. Thompson, "Thermal Ratings of Submarine HV Cables Informed by Environmental Considerations," in *9th International Conference on Insulated Power Cables*, 2015.

[7] F. Gyger, E. Rochat, S. Chin, M. Niklès and L. Thévenaz, "Extending the sensing range of Brillouin optical time-domain analysis up to 325 km combining four optical repeaters," Proc.SPIE 9157, 2014.

[8] Omnisens, "Malta – Sicily 120 km 220 kV AC interconnector monitored by Omnisens," 2015. [Online]. Available: http://www.omnisens.ch/ditest/doc-news.php?id=392.

[9] B. Marx, A. Rath, F. Kolm, A. Schröder, C. Buntebarth, A. Dreß and W. Hill, "Brillouin distributed temperature sensing system for monitoring of submarine export cables of off-shore wind farms," in *Proceedings Volume 9916, Sixth European Workshop on Optical Fibre Sensors*, 2016.

[10] E. Rochat, S. Chin and R. Fabien, "Using Brillouin distributed sensing to reduce installation risk and optimize cable operation of subsea power cable," in *CIGRE 2018*, Paris, 2018.

IOP Conf. Series: Journal of Physics: Conf. Series **1102** (2018) 012039 doi:10.1088/1742-6596/1102/1/012039

O&M Cost-Based FMECA: Identification and Ranking of the Most Critical Components for 2-4 MW Geared Offshore Wind Turbines

D Cevasco[1], M Collu[2] and Z Lin[2]

[1] Offshore Energy Engineering Centre, Cranfield University, Bedfordshire MK430AL, United Kingdom

[2] Naval Architecture, Ocean and Marine Engineering, University of Strathclyde, Glasgow G11XQ, United Kingdom

Email: d.cevasco@cranfield.ac.uk

Abstract. To date, the focus of the research on offshore wind turbines (WTs) has been mainly on how to minimise their capital cost, but Operation and Maintenance (O&M) can represent up to a third of the lifetime costs of an offshore wind farm. The cost for the assets repair/replacement and for the logistics of the maintenance operations are two of the biggest contributors to O&M expenses. While the first is going to rise with the employment of bigger structures, the latter can significantly increase dependently on the reliability of the components, and thus the necessity to performed unscheduled maintenance operations. Using the reliability data for a population of offshore WTs (representing the configurations most employed offshore), first, the share of the components failures to the O&M cost, together with an estimation of their dependency on some O&M parameters has been derived. Then, by following a cost-based Failure Modes Effects and Criticality Analysis (FMECA), and ranking the components through O&M cost priority number, the most critical components for O&M unplanned operations are identified.

1. Introduction

Over the past decades, the offshore wind energy sector has been growing significantly, with the employment of more than 1,500 units in UK only. Efforts are being made to minimize the overall cost of a wind farm (WF) from the capital cost point of view. On the contrary, to date, not enough attention has been given to the other cost driver of offshore structures: operation and maintenance (O&M). The two main contributors to O&M expenses are the cost for repairing/replacing the assets and the cost for logistics. It has been forecasted [1] that, with the employment of bigger power class turbines, installed further offshore, the share of the O&M expenses could reach the 39% of the lifetime costs of an offshore WF. Thus, it looks clear how much the understanding of when, where and how the wind turbine (WT) components fail is becoming crucial to improve current maintenance scheme and planning.

For this reason, and as a part of HOME Offshore project [2], the work of the authors first focused on the development of a methodology for identifying the components that are worth being further analyzed and developed in a holistic multiphysics modelling of the WT. A flowchart of the methodology applied for approaching this research questions is shown in Figure 1. In particular, the main contribution of this work is the ranking methodology, and the following identification, of the most critical components from an O&M point of view (for unplanned operations). By following a cost-based Failure Modes Effects and Criticality Analysis (FMECA), and using the reliability data for a population of offshore wind turbines (representing the offshore most installed configuration - see Figure 3), the share of the components failure to the O&M cost, together with an estimation of their dependency on some O&M parameters, is derived. In a parallel work [3], the advanced interpretation of the fault mechanisms for a specific failure mode, and its physics of failure, of the most critical component here identified, is presented, with the main aim of reducing the order of the WT model of dynamics (closing up with the last steps of the flowchart).

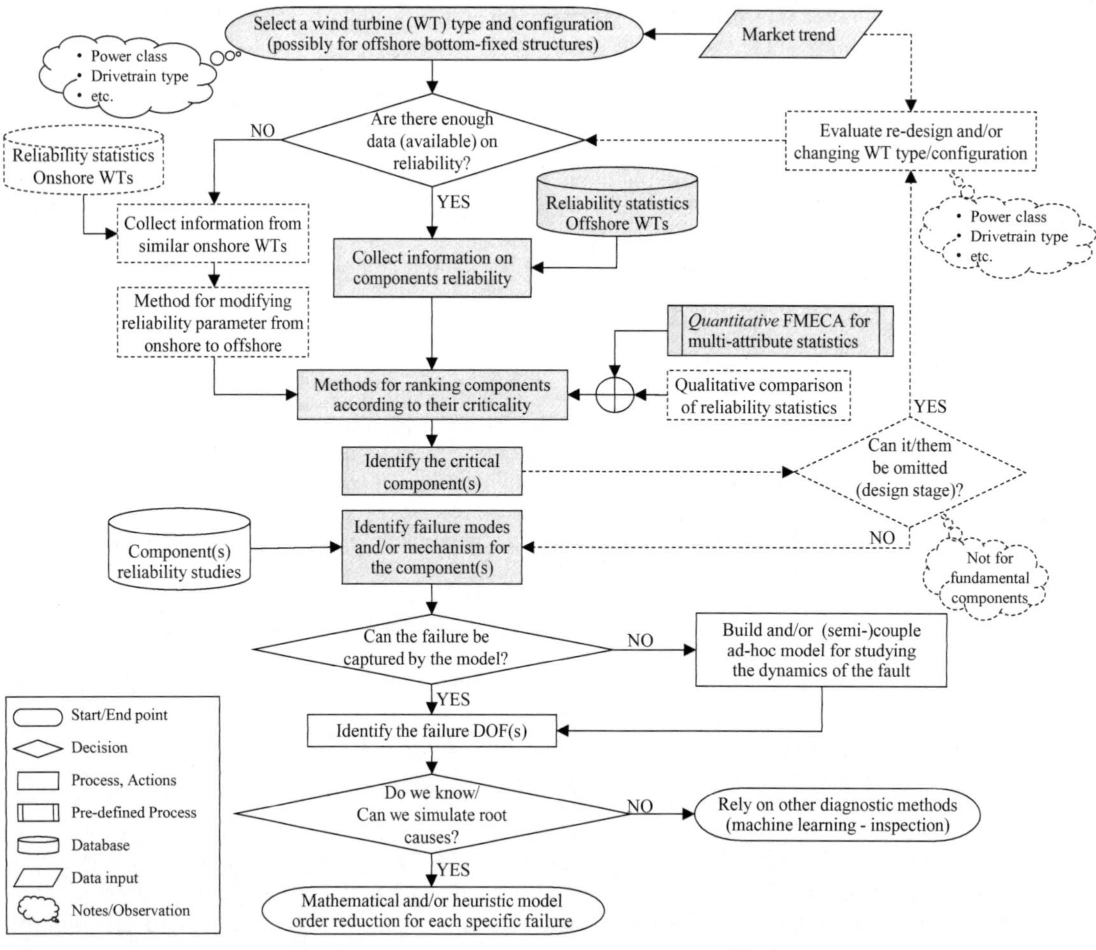

Figure 1. Overall methodology flowchart. In grey, the actions and processes presented in this work, while dashed is the loop for the additional review processes to include other drivetrain typologies.

2. Problem statement and aim

Reliability and maintainability data of WTs have been historically collected and analysed in databases, where maintenance logbooks, SCADA data and/or purchase/service bills, provided by owners and operators, are summarised in averaged results and statistics [4]. In particular, depending on the typology of data collected, more or less detailed analysis on the reliability of the components are possible.

Due to the lack of information available and/or accessible for offshore structures, the more comprehensive historical data for onshore WTs are first analysed (Figure 2). However, instead of answering the questions investigated, two main issues are identified:

- *Inconsistency of the statistics*. Due to either poorly documented collections or for confidentiality reasons, onshore databases generally report averaged results over broad populations, for varying characteristics of the units. By looking at Figure 2, where the most complete results for several onshore reliability sources are compared, a wide spread in both failure frequency and consequence (in terms of hour lost per failure) is observable.

- *Two-dimensional analysis only.* Frequencies and downtimes associated to the failure of the several components are the only information provided, at best. While they can be enough for the understanding of the criticality of onshore structure, offshore maintenance actions can be significantly affected by other logistic factors (such as number and typology of vessels and technicians required), as well as assets repair and replacement costs.

Therefore, despite the fact that drivetrain systems' faults are the highest contributors to the lost hours for the majority of the initiatives (Figure 2), it is not possible draw an immediate conclusion about its criticality when applied offshore, because of: the unjustified unlikely trend for others populations (e.g. Huandian statistics [5]), the possible influence of varying and disused typologies, and the lack of information about the severity of the maintenance action only over the time lost for the restoration of the system.

Learning from onshore databases shortcomings, recently launched offshore data collection international initiatives have been set up to obtain a more suitable and consistent set of data. The SPARTA (System Performance, Availability and Reliability Trend Analysis) initiative [6], initiated in 2013 by The Crown Estate, and under the supervision ORE Catapult, is gathering Key Performance Indicators (at the WF level) and reliability figures (at the subsystem level) from the participating operators. The output arc monthly benchmark reports, of which, to date, only partial and incomplete results (from April 2015 to May 2016) have been published [7]. The German equivalent, WInD-Pool (Wind-Energy-Information-Data-Pool), with Fraunhofer IEE as trustee, is analysing and benchmarking operational and maintenance data based on minimum data requirement [8]. This initiative can be seen as the successor of WMEP [9], where additional (but never published) information on the cost of the maintenance services were collected. In particular, it continues and merge the EVW [8] and Offshore~WMEP [10] research projects, gathering historic and newly collected data for both onshore and offshore WTs. Unfortunately, neither this initiative has published complete results yet (additional information in [11]).

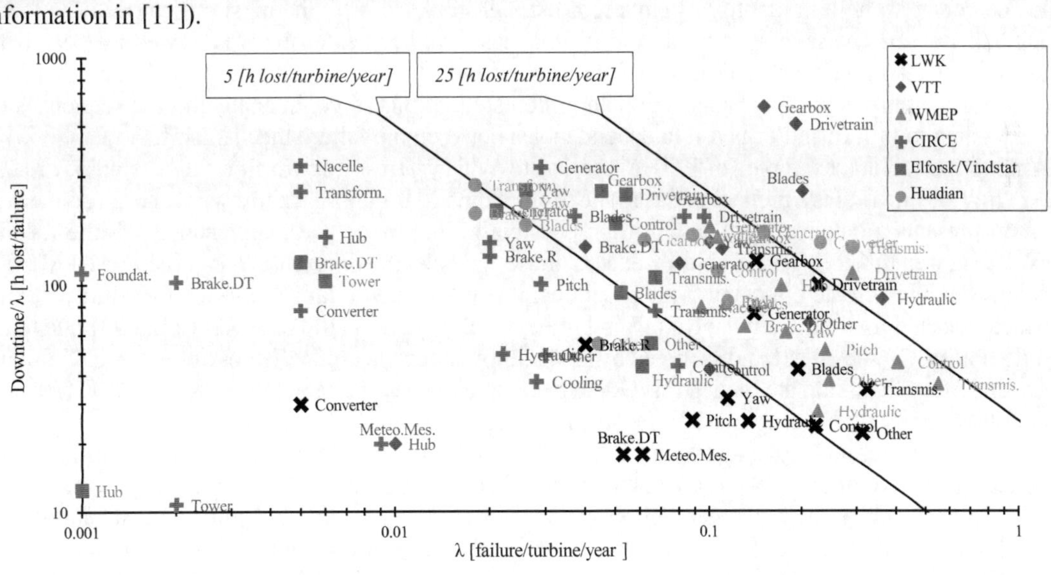

Figure 2. Comparison of the onshore initiatives based on the statistics reported by [11].

Consequently, publications from other independent authors are reviewed as well, giving a hint on the experience for various offshore WFs installed around Europe. Performance for UK offshore round 1 and 2, with evidences of WFs availability and/or capacity factor, are reported in [12], whilst maintenance records and operating issues of some specific WFs can be found in [13,14]. Outstanding among these "secondary" data collections, are the reliability and maintainability data, published by Carroll et al. [15], for a population of 350 offshore WTs. Despite the results presented are from a single manufacturer, the detailed definition of the failure and the maintenance statistics for the repair time (indicating the minimum downtime of the WT), material costs, and required technicians, per subassembly, are provided. For this reason and being representative for the majority of WTs installed offshore to date, these results - called from now on "Strathclyde" statistics - will be used in this paper.

Once accessed to this level details, however, another issues arises: how can we summarise this multi-attribute description of maintenance action, to identify the most critical failure modes? To answer this question, first reliability methods and/or O&M analysis tools, already applied for the wind sector, are briefly reviewed (section 3.1). Nonetheless, needing for a simple method, able to deal with a higher number of parameters in a consistent and systematic way, an ad-hoc FMECA approach is proposed, allowing to gather both performance and maintenance parameters.

3. Approach and methods

3.1. Methods to prioritise the failures

As suggested by the authors in [15], the purpose of reliability and maintainability data collection is to contribute to offshore wind O&M cost and resource modelling. Multi-agent systems simulation and modelling of O&M activities are continuously developed, to aid O&M planners and managers in decision making [16]. Although these tools (e.g. ECN O&M Tool [17]) are able to simulate both scheduled and unscheduled maintenance activities, eventually allowing to outline the most critical components from a cost point of view, they are generally employed for WF cost estimations. An application of the Strathclyde statistics for evaluating how the drivetrain technology choice, and the wind farm distance from shore influence availability and O&M costs, have been recently published by [18]. However, being usually based on several assumptions for both system reliability and maintenance actions, the semi-qualitative character of these tools does not match to the need of this analysis.

On the other hand, lower order analysis, by applying well-known reliability-based methods (and/or tools), can come to help in identifying, in a more straightforward way, the most critical failure modes for a given system. The several methods and approaches have been extensively reviewed by [19], with focus on the ones already applied to the offshore wind industry. In particular, a FMECA approach is selected for the purpose of this paper, not being interested, at this stage, in analyzing the sequences of events leading to failure and/or possible dependencies between the failure mechanisms.

Among the various examples of FMEA applied to WT systems, it is worth citing the work of [20] and [21]. While the first is a pioneer in the application of the FMEA approach to a wind energy converter, the latter presents a detailed breakdown of the possible failure mode and maintenance scheme for the 5 MW REpower turbine. To be able to extend these analyses to different WTs structures, FMECA software were then created. Among these programs, it is worth mentioning the Relex Reliability Studio software, selected for analyzing the ReliaWind project's WT models (R80 and R100) [22,23] due to its high flexibility. On the other hand, as noted by [24], although it is theoretically possible to find all failure modes and insert them into the model, this would necessitate excessive calculation work. For this reason, other authors suggested to employed of the so called correlation-FMEA [19,24].

Despite the different implementations, all the FME(C)A methods and tools above mentioned, are related to qualitative judgements of the authors. In a FMECA, indeed, the failures ranking is performed by the use of a risk priority number (RPN), which is derived from the multiplication of biased and unweighted risk factors. The RPN, difficult to be accurately determined for each single failure mode, has additionally a very little informative character when comparing different wind turbine. While, in [25], Braglia suggested a method to cope with the prioritization of the attributes, Kahrobaee and Asgarpoor [26] first, and Shafiee [27] after, proposed a semi-unbiased approach, by the use of a cost priority number (CPN). Computed based on the probability of failure, its consequences (in terms of cost), and the probability of non-detectability, this methods still includes both qualitative and quantitative measures. However, the economic measure given by the CPN allows a more realistic and quantitative comparison, with respect to criticality, of the different WTs systems.

3.2. Suggested approach: O&M cost-based FMECA

Starting from the work of these latter [26,27], but aiming to get rid of the last qualitative aspect of their risk-based FMECA analysis, an ad-hoc cost-based FMECA for offshore WTs is developed (section 3.2.1 to 3.2.4). In particular, relying on data statistics, the analysis of the failure modes is modified in an analysis of the components criticality first. The detailed study of the modes and mechanisms of failure will happen in a second phase (see flowchart in Figure 1,), based on additional information and details from the statistics, and/or by accessing components reliability studies.

3.2.1. Occurrence. The probability of failure is represented by the failure frequency or rate (λ). Differently from [26,27], who used the dimensionless ratio between the specific component failure and the total amount of failure of the turbine in statistics time period, this attribute has been kept with its dimensions (number of failures per turbine year).

3.2.2. Severity. The severity term accounts for the effect and consequences of the failure, in term of downtime (and/or repair time) and by considering all the other possible informative statistics provided by the database (average number of technicians, type of vessel employed etc.). To deal with all these different nature attributes the severity is translated in terms of cost.

3.2.3. Detectability. Considered only if detailed information are provided, it is defined by [26,27] as a probability of non-detection, ranging 0 to 1 (ratio of the number of failures and total number of occurred and potential failures). Similarly, it is here suggested to integrate this factor as a percentage curtailment in the failure rate. However, information on detected possible failures (usually achieved only after inspection) are difficult to access and/to hardly ever reported. For this reason, and due to the broader application of SCADA system/alarms, an approach based on the ratio between number of effective failures and number of SCADA alarm could be applied, in line with the detection analysis in [28].

3.2.4. O&M Cost Priority Number. As for the well-known RPN, the CPN (here specifically called O&M CPN) is derived by the multiplication of the occurrence, severity and detectability attributes. In particular, by maintaining the dimension of the failure attributes, the O&M CPN will provide information on the total annual cost, per turbine, for unplanned O&M actions (with details on the share per component).

4. Case study

4.1. Reliability and maintainability database

As anticipated in section 2, Strathclyde statistics [15] are the most complete data on offshore structures available to date. Although the WTs in the population have varying power class of 2-4 MW, and the average dimension of the EU installed offshore WTs is over 4 MW (weighted on the projects capacity), their characteristics can be related to the two currently most employed offshore configurations in Europe (Figure 3): geared WTs with an induction machine.

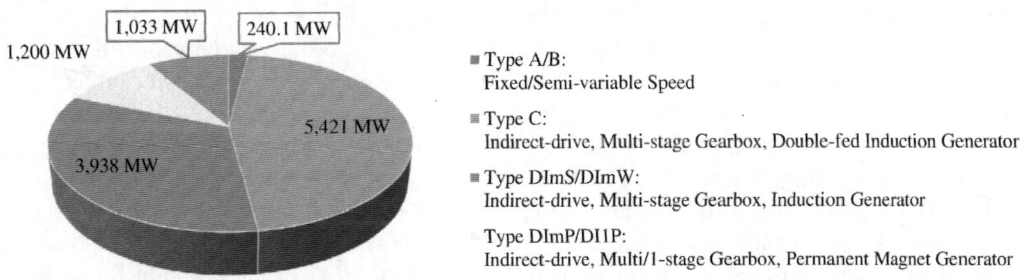

Figure 3. Drivetrain types, total EU installed capacity (for fully grid-connected wind farms, as estimated in [29]). The drivetrain types are described in [30]

Even though the taxonomy used by the authors do not conform to other internationally recognized [31], it has been judged clear enough for the purpose of this analysis. Additional details on the failure type are given by organizing the statistics in cost classes, either if no cost (*NC*) information were recorded, or according to the cost of the material for performing major replacements (*MR*), major and minor repairs (*Mr* and *mr*).

4.2. FMECA approach applied to the case study

Built on provided data, and aiming to maintain the analysis as unbiased as possible, the contributions of unknown possible cost is neglected (e.g. cost to equip the crew and the vessels, time to travel, and other possible logistic delays). Therefore, ignoring the cost of the service and the possible influence of weather days, the *O&M CPN* derived represents, for each component, a minimum forecast of the potential one. Furthermore, although Strathclyde statistics seem to be based on operational data coming from the collection of maintenance logs, no information are reported on potential vulnerabilities and near hit. Thus, to avoid adding qualitative judgement to the analysis, the detectability contribution is not considered either.

To not lose the details on the failure typology given by the division into classes of (material) cost to repair, the cost for O&M unplanned actions ($C_{O\&M}$) is defined per component (i-indexed) and cost classes (j-indexed), as from equation (1), weighting their contribution on the respective failure rate for determining the total averaged $C_{O\&M}$ per component, as shown in equation (2).

$$\begin{aligned}(C_{O\&M})_{ij} &= (C_M)_{ij}+(C_P)_{ij}+(C_L)_{ij} \\ &= (C_M)_{ij}+\overline{P}\cdot CoE\cdot \mathrm{MART}_{ij}+(N_{tech})_{ij}\cdot c_{tech}\cdot \mathrm{MART}_{ij}\end{aligned} \tag{1}$$

$$(C_{O\&M})_i = \frac{\sum_j (C_{O\&M})_{ij}\cdot\lambda_{ij}}{\sum_j \lambda_{ij}} \tag{2}$$

The C_M, C_P and C_L are, respectively, the cost for the repair/replacement material, the loss of revenue for the production loss, and the cost for the labor. While the first is given by reference [15], the latter are estimated as in shown equation (1). The $\overline{P}$, *CoE* and c_{tech} represent, respectively, the assumed average power output of the WTs population - as product of average capacity factor (*CF*) and power class -, the cost of energy and the cost hour per technician. The Mean Active Repair Time (MART) and the average number of technicians required to repair (N_{tech}) are taken from [15].

To understand the influence of the unknown parameters on the $C_{O\&M}$ and, consequently, the sensitivity of the *O&M CPN*, these constants are varied in a range compatible with the data of [12,32]: the *CF* is varied between 0.2÷0.6, the *CoE* between 30÷80 €/MWh, and c_{tech} between 50÷100 €/h.

5. Results and discussion

5.1. Sensitivity analysis: influence of performance and cost parameters

Although, as in [26], this analysis can be easily implemented in a spreadsheet, a preliminary study on the dependency of the results on the assumed parameters was performed in MATLAB. In the specific, a script was developed to calculate the $C_{O\&M}$ for varying scenarios (combinations of power class, averaged performance, and energy and labor cost) with respect to the several components and for the different cost class and types assigned. As stated in equation (1), linear relations between the $C_{O\&M}$ and the performance and cost parameters are expected.

Analyzing the scattered results of the total cost, it emerged that the effect of the *CoE* is significant only for high power rating, whilst the c_{tech} seems to be the biggest cost driver for major replacement and repair classes. In particular, the highest averaged cost is observed for major replacement actions, where the generally high cost of the material and long repair times lead to about €250,000 to €450,000 per failure. Investigating the share of the components to the total cost, its variation throughout the minimum, maximum and average *O&M CPN* scenarios is plotted in Table 1-2, preliminary neglecting the division in the *MR*, *Mr*, *mr* and *NC* cost classes.

Table 1. Percentage share of the components failure to the *O&M CPN*, and to the cost types identified (C_P, C_L and C_M) - graphical presentation - for the minimum, average and maximum possible scenarios.

	Case	*Gearbox*	*Generator*	*Pitch/Hydraulic Systems*	*Blades*	*Other Structural/ Mech. Components*
Share to the total cost	*Min*	75.27%	13.17%	2.46%	1.36%	1.09%
	Ave	71.49%	13.26%	3.37%	1.86%	1.56%
	Max	69.52%	13.16%	3.88%	2.16%	1.83%

Share to
Production Loss Cost
Labor Cost
Material Cost

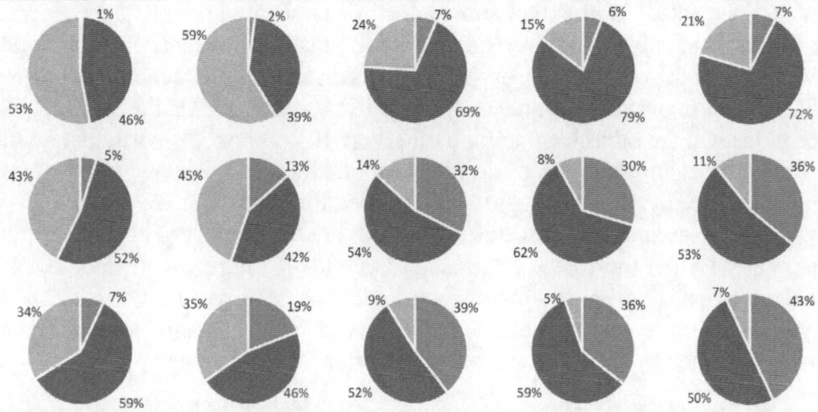

Table 2. Continuation of Table 1.

	Case	*Hub*	*Power Supply/ Converter*	*Controls*	*Breaker/Relay*	**O&M CPN (€/turbine/year)**
Share to the total cost	*Min*	1.03%	0.94%	0.81%	0.80%	**89,212**
	Ave	1.32%	0.99%	1.09%	1.00%	**116,533**
	Max	1.51%	1.00%	1.25%	1.11%	**151,667**

Share to
Production Loss Cost
Labor Cost
Material Cost

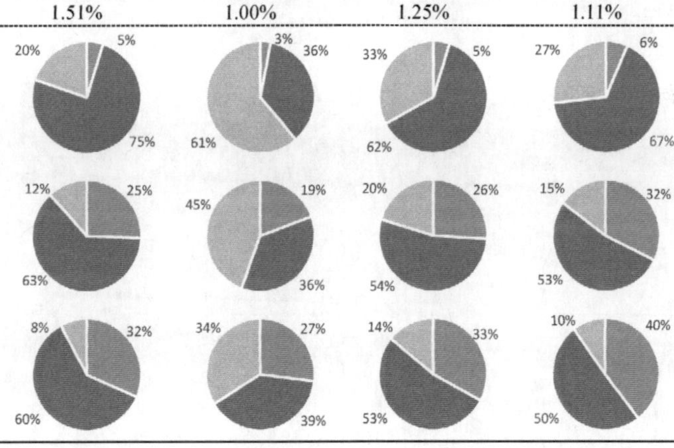

Table 3. Loss of production and technician cost factors for the O&M CPN minimum, average and maximum scenario

Scenario	$\bar{P} \cdot CoE$ **(€/h)**	c_{tech} **(€/h)**
Min	12	50
Ave	105	70
Max	192	100

It has to be noted that the average *O&M CPN* does not correspond to the midrange between the minimum and maximum value, being the *P* varied between 2-4 MW according to the actual capacity of the European installed offshore WTs [29]. The combinations of the performance and cost factors in the "*min*", "*ave*", and "*max*" scenarios are reported in Table 3.

Looking at Table 1-2, it is first noticeable that the top-six components do not change their ranking among the various scenarios. In agreement with what observed by [13], for the UK offshore round 1 structures, and tackled by the Gearbox Reliability Collaborative (GRC) project [33], the gearbox has the highest percentage of the total annual cost for unplanned maintenance operation. Its steady 70% share is caused by the elevated average cost of the material to repair, in the "*min*" scenario, and it is then overtaken by the cost of labor in the high cost scenario. The generator follows, with an almost constant 13% of the total cost, despite of the significant increase of the loss of production cost, among the scenarios. On the other hand, all the other top-10 components participate with values lower than 4% to the total cost. For this reason, despite of the reverse ranking of power supply (or converter), controls and breaker/relay, the average scenario is used for the final FMECA.

To understand, then, what type of failures are the main responsible for their high cost, the share per cost class for the top-three components is plotted in Figure 4. It is observable that both gearbox and generator cost are driven by major replacement cost. However, while the first show and additional high variance with the cost of the technicians, as a consequence of long average repair time, the latter is mainly affected by the high cost of the material for replacement.

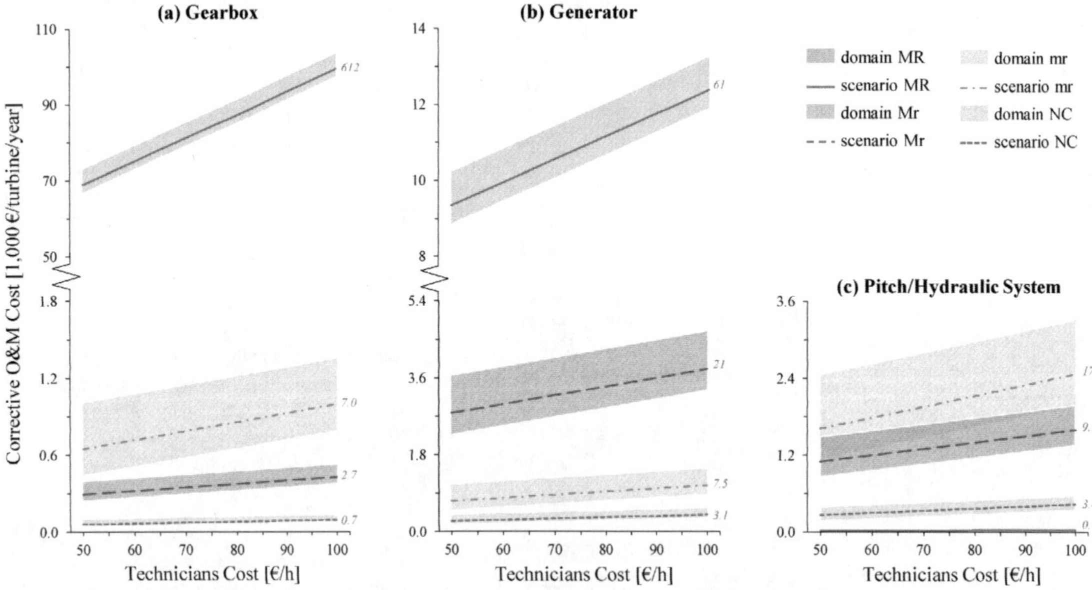

Figure 4. *O&M CPN* (in thousand € per turbine and year) vs c_{tech}, for the top-three ranked components: (a) gearbox, (b) generator, and (c) pitch/hydraulic system. The shaded areas represents the domain of the variation between the assumed *min* and *max* scenarios, whilst lines and the slopes show the trend against the c_{tech} for exemplificative scenario: P = 3.6 MW, CF = 0.36, CoE = 80 €/MWh.

5.2. Cost-based FMECA

The FMECA ranked results, already briefly and visually presented in [3], are shown in Table 4. In that first work the values of failure frequency and cost consequences are converted in dimensionless factor (ranged from 1 to 10), to obtain an expectable ranging of the final values. However, the authors following decided to keep the analysis in dimensional quantities, not to lose the sensitivity on the actual cost to repair, and to easily allow future comparisons between different WT systems. On the other hand, the use of average cost, instead of prioritising by distinguishing per cost class, ease the possible future integration of detectability factor.

Despite the relative long downtime for the reparation of the gearbox (with a weighted average MART of about 63 hours per failure), the share of the loss of production cost, compared to the material and the technicians' expenses, seems to be underestimated. More in general, a small contribution of the cost of energy loss to the total is observable for all the components. The following shortcomings of the presented analysis can possibly justify this trend:

- The use of the *CF* for calculating the lost energy might be appropriate for long downtimes, but it underestimates it for short downtimes. It has been already shown, indeed, that the λ can be related to the wind speed, and failures generally appear in periods with rather high winds [34].
- The logistics, technical and weather-day delays, not yet integrated in these results, can considerably stretch of both long and short downtimes.

Once identified the most critical component(s), a more in-depth analysis of the critical modes of failure has to be performed, following the approach suggested in Figure 1: by accessing the results of specific studies on the components reliability ([33,35]), and, if necessary, updating and completing the ranking by adding component failure-specific detection. A hint on the more likely frequency failure of the generator (either wound rotor or a caged one) and of the pitch/hydraulic systems are reported by [15], while the one for the gearbox have been investigated more in details by the GRC project [33].

Table 4. O&M Cost-based FMECA results (ordered) for the average scenario.

Systems and Sub-System	λ (failure/turbine/year)	Cost Severity (€/failure)				*O&M CPN* (€/turbine/year)
		C_M	C_P	C_L	$C_{O\&M}$	
Gearbox	0.633	56,184	6,534	68,807	131,524	83,255
Generator	0.999	6,908	2,083	6,455	15,445	15,430
Pitch/hydraulic system	1.076	490	1,163	1,981	3,633	3,909
Blades	0.52	351	1,213	2,587	4,151	2,159
Other components	1.005	199	637	967	1,803	1,812
Hub	0.235	771	1,654	4,116	6,541	1,537
Controls	0.428	449	948	1,569	2,966	1,269
Contactor/circuit/breaker/relay	0.43	549	692	1,452	2,692	1,158
Power supply/converter	0.18	2,847	1,227	2,312	6,387	1,150
Sensors	0.346	613	787	1,233	2,633	911
Electrical components	0.435	211	587	918	1,716	746
Pumps/motors	0.346	514	515	733	1,762	610
Grease/oil/cooling liquid	0.471	164	420	587	1,171	552
Heaters/coolers	0.213	458	549	886	1,892	403
Yaw system	0.189	281	636	1,026	1,943	367
Tower/foundation	0.185	599	371	568	1,538	284
Transformer	0.065	1,259	979	1,782	4,021	261
Safety	0.392	148	213	267	628	246
Service items	0.125	79	746	1,109	1,934	242

6. Conclusion

In this work, an unbiased approach for the ranking and identification of the most critical components of a population of currently installed WT systems has been proposed, based on an O&M cost estimation.

The main advantages of the proposed methodology are:

- Elimination of the qualitative nature of the severity and detection criticalities, opting for a quantitative conversion of reliability and maintenance statistics;
- Then, elimination of additional "weighting factors" for integrating the relative importance of the criteria.

On the other hand, the areas of improvement are:

- The need to add a second-stage FMEA, to complete the analysis with the components-specific failure modes and failures-specific detection criticality;
- This method can only be used when enough reliability data and/or cost information are provided.

As regards the results from the components ranking, it is clear that further effort is needed to improve the reliability and maintenance of the gearbox, the induction generator, and the pitch system, and to understand the root cause(s) and physics of failure of both their major and minor failures modes. However, the analysis still need to be completed with sensitivity on the service cost (component-dependent attribute, for the type of technicians and unit required) and possible weather day effect.

Acknowledgements

This work is supported by the UK Engineering and Physical Sciences Research Council (EPSRC) HOME-Offshore project (EPSRC Reference: EP/P009743/1). The first author is supported by grant EP/L016303/1 for Cranfield University and the University of Oxford, Centre for Doctoral Training in Renewable Energy Marine Structures - REMS (http://www.rems-cdt.ac.uk/) from the UK Engineering and Physical Sciences Research Council (EPSRC).

References

[1] GL Garrad Hassan 2013 *A guide to UK offshore wind operations and maintenance*
[2] EPSRC 2018 HOME Offshore: Holistic Operation and Maintenance for Energy from Offshore Wind Farms
[3] Lin Z, Cevasco D and Collu M 2018 Progress on the development of a holistic coupled model of dynamics for offshore wind farms, phase I: Aero-Hydro-Servo-Elastic Model, with drive train model, for a single wind turbine *Proceedings of the ASME 2018 37th International Conference on Ocean, Offshore and Arctic Engineering (OMAE2018)* Madrid (Spain) pp 1–9
[4] Kaidis C, Uzunoglu B and Amoiralis F 2015 Wind turbine reliability estimation for different assemblies and failure severity categories *IET Renew. Power Gener.* **9** 892–9
[5] Ma Z, An G, Sun X and Chai J 2015 A study of fault statistical analysis and maintenance policy of wind turbine system *International Conference on Renewable Power Generation (RPG)* p 4
[6] CATAPULT Offshore Renewable Energy 2018 SPARTA: A UK offshore wind performance data benchmarking platform, transforming wind turbine operational performance
[7] Fox A and Hill C 2017 *System Performance, Availability and Reliability Trend Analysis: Portfolio Review 2016*
[8] Fraunhofer IEE and Dresden IZP 2018 WInD-Pool (Windenergie-Informations-Datenpool)
[9] Faulstich S, Hahn B and Tavner P J 2011 Wind turbine downtime and its importance for offshore deployment *Wind Energy* **14** 327–37
[10] Faulstich S, Pfaffel S, Kühn P and Lyding P 2012 Monitoring offshore wind energy use in Europe - Offshore~WMEP *Energy Procedia* **24** 322–7
[11] Pfaffel S, Faulstich S and Rohrig K 2017 Performance and reliability of wind turbines: A review *Energies* **10** 1–27
[12] Crabtree C J, Zappalá D and Hogg S I 2015 Wind energy: UK experiences and offshore operational challenges *Proc. Inst. Mech. Eng. Part A J. Power Energy* **229** 727–46
[13] Feng Y, Tavner P J and Long H 2010 Early experiences with UK Round 1 offshore wind farms *Proc. Inst. Civ. Eng. Energy* **163** 167–81
[14] NoordzeeWind 2008 *Operations Report 2007-2009*
[15] Carroll J, McDonald A and McMillan D 2016 Failure rate, repair time and unscheduled O&M cost analysis of offshore wind turbines *Wind Energy* **19** 1107–1119
[16] Sahnoun M, Baudry D, Mustafee N, Louis A, Smart P A, Godsiff P and Mazari B 2015 Modelling and simulation of operation and maintenance strategy for offshore wind farms based on multi-agent system *J. Intell. Manuf.* 1–17
[17] Rademakers L W M M, Braam H, Obdam T S and Pietermen R P v. d. 2009 *Operation and Maintenance Cost Estimator (OMCE)*
[18] Carroll J, McDonald A, Dinwoodie I, McMillan D, Revie M and Lazakis I 2017 Availability, operation and maintenance costs of offshore wind turbines with different drive train configuration *Wind Energy* **20** 361–78

[19] Leimeister M and Kolios A 2018 A review of reliability-based methods for risk analysis and their application in the offshore wind industry *Renew. Sustain. Energy Rev.* **91** 1065–76
[20] General Electric Company 1979 *Mod 1 Wind Turbine Generator Failure Modes and Effects Analysis*
[21] Bharatbhai M G 2015 Failure Mode and Effect Analysis of Repower 5M Wind Turbine *Int. J. Adv. Res. Eng. Sci. Technol.* **2** 2394–444
[22] Hoseynabadi H A, Oraee H and Tavner P 2010 Failure Modes and Effect Analysis (FMEA) for Wind Turbines *Int. J. Electr. Power Energy Syst.* **32** 817–24
[23] Bueno Gayo J 2011 Final Publishable Summary of Results of Project ReliaWind 1–62
[24] Kang J, Sun L, Sun H and Wu C 2017 Risk assessment of floating offshore wind turbine based on correlation-FMEA *Ocean Eng.* **129** 382–8
[25] Braglia M 2000 MAFMA: multi- attribute failure mode analysis *Int. J. Qual. Reliab. Manag.* **17** 1017–33
[26] Kahrobaee S and Asgarpoor S 2011 Risk-based Failure Mode and Effect Analysis for wind turbines (RB-FMEA) *North American Power Symposium (NAPS)* pp 1–7
[27] Shafiee M and Dinmohammadi F 2014 An FMEA-based risk assessment approach for wind turbine systems: A comparative study of onshore and offshore *Energies* **7** 619–42
[28] Reder M D, Gonzalez E and Melero J J 2016 Wind Turbine Failures - Tackling current Problems in Failure Data Analysis *J. Phys. Conf. Ser.* **753**
[29] Cevasco D and Collu M 2018 Dataset for European Installed Offshore Wind Turbines (until year end 2017)
[30] Pinar Pérez J M, García Márquez F P, Tobias A and Papaelias M 2013 Wind turbine reliability analysis *Renew. Sustain. Energy Rev.* **23** 463–72
[31] Hahn B, Welte T, Faulstich S, Bangalore P, Boussion C, Harrison K, Miguelanez-Martin E, O'Connor F, Pettersson L, Soraghan C, Stock-Williams C, Dalsgaard Sørensen J, Van Bussel G and Vatn J 2017 Recommended practices for wind farm data collection and reliability assessment for O&M optimization *14th Deep Sea Offshore Wind R&D Conference, EERA DeepWind'2017* vol 137 (Trondheim, Norway: Elsevier B.V.) pp 358–65
[32] Maples B, Saur G, Hand M, van de Pietermen R and Obdam T 2013 Installation, Operation, and Maintenance Strategies to Reduce the Cost of Offshore Wind Energy *Tech. Rep. Nrel/Tp-5000-57403* 1–106
[33] Link H, Lacava W, Dam J Van, Mcniff B, Sheng S, Wallen R, McDade M, Lambert S, Butterfield S and Oyague F 2011 *Gearbox Reliability Collaborative Project Report : Findings from Phase 1 and Phase 2 Testing* 85
[34] Tavner P, Edwards C, Brinkman A and Spinato F 2006 Influence of Wind Speed on Wind Turbine Reliability *Wind Eng.* **30** 55–72
[35] Sandia National Laboratories 2015 Blade Reliability Collaborative

IOP Conf. Series: Journal of Physics: Conf. Series **1102** (2018) 012040 doi:10.1088/1742-6596/1102/1/012040

Multi-task distribution learning approach to anomaly detection of operational states of wind turbines

Stephan Vogt[1], Scott Otterson[1,2], Volker Berkhout[1]

[1]Fraunhofer Institute for Energy System Technology and Energy Economy (IEE), Königstor 59, 34119 Kassel, Germany

[2]TenneT TSO GmbH, Bernecker Straße 70, 95448 Bayreuth, Germany

stephan.vogt@iee.fraunhofer.de

Abstract. The detection of abnormal operation modes is of fundamental importance for both operational management and predictive maintenance of wind turbines. Anomaly detection approaches in this context should consider the additional information content that probabilistic models can provide. Instead of binary anomaly classification, the probabilistic information is necessary for proper decision making and risk assessment. Common models, such as quantile and distribution regression can provide probabilistic information. While they are appropriate in predicting the cumulative distribution function, they struggle to accurately describe the probability of an event to occur. In this article we present a new, multi-task learning based approach for a continuous distribution regression with deep neural networks. Using real-world data from an offshore wind turbine, we show that with this model we can better reflect the probability of observed events than with conventional methods. While the predicted cumulative distribution function has a similar quality and no significant differences are visible in the continuous ranked probability score, the probability density function will be substantially smoother. This is also reflected in a significantly lower ignorance score.

1. Introduction

Improving the performance of wind turbines by applying predictive maintenance techniques with data-driven approaches and machine learning is a field with very active ongoing research and high potential for increased yield and reduced operational cost. A key challenge for machine learning algorithms such as artificial neural networks is to detect time periods where a system, such as a wind turbine operates in an abnormal state. Methods for this anomaly detection problem in wind energy have been discussed for supervised [1-3] and semi- or unsupervised approaches [4,5]. Moreover, the assessment of a possibly detected anomaly which may depend on wide range of features within and outside the turbine should rather be expressed in a probabilistic way instead of giving a binary classification. This allows adjusting for the level of confidence with which an anomaly is detected and helps to reduce the number of false positives which may lead to unnecessary service and inspection efforts and low acceptance of a predictive maintenance system. Approaches to use probabilistic failure information in turbine maintenance have been described by Bangalore and Patriksson [6].

The paper will start with a description of the basic characteristics of artificial neural networks predicting the turbine state based on measured sensor values. The quantile and distribution regression approach will be described and shortcomings will be pointed out.

Then we present the concept of multi-task learning in the application for modeling uncertainties with conditional distributions. Based on this concept we develop a new approach and show improvements compared to quantile and distribution regression algorithms. Finally the paper will show the results of the newly developed algorithm when applied on a real field data set from an offshore wind turbine.

An intuitive and widely used technique for anomaly-detection in measured systems is to predict every sensor based on all remaining sensors. Most sensors show values that can be plausibilized using other sensors. Due to uncertainty in the measurements and because the state of the turbine can only be partially observed, each sensor should be modelled probabilistically. The estimated conditional distribution can then be used to quantify the plausibility of a sensor value given the state of the turbine with respect to other sensor measurements. Widespread state-of-the-art probabilistic models are quantile regression (QR) and distribution regression (DR). Both models can be combined with artificial neural networks to better represent non-linear behavior. The following shows how QR and DR can be applied to neural networks and how neural networks can be combined even more efficiently with distribution regression in a multi-task learning setup.

2. Model Configurations for Nonparametric Conditional Distributions with Neural Networks

Modern artificial neural networks (ANN) are composed by a set of partially differentiable operations, input variables and parameter matrices and vectors. The key concept is to predict values based on the operational data from the training period by means of advanced statistical regression.

Depending on the demanded task, a suitable loss function is used to evaluate the trained parameters to create an indicator for the quality of the prediction by the ANN. The goal here is for the regression models to represent the conditional distribution in the form of the cumulative distribution function (CDF) as a function of independent variables, such as other sensors. Any combination of independent sensors based on expert knowledge can be selected to estimate the conditional distribution of a different dependent sensor for each time step. By comparing the distribution with the actual observation of the dependent sensor, it can be evaluated whether the respective sensor shows anomalous behavior. In the experiment the model approach of quantile regression, distribution regression and the multi-task learning distribution regression, which is presented in this paper, were investigated in combination with an artificial neural network. The following paragraphs describe the configurations used and illustrate them in Figure 1.

IOP Conf. Series: Journal of Physics: Conf. Series **1102** (2018) 012040 doi:10.1088/1742-6596/1102/1/012040

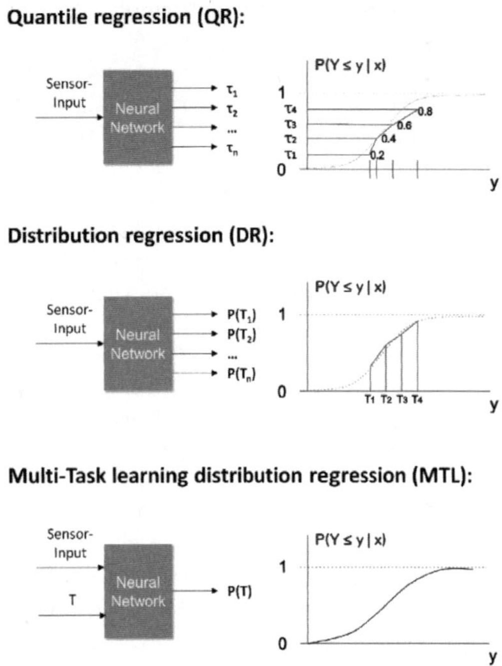

Figure 1. Schematic model structure for the quantile regression neural network (QRNN / top), the distribution regression neural network (DRNN / middle) and the multi-task learning distribution regression (mtlDRNN / bottom) with their respective discretization of the cumulative distribution probability function.

2.1. Quantile Regression Neural Network QRNN

The neural network in combination with quantile regression was first described by Taylor in 2000 [9]. The already known method of quantile regression, described by Koenker in 1978 [10], was applied and the gradient ∇L_τ of the quantile loss function

$$L_\tau = \begin{cases} \tau \cdot |y - \hat{y}_\tau| & if \quad y \geq \hat{y}_\tau \\ (1-\tau) \cdot |y - \hat{y}_\tau| & else \end{cases}, \tag{1}$$

where y is the observed power and $\hat{y}_\tau$ is the predicted value of the quantile τ, was used to tune the model parameters of the network. Ideally, with a well-trained model $P(Y \leq \hat{y}_\tau \mid \boldsymbol{x}) = \tau$ should apply, where Y is the target random variable and $\boldsymbol{x}$ is a vector of the independent sensors values. A further development and description of the corresponding toolbox in R has been provided by Cannon [11].

The quantile regression neural network (QRNN) used in this experiment has the following structure: An output unit was added to the neural network for each quantile between $\tau_1 = 0.05$ and $\tau_n = 0.95$ in 0.05 steps. For this purpose, a linear function is selected as an activation function, since the quantile predictions could be positive as well as negative real-valued values. Each output unit is trained with the same target value, but not with the same loss function, which is the quantile score described by Koenker [10]. It evaluates deviations upwards or downwards differently depending on the quantile. The gradients of the loss function are computed (backpropagation) and used to optimize the neural network weights with gradient descent. While the weights of the respective output unit only have to minimize the individual loss function per quantile, the weights of previous layers must be adjusted to optimize all loss functions at the same time, since they are shared by all output units.

2.2. Distribution Regression Neural Network DRNN

A description of the distribution regression by Peracchi can be found in [12]. Similar to the QRNN, the distribution regression neural network (DRNN) has several output units in the distribution regression. In contrast to QRNN, there is only one loss function. However, the target value differs from unit to unit. Instead of a real-valued target value, it is transformed with several threshold values into a higher-dimensional categorical target value. More precisely, 100 uniform thresholds are chosen between the slightly less than the smallest and slightly more than the largest target value, so that the thresholds cover 110% of the range of the target value. For each threshold value, an individual neuron predicts the probability whether the observation will be below this value. A sigmoid activation function is selected for each output unit to output a probability and thus a value between 0 and 1. The loss function used here, described by Brier in 1950 [13], is the Brier score

$$BS = \frac{1}{N}\sum_{i=1}^{N}(P_i(T) - O_i(T))^2, \tag{2}$$

which minimizes the averaged quadratic deviation between the binary target values $O_i(T)$ indicating that the observed power $y < T$ and the predicted probabilities $P_i(T)$ given a threshold T for all samples $i = 1, \ldots, N$.

In contrast to the QRNN, the DRNN usually requires a higher number of output units. The range covered by the threshold values of the DRNN often covers more than the individual conditional distributions would need. However, this is necessary in order to be able to make a prediction for every possible conditional distribution.

2.3. Multi-Task Learning Distribution Regression Neural Network mtlDRNN

Typically it is of interest to predict the probabilities of various thresholds in order to discretely approximate the conditional CDF. One rather time consuming way to achieve this is by training an individual model per threshold as in DR. Instead of training an individual neural network per threshold one single neural network can predict all threshold probabilities by using several output units. Since every threshold implies an individual loss function every output unit is concerned with an individual task. Applying various loss functions to one neural network is known as multi-task learning where every task is represented by a distinct loss. Multi-task learning has a regularizing effect, since a subset of the trainable parameters is combined with a larger set of tasks, where every task sets distinct requirements. They are jointly less free and therefore often lead to useful internal representations of the data.

One major drawback of distribution regression is that with many thresholds redundancy is generated, since similar thresholds will generate similar outputs. Since the probabilities are predicted independently it occurs, that a higher threshold has a lower predicted probability than a lower threshold. This is referred to as quantile crossing, which is also a disadvantage of the method and must be corrected by post-processing the model output. With a large number of thresholds, the probability of quantile crossing rises. Opposed to this, a low number of thresholds restrict the resolution with which the predicted distribution is scanned.

The newly proposed approach, multi-task learning distribution regression neural network mtlDRNN, takes up the idea of DRNN. At the same time, it eliminates the disadvantage of the many output units. Instead of giving the neural network more than one output unit, another input unit is added instead. The threshold value for which the prediction is to be made is not determined in advance. Instead, this is communicated to the neural network as an input. Like an ordinary CDF, the neural network can be evaluated for any (threshold) value. This makes it possible, for example, to estimate the distribution function of the power, while the neural network takes into account the dependency on the input variables, such as wind speed and wind direction. To train this model it is necessary to combine as many different threshold values as possible with the samples. In theory, this would result in an extremely large data set that would be very difficult to keep in memory. Instead, only the original record is kept in memory. In each epoch, each sample is combined with a new threshold value that can be drawn from a distribution. To ensure that this lies within the range of the target value, it is advisable to draw it from a uniform distribution within the same value range as the threshold values of the DRNN described before.

Our approach avoids the discrete representation of the CDF: A single output unit can be used to predict the probability of being less than any desired threshold. To achieve this, the desired threshold value is added to the set of input variables. The net is then trained in order to correctly predict the probability of falling below the given threshold. The trained net can therefore be evaluated for any threshold value and has thus directly approximated the CDF instead of discretely scanning it. The neural network is forced to perform not only a discrete but an infinite number of tasks. This increases the demand for multi-task learning and avoids redundancies in training. As a result, the number of parameters can be lower while reducing the risk of quantile crossing. This approach is referenced hereafter as multi-task learning distribution regression neural networks, short mtlDRNN.

2.4. Further ANN-Setup

A consistent deep neural network architecture across all model approaches was chosen to carry out the experiments. The selected deep neural networks each have an entrance layer with at least three dimensions for wind speed and the wind direction components. With 200, 100 and 50 units, the layers have a decreasing number of neurons. However, the number of units in the output layer depends on the respective approach and which requirements are made on the resolution of the conditional distribution. Deep neural networks often have the problem of vanishing gradients, which describes the effect of the initially low dependence of the loss function on the weights of the first layers and the associated difficulties of optimization.

This leads to a slow convergence and can sometimes lead to worse results than with a shallow net. The Leaky ReLU proposed by Maas [7] was selected for the activation function as it has good properties against this vanishing gradient problem. In addition, the weights were trained with the help of the ADAM optimizer, proposed by Kingma [8], which enables a fast training without having the need of carrying all data in the memory. The standard parameters of the ADAM optimizer were used for all networks. The training was conducted in 500 epochs in which the entire data set was processed 500 times. However, the hyperparameters of the neural network were not optimized, since the aim of the study was not to compare the neural networks, but to combine them with the approaches above. Both the optimizer and the ANNs have been implemented in Matlab and extend the usual range of available tools.

3. Data and Data Pre-Processing

Quantile regression, distribution regression and the proposed CDF multi-task learning approach are compared in a data case study. Operational SCADA-data from a wind turbine from a North Sea offshore wind farm is used. The data comprises 10 minute averages of the wind speed and wind direction measured at the nacelle. Wind speed is taken as a direct input without considering air density influence factors such as temperature and air pressure. To avoid the non-continuous step from less than 360° to above 0° degrees, the wind direction was converted into a direction vector with a sine and cosine component. The wind direction components and wind speed are scaled before use so that the mean value is zero and the standard deviation is one. As an example, the active electrical power was selected as the target value. Each trained neural network will therefore map the relationship between direction and magnitude of the wind vector into a power value and can thus be interpreted as a direction-dependent power curve, which additionally allows a conclusion about the uncertainty of the estimation. The experiment comprises data from the whole year 2016, using the first two quarters and the last quarter as training period, while the remaining quarter is withheld from model training for test purposes.

For this case study the training data was filtered in order to be able to model a failure-free production operation. Samples whose power measurements at wind speeds between 5 and 25 m/s fall below a power of 5% of the nominal power are removed. In addition, all samples in which the turbine was not in "production" mode at wind speeds greater than 5m/s were removed, for example, to prevent manual operation by maintenance personnel. Furthermore, the adjacent timestamps of the already excluded time periods were removed as a precautionary measure in order to eliminate reduced power samples which appear in the transition from production to shut down or vice versa. For timestamps between 5% and 95% of the nominal power and at wind speeds greater than 5 m/s, curtailment was also detected and removed. For this purpose, two conditions were checked. First, it is checked whether the variation in power over time is less than 30% of the cases that occur. Second, it is tested whether the wind speed is greater than the 99% quantile of the wind speeds that would occur at the measured power without curtailment. The result of the data cleansing is shown in Figure 2.

When monitoring other performance indicators the filtering has to be adjusted for the specific purpose or alternatively left out altogether which would mean that no prior expert knowledge has to be applied.

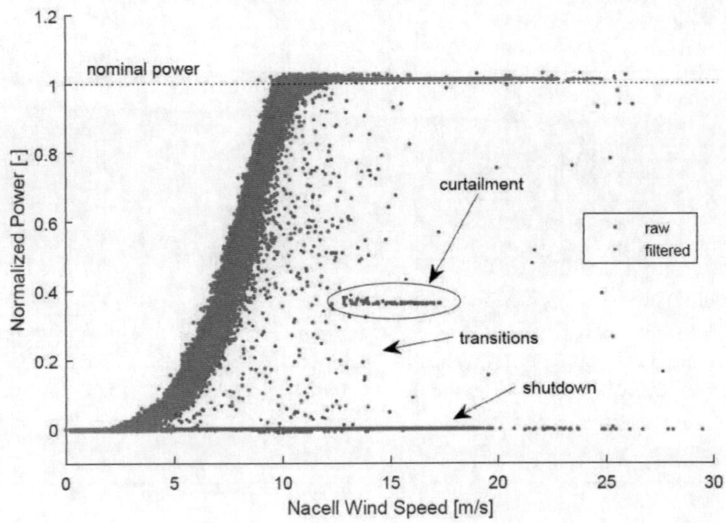

Figure 2. Visualization of the raw and filtered data by plotting the normalized power against the nacelle wind speed measurements.

4. Case-Study Results

Figure 3 shows the results of the models and the measured performance for a selected day in the test period. The output of QRNN, DRNN and mtlDRNN is displayed from the top to the bottom diagram. The blue line shows the measured power, which was normalized with the nominal power of the wind turbine. The orange area represents the predicted conditional distribution, which was converted into a probability density function. For reasons of better visibility, the color values were scaled for each point in time so that the highest probability density corresponds to constant maximum color intensity.

The temporal profile of all models seems to reproduce the temporal profile of the measurement more or less equally well. Between 04:10 a.m. and 01:50 a.m. the turbine reaches nominal power. Here a slight overestimation by QRNN and mtlDRNN and a very sharp prediction by the DRNN are visible. The overestimation by mtlDRNN is accompanied by a greater predicted uncertainty, which is why this is a rather minor problem. From about 02:00 p.m. the wind turbine is experiencing a pitch error. However, the turbine remains in productive mode for another hour until 03:00 p.m. and then changes the operating mode to failure. This point in time is marked with a black vertical dotted line. The difference between the predicted distribution and the observed power successfully detects anomalous behavior one hour before. By specifying the conditional density distribution, it is numerically possible to classify this deviation as unusual, while the deviations at the beginning of the day can be classified as in the range of possible. For anomaly detection also the sequence of time stamps with low probability should be taken into account which can be realized by adding a sequential probability ratio test (SPRT).

With regard to the density function, there are differences between the three models at different points in time. It can be seen that the vertical color gradient is not always continuous. This applies to the DRNN and especially to the QRNN, while the color gradient of the mtlDRNN makes a continuous appearance. In the case of the QRNN, the non-continuous behavior runs parallel to the power measurement, but in the case of the DRNN it seems to run parallel to the x-axis. The explanation can be found in the weaknesses of the output units. The individual outputs of the QRNN are the quantiles, which ideally run parallel to the performance. The output unit of the DRNN refers to one threshold value each. If individual output units are underestimated or overestimated, the strips described in the display appear. The vertical red dashed line marks August 7, 2016 at 01:00 am. For this point in time, the CDF and the corresponding probability density of normalized power are shown in Figure 4.

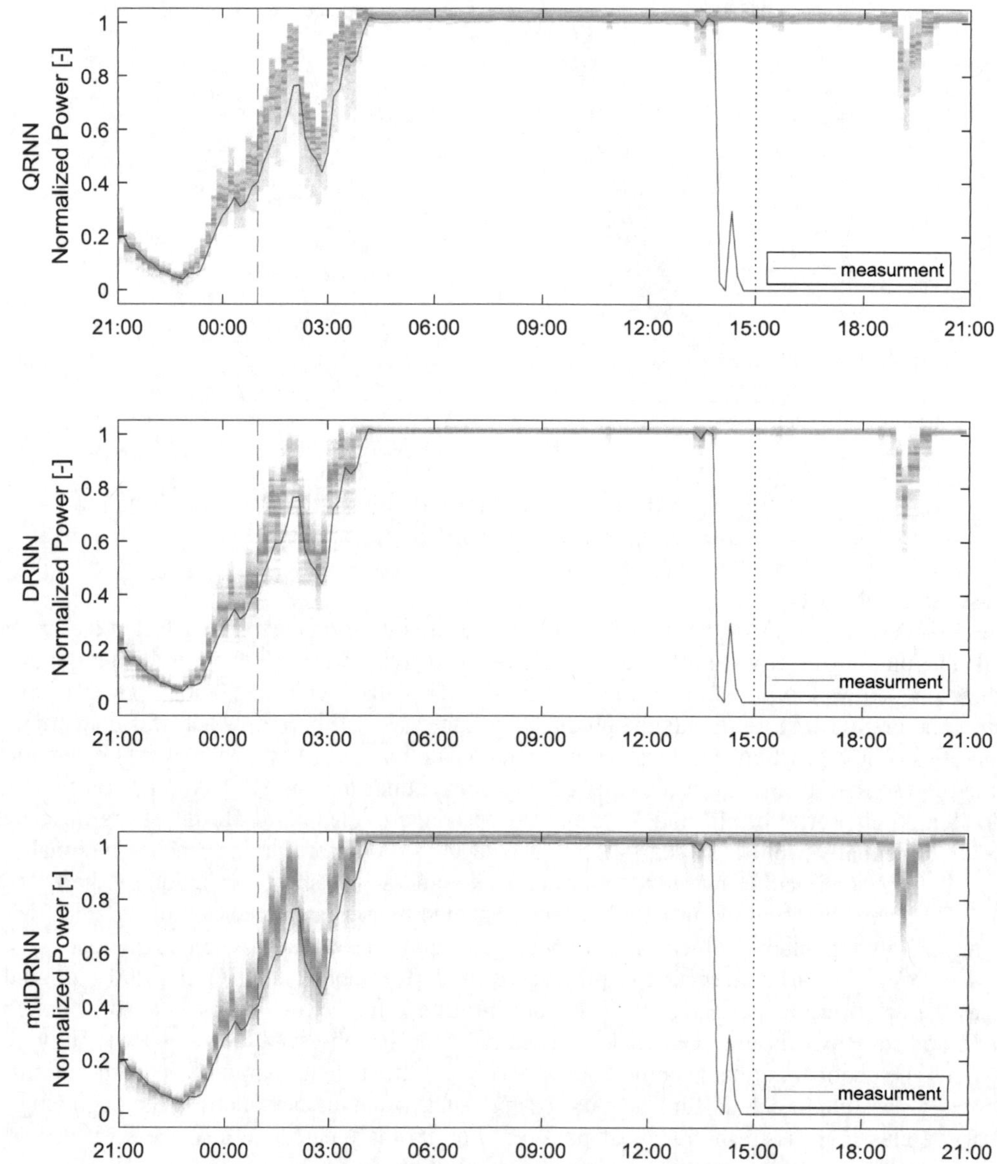

Figure 3. Progress of the power (normalized with the nominal power) and the predicted conditional density distribution.

While there are hardly any differences between the CDFs of the individual models, there are clear differences in the probability density function. This is even more surprising as the CDF is the integral part of the PDF. A closer look at the CDF reveals that in the case of the QRNN and the DRNN a slight jagged pattern is visible. This results in short successive changes in the gradient and thus leads to the large variations in the PDF. In the case of the mtlDRNN it comes to a clearly smoother development. This behavior is much more realistic, as it can be expected that a power measurement should have approximately the same probability of occurrence as a minimally larger or smaller power measurement. If anomalies on the basis of probability density are to be detected, in the worst case it may lead to a power measurement being marked as anomaly due to a low probability density, while a directly adjacent power value with a greater density is classified as normal.

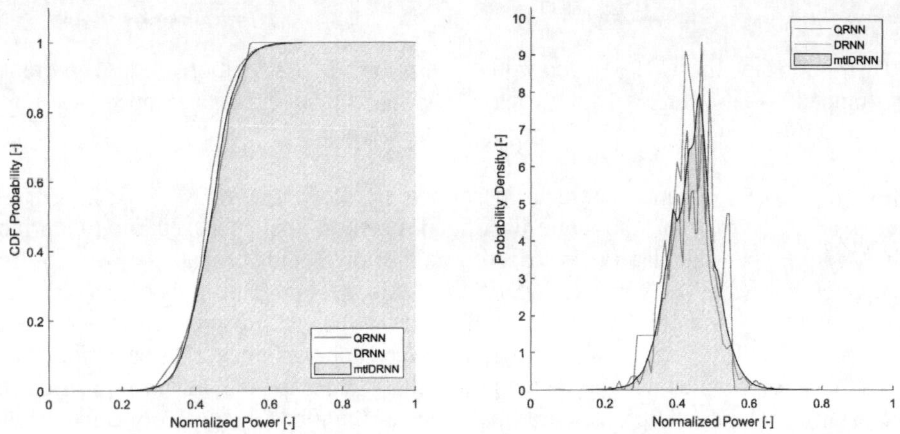

Figure 4: Cumulative distribution and probability density functions derived by the QRNN, DRNN and mtlDRNN for 01:00 a.m., 7th August 2016.

Probabilistic models of this type are usually evaluated using the continuous ranked probability score (CRPS) as suggested by Brown in 1974 [14]. This is a global score that compares the entire CDF with the sample and thus evaluates the distribution not only at the sample location, as a local score would. The CRPS calculates by modelling the CDF of the measurement as a Heaviside function. The difference between the predicted CDF and the measured Heaviside function is squared and the integral from minus to plus infinity is then evaluated as a measure for the sign-independent error area.

This process is repeated for each sample and the determined integrals are averaged. Figure 5 shows a boxplot with the CRPS values of the different models. The individual boxes contain 100 block bootstrap estimates of the test set CRPS with a block length of 7 days. The test period is divided into 7 days long sections, which are numbered consecutively. For each bootstrap CRPS, as many blocks are selected randomly from these blocks as are available in the test set. Then the CRPS is determined for each bootstrap set and this process is repeated 100 times. The distribution of the bootstrap CRPS indicates the uncertainty of the CRPS estimate. Since the distributions cover almost the same range, it can be concluded that there is no significant difference between the models in the CRPS.

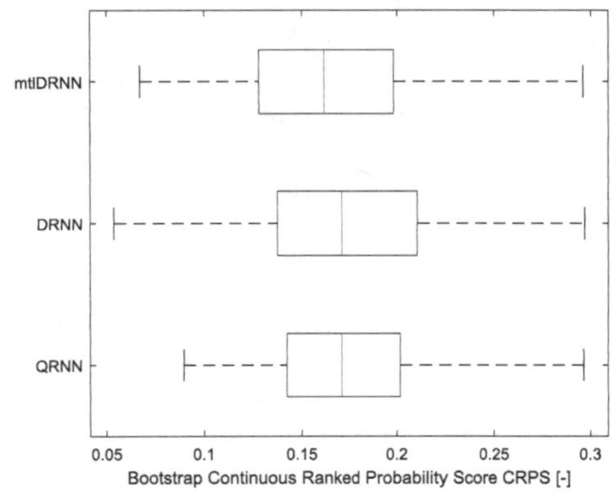

Figure 5. Distribution of the CRPS given 100 Block-Bootstrapped Test-Samples for each approach.

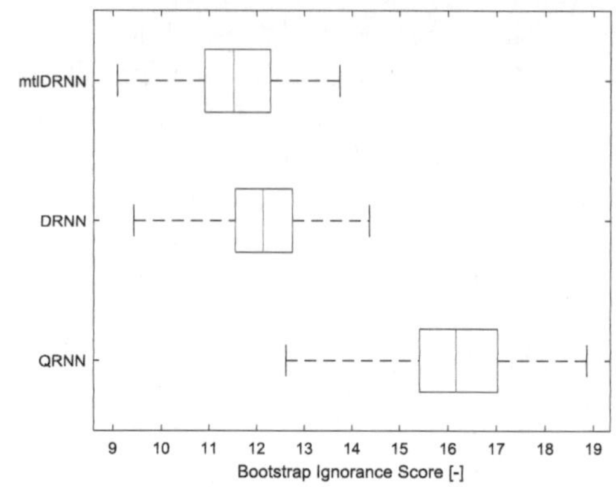

Figure 6. Distribution of the Ignorance Score given 100 Block-Bootstrapped Test-Samples for each approach.

A local score is the ignorance score as described by Roulston in 2002 [15]. It has a close relationship to cross entropy or the logic-likelihood function and is based on the principles of information theory. The calculation is based on the predicted probability of observing the corresponding measurement. The negative logarithm of this local probability is then averaged over all samples. To calculate the probability, the range of power values was discretized into 200 equidistant sections and the probability density was integrated over each sub-interval. The ignorance score is significantly stricter for samples with predicted probabilities close to zero, as they approach infinity in the negative logarithm. Since the predicted probabilities of the models repeatedly came close to zero, these probabilities were set to a thousandth of the smallest predicted probability of another model. Figure 6 shows the distribution of the 100 block bootstraps of the ignorance score of the respective models. The use of mtlDRNNs results in significantly fewer errors than in the case of DRNN. DRNN, on the other hand, clearly has smaller errors than the QRNN.

5. Conclusions

The methods investigated allow probabilistic models, based on deep neural networks, to be used for the detection of anomalies in wind turbines. We propose a new method for a continuous approximation of conditional distribution functions. This approach avoids typical weaknesses of models such as QR or DR: On the one hand, inconsistencies such as quantile crossing are avoided. On the other hand, high-dimensional and extensive data sets can be processed using mini-batch and artificial neural networks. This is made possible by avoiding redundant parameters in the model through the use of multi-task learning. The approach allows information to be exchanged between individual quantiles in order to simultaneously train a leaner model. It may be applied to monitor different kinds of performance indicators for predictive maintenance and is not limited to a specific time resolution. The approach advances the probabilistic output of the anomaly detection system by providing a more realistic probability density function and avoiding misdetections due to local effects. The model has been tested and approved by the use of real-world data from an offshore wind turbine.

In the example shown, information on wind speed and direction was used in the neural network. However, the application is not limited to these variables. Depending on the extent of the training data, many more variables can also be taken into account. Quantile and distribution regression in neural network applications are limited in resolution. These procedures sometimes have a partly questionable, volatile profile in respect of probability density. In addition, an increased number of output units is required in the ANN. Predicting multidimensional target values, would mean that the number of needed output units is proportional to the number of estimated quantiles or threshold values times the dimension of the target value. The approach presented here, on the other hand, could represent each additional target variable dimension with an additional output unit and an additional input unit, thus greatly limiting the complexity.

Acknowledgement

The work that has led to this publication was carried out within the research project ModernWindABS (FKZ 0324128), which is being funded by the German Federal Ministry for Economic Affairs and Energy.

References

[1] Janssens, Noppe, Devriendt, Walle, & Hoecke. (2016). Data-driven multivariate power curve modeling of offshore wind turbines. Engineering Applications of Artificial Intelligence, 55, 331-338.
[2] Carmona, M., & Sanz-Bobi, M. (2016). Normal power generation area of wind turbines for the detection of abnormal performance. Renewable Energy Research and Applications (ICRERA), 2016 IEEE International Conference on, 335-340.
[3] Sun, Li, Wang, & Lei. (2016). A generalized model for wind turbine anomaly identification based on SCADA data. Applied Energy, 168, 550-567.
[4] Zhou, Y., Hu, W., Min, Y., Zheng, L., Liu, B., Yu, R., & Dong, Y. (2017). A semi-supervised anomaly detection method for wind farm power data preprocessing. *Power & Energy Society General Meeting, 2017 IEEE, 2018*, 1-5.
[5] Zhao, Liu, Hu, & Yan. (2018). Anomaly detection and fault analysis of wind turbine components based on deep learning network. Renewable Energy, 127, 825-834.
[6] Bangalore, & Patriksson. (2018). Analysis of SCADA data for early fault detection, with application to the maintenance management of wind turbines. *Renewable Energy, 115*, 521-532.
[7] Maas, Andrew L., Awni Y. Hannun, and Andrew Y. Ng. "Rectifier nonlinearities A reference improve neural network acoustic models." Proc. icml. Vol. 30. No. 1. 2013.
[8] Kingma, Diederik P., and Jimmy Ba. "Adam: A method for stochastic optimization.", 3rd International Conference for Learning Representations , San Diego, 2015,arXiv preprint arXiv:1412.6980 (2014).
[9] Taylor, James W. "A quantile regression neural network approach to estimating the conditional density of multiperiod returns." *Journal of Forecasting* 19.4 (2000): 299-311.
[10] Koenker, Roger, and Gilbert Bassett Jr. "Regression quantiles." Econometrica: journal of the Econometric Society (1978): 33-50.
[11] Cannon, Alex J. "Quantile regression neural networks: Implementation in R and application to precipitation downscaling." Computers & geosciences 37.9 (2011): 1277-1284.
[12] Peracchi, Franco. "On estimating conditional quantiles and distribution functions." Computational statistics & data analysis 38.4 (2002): 433-447.
[13] Brier, Glenn W. "Verification of forecasts expressed in terms of probability." Monthey Weather Review 78.1 (1950): 1-3.
[14] Brown, Thomas A. "Admissible scoring systems for continuous distributions." Manuscript P-5235, The Rand Corporation, Santa Monica, CA, 22 pp. [Available from The Rand Corporation, 1700 Main St., Santa Monica, CA 90407-2138.]
[15] Roulston, Mark S., and Leonard A. Smith. "Evaluating probabilistic forecasts using information theory." *Monthly Weather Review* 130.6 (2002): 1653-1660.